SECURITY
OF MOBILE
COMMUNICATIONS

SECURITY
OF MOBILE
COMMUNICATIONS

Noureddine Boudriga

CRC Press is an imprint of the
Taylor & Francis Group, an **informa** business
AN AUERBACH BOOK

CRC Press
Taylor & Francis Group
6000 Broken Sound Parkway NW, Suite 300
Boca Raton, FL 33487-2742

First issued in paperback 2019

© 2010 by Taylor & Francis Group, LLC
CRC Press is an imprint of Taylor & Francis Group, an Informa business

No claim to original U.S. Government works

ISBN-13: 978-0-8493-7941-3 (hbk)
ISBN-13: 978-0-367-38541-5 (pbk)

Library of Congress Cataloging-in-Publication Data

Boudriga, Noureddine.
Security of mobile communications / Noureddine Boudriga.
p. cm.
Includes bibliographical references and index.
ISBN 978-0-8493-7941-3 (alk. paper)
1. Mobile communication systems--Security measures. 2. Wireless communication systems--Security measures. 3. Mobile computing--Security measures. I. Title.

TK5102.85.B68 2009
005.8--dc22   2008047407

Visit the Taylor & Francis Web site at
http://www.taylorandfrancis.com

and the CRC Press Web site at
http://www.crcpress.com

To my parents

To my wife

To my son, daughter, and son-in-law

Contents

SECTION III   SECURITY OF NETWORK-BASED SERVICES IN MOBILE COMMUNICATIONS

9   Inter-System Roaming and Internetworking Security 327

Preface

Mobile communications offer wireless connectivity that enables mobility and computing in diverse communication environments. The large demands from social markets are driving the growing development of mobile communications more rapidly than ever before. Consequently, a large set of new advanced techniques has emerged, brought about by a larger bandwidth, more powerful processing capability, and advances in computing technology. Many new services are provided or will be provided to potential users, and delivered with high-level quality by usage of GSM, 3G networks, and wireless local area networks in public, home, and corporate scenarios.

The exceptional growth in mobile and wireless communications gives rise to serious problems of security at the level of the customer, network operator, and service provider. The causes of such rise, typically due to the fragility of the wireless link nature, the mobility features, and the variety of the provided services, can be classified into the following six categories:

1. *The physical weaknesses and limitations of mobile communications*: The entirely exposed environment of the wireless radio links and devices provides more opportunities of being subject to malicious attacks. A high error rate and unpredictable error behavior due to external interference, interoperation, and mobility can influence the characteristics of the system security.
2. *The architecture limitations*: Mobile communication protocols are built to provide large-scale communication and high bandwidth. However, due to the scarcity of the frequency spectrum, these protocols are confined to use limited resources. In addition, due to the nature of the security problems, these protocols are not able to handle large protection mechanisms without interacting with each other. The interaction can provide suitable points of access to hackers.
3. *The complexity of services*: Applications are becoming more complex than ever. They are more distributed, involve large sets of users, and use transactions and/or interactions that are structurally more complex. Mobile applications and services in areas such as military, health care, and business are

requiring communication continuity, distributed information collection, and high QoS requirements. All these requirements may be affected by viruses and other security attacks.

4. ***The user requirements***: Services such as location-awareness services and context-based applications may bring mobile terminals easily in contact with possible threats of intruding privacy. Transactions, such as mobile payments, require strong security mechanisms that are able to provide authentication, authorization, and accounting.

5. ***The content of provided services***: Most services, such as multimedia-based applications, are valuable not only to mobile subscribers but also to network operators and providers. Therefore, secure protective measures are needed at access control, content browsing, and delivery. Obviously, the simple migration of securing methods from wired communications to mobile environments cannot satisfy the security requirements of anyone, besides hackers.

6. ***The evolution of hacking techniques***: Hacking techniques are evolving tremendously in structure, nature, and complexity, inducing therefore new vulnerabilities and threats to protected operational systems. One cannot say that a security solution is able to protect a system without updates. Operators and service providers should be able to protect their assets, connections, and services, even in the presence of infected terminals.

Many studies have carefully addressed mobile subscriber authentication, radio-path encryption, and secure mobility; however, the so-called *security of mobile communications* does not involve only these relatively independent domains. Indeed, it needs a more systematic approach to build up a framework layout capable of allowing risk analysis of the threats and vulnerabilities of a mobile communication system, the assessment of a mobile communication system in terms of provided QoS and security, the protection of a service provided via mobile communication systems, and the engineering and management of mobile communication security.

The major goals of this book target five objectives: (1) analyzing and discussing the security proposals made available by mobile communications systems; (2) highlighting the importance of security attacks, mobile viruses, and hacking techniques; (3) developing security policies, security practices, and security guidelines to help better address the security problem; (4) discussing the role of the network operator, service provider, and customer in securing mobile communications; and (5) analyzing the promises, requirements, and limits of service provision in terms of security needs. Advanced services such as IP mobility, voice over mobile IP, mobile payment, and support for roaming are also integrated in the overall framework provided by this book.

The most important issues developed through the following chapters include

1. The analysis of typical attacks and viruses against which a mobile communication system needs to be protected.
2. A detailed analysis of major mobile standards for mobile communication systems with respect to security needs.
3. The study of architectures that are able to provide authentication, data confidentiality, integrity, and privacy to mobile users in different wireless environments.
4. The mechanisms made available by service providers for specific services such as mobile payment, mobile commerce, and other mobile IP-based applications.

The book contains fifteen chapters classified into four major sections of expertise. The first part introduces the fields of mobile communication and discusses the major security concepts. Threats, vulnerabilities, attacks, malware, and risk analysis are discussed in this part, along with the major techniques used to provide encryption, digital signature, access control, authentication, and authorization in mobile communications. The security of SIM and USIM cards is also addressed as well as the major techniques that are used to provide security solutions in different wireless communication systems. Section I contains four chapters. Chapter 1 discusses issues related to the classification of mobile networks and presents the major features of infrastructure-based wireless networks and the infrastructureless networks. The chapter also introduces the security and privacy issues in mobile communications as well as the basic security requirements of the mobile communication systems. Several attacks are described and a classification of the malware targeting mobile systems is discussed.

Chapter 2 presents the main feature and major examples of symmetric and public key cryptography. It also defines the digital signature and discusses basic examples of generation and verification techniques in mobile communication systems. A review of the major authentication techniques deployed in mobile networks is addressed and common attacks against authentication in mobile wireless networks are detailed. In addition, authorization, access control, and key distribution management in mobile communication systems are detailed.

Chapter 3 discusses the most used techniques in mobile communications security. In particular, IPsec is described as a major technique addressing the security of network protocols. The limits of IPsec are discussed and several attacks targeting IPsec are detailed. The security of transport protocols and the limits of SSL, TLS, and SSH are studied in mobile environments. The security features of WTLS that are analyzed to provide security at the transport level are highlighted. Attacks against transport security services are presented. Finally, the role and function of mobile public key infrastructures are discussed.

Chapter 4 describes the major techniques provided for the use, protection, and development of smart cards. Then the utilization of smart cards in communications is developed through the analysis of the SIM and USIM cards. The chapter details a classification of the attacks targeting smart cards and gives the details of several among these attacks. In addition, the chapter develops techniques to protect the log file management in smart cards.

The second part discusses and analyzes the mechanisms and standards implemented by GSM, third generation networks, WLANs, and ad hoc networks to protect the communication services they provide. These networks use a representative class of the techniques currently available to protect mobile communication systems. Section II also contains four chapters. Chapter 5 discusses the infrastructure and GSM mobility scheme to present the major features needed to assess the GSM security solution. The security model of GSM is developed through the description of the major functions involved in the security solution. Several basic attacks on GSM are focused on. The encryption algorithms used in GSM are discussed in detail and the limits are emphasized.

Chapter 6 explains the 3G networks architecture and the security requirements that a 3G implantation should satisfy. The UMTS security architecture is discussed in detail. The authentication and key agreement, integrity protection of signaling messages, and UMTS major security functions are described. The security features of 3G networks are compared to GSM and several attacks against 3G networks are presented.

Chapter 7 starts with an introduction to WLAN communication. Then it discusses basic authentication and encryption schemes as well as several attacks that have defeated the WEP protocol. The vulnerabilities of WLANs and the major attacks targeting them are developed. The WiFi Protected Access is also analyzed and its vulnerabilities are addressed.

Chapter 8 discusses the challenges addressed by mobile ad hoc networks and the opportunities they have generated. Ad hoc networking and use are developed and the most representative routing schemes are analyzed. The chapter discusses the security of ad hoc networks, analyzes the various attacks that have been developed, and discusses the techniques developed to provide authentication in ad hoc networks.

Section III of this book discusses the security issues related to the provision of services using mobile communications. In particular, this part addresses issues related to the security of wireless sensor networks, satellite services, mobile e-services, and inter-system roaming and interconnecting systems. Chapter 9 discusses the concepts of inter-network roaming and internetworking and develops the major techniques used to provide handover among heterogeneous networks. The security provided to protect users and resources from attacks during roaming and handover are discussed. In addition, some attacks performed through roaming procedures are addressed.

Chapter 10 adapts the concept of e-services to mobile communications systems and then discusses the major operations to compose, design, promote, and deliver m-services. It introduces some basic m-services and analyzes their challenges. The m-government and m-commerce systems are particularly addressed and vulnerabilities are highlighted. The techniques to protect m-service messages are developed.

Chapter 11 talks about the specific features that make a wireless sensor network different from ad hoc networks. Then it discusses various issues related to resource management, trust management, and vulnerability protection. Several specific attack schemes are detailed. Challenges and security requirements of WSNs are addressed. Security measures and key distribution are also dealt with.

Chapter 12 presents a classification of satellite networks, develops the features of hybrid satellite networks, and discusses two special functions, mobility and handover. Examples of commercial satellite networks are also presented. Threats and attacks are discussed basically for LEO-satellite networks, as they represent a challenging class. The security techniques provided for satellite networks are then developed.

Section IV of this book consists of three chapters that discuss several issues related to mobile applications. In particular, it addresses the security of mobile payments, the security of mobile multimedia communication and the security of mobile VoIP. Chapter 13 focuses on a classification of mobile payment systems and models used to perform payment. Then, it discusses issues related to privacy and anonymity in electronic payment, performs an analysis of existing mobile payment systems, and discusses security payment transactions in untrusted hosts as well as the protection of micro-payment systems.

Chapter 14 discusses the basic techniques used in VoIP including a description and comparison of solutions offered by SIP and H323 to provide VoIP services. It also discusses some mobility and transport issues for VoIP users. The chapter then develops the main security issues in VoIP and discusses features and security solutions such as SRTP and Mikey.

Chapter 15 discusses transmission issues of mobile multimedia and techniques for securing copyright in mobile networks. The major watermarking techniques for image and video streaming protection are assessed. Several attacks against mobile multimedia are analyzed. Finally, the use of watermarking in providing security services such as key distribution is addressed.

Acknowledgments

I would like to thank my colleagues and students for the help they provided in the realization of this book. Particular thanks go to Neila Krichene for the good work she achieved in contributing to Chapter 7, and to M. Hamdi, S. Rekhis, and Y. Djemaiel for their careful reading of the chapters of this book. Many thanks also go to Rich O'Hanley for his support and professionalism. A final thanks goes to S. Ayari for her diligent work to put things together.

About the Author

Prof. Noureddine Boudriga is an internationally known scientist/academic. He received his Ph.D. in algebraic topology from University Paris XI (France) and his Ph.D. in computer science from the University of Tunis (Tunisia). He is currently a professor of telecommunications at the University of Carthage, Tunisia, and the director of the Communication Networks and Security Research Laboratory (CNAS). Dr. Boudriga is the recipient of the Tunisian Presidential Award in Science and Research (2004). He has served as the general director and founder of the Tunisian National Digital Certification Agency, and has been involved in very active research and has authored or co-authored many chapters and books including over 250 published refereed journal and conference papers. Prof. Boudriga is the President of the Tunisia Scientific Telecommunications Society.

BASIC TECHNIQUES FOR MOBILE COMMUNICATIONS AND SECURITY

I

Chapter 1

Threats, Hacking, and Viruses in Mobile Communications

1.1 Introduction to Mobile Communications

Wireless networking has witnessed a tremendous growth in recent years. Wireless networks offer attractive flexibility and coverage to network operators and users. Ubiquitous network coverage, for local and wide areas, can be provided without the cost of deploying and maintaining wired-based infrastructures. The current wireless networks are based on the concept of (radio) cells, which divide a telephone service zone into small areas for efficient use of low-power transmitters with minimum interference. The terminal devices are generally unintelligent terminals, meaning that the call management, routing, and other services are handled in the network. The terminals can move from one area to another, requiring their calls to remain uninterrupted during mobility. A simplified architecture of the wireless network is depicted in Figure 1.1. When a mobile terminal is turned on, it locates a nearby cell, which tells it what radio channels and what transmission power to use. As the terminal moves, it will repeat this process as it enters new cells. The mobile terminals identify themselves to the network using a unique identifier (ID), having two components that we denote, temporarily, by the electronic serial number (ESN) and the mobile identification number (MIN). While the first component physically

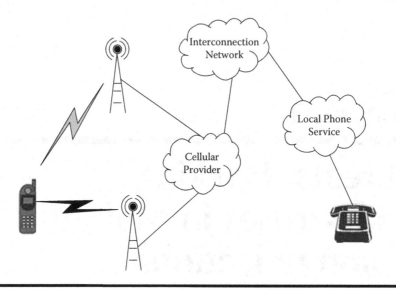

Figure 1.1   Simplified architecture of a mobile wireless phone system.

identifies the terminal, the second identifies the terminal holder. The second component is actually the telephone number. The ESN and MIN are continuously validated by the network. A common attack on the communication systems, to obtain fraudulent services, is called cloning attack, in which the ESN and MIN are duplicated in another terminal and submitted to the system by the other terminal.

To perform its tasks, a mobile communication system implements several fundamental functions, including call management, call billing, call routing, data communication, and data protection. For billing needs, the mobile communication system handling the communication connection is responsible for sending to the communication's service provider a billing record, translating the raw connection details, according to rates and plans, into particular charges (for the MIN holder). The signal carriers, involved in the connection establishment, set up accounts with each other for the services provided on the other networks. End users typically pay a combination of monthly service charges and per-minute charges, although prepaid plans with only per-minute charges are becoming more popular.

The various wireless network technologies, currently used to support mobile communication systems, have very different approaches to provide security. For example, one can notice that (a) the early analog cell phones have essentially no security capabilities; (b) the North American Digital Standards have voice privacy using an encryption that is commonly known as presenting various weaknesses; and (c) the European GSM standards support voice encryption, but the GSM networks are subject to a large range of straightforward attacks (as it will be shown in the subsequent chapters).

1.1.1 Security and Privacy in Mobile Communications

Nowadays, mobile communication systems are increasingly used for private discussions and business traffic involving sensitive data, business secrets, or personal information. Most of the time, the end users of these systems give little interest to the security of their communications, the privacy of their data, and the protection of their mobile stations (i.e., terminal systems). As the use of mobile communications devices has increased, many people have become more concerned with the privacy of communications. Most recently, location-based services have raised severe issues related to the privacy of user location. In particular, people realized that the mobile phones are easily tracked; indeed, the current architecture of mobile network systems essentially requires that the mobile stations be tracked. Additionally, many users have become more interested in the ways provided by the network operators to guarantee correct billing and authentication, especially when they are roaming under foreign networks.

As the mobile communications systems have evolved, successive improvements have not been addressed with security and privacy in mind. For example, some technological choices, related to deployment, have been made by communication network providers while making it possible for agencies (and perhaps for other administrations) to eavesdrop on conversations and obtain call history information.

Typically, various security issues are involved in a mobile communication. Five issues can be distinguished among the most trivial ones:

1. First, the communicating parties do not necessarily have a correct idea of the identity of the other communicating party. When they are calling users they know, they identify them by voice. But, beyond that, there is no real authentication in the system. Many people easily accept any identification given to them on the phone. Similarly, there is no certainty that calling a phone number gets to the right person or organization.
2. A user may be eavesdropping on a communication (or a conversation), whether by a wiretap or by intercepting a call broadcast by a cell phone. Existing standards for voice privacy on digital phones provide some protection, but not against a determined eavesdropping. In practice, the technology discourages simple scanning and opportunistic eavesdropping.
3. The billing records provide an audit trail held by the service provider of the communications made to and received from another user along with the mobile identification number of that party. These records are often used by law enforcement. The registered traces provide the digits of MINs dialed without actually storing the content of the connection (e.g., conversation). Such traces are done in real time.
4. The caller ID (CID) reveals the phone number of the caller to the recipient. While it is frequently useful, the CID does reveal information that the caller

may wish to keep private. In addition, having a single phone number used for multiple calls, or by different callers, may link together information that the recipient may wish to hide.

5. The weak authentication of devices in some systems makes fraud and masquerading possible. The mobile telecommunication's industry has spent large amounts of money to protect against attacks targeting authentication by putting more intelligence in the mobile network, for example.

It is, however, worthy to notice two facts. First, there are two types of authentication that need to be distinguished: the mobile station authentication and the authentication to set up between two mobile stations to help them authenticate each other, regardless of the terminals they are using. Billing issues are related to the first type. Second, solving many of the aforementioned security issues requires an end-to-end approach; this means that a clear security relationship should exist between the end users involved in the communication in addition to what can be established between the delivering nodes in the network. To explain this, let us consider the following example: a connection between two peers can be protected against attacks using two ways. In the first, encryption and decryption are performed at the user level. The second assumes that the traffic is encrypted between the mobile station and the network node that receives the traffic; the traffic is then decrypted and sent to the delivering node, and then, encrypted between that node and the final destination. The second approach is not end-to-end, because an intermediate system may have access to the clear traffic.

Addressing the aforementioned issues is fundamental to provide security and privacy in mobile communications. The effectiveness of a given solution can be measured by how efficiently it eliminates each threat. However, an efficient solution should be built on the communication architecture supporting the mobile system.

1.1.2 Basic Security Requirements

In addition to the very basic requirements for a secure and private mobile communications system, there are several essential requirements that have to be discussed from the perspective of the four actors involved in a mobile communications network, the mobile user, the network provider, the communication and security-related agencies, and the mobile service provider. Let us discuss the requirements from each point of view.

1.1.2.1 End User Perspective

The category of end users includes mobile users who initiate and receive communication calls, as well as Mobile Application Providers (MAP). A MAP provides some services over the mobile communication network based on applications connected to it. The simplest case of an application provider is just a call center offering commercial support to the clients of a given enterprise. The requirements necessary for the end users should include the following:

- No entity, except the authorized center, should be able to bill calls to an account user and have access to its private information.
- A stolen mobile station should be unable to place a call.
- The network should keep no record of calls sent or received. It can only store records needed for the correct billing. The user should be able to have full access to the stored information.
- No records about the uses of digital information services can be made.
- It should not be possible to record a clear copy of a conversation or data session.
- It should not be possible to discover the location of a mobile user, but the user should be able to release his location as desired.
- It should not be possible to identify the end user or the end device, unless the user or device accepts it.
- Location information is not available to unauthorized entities. The network does know the location of a mobile station that is transmitting at a particular time. The users can choose to release their location data to application providers. This information can automatically be released, for a given call, to emergency services.

1.1.2.2 Communication Provider Perspective

Mobile communication providers (or network operators) have many requirements related to the security and efficiency of the system they will deploy and operate. Deployment rules and security solutions must be scalable and reliable. Among the most important requirements that make it possible to operate the service as a business, a provider would need the following:

- Communication services provision should be paid correctly.
- Adequate measures should be selected and implemented against all types of fraud. Updating procedures of the measures should be provided.
- Mechanisms for correctly naming and addressing end devices should be implemented properly. Security of routing functions should be provided in the deployed infrastructure.
- The provider should be able to implement additional functions, such as voice-mail and call forwarding, while providing ad hoc security features to them.

1.1.2.3 For Governments

The primary requirements that are commonly seen for governmental agencies involved in the security of mobile communications systems are the following:

■ Provide correct location information for the emergency services.
■ Provide a robust infrastructure for use in emergencies.
■ Provide access to communications and information about communications for law enforcement.
■ Provide useful measures for a monitoring process made to protect essential assets and infrastructures.

While some of the aforementioned requirements are easy to satisfy, other requirements are very hard to provide or opposable. For example, some users strongly oppose the idea that everyone's privacy must suffer to make law enforcement easier to achieve. Some also believe that the benefits to society given by removing security and privacy from the network are outweighed by the risks of giving the agencies too much tools.

1.2 Basics of Mobile Communications

Mobility support is a significant feature of wireless networks that offer the users not only an anytime, anywhere network access but also the autonomy of roaming while communicating. Recent advances in mobile communications technology have provided increasing data rates comparable, in some cases, to their wired counterparts.

The dominant generations, which are nowadays largely used, are the 2G, 2.5G, and 3G networks explain this increase. While the circuit-switched GSM (2G network) provides very slow data rates (from 9.6 to 14.4 kb/s) to satisfy the needs of advanced applications, the packet-switched networks (2.5 G), which is based on the access network of the GSM, appeared with the promise of higher bit rates (offering a theoretical rate 172 kb/s), but the maximum bit rate achieved, in practice, is about 45 kb/s. On the other hand, the UMTS (3G network) achieves higher data rates. Typically, the UMTS offers data rates up to 384 kb/s, even if in theory a 2 Mb/s transfer rate is possible. Nevertheless, the actual performance of UMTS has still to be monitored during real operation conditions with heavy network loads. In addition, various other wireless communication systems such as the satellite-based systems, the wireless local area networks (WLANs), and the wireless personal area networks (WPANs), which are built based on wireless communication capabilities between devices located in reduced areas, have been deployed. These networks provide high data rates. In fact, versions of the WLANs specified by the series 802.11* can achieve a throughput up to 5 Mb/s (for the 802.11b) and rates approximating 25 Mb/s (for the 802.11a), with the perspective of reaching in the future rates comparable to 155 Mb/s.

On the other hand, enhanced IP networking technologies have been developed to provide a high level of control on transmitted traffic, offer a measurable constraining quality of service, and integrate current and future communication systems (wireless as well as wired) to a unified global network, enabling a truly seamless mobile Internet, beyond the simple wireless access to the Internet. For instance, the Internet Protocol IPv6 does not only offer virtually unlimited address space, but it also constitutes the technical foundation for increasing wireless networking capabilities, offering also interoperability and interconnectivity with respect to security, mobility, and Quality of Service (QoS).

Nowadays, three major categories of mobile communications can be distinguished. They all differ from each other in terms of communication architecture and security techniques. They are the following:

1. The systems based on a fixed communication infrastructure. These systems are very common systems and are not only limited to wireless local area networks (WLAN), where an "access point" is used to gain access to the network; they also include cellular systems such as the 2G and 3G networks.
2. The communication systems built on the collaborative work of autonomous nodes for the implementation of basic functions such as routing and relaying of traffic. Often, these systems do not use any fixed communication infrastructure. Examples of such systems include the ad hoc networks and the wireless sensor networks.
3. The hybrid networks, which combine the concepts used by the previous two categories. Two types of hybrid networks are of major interest: the heterogeneous networks and the provider mediated. While the former type provides different infrastructure-based networks and infrastructureless networks interconnected into a backbone, to which various access networks are connected, the latter is a virtual network provided by a mediator based on direct connections negotiated between autonomous nodes occurring on heterogeneous networks. Both networks are characterized by a dynamic topology, variable links, and heterogeneous components. Variability is observed in the first type because ad hoc networks can serve as access networks; it occurs also in the second type because of the use of infrastructure for mediation purposes (such as routing decisions and the provision of security services).

In the following, we will discuss the main features of the three categories and prepare the discussion related to the need of security services as part of the quality of service provided by any mobile communication system.

1.2.1 Infrastructure-Based Wireless Networks

To achieve the communication needs of the users (e.g., subscribers, in the case of public networks), an infrastructure-based wireless network is assumed to provide some access points, over which a mobile user can connect, to a destination, and provide the requested information or route data to establish a connection to destination. To do so, the mobile user (or its station) establishes a wireless connection with an access node. This connection has to be secured, in some sense. Typically, this kind of setting can be performed by two communication scenarios: (a) a mobile station (such as a portable PC) connecting to a WLAN access point and (b) a mobile station (such as a mobile phone) connected to a cellular base station.

In the following, we discuss the major security issues related to the networks belonging to the first category.

1.2.1.1 Cellular Network Security

Several cellular networks have been developed to provide wide access to a large number of users. The ubiquity of the cellular standards has been advantageous to both consumers (who have the ability to roam and switch carriers without switching mobile stations) and network operators (who can acquire and deploy equipment from any vendor implementing the cellular). Mobile cellular networks guarantee, at least, the authentication of users and provide the confidentiality of the exchanged data between the users and their access points when needed. Second generation cellular networks (such as the GSM), for example, were designed with a moderate level of protection. They were intended to authenticate the subscriber using a pre-shared key and challenge-response. Communications between the subscriber and the base station can be encrypted. The development of 3G networks (such the UMTS) introduced an optional Universal Subscriber Identity Module (USIM), that is able to (a) store user subscriber information; (b) authenticate information and provide storage space for text messages and phone book contacts; and (c) use a longer authentication key to give greater security as well as mutually authenticating the network and the user. Therefore, the security model used in the 3G networks offers confidentiality and authentication. However, it allows limited authorization capabilities, but no non-repudiation.

The GSM uses several cryptographic algorithms for security, namely the A5/1 and A5/2 stream ciphers. These algorithms are used for ensuring over-the-air voice privacy. Despite the fact that A5/1 was developed as a strong algorithm and A5/2 as weaker, severe weaknesses have been found in both algorithms: for example, it has been shown that it is possible to break A5/2 in real-time using the so-called ciphertext-only attack, and that A5/1 can be broken with an attack called rainbow table attack (Chapter 5 will discuss in details the weaknesses of these algorithms).

1.2.1.2 WLAN Security

Several schemes to protect the traffic sent over a wireless LAN have been developed. The first solution came with the IEEE Standard 802.11. It is called the Wired Equivalent Privacy (WEP) and was soon proven to be insecure, because of the use of too short keys (due mainly to US export restrictions) and fundamental problems inherited with the cryptographic algorithm it uses. WEP2 has attempted to cope with these limitations. It just required a minimum key size and did not solve the algorithmic problems. The currently recommended solution for securing WLAN is the so-called Wi-Fi Protected Access (WPA). It was introduced with the IEEE 802.11i standard (for details see Chapter 7).

It is worth noting that all of the aforementioned methods were only designed to secure the communication between the mobile nodes and the access points in a WLAN. They all assume that the access points have to be trusted. This assumption occurred by the past to be acceptable, since the WLANs are run in a reduced and controlled zone (for example, at home for the need of family communications). However, with today's use of WLAN, public or private institutions have made available many access network based on WLANs, particularly in public zones where protection cannot be achieved properly. This offers the opportunity for attackers to intercept the network traffic behind the access points. In addition, the usage of WLAN encryption is not feasible anymore when WLAN access is provided to the public users.

1.2.1.3 Virtual Private Networks

A major observation can be made on the infrastructure-based wireless networks: The security services provide mainly a protection of the radio link between the user and the access point to which it is connected. This allows the visibility of the clear form of the exchanged messages at different points in the network. In addition, various opinions made on the Internet require that services based on a peer to peer relationship should be established for the cases where the responsibility of the partners need to be established, or when the connection relating the partners flows through untrusted environments.

In fact, security has to be provided until the home network; for instance, a company network may desire to allow employees a secure access from everywhere, while protecting their communications from external exposure for business reasons. To solve these problems, various solutions have been developed to provide the so-called Virtual Private Network (VPN). A VPN allows a mobile node to be connected through a secure tunnel, over the public network, to a home network. IPsec is a protocol that adapts the protocol IP to provide such a tunnel, so that real end-to-end security can be achieved regardless of the occurrence of wireless links. While the IPsec extension supports tunneling natively and allows encryption to take place on the network layer, the WEP and WPA provided the security on the link layer. VPN implementations are widely used and the concept is well tested. However, several solutions suffer from important weaknesses. The limitation of IPsec will be discussed in Chapter 3.

1.2.1.4 Mobile IP

Usually, an IP address is assigned to a mobile station depending on the network link, on which it is connected. This means that, if a node changes to another network, its IP address may change. Using a protocol such as IPv4 or IPv6 would require mechanisms that are hard to implement to send a message to such a moving node. Nevertheless, wireless networks should provide this feature; in particular, many open WLAN access points, that are available for public use, should allow a moving user to attach its mobile device to any access point and switch between them.

To overcome these problems, different works have proposed IP mobility support for IPv4 and IPv6. Mobile IP is a standard communication protocol that is designed to allow mobile device users to move from one network to another while maintaining a permanent IP address. In Mobile IP every user has a home address, which is associated to the mobile node, when it is in its home network and a care-of-address is associated to it when being somewhere else. Mobile IP allows each mobile node to be identified by its home address ignoring its current location in the Internet. While away from home, a mobile node is associated with a care-of address that gives information about its current location. Mobile IP specifies how a mobile node registers with its home network (via a dedicated home agent) and how the home network routes messages to the mobile node through a tunnel, when he is not in his home network. Mobile IP provides an efficient and scalable mechanism for roaming within the Internet and allows users to maintain transport and higher-layer connections while moving. The other end of this forwarding tunnel is called a foreign agent. Then, this foreign agent knows how to reach the mobile node. When the mobile node sends packets to another mobile, it can send them directly over the foreign agent or also tunneled over the home-agent. The latter has the advantage of not revealing the mobile node's location.

Virtual private networks and mobile IP look similar in some sense: both can provide a possibly secure tunnel to a home network and also location transparency. Mobile IP can use IPsec to support encryption on the network layer. Besides IPsec, some libraries and applications can be used by a network implementing mobile IP to provide cryptography on the transport layer. SSL/TLS is an example for a library providing end-to-end security on the network layer. The secure shell (SSH) is another application. However, adding cryptographic capabilities at a later point in time after the development and deployment of a system is easily realized with traditional communicating nodes; but applying it to small mobile devices is really questionable, because of the special environmental restrictions of the mobile device. Chapter 3 will discuss the major features and uses of protocols IPsec, SSL, TLS, and SSH. It also addresses some of their flaws.

1.2.2 Infrastructureless Networks

When the basic wireless communication capabilities are integrated into mobile stations desiring to collaborate to realize a communicating system, then with little overhead the node can establish a data network without using the expensive service-provider network and without having a previously set up infrastructure.

The achieved networks are called ad hoc networks. However, in such a scenario it is quite hard to use security mechanisms that are as efficient as what we could expect, for instance, in wired systems. Basic reasons that explain why the security service provision in ad hoc networks is hard to achieve include the following:

1. The communicating entities can be heterogeneous; thus, while a large capacity node is easily able to deal with public-key cryptography, less equipped nodes could just have too small calculating capability and reduced resources.
2. Sensor networks, which can be considered as a special case of ad hoc networks, consider that the energy aspect is critical for the lifetime of the sensor nodes. Developing an energy efficient cryptographic is a major concern for the availability of wireless sensor networks.
3. A central public-key infrastructure (PKI) is unachievable in ad hoc networks, since node implementing the PKI would represent a single point of failure and the method would suffer from scaling problems. A hierarchical solution to deploy a PKI seems to scale better. However, the autonomy of the ad hoc nodes and their mobility over several hops might be error prone.
4. The redundancy and distribution, which are the major concepts in the development of ad hoc networks to prevent central point failures, may present more opportunity for the attacker to launch effective attacks on the basic functions provided by the ad hoc network.

1.2.2.1 Distributing Encryption Keys in Infrastructureless Wireless Networks

Public keys and security keys need to be distributed to mobile users. They can be used to authenticate the users, check their signatures, or decrypt their messages. Public key infrastructures (PKI) represent a good solution to provide key distribution. Even though a centralized PKI is not viable in typical infrastructureless networks, they are useful to achieve the security services as authenticity, integrity, and non-repudiation manner in infrastructureless systems as ad hoc networks. A distributed PKI is able to implement efficiently tasks such as keeping the private key secure, key distribution, and certificate issuance, as well as key revocation (Chapters 2 and 3 will detail these tasks).

Since the availability of the network is not always guaranteed, the deployment of a central PKI, or partly distributed PKI (where only a few nodes perform the basic PKI functions), presents some shortcomings. The new mobile nodes are not always able to get in the ad hoc network to deliver their public key so that a certificate can be issued for them; in addition, mobile nodes desiring to revoke their certificates will not be allowed always to do it. Network entities could possibly fall back to using unsecured communication or compromised keys. A solution to overcome this problem tries to create a fully distributed public-key infrastructure based on the central following idea: the threshold secret sharing, where n nodes are selected and each one is assumed to hold a part of a secret in such a way that whenever k $(k < n)$ nodes put their secret parts together, the secret can be recovered. The k nodes can then cooperate in a way to generate any certificate and allow key distribution. They can issue parts of the certificate with partial signatures, out of which the new node can compute its final certificate. This is done without revealing the actual secret key (the shared secret) at any time. This eliminates the burden of having predefined (central) points of failure. In fact, an attacker willing to obtain the shared secret needs to get control of k different nodes among the n nodes. In addition, if the attacker wants to destroy such a distributed PKI, he has to destroy $n - (k + 1)$ entities, at least.

Certificate revocation is realized in the aforementioned approach following two methods: implicit and explicit revocations. Implicit revocation is done by defining relatively short certificate life-times. Once a certificate has expired, a new one has to be issued (in the way defined above). Explicit revocation works using a counter for certificates; an entity creates a special signed message and floods it over the network. Others remember this revocation just as long as the revoked certificate would anyway still be valid. The two methods generate some unacceptable overhead: a lot of calculations are needed and the nodes are required to continuously renew their certificates. This calls for a tradeoff between simplicity and insecurity, saving battery life on one side and providing strong cryptography combined with highly robust protocols, on the other side.

The whole distributed public-key infrastructure can be used for securing the connections as well as for providing the certification services for any type of infrastructureless networks. Signing electronic contracts using this security platform can be handled because the basic requirements for contract signing can be satisfied. Therefore, authenticity, integrity, and non-repudiation are achievable. However, the whole solution as distributed public-key infrastructure probably will not satisfy the extreme high requirements to provide infrastructure for legal contracts. Related to this issue, some countries would require the certification authority to be independent from any network.

1.2.2.2 Routing Security in Infrastructureless Wireless Networks

Since there is no preexisting infrastructure, paths between nodes are formed spontaneously and cooperatively, and no one can guarantee that a path is free of malicious or misbehaving nodes. One can see the existence of a large spectrum of attacks on the routing protocols used in the infrastructureless wireless networks. In that sense, it is essential to make routing protocols more robust so that the misbehaving nodes can be put in quarantine.

Various secure routing protocol have been proposed in the literature (Chapter 8 describes the most important ones). These protocols use cryptography in order to secure establishing a route up to some extent. They guarantee that illegal route replies during route discovery can be detected and ignored, so that nodes trying to artificially attract routes will fail in most cases. In addition, a large set of secure routing protocols requires specifically the use of public key infrastructures so that any pair of nodes can authenticate each other, preserve information authenticity and integrity between them, establish a security association (or a shared secret) using public-key cryptography, and use the SA to efficiently protect message integrity and authenticity. On the other hand, intermediate nodes on the path kinking the pair of communicating nodes do not need to care about encryption.

To find a route in the infrastructureless wireless network, a requesting node (or source) constructs a protected request-message that carries source and destination address and a nonce (or a sequence number) to prevent replay attacks. Nodes receiving the request just forward messages (if they are not the destination or do not know how to get to destination) so that a message finally arrives at the node carrying a list of all nodes on a path leading to destination. The latter node will check the integrity of the request and then send back this list to the source (via the same path it has used to come). The path information sent back is also protected. Upon arrival back at the source node the integrity of the reply is checked again and the list of nodes on the return path compared with the route taken on the way to the destination.

Various attacks have been developed against infrastructureless networks and the routing they implement. Some of these attacks use forged or modified routing information during route discovery. Such attacks can be prevented using the previous approach. Another approach to protect against these attacks assumes that the nodes in the network maintain a priority list of neighbors to determine which neighbor sends or forwards more messages. A node gets higher ranking when it produces less requests. That way the malicious nodes flooding the network with invalid requests are automatically ranked as being inappropriate nodes for routing.

1.2.3 Heterogeneous Wireless Networks

Integrating the advantages of controlling procedures in infrastructure-based networks with the advantages of the ad hoc networks has led to the emerging of hybrid networks. Leading factors in the integration include the cost of network deployment, the need to have existing technologies interoperating and cooperating to achieve seamless roaming of mobile users, and the requirement to provide better provision of quality of service. Various solutions have been proposed to provide integration.

A particular solution created is the so-called Cellular Aided Mobile Ad hoc Network (CAMA; Bhargava, 2004). A typical CAMA architecture is depicted in Figure 1.2. It can be deployed and operated in areas where one (or more than one) mobile ad hoc network overlaps a cellular network covered area. The nodes that are in charge of operating CAMA network called CAMA agents, are deployed under the coverage of the cellular network and connected with the cellular network. Each CAMA agent covers a number of cells and knows which ad hoc network user is

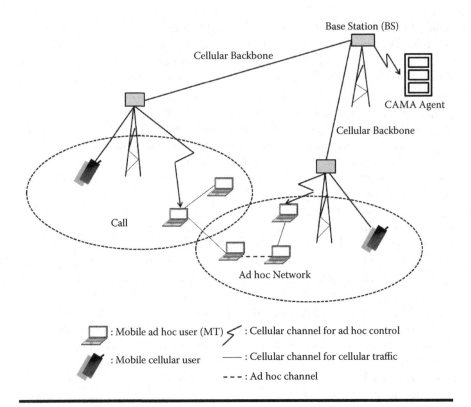

Base Station (BS)

Cellular Backbone

CAMA Agent

Cellular Backbone

Call

Ad hoc Network

⬚ : Mobile ad hoc user (MT)   ⌇ : Cellular channel for ad hoc control

◣ : Mobile cellular user   —— : Cellular channel for cellular traffic

- - - : Ad hoc channel

Figure 1.2   CAMA architecture.

registered as a CAMA user. Agents collect information for the entire ad hoc network and are involved in performing its major functions, including authentication,
routing, and security. A mobile station may contact the CAMA agents through the
cellular network's radio channels to exchange the control information. Also, position information can be retrieved (via GPS) and used to calculate optimal routes
in CAMA based on a central position database. Thus, CAMA presents no need for
route discovery. Instead, nodes could just ask for routes from the CAMA agent.

When a node needs to know the route to another node, two basic approaches
can be used. In the first approach the CAMA agent at the provider side computes
the route and transmits it directly to the mobile node wanting to transmit something. The second approach allows the mobile node to determine the route itself.
On the other hand, power can be saved in CAMA by letting mobile nodes move
to an idle state and just listen to the cellular channel. Whenever the CAMA agent
gets a routing request and makes a routing decision, it wakes up the nodes involved
in the routing process. After transmitting all the packets these nodes can get back
to the idle state.

Solutions for 3G hybrid networks have been proposed following similar ways. A more general solution considers the 4G system, an abbreviation for Fourth-Generation Communications System. This is a concept used to describe the next step in wireless communications. A 4G network will be able to provide a complete IP solution where voice, data, and multimedia can be provided to mobile users on an "Anytime, Anywhere" basis, and at higher data rates than the 3G systems. Nowadays, there is no finalized specification for what 4G system is (Zarai, 2007). 4G systems aim at providing different objectives including (a) they should be fully IP-based integrated systems; (b) they should be capable of providing between 100 Mbit/s and 1 Gbit/s speeds (both indoors and outdoors); and (c) they should be able to offer all types of services at an affordable cost.

The major challenge in a heterogeneous networking environment based on three different cellular access networks (GSM, GPRS, UMTS), WLANs, and ad hoc networks is to exploit the advantages of WLAN systems focusing on their seamless integration in composite radio environments, ease of deploy, and limitation of installation cost. Mobile nodes that are connected to the cellular backbone can use the ad hoc network to exchange data with each other. The usage of cluster heads to route traffic via the ad hoc networks can reduce the load of the cellular networks.

1.3 Wireless Vulnerabilities and Threats

A vulnerability can be defined as a weakness (or fault) in the communication system, its components, the medium it uses, or the protocol it implements that allows compromising the security of the network (or one of its components). Most of the existing vulnerabilities in the wireless networks are caused by the medium. Because transmissions are broadcast, they are easily available to anyone who can listen to them. Particular threats of the wireless communication are device theft, malicious hacker, malicious code, and theft of service (Boncella, 2006). Multiple wireless vulnerabilities and threats have been studied in the literature for the purpose of detecting attacks, exploiting them, and providing appropriate reactions. They can be addressed following two dimensions, the environment where they can be activated and the assets they target. In the first dimension, we distinguish two groups of vulnerabilities and threats: those existing in a LAN-like wireless network (WLAN) and those existing in cellular-like wireless networks (Hutchison, 2004). In the second dimension, three groups of threats can be characterized: the application-based threats, the content-based threats, and the mixed threats.

In the following, we discuss the basic features of the five groups of vulnerabilities and threats along with security policy requirements that help reduce vulnerabilities and attack damages. The basic attacks exploiting the discussed vulnerabilities will be presented in the following section of this chapter.

1.3.1 WLAN Vulnerabilities and Threats

The following represent the typical vulnerabilities witnessed at the main component of WLAN, namely the access point (AP).

■ *The easy installation and use of an AP.* This vulnerability allows any individual to introduce an unauthorized wireless network in unauthorized areas. The easy installation and configuration of the AP make this feasible for legitimate or illegitimate users.

■ *The AP configuration.* If the AP is poorly configured or unauthorized, then it can provide an open door to attackers. This is caused by using a default configuration that annihilates the security controls and encryption mechanisms that the AP is able to provide in normal use.

■ *Physical security of an authorized AP.* Because most APs are deployed by default, their placement and ease of access are critical. An AP has to be correctly placed and physically protected in order to avoid accidental damage (made, for example, by a direct access to the physical cable attaching the AP). Many solutions were proposed to physically protect the access to the AP, but all of them require a mandatory policy.

■ *Signal range of an authorized AP.* This vulnerability is characterized by the possibility that the AP signal strength extends beyond a given perimeter (the perimeter of a building, for example). Consequently, the AP's placement and the signal strength have to be closely studied to make sure that the transmitting coverage of the AP is just enough to cover the required area and does not extend out of this area.

■ *Rogue AP.* This vulnerability allows an attacker to place an unauthorized (or rogue) AP on the network area and configure it to look legitimate to the network users to gain access to a wireless user's sensitive data. The vulnerability is represented by the criteria of AP selection implemented within the mobile stations. Indeed, the user's devices need to be connected to the strongest available AP signal.

■ *Protocol weaknesses and capacity limits on authorized APs.* These vulnerabilities can cause Denial of Service attacks (DoS) from malicious users utilizing unauthorized APs when they can flood authorized AP with traffic forcing them to reboot or deny accesses.

1.3.2 Cellular System Vulnerabilities and Threats

Vulnerabilities and threats commonly observed in cellular communication systems contain the following four major categories (Randall, 2002):

- *Service interruption vulnerabilities*: The increased capacity offered by the high-speed communication technologies has resulted in the reduction of cable routes necessary to meet traffic capacity requirements. Consequently, this has decreased the number of switches and enhanced their capacities, and increased the vulnerability of telecommunication infrastructures.
- *Natural threats*: These threats comprise a large category of natural events such as the climatic, geological, or seismic events. Severe damages resulting from natural disaster can cause long-term damage to the wireless communications infrastructures.
- *Handset vulnerabilities*: Unlike computer systems, handsets are limited regarding the security features. The implementation of security mechanisms can present some weaknesses allowing attackers to launch successful attacks.
- *Radio link protection-only vulnerability*: Because wireless messages travel through the air, between the handset and the access node, for transmission to the receiver, messages may need to be changed to another protocol. Such change can be done at a gateway, for example, to allow a wireless transport layer security (WTLS) message to be changed to a secure socket layer (SSL) message. This operation presents some vulnerability, because anyone may attempt to access the network at this moment and get the message during transformation.

1.3.3 Application-Based Threats

Application-based threats are roughly posed by executable malicious codes that are inserted into existing or new wireless applications. They are potentially present anytime a software program is downloaded to (or executed) on a wireless terminal. This is particularly true when the program is downloaded or received from an unknown source. These threats are equivalent to the earliest type of computer viruses that attacked executable programs.

The first malicious application-based program that specifically targeted the operating system used in personal digital assistants (PDAs) was called the Liberty Crack. The free software, which could be downloaded from a Web site or accessed via Internet relay chat rooms, pretended to convert the shareware specific game program into a registered version. When the program is executing, the user cannot see that the program is simply deleting all executable applications in the handheld device. Fortunately, Liberty Crack did not affect the underlying operating system of the embedded applications.

1.3.3.1 Content-Based Threats

In content-based threats, the content is the threat (e.g., derogatory messages) or the malicious use of the content is the threat (e.g., spamming of email). Networks have been known to crash under the weight of spam attacks. While email is one of the key features of the wireless world, email is also one among the most vulnerable to attacks. Hence, the most common content-based threats to the wireless infrastructure occurred through infected email or spam mail. The first content-based threat against wireless devices occurred in June 2000 with the so-called Visual Basic Script (VBS). The related attack proliferates by sending infected email messages from affected computers. When an infected email reaches a PC, it uses Microsoft Outlook to send a copy of itself via infected emails to all addresses in the MS Outlook address book.

Although the program reached out into the wireless world, it was benign and caused little damage because it propagated via PCs and emails, not directly from mobile phone to mobile phone. Nevertheless, the attack demonstrated the ability of a malicious code to hit the wireless infrastructure and spread with a considerable speed. The attack has shown the potential to flood the wireless network with messages, reducing its performance or even reducing its ability to meet expected load. Worse, it has demonstrated its ability to impact the billing features. In fact, wireless users billed on a per-message basis may need to support the most of receiving spams.

1.3.3.2 Mixed Application

The third type of threat offers a greater potential for damage than the previous two types of known threats. While still considered to be theoretical, a mixed application threat would integrate techniques from content-based and application-based threats. Considerable damages can be achieved by such threats. For example, an attack could involve the unwitting download of a sophisticated malicious code attached to a shareware program that deletes wireless device applications and propagates itself rapidly across the wireless infrastructure via address books of email. This attack can cause damage to any mobile station it visits, and spreads across very large areas over a limited period of time.

It will be shown in the sequel that some mixed attacks have been created (with Nimda.A, for example) that replicate and spread rapidly. Consequently, it appears that some type of highly destructive and rapidly spreading wireless mixed threats will inevitably surface and that an adequate comprehensive wireless infrastructure protection against it is needed.

Nowadays, cellular phones are used almost exclusively for voice communication. However, cellular communication technology is already merging with the platform-independent programming models and new technologies such as Bluetooth. In a near future, cell phones will be able to send and receive data and applications, directly to another wireless device cell phone. Unfortunately, this expected wireless environment is unlikely to come without the price of increasingly sophisticated wireless mixed threats utilizing high capabilities of connectivity, functionality, and speed.

1.4 Attacks in Mobile Environments

Detecting a large set of attacks by a wireless intrusion detection system (WIDS) requires studying and discovering the attacker's methods and strategies. We discuss in this subsection the typical attacks and malicious events that can be detected by a WIDS (Valdimirov, 2004). We then discuss in a following section the typical application-based attacks and give a picture of the basic detection and protection techniques.

1.4.1 Typical Attacks

Basic attacks can be classified into four major classes, namely the illicit use, the wireless spoofing, the man-in-the-middle attacks, and the denial of service attacks. A description of the features of the typical attacks is given as follows.

1.4.1.1 Class of Illicit Use Attacks

Illicit use is a passive attack that does not cause damage to the physical network. It involves an attacker that is placed close to AP (or BS) and gets illicit information extracted from the traffic it can listen to. Illicit use includes the following attacks (Mateli, 2006):

■ *Wireless network sniffing*: When wireless packets traverse the air, attackers equipped with appropriate devices and software can capture them. Sniffing attack methods include the following:
  - *Passive scanning*: This attack aims at listening to each wireless communication channel and copying, for future analysis, the traffic flowing through it. It can be done without sending information and can use some tools such as the radio frequency monitors, which allow copying frames on a channel.
  - *Identity detection*: This attack consists in retrieving the identity of important entities occurring in a wireless network (such as the identity of the AP, in WLAN) by scanning specific frames such as the frames of the following types: beacon, probe requests, probe responses, association requests, and re-association requests.
  - *MAC address collection*: To construct spoofed frames, the attacker has to use legitimate MAC addresses. These addresses can be utilized for accessing active AP by filtering out the frames with non-registered MAC addresses.
■ *Probing and network discovery*: This attack aims at identifying various wireless targets. It uses two forms of probing: active and passive. Active probing involves the attacker actively sending probe requests with no identification using the SSID (Service Set Identifier) configured in order to solicit a probe

response with SSID information (and other information) from any active AP. When an attacker uses passive probing, he listens on all channels for all wireless packets.

- *Inspection*: The attacker can inspect network information using tools such as Kismet and Airodump (Low, 2005). He could identify MAC addresses, IP address ranges, and gateways.

1.4.1.2 Wireless Spoofing

The spoofing intent is to modify identification parameters in data packets for different purposes. Typical spoofing attacks include the following:

- *MAC address spoofing*: MAC spoofing aims at changing the attacker's MAC address by a legitimate MAC address. This attack is easy to launch because some client-side software allows the user to manipulate their MAC addresses.
- *IP spoofing*: IP spoofing attempts to change the source or destination IP addresses by talking directly with the network device, for example.
- *Frame spoofing*: The attacker injects frames with spoofed content. When the network lacks authentication, spoofed frames cannot be detected.

1.4.1.3 Man-in-the-Middle Attacks

This attack attempts to insert the attacker in the middle (MITM attack) of a communication for purposes of intercepting client's data and modifying them before discarding them or sending them out to the real destination. To perform this attack, two steps have to be accomplished. First, the legitimate AP serving the client must be brought down to create a "difficult to connect" scenario. Second, the attacker must set up an alternate rogue AP with the same credentials as the original for purposes of allowing the client to connect to it. Two main forms of the MITM exist: the eavesdropping and manipulation MITM attacks. Eavesdropping can be done by receiving radio waves on the wireless network, which may require sensitive antenna. Manipulation requires not only having the ability to receive the victim's data but then be able to retransmit the data after changing it.

1.4.1.4 Denial of Service Attacks

Denial of service (DoS) attacks aim at denying or degrading the quality of a legitimate user's access to a service or network resource. It also can bring down the server offering such services itself. DoS attacks can be classified into two categories:

- *The disabling services attacks*: A DoS attacker makes use of implementation weaknesses to disable service provision. Weaknesses that are used with these attacks include buffer overflow.
- *Resource undermining*: Undermining can be achieved by causing expensive computations, storage of state information, resource reservations, or high traffic load.

The techniques used in DoS attacks can be applied to protocol processing functions at different layers of the communication architecture. DoS attacks can threaten the services offered to mobile users (e.g., servers offering specific information, or servers of specific companies) and the communication infrastructure itself. Especially, specific access resources such as bandwidth can represent a serious problem (since it most likely will remain a scarce resource in access networks). DoS attacks can target different network layers as explained in the following:

- *At the application layer*: DoS occurs when a large amount of legitimate requests are sent. It aims to prevent other users from accessing the service by forcing the server to respond to a large number of request transactions.
- *At the transport layer*: DoS is performed when many connection requests are sent. It targets the operating system of the victim's computer. The typical attack in this case is a SYN flooding.
- *At the network layer*: If the network allows associating clients, an attacker can flood the network with traffic to deny access to other devices. Typically, this attack is performed by allowing one among the following three tasks:
  - The malicious node participates in a route but simply drops several data packets. This causes the deterioration of the connection.
  - The malicious node transmits falsified route updates or replays false updates. These might cause route failures, thereby deteriorating performance.
  - The malicious node reduces the time-to-live field in the IP header so that packets never reach destinations since they are dropped by other nodes before destination.
- *At the data link layer*: DoS targeting the link layer can be performed as follows:
  - Since we assume that there is a single channel that is reused, keeping the channel busy in the node leads to a DoS attack at that node.
  - By inducing a particular node to continually relay spurious data so that the battery life of that node may be drained. An end-to-end authentication may prevent these attacks from being launched.
- *At the physical layer*: This kind of DoS can be executed by emitting a very strong RF interference on the operating channel. This will cause interference to all wireless networks that are operating at or near that channel.

1.4.2 Distributed DoS Attacks in Mobile Communications

To make DoS threats worse, recent reports indicate that attackers have developed effective tools to coordinate distributed denial of service (or DDoS) attacks that can be launched and coordinated from a large number of sites. A DDoS attack is distinguished from a common DoS attack by its ability to launch its actions in a distributed manner over the wireless communicating system and to aggregate these forces to create dangerous traffic. According to different reports including the annual CSI computer crime and security report, the DDoS attacks have induced large financial costs to companies in recent years (Richardson, 2007). In addition, they caused large damage to consumer confidence in e-commerce.

There are various types of DDoS attacks. They all share the same typical structure that is depicted in Figure 1.3. The attacker, in a DDoS, first gains control of several master computers connected to the wireless network by hacking into them, for example. Then the master computers gain control of more computers (referred to as zombie computers) by different means. Finally, a message is sent by the attacker to synchronize all zombies to send the required traffic to the victim. In the following we first describe two examples of mobile systems that are targeted

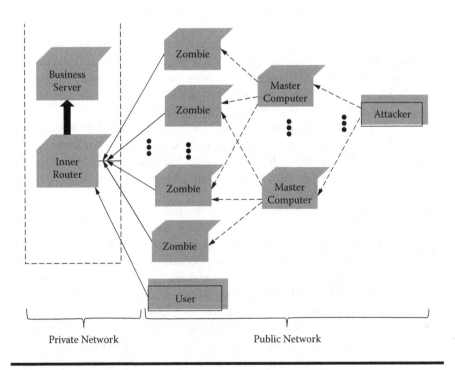

Figure 1.3 Typical DDoS structure.

by DDoS attacks. Then we present some of the countermeasures that should be provided to protect against DDoS.

1.4.2.1 Targeted Environments

Two wireless communication systems are of interest to DDoS attackers, the wireless extended Internet-based networks (WEIN), where wireless technology is used only for the last mile, and the ad hoc networks (AHN), which represent, in the opinion of a large number of experts, the best architectures against DDoS attacks, since they have no central nodes and may implement severe admission policies making it very difficult for malicious users to enter into the communication infrastructure. An example of WEIN is a network that is able to connect mobile devices to fixed networks via radio frequency (RF) channels using the traditional Client/Server architecture and the existing transport layer protocols (e.g., TCP). All the DDoS attacks achievable in the wired Internet are still feasible in the WEIN.

DDoS targeting WEIN and mobile ad hoc networks include, but are not limited to, the following attacks:

- *Attacking the wireless Internet content servers*: Since mobile devices have little computation and communication capabilities, a DDoS attack, even launched by a small number of powerful fixed terminals, can effortlessly disable a large range of mobile devices. Wireless Internet content servers, such as the WAP server, the wireless game servers, and the mail server, are often optimized for small throughput and timely response. They are particularly vulnerable to DDoS attacks compared with traditional wired servers. Furthermore, new forms of DDoS attacks may emerge taking advantage of the attractive features presented by the WEIN and ad hoc networks.
- *DDoS attacks on radio spectrum*: Often, the limited availability of radio spectrum is the bottleneck in a mobile network. Even if license-free RF bands are used and pico-cell-based (or reduced area) technologies are employed to expand transmission rates, the radio spectrum is still a scarce resource as the number of users and the demand for bandwidth is increasing tremendously. A DDoS attack can deliberately coordinate mobile devices to send out synchronized traffic to easily consume all spectrum resources or (at least) significantly reduce the capacity of any communication channel offered by the networks.
- *Attacks aiming at avoiding tracing back DDoS*: Some of the WEINs, such as the mobile IP protocol based networks, present weaknesses that a DDoS attacker can use to launch attacks. For example, the Mobile IP protocol requires two IP addresses: the home address and the care-of address. The home address is permanently assigned to a mobile device, while the care-of address is temporarily assigned by the visiting foreign network. This allows a mobile device to send IP packets using its fixed home address, even when it is roaming, while

applying the Non-Disclosure Method (NDM), which gives mobile users control over the revelation of their location information. Consequently, victim sites will find it hard to trace sources of DDoS attacks.

■ *DDoS attacking devices using aggregated traffic*: Although the bandwidths used for the transmission in WEIN and ad hoc networks are much lower than those in wired networks, potential DDoS attacks are feasible mainly because of the fact that a large set of mobile devices can be involved. In particular, any wireless data packet traffic is a potential path for DDoS attacks.

1.4.2.2 Defending against DDoS Attacks

In the event of a typical DDoS attack, the victim alone cannot effectively defend itself. Cooperation among all involved parties is necessary. Typical methods to protect against DDoS attacks focus on effective coordinated technological solutions. There are three major types of coordinated technological solutions: (a) improving the security of all relevant devices; (b) enhancing the User-level traffic control; and (c) coordinating filters and tracing back methods (Geng, 2000).

1.4.2.2.1 Improving the Security of the Relevant Devices

Before initiating an effective DDoS attack, the attacker needs to involve enough zombie devices to secure the ability to generate sufficient traffic. An ineffective and direct countermeasure is to secure all devices to make it difficult for the attacker to install and take control of a large number of zombies. An alternative and effective solution would be to selectively secure those devices that have high traffic throughput, such as routers in the WEINs or the clusterhead nodes in the ad hoc networks.

1.4.2.2.2 Mobile User-Level Traffic Control

The traffic control, at the user-level, can be achieved by a set of traffic control rules. For example, the mobile user can set up a rule that fixes a daily traffic limit that is high enough not to disturb the normal activity of the user, while the unusually large traffic is stopped and may trigger an alarm (to the user or to a network administrator) for a subsequent diagnosis. The traffic control rules can also describe the data to be dropped or delayed if the network is experiencing congestion. An alternative solution can use a timestamp model to control traffic even when user devices are hacked. This technique, however, experiences some drawbacks, including the fact that user-level traffic control rules for a specific network device need to be protected more securely than the network device itself. Edge routers in the WEINs are the perfect hosts for coordinating user-level traffic control rules. On the other hand, the designation of a host for traffic control rule coordination is more complicated in a mobile AHN, since no node (including the clusterheads) is more likely to be in a central position than another to host the rules.

1.4.2.2.3 Coordinated Filters and Tracing Back

Wireless Internet service providers in the WEINs can try to overwhelm the DDoS attacks by identifying the attacking traffics and stopping them using coordinated filters, whose aim is to stop the traffic as early as possible, along the attacking paths, to prevent the damage from aggregated traffic. For a mobile AHN, the filtering is not directly applicable due to the symmetric structure of the AHN. However, a dynamic voting mechanism may play an essential role to select those in charge of performing this function.

Cost-effectiveness arises as a crucial issue in the defense against DDoS attacks, because it may require the update of the current network infrastructure. Several advanced network management technologies have been proposed to address the traffic control problem. The use of these technologies will significantly reduce the costs and risks in designing future WEINs. In particular, Policy Based Networking (PBN) represents a promising technology for implementing usage-based fees practices to deal with DDoS attacks (Yavatkar, 2000). PBN provides rules that describe actions to take when specific conditions occur. These rules are able to control critical network resources such as bandwidth, QoS, security, and Web access across heterogeneous networks. Thus, it allows congestion to be under the control of a globally coordinated structure.

In the PBN typical scheme, two components can be distinguished for the traffic control: the Policy Enforcement Point (PEP) and the Policy Decision Point (PDP). While the Wireless Location Register/Authentication Center is a PDP with additional functionality such as accounting and policy information storage, the PEP accepts or denies requests appropriately, at the network border points. PDPs and PEPs can exchange policy information through secure and reliable channels to achieve efficiently their roles.

1.5 Mobile Malware

1.5.1 Basics on Malware

Malware (or Malicious Software) can be any malicious, unauthorized, or unexpected program (or code) that aims at realizing unauthorized actions on a computer, network components, or a mobile terminal. Some examples of the actions a malware can perform include spying on wireless traffic, recording private communications, stealing and distributing private and confidential information, disabling computers, and erasing files. Malware can be divided into eight different categories:

1. *Worms*: A worm is a program that makes copies of itself (by various means including copying itself using email or another transport mechanism). A worm may damage and/or compromise the security of the visited (or infected) computer by executing special actions.

2. *Zombies*: A zombie is a program that secretly takes over another Internet-attached computer and then uses that computer to launch attacks that are difficult to trace to the zombie's creator. Zombies can be used to launch denial of service attacks, typically against targeted Web sites. The zombies can be installed on hundreds of computers belonging to unsuspecting third parties. They are then used synchronously to overloading the victim target by launching an overwhelming onslaught of Internet traffic.

3. *Viruses*: A virus is a sequence of code that is inserted into another executable code, so that when the regular program is run, the viral code is also executed. The viral code causes a copy of itself to be inserted in one or more than one program. Viruses are not distinct programs; they cannot run on their own and need to have some host program, of which they are a part, executed to activate them.

4. *Trojan Horses*: A Trojan is a malware that performs unauthorized, often malicious, actions. The main difference between a Trojan and a virus is the inability to replicate itself. Like a virus, a Trojan can cause damage or an unexpected system behavior, and can compromise the security of the visited systems; but, unlike viruses, it does not replicate. A Trojan looks like any normal program, but it has some hidden malicious code within it.

   Often, a Trojans is composed of two parts, a client part and a server part. When a victim executes a Trojan server on his machine, the attacker then uses the client part of that Trojan to connect to the server and start using it based on TCP or UDP, for example. When a Trojan server runs on a victim's computer, it (often) tries to hide somewhere on the computer; it then starts listening for incoming connections from the attacker on one or more ports; then attempts to modify the system registry or use some other auto-starting method. Most Trojans use an auto-starting method that allows them to restart and grant an attacker access to the infected machine.

5. *Logic Bombs*: A logic bomb is a programming code inserted secretly or intentionally. The bomb is designed to execute (or explode) under special circumstances, such as the amount of time elapsed since an event has occurred. It is in effect a delayed-action computer virus or Trojan. A logic bomb may be designed to display a fake message, delete data, corrupt data, or have other undesirable effects, when executed.

6. *Trap Doors*: A trap door, sometimes called back door, is a secret entry point into a program that allows someone that is aware of the trapdoor to gain access without going through the usual security access procedures. The difference between a trap door and a Remote Access Trojan (RAT) is that the

trap door only opens a port, often with a shell. The RAT is designed with a client-server architecture.

7. *Phishing Scam (PS)*: A PS is a fraudulent Web page, an email, or a text message that attracts the unsuspecting users to reveal sensitive information such as passwords, financial details, or other private data.

8. *Spyware*: A spyware is a software that reveals private information about the mobile user or its computer system to eavesdroppers.

The first example of a mobile malware for mobile cellular phones was built in June 2004, for the Symbian operating system. The antivirus companies now have hundreds of Trojans and worms for mobile phones in their antivirus databases, and new malicious programs have become a constant stream. Nowadays, about ten viruses are added every week. The worst thing in that information is that worms and Trojans for mobile devices are spreading so fast, causing more and more outbreaks. Two major reasons can explain this spreading: First, one can notice that the typical mobile phone user is less security conscious than the average Internet user. Second, for a long time mobile users had treated mobile malware as a problem that is not a real concern to them.

Let us also notice that today's mobile malware are very similar to computer malware in terms of the techniques they can use. However, while it took computer viruses over two decades to progress, the mobile viruses can cover the same ground in a few years. No doubt, mobile malware is the most quickly evolving type of malicious code. A short list of the actions that a mobile virus can do includes, but is not limited to, the following actions:

■ Block memory cards;
■ Combat antivirus programs;
■ Infect personal files;
■ Modify icons and system applications;
■ Install "false" or non-operational fonts, applications, and malicious programs; and
■ Steal data and send messages to other users.

A study made by McAfee, Inc., has revealed that mobile operators globally are experiencing more mobile malware attacks than ever before, and that they are spending more time and money on recovery from these attacks. The study demonstrated that nearly half of the network operators who have experienced mobile malware outbreaks have had one within the last three months prior to the study. Twice as many mobile operators spent over $200,000 on mobile security in 2006, compared to 2005. The purpose of the study was to discover to what extent mobile operators are affected by mobile threats. The findings of the study revealed that (McAfee, 2008):

- 83 percent of mobile operators questioned have been hit by mobile device infections.
- The number of reported security incidents in 2006 was more than five times higher than it was in 2005.
- The number of mobile operators in Europe reporting incidents affecting more than 1,000 devices more than doubled in 2006.
- 100% more mobile operators spent over $200,000 on mobile security in 2006 compared to 2005.
- The number of mobile operators estimating that the cost of dealing with mobile threats is more than 1,000 hours and increased by 700 percent.

Furthermore, almost three operators out of ten stated that subscriber satisfaction had suffered more than any other factor and that the second most serious impact from mobile malware infections was on the network performance.

1.5.2 Examples of Mobile Malware

Nowadays, mobile phones are equipped with well performing operating systems (OS) such as the Symbian OS, Microsoft Mobile OS, and Palm OS. These OSes present interesting features such as built-in cameras, high-resolution color screens, wireless data access, MP3 players, email services, and useful tools such as calendars and address books that can be linked wirelessly with a computer. Some mobile phones are also equipped with Bluetooth or/and other wireless technologies, making them directly accessible from computers. It is expected that mobile malware will represent a new mobile threat and will become more serious in the near future.

In the sequel, we will show that Bluetooth and Symbian OS represent fine examples of how the technology can be abused to distribute mobile malware. Let us recall first that Symbian OS is the advanced operating system licensed by the world's leading mobile phone manufacturers (e.g., Nokia, Motorola, and Sony Ericsson). Designed to comply with the specific requirements of advanced cellular communications (e.g., 2G, 2.5G, and 3G), Symbian OS combines the power of an integrated applications environment with mobility, bringing advanced data services to the public. Mobile phones that run on the Symbian platform are relatively large and includes, among others, Nokia 6600 and 7610, Sony Ericsson P900 and P910, and Motorola A925.

Packaged with an OS and multiple applications, the mobile phones—just like computers connected to a wireless network—are vulnerable to security threats malware such as worms and Trojan horses. Recently, security experts and antivirus companies have identified the different malware that have emerged on mobile phones. We list below some among the most known malware:

- **Cabir:** Cabir is a worm. It was the first identified malware for cellular phones. It uses Bluetooth to infect the phones and to transfer itself to a new host as a file. Two new versions of Cabir worm, namely Cabir.H and Cabir.I, have been created. They are able to search for (and find) new mobile targets. They spread faster between mobile phones using a specially formatted Symbian Installation System (SIS) file.

When infected by Cabir worms, the mobile phone's OS is modified so that Cabir is executed each time the phone is switched on. The infected mobile phone also will have to scan for vulnerable phones using Bluetooth. Finding a target, the phone will send the so-called "velasco.sis" file including the Cabir worm. The versions Cabir.H and Cabir.I do not destroy data on the mobile phones they infect. Instead, they block legitimate Bluetooth wireless connections and rapidly consume the phone's battery.

■ **Cardtrap.A:** This Trojan has the capacity to infect computers when users transfer data from their infected mobile phones to computers. It may have a built-in mechanism that places several worms on a mobile device's memory card with the final objective of infecting a computer. In addition to placing two traditional worms (namely, the Win32/Padobot.Z and Win32/Rays worms), the malware Cardtrap.A also creates an autorun file on the card so that when the card is inserted into a computer, it automatically installs and runs the malware. It also overwrites normal applications installed on the infected mobile, preventing them from working properly.

The basic objective of the Cardtrap.A virus is most likely to cause the user to infect his computer with worms in the action of attempting to disinfect the phone. A typical damage would be achieved when a user, who discovers the infection made, would be to insert the phone memory card into the PC to copy the file manager or disinfection tool to the card. But this is precisely what would create the computer infection. Cardtrap.A presents a low risk to most mobile users, since it has not spread very far at this point. It can infect devices running some versions of the Symbian OS, including different products from Nokia, Panasonic, Sendo, and Siemens.

■ **Commwarrior:** This is the first worm to spread via MMS. Like Cabir, it can spread via Bluetooth. MMS is the main method used, making Comwar potentially extremely dangerous, since Bluetooth operates within a distance of about 15 meters and any device can be infected if it is within this range. MMS has no boundaries and can be instantly sent even to handsets in other sites. Currently, more than seven modifications of this worm can be distinguished. One of these variants (Comwar.g) was the first to include a file infector functionality. The worm looks for other sis files in the phone's memory and appends its code to these files. This provides an additional propagation method compared to the traditional MMS and Bluetooth.

■ **MetalGear:** This Trojan horse combines several malicious mobile phone programs that work on the infected phone to spread over Symbian-based phones. Like a fake version of the Metal Gear Solid game, it disables antivirus programs and other programs. It then installs the Cabir worms. This installer adds code that disables the handset's Menu button.

■ **FlexiSpy:** This malware was discovered in March 2006. It is a spyware that is typically installed by someone other than the phone owner. It sends a log of phone calls and copies of texts and MMS messages to a commercial Internet server for viewing by a third party.

Very few mobile viruses are truly original. Cabir served as the basis for a number of its own variants, which differ only in terms of the file names and the contents of the sis installation files. Cabir was also used as the basis for new families such as Lasco and Pbstealer. Malware from the Lasco family are capable of infecting files in the phone memory. Pbstealer, which is the first Trojan spy for Symbian, is another Cabir-like malware. It searches for the phone's address book and sends data contained in it via Bluetooth to the first device found in that list. The name Pbstealer stands for "Phonebook Stealer." Until the construction of Pbstealer, the cybercriminals used various vulnerabilities in the Bluetooth protocol to steal such information (e.g., BlueSnarf).

To date, some of the mobile malware have failed to spread. Users can prevent attacks by disabling Bluetooth and declining to accept and install any new software from the networks, especially pirated software. On the other hand, according to several experts, users most likely to be hit by Trojan horses, such as MetalGear, are typically those who like to download new software from Symbian freeware sites or peer-to-peer networks. However, the only way to protect against Trojans is to reset the infected phone to its default factory setting. Unfortunately, this means that all the data and configuration stored in the infected phone will also be lost.

1.6 Prevention Techniques in Mobile Systems

Two key mechanisms are used by the enterprises to provide access control to their resources via a wireless means, namely the intrusion prevention and intrusion detection. Intrusion prevention is a first line of defense. It is carried out through encryption and authentication (e.g., using passwords or biometrics), anti-virus applications and firewall, whereas intrusion detection is performed by mechanisms that are able to recognize misuse and/or anomaly in communication activities and use of resources.

1.6.1 Firewall in Mobile Communication

A firewall can be defined as a communication device placed between a network (the protected network) and another network (the public network) that is able to filter access to the protected network. The firewall observes all traffic routed between the two networks to check whether it meets specific criteria. If a criterion is met, the traffic is routed between the networks; otherwise, it is stopped. The firewalls can be used to keep track of all attempts to enter the protected network and trigger alarms when hostile or unauthorized actions are attempted. They can filter packets based on the content of their fields to achieve address filtering (using source and destination addresses, port numbers) and protocol filtering (using specific types of network traffic).

A special firewall is called the personal firewall; it can be defined as a computer having the ability to filter its incoming and outgoing traffic. A personal firewall is a piece of software or hardware whose responsibility it is to protect the machine it is installed on. In addition to the already implemented functions in it, the personal firewall can be post-configured by the user. To this end, the user sets special rules for handling the traffic. Based on these rules, the personal firewall can deny traffic coming from untrusted source or allow it. On the other hand, the personal firewall is able to monitor every application (implemented in the machine the firewall is installed on) that attempts a connection to the Internet or external network. In particular, a personal firewall allows outgoing traffic from applications that are on trusted application list. This is an important measure for preventing Trojan horse programs from communicating with the Internet.

The personal firewall is able to stop Trojans that seek to control the system from an incoming connection, as well as scripts that attempt to send emails using the user's name. It allows the user to control which peers can view and access shared folders, and even prevents others from detecting a computer's presence on the network. However, some spyware programs are getting smarter, for example, knowing that certain personal firewalls look at the name of the application to decide whether its outgoing traffic is allowed, and the spyware has the ability to rename listed names. This is why detecting outgoing traffic is an important feature in personal firewall, and outgoing traffic should be based on additional control such as the checksum of the entire application, instead of just its name.

Personal firewalling in a mobile station can be responsible for the scan for patterns of network traffic that indicate a known attack attempt. It may even have a maintainable list of patterns (or intrusion signatures) to respond to newly discovered attacks methods. Some of the functions in the personal firewall could be useful in preventing mobile phones from being infected by the mobile malware. Among these functions, one can mention (a) the monitoring of incoming and outgoing traffic; (b) the detection of signs or attempts of attacks; and (c) the detection of active content nuisance. Monitor incoming traffic should block programs and packets that may include insecure and unusual content. Indeed, the personal firewall can detect any abnormal behavior from programs installed in the mobile station or unusual content in packets processed by it. It can also allow only certain trusted servers, applications, and specified sources to issue traffic to the mobile user.

In addition to monitoring the incoming (and outgoing) traffic and filtering and blocking traffic in a mobile station, the personal firewall can provide anti-virus protection and should have secure software built in phones. It should also provide functionality for anti-virus detection and email virus detection. To prevent file inflecting viruses or Trojans, the mobile phone should include monitor outgoing traffic function in personal firewall. Every time the mobile phone starts to send traffic, the user should get a warning on the display. When this function is enabled, the user is aware of what program or content mobile phone is trying to broadcast. In addition to the mentioned functions, the personal firewall should include detection intrusion attempts and active contents nuisance. This will make the user more secure when it is connected to Internet. The list below summarizes the main reasons why personal firewall is important.

■ The personal firewall prevents a wide range of attacks coming from the network including address IP spoofing, port scanning, and denial of service. It can contribute indirectly to the provision and conservation of quality of service for all users on the network.

■ The personal firewall prevents billing attacks, in which the attacker can run another user's bill simply by involving them in the exchange of IP traffic.

■ The personal firewall contributes in preventing the mobile stations from consuming extra processing power and draining their batteries. This can be done by adding, to the firewall, filtering rules to reduce unnecessary outgoing and incoming traffic and stop leaking information.

■ The personal firewall supports the efficiency of mobile communication functions such as the peer-to-peer services over IP. It also helps protect communication protocols such as the WAP and HTTP.

■ The personal firewall protects mobile terminals from being infected by mobile viruses and insecure content in downloaded games and applications from the Internet. This can be done by personal firewall–specific functions such as the active content nuisance, the JavaScript Pop-Ups blocker, and the collection of behavioral patterns.

1.6.2 Encryption-Based Mechanisms for Prevention

While the need for encryption is accomplished by making use of suitable cryptographic mechanisms, authentication functions and protocols are implemented at different layers of the communication architecture. Their aim is to produce an authenticator, a value that is used to verify the identity of an entity. The authentication protocols fall into one of two categories: First, the protocols authenticating devices are typically implemented at the link layer using hardware or software. Second, higher-layer protocols authenticate users and provide other security services. Examples of the second category include end-to-end encryption and non-repudiation. These security protocols can be implemented using a combination of hardware, firmware, and software. Examples of authentication protocols include the authentication protocols using message authentication code, such as the Authentication Header (AH) or Encapsulating Security Payload (ESP) occurring in IPsec.

Unlike IPsec, which can provide a generic authentication solution that is transparent to end users and applications, protocols such as the Secure Sockets Layer (SSL) and the Transport Layer Security (TLS) have been designed to mitigate the increased risk associated with Web-based applications. Typically, these protocols use the transport protocol TCP to provide point-to-point security services, namely authentication and confidentiality, between the client and the server. As a result, application-level protocols, such as the HyperText and the file transport protocols, can use these security services. Like in IPsec, the authentication and encryption in SSL/TLS are based on a shared secret key that is established using a handshake protocol (which is one of the SSL/TLS suite). The handshake allows the negotiation of a cipher suite, which negotiates the key exchange methods and the specific cryptographic algorithms to use by the client and the server. However, digital certificates are exchanged for the need of initial authentication and the exchange of the shared secret key.

In the following, we identify specific flaws of the wireless communications and discuss the current state of intrusion prevention in wireless communications, by examining the major authentication protocols that play an important role in the front door security. The protocols provide the means to authenticate users, processes, hosts, and devices. In particular, the weaknesses shown by these protocols are presented and some among the attacks that exploit these weaknesses are described. On the other hand, the following chapters will analyze in more detail the security attacks and threats targeting mobile communications.

1.6.2.1 WLAN Authentication Flaws

A support for unilateral authentication of wireless devices is provided, for WLANs, using typical open authentication and shared key authentication schemes. The WEP specification, for example, accommodates the need for user confidentiality by allowing a pseudorandom generator to create a key sequence that is XORed with the payload of each frame needing encryption. However, despite its fundamental goal (which is preventing casual eavesdropping), the WEP allows the null-authentication option to permit all users to access the WLAN and uses small sized keys. Proceeding that way has created unacceptable flaws in the WEP implementations. More flaws will be discussed in Chapter 7 for the WEP protocols.

More secure versions of the authentication protocols implemented in WLANs use larger key sizes. Authentication of devices is based on the challenge-response mechanism and symmetric encryption. Typically, these algorithms assume that, when a mobile node or wireless device attempts to connect to an AP, it sends an authentication request management frame, specifying the use of shared-key authentication that contains a challenge, its identity, the identity of the access point, and a random value called initial vector.

While the strength of the authentication protocols in WLAN is based on the difficulty of discovering the secret key, generally through brute-force attacks, one of the most significant flaws with the authentication protocols is the use of stream cipher for symmetric encryption that is not robust. In addition, another flaw can be noticed: Some of these protocols allow the derivation of the key stream, since they use very small initial vectors (IV). Once the key stream is known, for a given IV, the attacker can respond to future challenges until the IV is changed. At that moment, a new key stream must be derived.

On the other hand, as the standard does not specify a mechanism (automated or manual) for the distribution of keys to different devices in practice, most WLAN installations not only use a single shared key but fail to change it periodically. The consequences of this action are quite severe: Attacks, associated with key stream reuse, become more feasible since the initial vector space is often exhausted in less than a day. In addition, the IV that is used to randomize the key stream is not only too short, but the standard does not require that a different value be used in each frame. The small space of IV values, combined with a static nature of shared key, increases the feasibility of constructing a decryption dictionary.

1.6.2.2 Wireless Ad Hoc Authentication Flaws

As ad hoc wireless networks are characterized by the lack of infrastructure, each mobile node (MN) will need to communicate with the other nodes using a multi-access control to shared medium such as the Carrier Sense Multiple Access with Collision Avoidance (CSMA/CA) protocol. This protocol, along with other protocols implemented to allow ad hoc nodes to communicate, induces several important flaws in the ad hoc network. To discuss the authentication flaws, we consider the case of bluetooth networks (BT), one of the most well-known examples of ad hoc networking. The term Bluetooth refers to an open specification that enables short-range peer-to-peer wireless communications of voice and data, based on proximity networking. The BT is a preferred choice for deploying personal area networks.

Mutual authentication of BT devices is realized by the link manager, while encryption of the packets, at the link layer, is carried out using a stream cipher. It is based on the use of a secret key that is shared by the pair of participating devices (called the master and slave). The key length is made variable to accommodate the security requirements of different applications. It is negotiated between the applications that reside on the participating devices. Authentication of a BT slave by a master is based on a challenge-response mechanism, which requires the address of the BT device, a shared secret key or link key, and a random number. In order to generate a link key, the devices undergo a pairing process (Barbeau, 2006). The end of this process is marked by the creation of the initialization key. This key is used subsequently by the slave and the master to encrypt data during the link key generation process. Thus, practical studies on the security of BT have concentrated on the keys discovery. The most appropriate moment to launch these attacks was shown to be the time interval of the initial pairing process.

One of the problems, revealed by users, is the need to enter a PIN twice, every time two devices are involved in a communication. This gave, for some users, the choice to use the shortest PINs possible. As the specification does not define a precise mechanism for the distribution of PINs, various strategies have been adopted. One option was to transmit the PIN in clear. This is obviously a poor choice since it can be captured by an attacker. However, even if it is encrypted, using application level encryption, before being sent, it can still be discovered.

It has been shown that the keys discovery process can be further simplified by the fact that most users often use PINs with length smaller than 5 digits and that a large percentage of users utilize a PIN set to 0000. Thus, it has been recommended that using longer PINs would minimize the discovery vulnerability. Moreover, it has been suggested that a PIN of more than 64-bits should be better securing. While this suggestion is technically feasible, it appears unpractical, given the fact it has to be inserted twice (Jakobsson, 2001).

1.6.2.3 Cellular Authentication Flaws

Access control decisions are made in cellular communication networks based on two elements of subscriber data (including the identity of a subscriber) and a secret key inserted securely in the smart card attached to the mobile station. The acquisition of the private information permits an intruder to impersonate the legitimate subscriber that holds the smart card. Unfortunately, it is the victim who is forced to accept the financial costs and other related costs resulting from an impersonation attack. The authentication in cellular networks present some flaws that depend on the technique used between the mobile station and the node to which the mobile station connects.

For the sake of clarity, let us consider the case of the GSM network. The lack of mutual authentication represents the most severe flaw in the GSM authentication system. That is, there is no mechanism implemented on the mobile station that allows the subscriber to verify the credentials of the network it connects, either implicitly or explicitly. This weakness makes the mobile station vulnerable to impersonation attacks. In addition, according to GSM specifications, neither authentication nor data encryption is carried out internally to the network to protect the exchange of data needed for authenticating the subscriber. The components involved in that exchange do it under the support of a mutual trust relationship.

Of course, when the exchange is performed in the network where the subscriber is permanently registered, this level of trust can be justified and accepted. However, when international roaming is taken into consideration, there is a need for an enhanced security; such need can involve strong cryptographic methods.

In any situation, the path connecting the ingress node (to which a user is requesting authentication) to the home registering node (of that user) remains vulnerable to third-party attacks. Moreover, the link "mobile station-ingress node" remains equally vulnerable. The lack of mutual authentication and the absence of data encryption, during the initial phase of the authentication process, make this link insecure. As a matter of fact, data encryption cannot be initiated in GSM until the authentication process has been completed.

1.7 Intrusion Detection in Wireless Communications

This section discusses the major security techniques provided for wireless networks for the detection of intrusions. In particular, the cases of WLAN and ad hoc networks will be addressed. The discussed methods include four among the major techniques used for that purpose, namely, the radio frequency fingerprinting, cluster-based detection, mobile devices monitoring, and mobile profile construction.

Wireless intrusion detection protects wireless networks against attacks, by monitoring traffic and generating alerts when signs of attack or actual attack attempts are detected. Two classes of detection techniques can be distinguished: signature-based and anomaly-based approaches. The first category aims at detecting known attacks by looking for their signatures (or known patterns). The main disadvantage of such approaches is their limitation to detect only known attacks. The anomaly-based approaches look for abnormal behavior in the traffic, related to resource uses or user behavior. They are not often implemented, mostly because of the high amount of false alarms that have to be managed and the large amount of time they may waste in processing. Anomaly-based detection develops a database of profiles characterizing normal behaviors or traffic. When an abnormal traffic is detected, an alert is generated. The main advantage of the anomaly-based approaches is their capacity to cope with unknown attacks.

To benefit from the advantages of the previous two approaches, hybrid approaches implement in the same system the two approaches simultaneously. However, an essential issue has to be addressed to provide efficiency for the hybrid intrusion detection approaches: detection has to run online and be real-time reactive. Otherwise, these approaches will only be useful for audit or postmortem digital investigation. In addition, the real-time intrusion detection has to be able to collect data from the network in order to store, analyze, and correlate them. This can, however, decrease the network performance (Hutchison, 2004).

1.7.1 Wireless Detection Approaches

The main objective of wireless detection is to protect the wireless network by detecting any deviation with respect to the security policy. This can be done by monitoring the active components of the wireless network, such as the APs. Generally, the wireless intrusion detection systems (WIDS) are designed to monitor and report on network activities between communicating devices. For this, they have to capture and decode wireless network traffic. While some WIDS can only capture and store wireless traffic, other WIDS can add traffic analysis and reports generation to these functions. Other WIDS are able to analyze signal fingerprints, which can be useful in detecting and tracking rogue AP attacks. The following intrusion detection techniques are of utmost interest for WIDS design and analysis. A classification of these techniques can be made according to several dimensions: (a) the approach, which can be signature-based or anomaly-based; (b) the monitored system, which can network-based, radio, or host-based; and (c) the way of response performed by the WIDS, which can be active or passive.

1.7.1.1 Mobile Profiles Construction

The main objectives when using the anomaly-based approach are to define the user profiles, application profiles, and user mobility profiles, and to design an efficient mechanism that permits the detection of any deviation with respect to the stored profiles. The construction of a profile begins with the collection and processing of the related data. The user mobility is constructed by collecting the user locations and deduces the coordinates of these locations in a way that reduces the granularity of the location data in order to accommodate minor deviations or intra-user variability between successive location broadcasts. Then, useful features of the successive locations are extracted. The set of chronologically-ordered features are subsequently concatenated to define a mobility sequence (Hall, 2005). This process continues until the creation of the mobility sequences. A training process can be organized on the patterns collected, characterizing the user mobility behavior and other user-related information. During the classification phase, a set of user mobility sequences are observed and compared to the training patterns in the user's profile to evaluate a similarity measure to profile parameter. If the average value of this parameter exceeds predefined thresholds, then the mobility sequences can be considered abnormal and an alert is generated (Hall, 2005).

1.7.1.2 Monitoring Wireless Devices

Using a signature-based approach, the WIDS bases its processing on the recognition of intrusion patterns from the traffic outputs. This requires monitoring several parameters on the AP outputs and the wireless client's station. Monitoring APs is about monitoring their respective identities, MAC addresses, and channel information. This requires listening to wireless frames such as beacons, probe response and authentication/association frames (on the wireless link and at the access node), and compare them to the predefined attack signatures.

Because authorized clients cannot be listed, the information that may help in detecting an attack cannot be totally available; nevertheless, the following aspects can be monitored (Low, 2005):

■ A predefined "blacklist" of wireless clients can be checked against all connecting clients. Any client within this list trying to access the network would be automatically denied and an alert can be sent to the right partner.
■ All wireless clients with an "illegal" MAC address (MAC address ranges, which have not been allocated) are automatically denied access and an alert is sent off.
■ A wireless client, who just sends out probe requests or special distinguishable data packets after the initial probe request and has not been authenticated, can be flagged out as potential source of a network discovery attack.
■ When a sequence number (SN) is implemented on the exchanged data, the potential impersonators could be identified by simply monitoring the SN. When impersonation attacks are ongoing, the attacker will not be able to continue with the SN used previously by the victim.

1.7.1.3 Radio Frequency Fingerprinting (RFF)

The RFF is defined as the process identifying a cellular phone by the unique "fingerprint" that characterizes its signal transmission. It is used to prevent cloning fraud, because a cloned phone will not have the same fingerprint as the legal phone with the same electronic identification numbers. This process aims to enhance the anomaly-based wireless intrusion detection by associating a MAC address with the corresponding transceiver profile. The fingerprint of a signal is generally represented by a set of values including amplitude, phase and frequency of the signal, and some other values deduced mathematically from the signal during a period of time (by applying the Discrete Wavelet Transform, for example).

The architecture of the corresponding WIDS is shown by Figure 1.4, where the main objective is to classify an observed transceiver fingerprint as normal when it belongs to the transceiver of a device with a legitimate MAC address, or anomalous when it belongs to another transceiver (Hall, 2005; Barbeau, 2006).

As illustrated in Figure 1.4, the process starts by converting the analog signal to a digital signal. This is done by the converter component. Second, the features extractor extracts components such the transient portion from the digital signal. Then, the amplitude, phase, frequency, and other parameters defining the transceiverprint are extracted by the feature extraction component. These features are compared to the transceiver profiles stored in the base of fingerprint handled by the

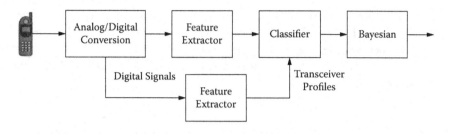

Figure 1.4   The enhanced architecture of WIDS based on fingerprinting.

WIDS. This operation is performed by the classifier component. To decide about the status of the transceiverprint, the Bayesian filter is applied because of the noise and interference, which are special characteristics of the wireless environment. The Bayesian filter has to estimate the state of the system from noisy observations. This process requires extracting predefined transceiver's profiles.

To classify a signal as anomalous, the probability of match has to be determined for each transceiver profile. Therefore, a statistical classifier using neural networks can be used, where the set of extracted features represents a vector and the outputs are a set of matching probabilities.

1.7.2 Cluster-Based Detection in Ad Hoc Networks

Intrusion detection in ad hoc networks provides audit and monitoring capabilities that offer local security to a node and helps to assign specific trust levels of the other nodes (Ejaz, 2006; Kashan, 2005). Clustering protocols can be taken as an additional advantage in these processing constrained networks to collaboratively detect intrusions with less power usage and minimal overhead. Because of their relation with routes, existing clustering protocols are not suitable for intrusion detection. The route establishment and route renewal affect clusters. Consequently, processing and traffic overhead increase, due to instability of clusters. Ad hoc networks present battery and power constraint. Therefore, the monitoring node should be available to detect and respond against intrusions in time. This can be achieved only if clusters are stable for a long time period. If clusters are regularly changed due to routes, the intrusion detection will not be efficient. Therefore, a generalized clustering algorithm, detailed in Ejaz, 2006, has been discussed. It is also useful to detect collaborative intrusions (Kashan, 2005).

In the other hand, clusters are formed to divide the network into manageable entities for efficient monitoring and low processing. Clustering schemes result in a special type of node, called the "Cluster Head" (CH), to monitor traffic within its cluster. It not only manages its own cluster but also communicates with other clusters for cooperative detection and response. It maintains information of every member node (MN) and neighbor clusters. The cluster management responsibility is rotated among the cluster members for load balancing and fault tolerance and must be fair and secure. This can be achieved by conducting regular elections (Kashan, 2005). Every node in the cluster must participate in the election process by casting their vote showing their willingness to become the CH. The node showing the highest willingness, by proving the set of criteria, becomes the CH until the next timeout period.

References

E. Ahmed, K. Samad, and W. Mahmood, Cluster-Based Intrusion Detection (CBID) Architecture for Mobile Ad Hoc Networks, *Proceedings of AusCERT Asia Pacific Information Technology Security Conference (AusCERT)*, Asia, 2006.

M. Barbeau, J. Hall, and E. Kranakis, Detection of Rogue Devices in Bluetooth Networks using Radio Frequency Fingerprinting, *Proceedings of the 3rd IASTED Int. Conf. on Communications and Computer Networks*, Lima, Peru, 4–6 October 2006.

B. Bhargava, X. Wu, Y. Lu, and W. Wang, Integrating Heterogeneous Wireless Technologies: A Cellular Aided Mobile Ad hoc Network (CAMA), in *Mobile Networks and Applications*, v. 9 n. 4, pp. 393–408, August 2004.

R. J. Boncella, *Wireless Threats and Attacks, Handbook of Information Security*, Hossein Bidgoli, John Wiley & Sons publisher, pp. 165–175, 2006.

X. Geng and A.B. Whinston, Defeating distributed Denial-of-Service Attacks, *IEEE IT Professional*, July/August 2000, 36–41.

J. Hall, M. Barbeau, and E. Kranakis, Using Mobility Profiles for Anomaly-Based Intrusion Detection in Mobile Networks, *The 12th Annual Network and Distributed System Security Symposium*, San Diego, CA, 3–4 February 2005.

K. Hutchison, Wireless Intrusion Detection Systems, GIAC Security Essentials Certification (GSEC) Practical Assignment Version 1.4c, SANS Institute, http://www.sans.org/reading_room/whitepapers/wireless/, 18 October 2004.

M. Jakobsson and S. Wetzel, Security Weaknesses in Bluetooth, *Lecture Notes in Computer Science*, pp. 176–191, 2001.

C. Low, Understanding Wireless Attacks & Detection GIAC Security Essentials Certification (GSEC) Practical Assignment Version 1.4c. http://cnscenter.future.es.kr/resource/hot-topic/wlan/1633.pdf. Accessed February 3, 2009.

P. Mateli, Hacking Techniques in Wireless Networks, *Handbook of Information Security*, Hossein Bidgoli, John Wiley & Sons publisher, pp. 83–93. USA, 2006.

McAfee, Mobility Security Report 2008, available at http://www.mcafee.com/us/local_content/reports/mcafee_mobile_security_report_2008.pdf. Accessed February 3, 2009.

R. K. Nichols and P. C. Lekkas, *Telephone System Vulnerabilities*, McGraw-Hill Publications, USA, 2002.

R. Richardson, *The CSI Computer Crime and Security Survey*, Computer Security Institute, available at: http://www.gocsi.com/forms/csi_survey.jhtml

K. Samad, E. Ahmed, and W. Mehmood, Simplified Clustering Approach for Intrusion Detection in Mobile Ad Hoc Networks, 13th Int. Conference on Software, *Telecommunications and Computer Networks (SoftCOM 2005)*, Split, Croatia, 15–17 September, 2005.

A. A. Vladimirov, K. V. Gavrilenko, and A. A. Mikhailovsky, Counterintelligence: Wireless IDS Systems, in *WI-Foo: The Secrets of Wireless Hacking*, chapter 15, pp. 435–456, Pearson/Addison Wesley publisher, 2004.

R. Yavatkar et al., A Framework for Policy Based Admission Control, *IETF RFC* 2753 (January 2000).

F. Zarai and N. Boudriga, Provision of Quality of Service in Wireless Fourth Generation Networks, *Int. J. of Commun. Syst.*, Vol. 20(11)(2007): 1215–1243.

Chapter 2

Access Control and Authentication in Mobile Communications

2.1 Wireless System Security

A secure system can be defined as a system that performs exactly what its designers have envisioned it to do, performing what should be authorized, and interdicting what should not be allowed. Also, it should not show any unexpected behavior, even when attackers try to make the system act differently. As complete security is impossible to achieve in a communication system, a study about the cost/benefit balance must be established for any solution to deploy. It must be reminded that enforcing security requires that the defender covers all issues of possible attacks, whereas it is sufficient for the attacker to concentrate efforts on a few vulnerabilities to launch an attack. Thus, a system is only as secure as its less reliable security asset. The less reliable security assets in a mobile network include the radio link and the mobile terminals.

The major security goals are defined in terms of the security services provided for the communication networks. The basic security services and their goals are defined as follows:

- *Confidentiality*: This means that the transmitted information is only disclosed to the authorized parties. Sensitive information disclosed to an adversary could have severe consequences.

- *Integrity*: This assumes that a message is not altered in transit between sender and receiver. Messages could be corrupted due to network malfunctioning or malicious attacks.
- *Non-repudiation*: This means that the source of a message cannot deny having sent the message. An attacker could generate a wrong message that appears to be initiated from an authorized party, with the aim of making that party the guilty one. If non-repudiation is guaranteed, the receiver of a wrong message can prove that the originator has transmitted it, and that, therefore, the originator misbehaved.

Other security goals, which are of large interest, may be more difficult to achieve, since the attacks can be combined. For example, the intruder may get into the communication system to prepare a denial of service from inside or it may perform an eavesdropping attack with the purpose of gaining unauthorized access later. Among these goals, we mention the authentication, access control, and network availability.

- *Authentication*: Authentication guarantees the identity of the entity with which communications are established, before granting it the access to the resources of the network. In the absence of authentication mechanisms, an attacker could masquerade as a legitimate entity and attempt to violate the security of the network.
- *Access control*: Access control means that only authorized parties can be allowed to access a service on the network, use a resource, or participate in the communications; any other entity is denied access. The access control assumes the authentication of the entity trying to get access to the network.
- *Network availability*: Availability ensures that all resources of the communications network are always utilizable by authorized parties. An attacker may launch a Denial of Service (DoS) attack by saturating the medium, jamming the communications, or keeping the system resources busy in any other way or by any other means. The aim here is just to slow down or stop authorized parties from having access to the resources, thereby making the network unusable.

Many security mechanisms were made available to provide authentication and access control in mobile wireless networks. Some other countermeasures are attempting to address attacks against availability. The major techniques used in the aforementioned mechanism include cryptography (Schneier, 1995), digital signature, trust management, and security policy-based authorization.

2.2 Cryptography Basics

Encryption is the process of masking a message in such a way that it keeps secret its content; the operation aims at transforming the message from its original form (called *plaintext*) to an unintelligible form (called *ciphertext*). The reverse process is called decryption. Cryptographic techniques are carried out using a cryptographic algorithm (or *cipher*) and a specific key. It is often useful to apply more than one technique to protect a message; for example, the message can be encrypted and then digitally signed. As it will be explained in the following, signing a message means to add a sequence of bits, referred to as a *digital signature*, to the message in order to identify its authentic originator. With respect to the aforementioned security goals:

■ The encryption provides confidentiality, because the messages are transmitted in ciphertexts that only the owner(s) of the encrypting key can decrypt.
■ The digital signature provides non-repudiation, as only the owner of the key could have generated it. It also guarantees integrity of the message, while flowing in the network.

While service availability is not the concern of cryptography, authentication and access control are more complicated to obtain and require the use of more advanced cryptographic primitives. In fact, it is likely that information that was true at some time in the past may not be true anymore in the present. In addition, even assuming that a signature is successfully checked, previously transmitted messages can be sent again by an attacker; that is, an intruder may record a sequence of messages and re-send them some time later (or replay them). If these messages cannot be identified as old, they will be accepted as valid because they were properly signed. To protect against replay attacks, messages usually embed a representation of time, called timestamp, describing the time at which the message was generated. The timestamp is included in the computation of the signature.

Two classes of cryptography can be used: symmetric cryptography and asymmetric cryptography. Each is useful to perform different functions.

2.2.1 Symmetric Cryptography

Symmetric cryptography (also called secret key cryptography) is based on symmetric key algorithms, meaning that the algorithms use the same key for the encryption and the decryption (or, more broadly, the encryption key can be easily computed from the decryption key and vice versa). The sender and the receiver of a message must exchange (or agree on) a secret shared key, which will henceforth be used to encrypt and decrypt the exchanged messages. Formally, the symmetric cryptography process operates as follows. Let A and B be the set of plaintexts and the set of ciphertexts, respectively. Let also Λ be the set of keys and $E_k: A \rightarrow B$ and $D_k: B \rightarrow A$ be two applications satisfying the following:

$$D_k(E_k(x)) = x, \forall x \in A; \text{ and } E_k(D_k(y)) = y, \forall x = y \in B$$

$E_k(x)$ and $D_k(y)$ are computationally feasible, given x and y, respectively.
Assuming that it is computationally unfeasible

- to find x knowing $E_k(x)$,
- to find k knowing a subset of $\{E_k(x)|x \in A\}$,

then one can use $E_k(-)$ and $D_k(-)$ to provide the encryption and decryption schemes, respectively.

The key space Λ plays an important role in the protection of the symmetric cryptosystem. In fact, a necessary, but not sufficient, condition for an encryption scheme to be secure is that the key space should be large enough to prevent exhaustive attacks. Knowing the decryption transformation $D_*(-)$, an exhaustive set of attacks targeting the extraction of the key used to produce a ciphertext, say $E_k(m)$, can compute $D_j(E_k(m))$, $j \in A$ until m is found. If every computation takes 10^{-6} seconds, the exhaustive attack will take more than 5×10^{24} years, on average, to succeed, if the set of keys has 2^{128} different keys.

2.2.1.1 Classification of Symmetric Cryptosystems

Consider a symmetric encryption scheme consisting of the sets of encryption and decryption transformations:

$$\{E_k, k \in \Lambda\}, \{D_k, k \in \Lambda\}$$

where Λ is the key space. The encryption scheme can be classified into two classes: block cipher scheme and key stream cipher. A block cipher scheme breaks up the plaintext messages to be transmitted into segments (or blocks) of fixed length $b > 1$ and encrypts one block at a time. Key stream ciphers are very simple block ciphers acting using blocks with a length equal to 1. Cipher schemes are very special for two reasons: (a) the encryption transformation can change with each symbol of the plaintext to encrypt and (b) a special method can be used to change the symbols used for the key with each symbol of the plaintext. In this case, the key space for a set of n encryption transformations (n is often the length of the plaintext) is the set of sequences of the form $e_1 e_2 \ldots e_n$ called keystreams.

> **Example:** A one-pad cipher is a stream cipher defined over binary messages. It transforms a message $m = m_1 m_2 \ldots m_n$ using a binary key string $k = k_1 k_2 \ldots k_n$ of the same length to produce a ciphertext $c = c_1 c_2 \ldots c_n$ given by
>
> $$c_i = m_i \oplus k_i, \ 1 \le i \le n$$
>
> The decryption is operated by computing the following objects
>
> $$m_i = c_i \oplus k_i = (m_i \oplus k_i) \oplus k_i, \ 1 \le i \le n.$$

The one-pad cipher stream can be shown as highly secure. That is, if the hacker has $c = c_1c_2...c_n$, he can do no better than guessing at the plaintext being any sequence of bits of length n. Since a block cipher processes messages by decomposing them into segments, different modes of encryption have been developed. They all aim at adding several desirable features to the ciphertext such as including some form of randomness. We describe, in the following, two among these modes of operation: the cipher block chaining mode (CBC) and the output feedback mode (OFB). Other modes such as the electronic codebook mode (ECB) and the cipher feedback mode (CFB) are also largely used.

The CBC mode performs encryption of a message $m = m_1m_2...m_n$, using a special input block known as the initialization vector (IV), produced as a random n-bit string, the encryption mapping E, and a key k. The ciphertext $c = c_1c_2...c_n$ is computed as follows:

$$c_0 = IV,$$

$$c_1 = E_k(m_1 \oplus IV),$$

$$c_{i+1} = E_k(m_{i+1} \oplus c_i), i \geq 1$$

The decryption process applies as follows upon receiving $c_0, c_1, ..., c_n$

$$c_0 = IV,$$

$$m_1 = D_k(c_1) \oplus IV,$$

$$m_{i+1} = D_k(c_{i+1}) \oplus c_i, i \geq 1$$

A new initial vector IV is used with each new session. It does not need to be secret; it is transmitted in clear as a ciphertext block, and it needs to be unpredictable. On the other hand, the OFB mode requires an initial vector IV and uses an intermediate sequence and performs the following operations $I_1, I_2, ..., I_n$ for a plain text of the form $m = m_1m_2...m_n$. The OFB encryption is done by

$$I_1 = IV, c_1 = m_1 \oplus E_k(I_1)$$

$$I_j = E_k(I_{j-1}), c_j = I_j \oplus m_j$$

The OFB mode does not need IV to be transmitted secretly. The OFB mode encryption and decryption are identical.

2.2.1.2 Encrypting with AES

Various symmetric algorithms for encryption have been produced including DES, 3-DES, AES, IDEA, Twofish, Serpent, etc. To this class of algorithms also belong ancient algorithms such as the substitution ciphers, such as Caesar, Vigenere, and Playfair. These ciphers are not used anymore because they are easy to break. We present here the major features of AES.

AES is a block cipher. Originally named the *Rijndael algorithm*, it was selected as the algorithm for the *Advanced Encryption Standard* in 2001. AES applies a number of rounds, and each round makes a series of transformations on the state of the block to encrypt, using a round key derived from the encryption key. The number of rounds depends on the block and the key sizes. An encryption of a block starts with a transformation, called *AddRoundKey*; this is followed by an odd number of regular rounds, and ends with a different final round. Unlike other symmetric ciphers, AES uses invertible transformations, which make the decryption feasible. AES operates on a state that is initialized with a plaintext block, and after encryption this represents the ciphertext. The state can be pictured as a rectangular array of bytes. It consists of four rows and a number of columns defined by the block size in bytes divided by four. For example, a block size of 128 bits would require a state of four rows and four columns.

2.2.1.2.1 The Round Transformations

AES uses four transformations. They are:

1. *AddRoundKey*: This is an XOR operation between the state and the round key. This transformation is its own inverse.
2. *SubBytes*: This is a substitution of each byte in the block independently of the positioning in the state. It is called an S-box. It is invertible and characterizes a non-linear transformation. The S-box is proved to be optimal with regards to non-linearity.
3. *ShiftRows*: This is a cyclic shift of the bytes in the rows in the state and is clearly invertible (simply by applying a shift in the opposite direction by the same amount).
4. *MixColumns*: Each column in the state is considered a polynomial with the byte values as coefficients. The columns are transformed independently by multiplication using a special polynomial $c(x)$, which has an inverse $d(x)$ that is used to reverse the multiplication by $c(x)$.

The functions *Rnd* and *RndF* implementing a round and a final round are respectively given by:

$$Rnd(state,RoundKey) = MixColumns(ShiftRows(SubBytes(state))) \oplus RoundKey$$

$$RndF(state,RoundKey) = ShiftRows(SubBytes(state)) \oplus RoundKey$$

2.2.1.2.2 The Round Key

The Round keys are obtained by expanding the encryption key into an array holding the round keys one after another. The expansion works on words of four bytes. Let

N_k be a constant defined as the number of four bytes words in the key k. Roughly speaking, the encryption key is filled into the first N_k words and the rest of the key material is defined recursively from preceding words. The word in position i, say $W[i]$, except the first word of a RoundKey, is defined as the XOR between the preceding word, $W[i-1]$, and $W[i-N_k]$. The first word of each RoundKey, $W[i]$ (where $i = 0 \bmod N_k$) is defined as the XOR of a transformation on the preceding word, $T(W[i-1])$ and $W[i-N_k]$. The transformation T on a word, w, is w rotated to the left by one byte, XORed by around constant and with each byte substituted by the S-box.

2.2.2 Asymmetric Cryptography

The asymmetric cryptography (also called public key cryptography) assumes the existence of set Λ of pairs of keys (K,k), a key for encryption (public key, k) and another key for decryption (private key, K). An asymmetric cryptosystem with pairs of keys in Λ is defined by:

$$(E_k : A \rightarrow B, D_K : B \rightarrow A), \; (K,k) \in \Lambda$$

such that:

$$D_K(E_k(x)) = x, \forall x \in A; \text{ and } E_k(D_K(y)) = y, \forall x = y \in B$$

$E_k(x)$ and $D_K(y)$ are computationally feasible; and assuming that it is computationally unfeasible:

- to find x knowing $E_k(x)$ or find k knowing a subset of $\{E_k(x)|x \in A\}$;
- to find y knowing $D_K(y)$ or find K knowing a subset of $\{D_K(y)|y \in B\}$;
- to find K knowing k and find k knowing K.

Therefore, the encryption and decryption schemes are provided respectively by $E_k(-)$ and $D_k(-)$. An entity (or individual), say α, can leave its public key available to everyone, by publishing the key in a public directory, for example. The private key K_α of α needs to be kept undisclosed. All public key exchange may be done over an insecure channel, which can be any channel built on a network that may be subject to eavesdropping. Any other entity, say β, can therefore use the public key of α, encrypt messages, and send ciphertexts to α. Only entity α will be capable of decrypting the ciphertexts. However, public key cryptography requires an additional tool that allows β to authenticate the relationships between α, K_α and k_α. Additional tools can be needed to help generate the key pairs, distribute the public keys (if needed), and update and revoke the public keys. This can be performed by the Public Key Infrastructure (PKI), for example. In the following section, we will discuss the major functions for a PKI.

One of the most important concerns in public cryptosystems is about how to attach a public key with its legitimate owner – that is, how to be guaranteed that a specific public key is owned by a given entity and not by another, which would then be able to decrypt messages apparently sent to that entity. If two users, say α and β, want to exchange their public keys, they could do it over the same insecure channel (such as a radio link) that is used afterward to exchange their encrypted messages. However, if an adversary, say γ, is able to listen over the communication channel, it can make the protection ineffective by providing a public key to α (as if it comes from β) and a public key to β (as if it comes from α). This is a sort of *double identity spoofing*, called man-in-the-middle attack, in which an adversary takes place in the communication channel between two parties and acts with an entity as the other entity.

More explicitly, the man-in-the-middle attack is performed as follows. The adversary γ generates two public/private key pairs (K_γ^1, k_γ^1) and (K_γ^2, k_γ^2). User α transmits his public key k_α to β, but the adversary intercepts it, substitutes the legitimate key with its public key, k_γ^1, and sends it to user β. Similarly, β sends his public key k_β to user α, but the adversary γ intercepts and substitutes it with key k_γ^2 that is transmitted to user α. As a result, user α mistakenly believes the β's public key to be k_γ^2, and user β erroneously believes α's public key to be k_γ^1, while both keys are owned by the adversary.

From this point on, the adversary intercepts any message sent by user α to user β, decrypts it with K_γ^1, reads it, re-encrypts it with k_β, and sends the message to user β, who will decrypt it with his private key K_β. In the opposite direction, the adversary intercepts any message sent by user β, decrypts it with K_γ^2, reads it, re-encrypts it with k_β, and sends the message to user α, who will decrypt it with his private key K_β. Therefore, the adversary is able to read any message exchanged between α and β, while they are unaware of the adversary's presence and think their communications are kept confidential.

A particular solution against this attack involves a Trusted Third Party (or TTP), which must be trusted by all parties. The TTP stores the public key of all participants and guarantees the identities of the owner of each key. Many techniques can be used to implement the TTP role. This can be a Key Distribution Center (KDC) connected to the participants with secure channels or an entity, called Certification Authority (CA), which is able to deliver to any participant a digital certificate containing the identity of the key's owner, its public key, the certificate validity dates, and other useful information. Common certificates follow the X.509 standard (Garfinkel, 2001) developed by the International Telecommunication Union. However, the existence of a TTP may represent a point of weakness of the whole public key cryptographic system. For example, if the delivery of public keys is done on demand, an adversary could affect the availability of the whole system by launching a denial of service attack against the TTP. Furthermore, by compromising a certification authority the attacker can issue a forged certificate for any identity he wishes to perform spoofing and man-in-the-middle attacks. The solution based on a certification authority is largely utilized in wired networks and wireless communication networks. Protocols such as HTTPS, IPsec, and SSL are based on the use of digital certificates. For example, SSL certificates follow the X.509 standard and can be delivered by many commercial CAs such as VeriSign. In addition, public administrations and government agencies may have to create their own CAs, too.

To securely transmit a message, the originator of the message retrieves the destination's public key, encrypts the message using the public key of the destination, and transmits it to the destination, which can decrypt it with its private key. Examples of asymmetric cryptosystems include RSA (Rivest; Rivest, 1978), Knapsack, ElGamal (ElGamal, 1985), and the elliptic curve cryptography (Hankerson, 2003).

2.2.2.1 Encrypting with RSA

The RSA cryptosystem assumes the following mathematical results: Let $n = p.q$ be the product of two prime numbers and $\varphi(n)$ be the number of integers that are prime to n. Then

$$\varphi(n) = (p - 1)(q - 1)$$

If a pair of numbers (e,d) satisfies $e.d = 1 \pmod{\varphi(n)}$ then for all m, we have

$$(m^d)^e = 1 \bmod n$$

The RSA encryption proceeds with two steps: key generation and ciphertext production. During key generation, an entity chooses two random large prime numbers p and q, computes n and $\varphi(n)$, chooses a random integer $e < \varphi(n)$ such that $\gcd(e, \varphi(n)) = 1$ and computes the integer d such that

$$ed = 1 \pmod{\varphi(n)},$$

and publicizes (n,e) as the entity's public key. The entity is sure to find d by applying the extended Euclid algorithm. The entity should delete p, q, and $\varphi(n)$ for safety and keep d secret as its private key.

During encryption, an entity needing to send a confidential message $m < n$ to the owner of public key (n,e) should create the ciphertext c as follows:

$$c = m^d \pmod{n}$$

To decrypt c, the receiving entity can simply perform the computation $c^e \bmod n$.

Example: Assume that user α chooses a modulus n such that $n = 7 \times 13 = 91$. Then $\varphi(91) = 72$. By applying the extend Euclid algorithm, one can find that

$$5 \times 29 + 72 \times (-2) = 1$$

That is $5 \times 29 = 1 \pmod{72}$. Therefore, user α has computed 29 to be its private key decryption exponent. The user can publicize (91,5) as his public key. If another user wants to send to α the message $m = 7$, then he performs encryption by computing

$$c = 7^5 = 63 \pmod{91}$$

To decrypt the ciphertext, user α computes

$$m = 63^5 = 7 \pmod{91}$$

2.2.2.2 Encrypting with ElGamel

The ElGamel cryptosystem is an interesting application of the Diffie-Hellman one-way trapdoor function. It integrates three parts: key setup, plaintext encryption, and ciphertext decryption. During encryption phase, user α chooses a random prime number p, computes a random multiplicative generator element g modulus p non null ($\in Z_p$), chooses a random number x ($\in Z_{p-1}$), computes

$$y = g^x \pmod{p},$$

and publicizes the 3-tuple (p,g,y) as the α's public key while keeping x as the corresponding private key.

To send a confidential message $m < p$ to user α, user β chooses a number k ($\in Z_{p-1}$) and computes the following pair (c_1,c_2) of ciphertexts:

$$\begin{cases} c_1 = g^k & \pmod{p} \\ c_2 = y^k \times m \pmod{p} \end{cases}$$

The decryption process is obtained upon receiving the pair (c_1,c_2) of ciphertexts by computing

$$m = \frac{c_2}{c_1^x} \pmod{p}$$

The above equality holds because

$$c_1^x = (g^k)^x = (g^x)^k = y^k = \frac{c_2}{m} \pmod{p}$$

2.2.3 Symmetric vs. Asymmetric Cryptography

Symmetric and asymmetric cryptographies have both weak and strong points as it will be shown in the next section. Symmetric cryptosystem share the following advantages compared to asymmetric cryptosystems:

■ The data throughput rate is higher with symmetric ciphers and needs less computation power.

■ To provide the same level of security, the key size is significantly smaller with symmetric ciphers.

On the other hand, asymmetric cryptosystems present a better behavior in some perspectives:

■ Unlike symmetric cryptography, where the shared key must be maintained as secret, asymmetric cryptography requires only that the private key be kept secret. The public key can (and should) be publicly released.

■ To some extent, key management and key distribution are easier in asymmetric cryptography. To handle a secured message exchange between n parties, the number of symmetric keys to manage is very high (as there are at least $n(n-1)/2$ symmetric keys counting only direct links, while other keys are needed for group security). On the contrary, using asymmetric cryptography, the number of keys to manage is just $O(n)$-complex, since only n public keys are needed to the TTP.

■ A public/private key pair may remain unchanged for many sessions. In the opposite, symmetric keys should be renewed more often (even once per session) to guarantee the same level of security.

The advantages of asymmetric cryptosystems, however, are closely related to the trust level assigned to the TTP handling the public keys. If these keys are committed to a TTP, then it must be unconditionally trusted, particularly when it handles the private keys on behalf of the user. In the latter case, the TTP is theoretically able to encrypt and decrypt any message from or to any user.

A designed cryptographic application would take advantage of both schemes: a public key exchange could be used to establish a symmetric key between two parties, while further communications would be encrypted using the symmetric key. In addition, public key cryptography provides a good means for key distribution. A largely used combination of public key and symmetric key cryptosystems is the so-called *digital envelop* technique. Using this technique, the initiator of a communication request (let's call him α) needs first to download the public key k_β of the other user (let's call him β); then α will generate a random session key, envelops (or encrypts) it using k_β, and sends the envelope to β. After β has decrypted the ciphertext and retrieved the session key, the two parties can then use it to exchange confidential messages during the session.

The above approach presents, nonetheless, two limitations. First, the process uses a session key that is created by one party (the initiator) and the other party will have to entirely rely on the initiator's truthfulness in key generation. Second, an eavesdrop, who is able to force the receiver to reveal his private key, can recover the plaintext message. The aforementioned limitations may be reduced if the public key cryptographic part in the scheme uses a technique such as the Diffie-Hellman key exchange protocol, where the session key is obtained cooperatively (under the form g^{ab} (mod p) for the Diffie-Hellman technique, where a and b are the random values selected independently by α and β, respectively). Users α and β are equally involved in the construction of the session key.

2.3 Insecurity of Cryptosystems

Among the cryptographic algorithms, some have experienced security problems that can potentially make them unsafe to use. These problems are mainly linked to the size of the keys used in the algorithm or the parameters used to construct the keys. For the sake of simplicity in the description of the insecurity problems, we will consider in the following the security limitations of DES and the security problems of RSA. While the limitations of DES have conducted to the conclusion that the keys used are too short, the problems of RSA have only been conducted to the establishment of rules to select the prime components used to build the pair of keys.

2.3.1 Insecurity of Secret Cryptosystems

The brute-force attack should not be considered as a real attack, since it applies to all ciphers. The designers of ciphers have expected it and have wanted this attack to be the only means for an attacker. One can say, therefore, that during the 1970s, the DES was very successful, given the computation level of technology in that decade. The short-key weakness of DES has been discovered prematurely, since a key search machine, called the DES crackers, has been built with lower cost and with the capability of successfully finding the key in 56 hours (EFF, 1998). In addition, it was announced in 1993 that a special purpose VLSI DES key search can be built and can find the secret DES rapidly, knowing the pair of plaintext and ciphertext.

One solution that has been proposed to overcome the limitations of DES is to extend the size of the secret keys by applying DES several times. Using a key $k = (k_1, k_2, k_3)$ of size 168 bits, the solution is represented by encryption and decryption functions E' and D' using the encryption and decryption function E and D of DES as follows:

- Encryption: $E'_k = E_{k_1} D_{k_2} E_{k_3} \quad k = (k_1, k_2, k_3)$
- Decryption: $D'_k = D_{k_1} \circ E_{k_2} \circ D_{k_3} \quad k = (k_1, k_2, k_3)$

The solution is called 3-DES. It is compatible with DES. Now let us consider more generally a sth encryption E as defined by:

$$E_k^{(s)} = E_{k_1} \circ E_{k_2} \circ \ldots \circ E_{k_s} \qquad k = (k_1, k_2, \ldots, k_s)$$

where E_* is either the encryption or the decryption function of a given block cipher using key *. A naïve exhaustive attack on $E^{(s)}$ tries the $2^{s|k|}$ key pairs ($|k|$ is the key length of E_{k_i} for all i). An attack called *meet-in-the-middle attack* is able to overcome double encryption ($s = 2$) using $2^{|k|}$ encryption operations and $2^{|k|}$ decryption operations. This can be achieved as follows: with $,c$) a pair of plaintext and ciphertext, one can compute $c_i = E_i(m)$ for all possible key value i and store all the pairs (i,c_i). Then one can decipher c using all key values j. For each computed deciphertext m_j, we check for a match with a c_i. When the match is obtained, the pair of keys is determined as (i,j).

The meet-in-the-middle attack can be easily generalized to reduce the brute-force attack complexity on $E^{(n)}$, for $n > 2$.

2.3.2 Insecurity of Public Cryptosystems

For the sake of clarity, we consider in this subsection some of the weaknesses discovered on RSA cryptosystem during the last two decades. The attack that is most often considered for RSA is the factoring of the integer n public key (n,e). If factoring is achieved, all messages encrypted with the public key can be decrypted. Since the public key is chosen very large, the factoring operation takes an unreasonable amount of time. However, a common attack can be easily performed, provided that the modulus n, which is generated randomly, belongs to two different users of the public cryptosystem. This attack is called common modulus attack.

The objective of this subsection is to describe the major attacks on RSA that decrypt messages without directly factoring the RSA modulus n (Boneh, 1999).

2.3.2.1 Attacks Based on Small Private Keys

To reduce the decryption time, one may wish to use a small value of d rather than allow a random d. Since modular exponentiation takes time linear in $\log_2 d$, a small d can improve the performance of the decryption algorithm. Unfortunately, an interesting attack due to (Wiener, 1990) shows that a small value of d results in a total break of the cryptosystem. More formally, let $n = pq$ such that $q < p < 2q$ and assume that d satisfies the following inequality

$$d < \frac{n^{1/4}}{3},$$

then it can be stated that

$$\left|\frac{e}{n}-\frac{k}{d}\right| \leq \frac{3k}{d\sqrt[4]{n}} \leq \frac{1}{2d^2}$$

The number of fractions of the form k/d with $d < n$ approximating e/n is bounded by $\log_2 n$. An attack can be performed aiming at determining d. Its complexity is controlled by $\log_2 n$. Therefore, to avoid such attack, d needs to be higher than $n^{1/4}/3$.

2.3.2.2 Attacks Based on Small Public Exponents

To reduce encryption time, it is assumed that one can use a small public exponent e. The smallest possible value for e is equal to 3. The so-called *Franklin attack* operates as follows. Let (n,e) be an RSA public key and m_1 and m_2 be two messages satisfying $m_1 = P(m_2)$ for a polynomial function $P(x) = ax + b$, with coefficients in Z_n. and b non null. Then, let c_1,c_2 be defined by:

$$c_j = m_j^e, \; j = 1,2$$

Then m_2 is a zero of the flowing two polynomials

$$Q_1(x) = P(x)^e - c_1, \; Q_2(x) = x - m_2$$

If $e = 3$, one can use the Euclidean algorithm to compute $\gcd(Q_1(x),Q_2(x))$. This is a polynomial function of degree one. This allows the determination of m_2 and m_1.

2.3.2.3 Cycling Attacks

Let m be a plaintext and c be the related ciphertext with respect to a pair of private and public keys $(d; n,e)$; i.e., $c = m^e \pmod{n}$. Then there is a positive integer k such that $c = c^{e^k} \pmod{n}$. Thus, we conclude easily that $m = c^{e^{k-1}} \pmod{n}$. Based on this fact, an attacker can compute the sequence

$$c_k = c^{e^k} \pmod{n}, \text{ for } k = 1, 2, 3, \ldots$$

until c is obtained for the first time and then m is deduced.

2.3.2.4 Message Concealing

A plaintext message m is called unconcealed message for a public key (n,e) if it encrypts to itself ($m = m^e \pmod{n}$). It has been shown that the number of unconcealed messages is exactly given by:

$$[1 + gcd(e - 1, p - 1)] \times [1 + gcd(e - 1, q - 1)].$$

Thus, the number of unconcealed messages is always equal to 9 at least (because $e - 1$, $p - 1$, and $q - 1$ are all even). Since e is selected at random, this may constitute a security problem. However, if e is chosen to be a small number such as $e = 3$ or $e = 2^{16} + 1$, then the proportion of messages that are unconcealed by RSA encryption will be negligibly small, and thus unconcealed messages do not pose a threat to the security of RSA encryption.

2.4 Digital Signature

The digital signature is a cryptographic technique that is fundamental to authentication, access control, authorization, and non-repudiation. The object of a digital signature is to provide a means for an entity to attach its identity to a piece of information. The process of signing a message involves transforming the message and some secret information held by the entity into a mark called a signature. A basic definition of a digital signature is as follows (Menezes, 1996). Let:

- M be the set of messages to be signed,
- S be a set of elements called signatures (they are typically binary strings of a fixed length),
- Sg_α: $M \rightarrow S$ be a transformation from the set M to the set S, called a signing transformation for entity α. Transformation Sg_α is kept secret by user α and will be used to create the signatures related to the messages in M.
- V_α is a transformation from the set $M \times S$ to the set $\{1,0\}$, called verification transformation for α's signatures. It is used by other entities to verify signatures created by α.

The transformations Sg_α and V_α define a digital signature scheme for user α. Two processes can be associated with a signature scheme for user α, the signing procedure and the verification procedure. With the signing procedure, the signer generates a signature for a message m in M by computing $Sg_\alpha(m)$ and transmitting the pair $(m, Sg_\alpha(m))$ to user β. The verification of signature $Sg_\alpha(m)$ on a message m, an entity β verifier starts by getting V_α, then he/she computes $\varpi = V_\alpha(m, Sg_\alpha(m))$. Then, entity β accepts the signature as having been created by α if $\varpi = 1$ and rejects if $\varpi = 0$.

The properties required for a signature scheme (Sg_α, V_α) should include the following:

- $Sg_\alpha(m)$ is a valid signature of α on message m if and only if $V_\alpha(m, Sg_\alpha(m)) = 1$.
- It is computationally unfeasible for any entity other than α to find, for any m in M, a signature s in S such that $V_\alpha(m, s) = 1$.

- ■ The signature must depend on the message being signed and should prevent forgery.
- ■ The signature should be relatively easy to compute and relatively easy to verify.

The signature must be easy to store in terms of structure, size, and link to the related message. In the following subsections we consider two special examples of digital signatures based on the use of one-way trapdoor hash functions and public-key cryptography. They are: the RSA signature scheme and the *Digital Signature Standard* (DSS, 2000), which includes DSA (Digital Signature Algorithm).

The signature generation and verification schemes work as follows with a public cryptography: During signature generation, an entity wishing to sign a message m selects a public key cryptosystem and uses his private key to associate a hash value to m and encrypt that value. He then sends the message and the encrypted value to the recipient. During verification, the recipient re-computes the associated hash value of m and checks whether it matches the result obtained after decrypting the received encrypted value. Therefore, digital signatures involve only two communicating parties (signer and verifier) and assume that the verifier has access to (or knowledge of) the signer public key. A digital signature guarantees the integrity of the signed message. It fails, however, to provide confidentiality since the message is sent in clear text. Confidentiality can be provided by encrypting the entire message (including the encrypted value) with either a secret key shared with the destination or with the destination's public key.

The efficiency of a public key-based digital signature scheme relies on the security of the sender's private key and the guarantee of public key ownership. If a sender later wants to deny sending a message, he can claim that his private key was lost or that the public key used to verify the signature does not belong to him. Administrative controls related to the security of private keys and ownership of public keys can be employed to thwart this situation. Examples of control aim at insuring that (a) every signed message should include a timestamp; (b) every loss of keys should be immediately reported to the central authority handling the public keys; and (c) every public key is stored in a trusted directory after verification of its link with the related owner and private key.

2.4.1 The RSA Signature Scheme

The plaintext space and ciphertext space for RSA public key cryptosystem are equal to the set $\{0, .., n\}$ where n is a large composite number. Number n is assumed to be different from one user to the other. RSA assumes that the sender and the receiver are in possession of the public key of the other user. During the key generation process, the user selects randomly two large prime numbers p and q, and states $n = p.q$. The user then selects two random numbers such that

$$e.d = 1 \ (\text{mod} \ (p-1)(q-1))$$

The sender finally states that his private key is d and the public key is (n, e). To sign a message m, the user computes the hash value $h(m)$ using a one-value function (or hash function) h. Then he computes the signature $Sg(m)$ of m by

$$s = Sg(m) = (h(m))^d \ (\text{mod} \ n)$$

On the reception of the message m and signature $s = Sg(m)$, the recipient decrypts s using the sender's private key (n,e). This is given by $s^e(\text{mod} \ n)$. The verification process aims at checking whether $h(m) = s^e \ (\text{mod} \ n)$. The signature is considered valid if the equality is true.

Similar to any public key-based digital signature scheme, the RSA scheme is vulnerable to some form of replay attacks since an adversary, say β, can compromise the system by capturing a copy of signed message $<m,s>$ and resend it inducing damages (if the message is a transaction, for example). β can also damage the system if he compromises the signer's private key. To overcome the weakness associated with replay attacks, a modification can be made to the signature scheme to include a timestamp t in the message signature s, say

$$s = (<h(m),t>)^d \ (\text{mod} \ n)$$

In that case, the verification process should invalidate any received message that has a signature that has been previously received.

2.4.2 The DSA Signature Scheme

The DSA is based on the discrete logarithm problems. It uses a scheme called the *ElGamel scheme* (ElGamel, 1985). The plaintext space and ciphertext space are $\mathbf{Z}_{p-1}$, where p is a large prime number. Let consider two other parameters of interest for DSA, a large prime number q and a natural number g such that

$$q \ \text{divides} \ (p-1) \ \text{and} \ g = k^{(p-1)/q} \ (\text{mod} \ p)$$

where k is a number satisfying: $0 < k < (p-1)$.

During signature generation, the signer first chooses randomly a private key x and computes the related public key (p,q,g,y) by setting $y = g^x \ (\text{mod} \ p)$. Given the public key y, it is computationally unfeasible to extract the private key x (this is the discrete logarithm problem). To sign a message m, the signer determines two numbers from m, called r and s, using the following expressions:

$$r = g^k \ (\text{mod} \ p), \ s = k^{-1}(h(m) - xr) \ (\text{mod} \ p-1)$$

where $h(-)$ is a one-way function (or hash function), k is an additional number selected randomly by the sender such that $0 < k < p - 1$, and k^{-1} is the inverse of k in $\mathbf{Z}_{p-1}$ (i.e., $k.k^{-1} = 1 \bmod (p - 1)$). Moreover, k should be unique for every signing operation. The signature of message m provided by DSA is defined by pair

$$(r, s).$$

On reception, the receiver of $<m, (r,s)>$ starts the verification process. He (she) computes

$$y^r \times r^s$$

and checks whether the result is equal to $g^{h(m)} \bmod p$. The signature is valid, if the verification is successful, since the following holds:

$$y^r \times r^s = g^{h(m)} \,(\bmod\, p)$$

2.4.3 Message Digest

As seen in the previous subsection, digital signature based on asymmetric cryptography makes use of one-way functions (or hash functions) to provide a message digest (Menezes, 1996). A hash function h maps a bit-string of arbitrary finite length to another bit-string of fixed length n, where n depends on h. Thus, a hash function outputs a hash value, which is a condensed representative image of the bit-string fed in input. Hash function should comply with the following feature: Changing just one bit of the input string results in a very different hash value in output; this is known as the avalanche effect.

A hash function h should have the following properties:

- It is pre-image resistant, i.e., given an output y, it is computationally unfeasible to find an input x such that $h(x) = y$;
- It is collision resistant, meaning that given an input x it is computationally unfeasible to find another input x' (different from x) such that $h(x') = h(x)$;
- It is fully collision resistant, meaning that it is computationally unfeasible to find two different inputs x and x', such that $h(x) = h(x')$.

The major hash functions in use today include Message Digest (MD5), Snefru, RIPEMD-160, and the Secure Hash Algorithms that include SHA-1 and SHA-256. The cryptographic literature often references a random oracle, which is a theoretical model of a "perfect" hash function, which returns an answer uniformly selected among all possible answers. A hash function may be used in conjunction with a secret shared key to construct a keyed hash function. In this case, the digest is more often called Message Authentication Code (MAC). The resulting keyed hash function is called with a name that depends on the hash function used, for instance HMAC-MD5, HMAC-RIPEMD, or HMAC-SHA1.

2.5 Authentication

Entity authentication can be defined as the process through which the identity of an entity (such as a mobile user, a computer, an application, or a network) is demonstrated. Authentication involves two parties, a *claimant* (also called *prover*) and a *recipient* (also called *verifier*). The *claimant* presents his identity and a proof of that identity. The verifier checks that the *claimant* is who he claims to be, by confirming the proof. Authentication is a very special concept. It is distinct from identification, which aims at determining whether an individual is known to the system he accesses. It is also different from authorization, which can be defined as the process of granting the user access to specific resources, in the system, based on his profile and the policy controlling the resource access. In the following sections, however, we will use the terms *identification* and *authorization* to designate the same concept.

Message authentication, on the other hand, provides the assurance that a message has not been modified during its transmission. Two main differences can be observed between entity authentication and message authentication, as provided by the techniques described in this chapter: (a) message authentication does not provide time-related guarantees with respect to when the message has been created, signed, sent, or delivered to a destination and (b) entity authentication involves no meaningful information other than the claim of being a given entity.

Examples of authentication systems (or authenticators) include biometrics, where an individual's identity can be proved using biological features such as fingerprints, hand geometry, retinal patterns, voice recognition, or facial recognition. Most of these forms of authentication are being used widely. Digital signatures give another type of authenticators. They are beginning to replace passwords as an efficient means of authentication and access control to networks and information systems as well. Digital signatures allow users receiving data over a communication network to correctly determine the origin of the information and check whether it has been changed during transmission.

Because authentication has started to be an essential component of business and consumer activity, all the technologies being developed to provide authentication services will have to prove their efficiency and security capabilities. However, the functions used to produce an authenticator may be grouped into three categories: (a) message encryption-based authenticators, which use encryption so that the produced ciphertext of the entire message serves the authentication of the message; (b) cryptographic checksum-based authenticators, which use public checksum-like functions of the message and a secret key that produces a fixed-length value to serve as an authenticator; and (c) hash function-based authenticators, which use public one-way functions to map a message of any length into a fixed-length value that serves as the authenticator of message.

An entity authentication protocol is a real-time process that provides the assurance that the entity being authenticated is operational at the time when that entity has carried out some action since the start of the protocol execution. From the point of view of the verifier, the result of an entity authentication protocol is the acceptance of the prover's identity as authentic or its non-acceptance. The authentication protocol should fulfill the following objectives:

- The probability that any third party different from the prover, using the authentication protocol and impersonating the prover, can cause the verifier to authenticate the third party as the prover, is negligible.
- The verifier should not be able to reutilize the information provided by the prover to impersonate him to a third party.
- A signed message cannot be recovered from the signature code during signature verification.

A digital signature can be defined as a cryptographic message enhancement that identifies the signer. It authenticates the message on a bit basis (i.e., every bit is authenticated) and allows anyone to verify the signature, with the restriction that only the signer can apply it. The digital signature is different from the authentication. It identifies the signer with the signed message. By signing a message, the signer marks it in his own unique way and makes it attributed to him. The concepts of signer authentication and document authentication encompass the non-repudiation service, which provides a proof of the origin or the delivery of data in order to protect the sender against false denial by the recipient that has received the data, or to protect the recipient against a false denial by the sender that the data has been sent. Typically, a digital signature is attached to its message and stored or transmitted with this message. However, it may also be sent or stored as a separate data element since it maintains a reliable association with its message.

In addition to the above features, authentication and digital signature can be classified into weak and strong techniques, while entity authentication methods can be divided into three classes, depending on which paradigm they are based on. They are the following:

1. *Known-object-based authentication*: Methods in this class use as an input what the user presents (through challenge-response protocols) for authentication. Examples of known object-based methods include the standard password, the personal identification numbers (PIN), the secret keys, and the private keys.
2. *Possessed-object-based authentication*: Techniques in this class use physical devices to authenticate an entity. Examples of such techniques include the

credit card, the smart card, and the hand-held customized devices that provide time varying PINs.

3. *Biometric-object-based authentication*: Methods in this class use biological characteristics to achieve user authentication. Examples of characteristics include fingerprints, retinal patterns, voice, and hand geometry.

Techniques used in possessed-object-based authentication are typically non-cryptographic and will not be extensively considered in this book. The techniques belonging to third category will be discussed in Chapter 5. The current chapter discusses the techniques used by the known-object-based authentication.

2.5.1 Weak Authentication Schemes

Weak authentication typically fails to provide a complete and efficient authentication. Two classes of weak authentication schemes are of particular interest since they are commonly used. They are the password-based authentication and the PIN-based authentication. We discuss in this subsection the main features and drawbacks of weak authentication techniques.

2.5.1.1 Password-Based Authentication

Password authentication is perhaps the most common way of authenticating a user to a communication system. To be authenticated, the system compares the password entered by the user, after his login, against the expected response. Typically, a password-based authentication falls into the category of secret-key methods as it uses conventional password schemes that are time-invariant passwords. Passwords as associated with an entity are strings of characters (usually with a size larger than 8). They serve as a shared secret key between the entity and the system to be accessed. To be authenticated, the entity provides a (login, password) pair, where login is a claim of identity and password is the shared password that is used to support the claim. The system then checks whether the pair matches the entity's identity and the secret it shares with the entity.

Generally, the password schemes differ by the technique used to perform the verification and store the information providing password verification. For non-time-varying passwords, the system can store the entity's passwords in a system file, which is protected against read and write operations. In that case, the system does not use any kind of cryptographic object. Therefore, this method provides no protection against the privileged system users and subsequent accesses to the file after backup operations. To overcome such drawbacks, the verifying system can apply a one-way function to the passwords and store the resulting values. In such a situation, to verify the password provided by an entity, the system computes the one-way function on the entered data and checks whether the result matches with the entry it has stored.

Non-time-varying password schemes present various security weaknesses. An adversary, for example, can perform various attacks to get control of a user's password. He can observe it as it is introduced by the user, or during its transmission; and then he can use the captured password for a subsequent impersonation. He also can perform password-guessing and dictionary attacks. Finally, in the case of the use of one-way functions (as a means to protect passwords), an adversary may attempt to break the list of passwords by providing arbitrary passwords, one by one, and comparing their values to the passwords in the file.

While there are considerable problems with non-time-varying password-based authentication, it should be noticed that the passwords are very familiar and offer a high level of user acceptability, ease of use, and convenience. Added to this, administrative rules can be used within an enterprise to ensure the user-chosen passwords satisfy certain criteria for acceptability (e.g., size and use of digits) and make mandatory a periodic modification of the passwords under use.

A natural enhancement of the fixed-password schemes is given by time-varying password schemes such as the "one-time password" schemes, which ensure that a system user utilizes a new password at each new access. A list of passwords is managed for each user with any one-time password scheme. Each password in the list is used only once. The passwords can be in three forms: they are pre-written in the list; sequentially updated; or sequentially computed. With the first form, the list is not used sequentially, but the entity and the system agree to use a challenge-response table containing n pairs of the form

$$<i, password_i>, i < n.$$

On access request, the system challenges the users with a value for i, and waits for the right password. The second case considers an initial unique shared password.

The user is assumed to create and transmit the new password, say $password_{i+1}$, to the system during authentication period covered by $password_i$. The last case assumes that the users have an initial password, say pwd_0. The user utilizes a one-way function h to define the password sequence so that, for $i < n$, the ith password pwd_i is given by

$$pwd_i = h^i(pwd) = h(h(..(pwd)..)).$$

On the occurrence of problems (or on reaching the value n), the system is assumed to restart with the shared initial password. Even though they provide better security than the non-time-varying schemes, the one-time schemes, however, present several drawbacks. An active attacker, for example, can attempt to intercept unused one-time passwords and impersonate the system.

2.5.1.2 PIN-Based Authentication

Personal identification numbers (PIN) schemes can be classified as special time-invariant passwords since a physical device, such as a banking card or a token, is used to store the PIN. Typically, the PINs are short strings of digits (from 4 to 10 digits). PIN-based authentication represents a vulnerability that should be covered by additional measures. Examples of protections include invalidating the physical device when more than a pre-specified number of incorrect PINs are attempted by an adversary (or the user himself).

In an authentication system that uses PINs, a claimed identity accompanied by a user-provided PIN may be verified (on-line) by comparing it with the PIN stored for that user in a system database. An alternative approach, called validation off-line, does not use a centralized database and considers that the verification is performed based on information stored on the device itself. In such situation, the PIN may be defined to be a function of a secret key and the identity of the user associated with the PIN. Moreover, the device should contain additional information allowing the token (and therefore associated user) to be authenticated. This, however, requires the user to possess the device and remember the PIN.

2.5.2 Strong Authentication Schemes

Typically, a strong authentication scheme is based on the concept of cryptographic challenge-response protocol, which works generally as follows: A user wishing to access a service (or use a resource) should prove his identity to the verifier by demonstrating knowledge of a secret information known to be sufficient to authenticate him. The demonstration is usually made without revealing the secret information to the verifier. It is typically achieved by providing the right response to a time-varying question (or challenge) related to the secret information.

Typically, a cryptographic challenge-response protocol is built on secret key cryptosystems, public key cryptosystems, and zero-knowledge techniques. It often uses time-variant parameters to uniquely identify a message or a sequence involved in the process, and thus protect against replay and interleaving attacks. Examples of time-variant parameters include the timestamps, the random numbers, and the sequence numbers. Combinations of these parameters may be used to ensure that random numbers are not repeated, for example. Often, random numbers are used as follows: An entity can include a random number in a transmitted message. The next received message, whose construction has used the random number, is bound to that number, which tightly links the two messages. Various drawbacks can be observed with the protocols using this technique, including the utilization of pseudorandom number generators and the generation of additional messages.

A sequence number, such as serial number, transaction number, and counter value, serves as an identifier of a message within a series of messages. Sequence numbers must be associated with both source and destination of a message. The association can be explicit or implicit. Parties using a sequence number scheme agree on the fact that a message is accepted only if the sequence number contained in the message conforms to a well-defined policy. The least policy should define the starting values of the sequence, window time, and monotonicity form of the sequence. Several problems can limit the use of sequence numbers including the delays experienced at the verifier's side.

Timestamps can be used to provide timed guarantees and prevent message replays. They can also serve to implement access privileges. The timestamps are used as follows. An entity originating a message inserts in it a timestamp that is cryptographically bound to it. On the receipt of the timestamp, the destination computes the difference δ between its local time and the received timestamp. The message is accepted if δ is within the acceptance period of time and no other message with the same timestamp has been previously received from the originating entity. However, the security of the timestamp scheme relies on the use of a synchronized clock.

Various classes of strong authentication schemes are built as follows:

1. **Challenge-response by cryptosystems:** Challenge-response mechanisms-based cryptosystems expect the entity requiring an access to share a secret key with the verifier (in the case of secret key cryptosystem) and to secure his public key (in the case of public key cryptosystem).

   ■ *Secret Key cryptography*: The general model of a challenge-response mechanism using secret key cryptography (and a random number) can be described as follows: Let r denote a random number and E_k denote the encryption transformation using a secret key shared between user α and β. Assuming that each user is aware of the identity of the other entity, the authentication of user α starts by having the verifier β sending a random number r. On receiving r, user α computes $E_k(r,m)$ and sends the result back to user β. Message m is optional and is used to prevent replay attacks. Then, the verifier decrypts the received random number and checks whether it is the random number that was provided before. The above-mentioned mechanism may include a one-way function to provide a more efficient challenge-response mechanism.

     In the case of mutual authentication, three messages can be used between α and β:

$$\beta \rightarrow \alpha: \ r;$$

$$\alpha \rightarrow \beta: \ <r', E_k(r,r',m)>$$

$$\beta \rightarrow \alpha: \ E_k(h(<r,r',m'>))$$

where r is a random number generated by β, r' is a random number generated by α, m and m' are optional messages, and h is the one-way function.

■ *Public Key cryptography*: Public key techniques can be used to provide challenge-response authentication, with the originator of a request demonstrating knowledge of his private key. The originator can use two approaches to achieve this: either he decrypts a challenge that the verifier has encrypted using the originator's public key or he digitally signs a challenge.

Challenge-response based on public keys operates typically as follows: User β generates a random number r. Then, he computes its hash value $h(r)$ using a hash function, encrypts r along with a general message m, using α.'s public key, and sends to user α the value $h(r)$ appended to message m and the encryption result. User α. decrypts the received message to recover the first component r' and $h(r)$. If $h(r) = h(r')$, α. sends r to β; otherwise he exits. Formally, the exchange is done by

$$\beta \to \alpha.: E_{k_\alpha}(<r,\ h(r),m>)$$

$$\alpha. \to \beta: r \text{ if reception is valid, else exit}$$

where E_{k_α} and k_α stand for the public-key encryption function and the public key of α., respectively.

Challenge-response based on digital signature typically assumes that a request originator, upon receiving a random number r from the verifier, generates a random number r' and an optional message m (which may be equal to the identity of the verifier). Then he signs $<r,r',m>$ and sends the result to the verifier of the digital signature appended to r' and m. If β is the request generator, then the verifier result received by α. has the following form:

$$<<r,r',m>,\ E_{K_B}(h(r,r',m))>$$

where K_B stands for the private key of β.

2. ***Challenge-response by zero-knowledge techniques:*** The above-mentioned challenge-response mechanisms might reveal part of the secret information covered by the user wishing to be authenticated by a verifier. A malicious verifier, for example, may be able to submit specific challenges to obtain responses capable of recovering such parts of the secret information. Zero-knowledge was proposed in the literature to overcome such drawbacks by allowing a user to prove knowledge of secret information while revealing no information of some interest to the verifier whatsoever.

Roughly, the zero-knowledge technique is a form of interactive proofs, during which the originator (or prover) and the recipient (or verifier) exchange various messages and random numbers to achieve authentication. To perform such interactive proofs, the concept of proof is extended to integrate some forms of probabilistic features. The general form of a zero-knowledge scheme is given by a basic version of the Fiat-Shamir algorithm (Feige, 1977). We present this algorithm in the following. However, we notice that more efficient versions of this algorithm are now under use within various solutions. In these versions, multiple challenges may be used.

In the basic version of the Fiat-Shamir authentication, user α proves to β knowledge of a secret s in n executions of a 3-phase process:

■ **Phase 1**: *Secret generation.* A trusted third party (TTP) selects two large prime numbers p and q. Then he publishes $n = p.q$ while keeping secret p and q. A user, say α, wishing to be authenticated by user β, selects a secret s relatively prime to n, $1 < s < n$, computes a public key c by

$$c = s^2 \bmod n$$

and registers the public key with the trusted third party.

■ **Phase 2**: *Exchanging messages.* α generates a random number r, $0 < r < n$, and participates in the following three actions:

$$\alpha \rightarrow \beta : x = r^2 \bmod n$$

$$\beta \rightarrow \alpha : \textit{random Boolean number } b$$

$$\alpha \rightarrow \beta : y = r.s^b \bmod n$$

The above actions are repeated t times ($t < n$). At the end of the t rounds, we can say that we have (a) a number of $3t$ messages that have been exchanged; (b) user α has selected t random values $r_1, .., r_t$; and (c) α has computed t values $x_1, .., x_t$, and determined t numbers $y_1, .., y_t$, while β has selected t random Boolean values $b_1, .., b_t$ and received all x_i, and y_i, for $i = 1, .., t$, such that

$$x_i = r_i^2 \bmod n, \, y_i = r_i.s_i^{bi} \bmod n$$

■ **Phase 3**: *Verification.* User β accepts the proof if, for all $i < t$, the equality

$$y_i^2 = x_i.c^b \bmod n$$

is satisfied and that y is not 0. Both terms of the equality take the form $r_i^2 . s_i^{2bi}$ in the case of success.

By sending challenge b, the verifier aims to first check whether user α is able to demonstrate (t times) that he has knowledge of the secret s, and to deny actions performed by an adversary impersonating user α such as selecting any random number r and sending $x = r^2/c$ to verifier β. On receiving $b = 1$, for example, the adversary will only answer by sending r, which is enough to satisfy the above equality. However, this will not work for $b = 1$. The response $y = r$ (response for challenge $b = 0$) is independent of the secret s, while the response $y = r.s$ mod n (made when $b = 1$) provides no information about s because r is a random number.

3. ***Device-based authentication:*** Normally, many user authentication protocols are susceptible to masquerading, spoofing, interception, and replays authentication messages. Some current approaches address these exposures using authentication devices with limited processing capability. The devices contain a robust cryptographic algorithm, which aims to help user authentication in a hostile environment. The device's key is randomly selected out of the key space of the embedded cryptographic algorithm. Since the algorithm is robust and the key is large, the probability of success of a brute-force attack is almost null. The activation procedure of the device operation takes place directly between the user and the device and is performed by the user using a weak initial secret (such as a password or a PIN).

Current device-based authentication methods differ in many aspects. The following features characterize some dimensions of that difference:

- *Device-workstation interface*: The device and workstation need to communicate. They can communicate through an electronic interface, such as a card reader, or via the user himself, who may enter manually the information provided by the device.

- *Clock availability*: An internal clock may be needed by the device for the generation of the necessary parameters (e.g., generated random numbers, timestamps) or for the computation of other useful parameters (e.g., one-way function values).

- *Storage usage*: Non volatile read-only storage may be needed to store sensitive information and computational procedures within the device (e.g., cryptographic key, random numbers generator)

- *Encryption capabilities*: Specific devices implement public key cryptography. However, for the sake of complexity reduction, the devices may not have to implement the decryption algorithm or may only need to perform a one-way function value computation.

- *Exchanged information*: The structure and size of the exchanged information between a device and the system may differ from one device to another based on the methods used. The way the information is passed is also an important factor to consider

With the current methods, the user-to-device relationships developed for the needs of authentication is based on the delegation concept, where the device performs an authentication procedure on behalf of the user. Unfortunately, this may induce a major drawback, since a potential of masquerading can be performed using stolen devices.

2.6 Attacks against Authentication in Mobile Wireless Networks

Nowadays, malicious adversaries can attempt to defeat authentication schemes in mobile communications and to perform a set of damaging attacks.

2.6.1 Common Attacks

The basic attacks targeting authentication in mobile networks include the following non-exhaustive list:

- *Impersonation attacks*: The adversary can attempt to capture information about an authorized mobile user and impersonates that user. Impersonation is easy to perform if the adversary succeeds to compromise the keying material (such as password, PIN, or key) of the mobile user.
- *Replay attacks*: The adversary can attempt to capture the authentication related information and replays it to impersonate the mobile user originating the information with the same or a different verifier (Syverson, 1994). Assume, for example, that a user U authorizes the transfer of funds from a banking account to another by only signing the request by a signature key known only to him. To get this transaction done, user U sends it to the bank system, which checks the signature and executes the transactions. An adversary H, wishing to have the same request repeated without having U's authorization, would need only to produce the signed transfer, provided that anti-replay measures have not been provided.
- *Forced delay attacks*: The adversary executes a forced delay attack when he can intercept a message, drops it from the network, and relays it to its destination after a certain period of time. This attack is different from the replay attacks since the interception stops the original message so that it cannot reach the verifier. Delaying signed transactions may induce serious damages to e-business trust.
- *Interleaving attacks*: The adversary can involve selective combination of information from one or more authentication processes, which may be ongoing processes (Tzeng, 1999). Examples of interleaving attacks include the oracle session attack and the parallel session attack. To explain the basics of these attacks, we consider the case of the two-way authentication protocol, which can be described formally as follows:

$$\alpha \rightarrow \beta : E_k(r)$$

$$\beta \rightarrow \alpha : <r, E_k(r')>$$

$$\alpha \rightarrow \beta: r'$$

where k is a shared secret key, and r and r' are random numbers generated by α and β, respectively.

- *Oracle session attack*: This attack is performed as follows. An intruder H starts a session with a mobile user β, in which he impersonates a mobile user α. Intruder H generates a random number, which is considered to be $E_k(r_1)$, intruder H has no knowledge of r_1. He sends it to β, who assumes that its source is α. User β decrypts the "supposed" encrypted random number, r_1, with the secret key k that he shares with user α and he generates a random number r_2 that he encrypts with the same secret key and sends the random number r_1 and the encrypted random number $E_k(r_2)$ to user α. Then, the intruder intercepts the message and starts a new session with α, in which he impersonates β. He sends the encrypted random number $E_k(r_2)$ to α. User α decrypts the encrypted random number, generates a new random number r_3, and encrypts it. Then he sends the random number r_2 and the encrypted random number $E_k(r_3)$ to user β. The intruder intercepts the message and sends the decrypted random number r_2 to β. He then quits the session with α. User β receives the random number r_2, which is the same as the one he generated, and so he thinks that he is communicating with user α. Finally, the intruder has untruly authenticated himself as user α to user β. He has used user β as an oracle to decrypt the encrypted random number that β has generated.

  A security measure to prevent an adversary from performing an oracle session attack on the aforementioned two-way authentication protocol is to transform the second step as follows:

$$\beta \rightarrow \alpha : <E_k(r), E_k(r')>$$

- *Parallel session attack*: This attack works as follows, when performed on the enhanced two-way authentication protocol. Assume that user α wishes to start a session with β. He generates a random number r_1 and sends it to β. An adversary H intercepts the message and starts a second session with α, in which he impersonates β and sends r_1 to α (as if he has generated it). User α assumes that the source of the random number is β and encrypts it with the secret key k he shares with β. Then α generates a random number r_2, encrypts with k, and sends the encrypted numbers to β as part of the parallel session. The adversary H intercepts the encrypted random numbers and sends them back to α as part of the first session.

Assuming that he is receiving the encrypted random numbers from β, α decrypts them. The received number r_1 is the same as the one he has generated for the first session, so he believes that he is communicating with β. Therefore, H has now authenticated himself falsely to α as β. After that authentication α sends r_2 back to β to complete the first session. Us_H intercepts the message and sends r_2 back to α as part of the parallel session. α receives the random number r_2 that is the same as the one he has generated for the parallel session and believes that he is communicating with β. Hence, H has finally authenticated himself two times falsely to user α as user β. Adversary H has used α as an oracle to encrypt the random numbers that he generated. Hence, two authentication sessions took place.

2.6.2 Common Guidelines for Protection

The above-mentioned authentication attacks can be avoided by applying several actions. Replay attacks can be stopped by the challenge-response techniques using a sequence number. Interleaving attacks can be avoided by linking together all messages using a sequence number. A protection against forced delay attacks can combine the use of random numbers with reduced time-windows. Nevertheless, the efficient security solutions that provide protection to the authentication protocols should consider how the protocols operate and suggest actions appropriately. This makes the guidelines to protect a communication system very useful. The following general security guidelines must be observed to protect authentications schemes:

- When a weak authentication based on passwords is used, a password policy should be stated. It should include rules connected to password complexity, length, aging, reusability, and timetables access. Such parameters depend heavily on the context of the accessed services and the frequency of the server's use. They depend also on account types to access the services, and the risk associated with passwords compromise.
- A security policy should be made available to describe under what conditions the accounts of mobile users are created, modified, and deleted. A set of administrative procedures should be referred to specify obligations regarding users' usage of authentication materials.
- A risk mitigation strategy is highly recommended to implement in order to study the possible attack scenarios, provide efficient authentication, and reduce the cost attached to the security measures that can be applied against threats.
- If authentication needs to be valid during a connection lifecycle, authentication should be re-performed periodically in a way that an adversary cannot benefit from the duration of the process.

- If the authentication process is linked to any active integrity service, the authentication process and the integrity service should use different keying materials.
- If timestamps are used in an authentication scheme, the working systems that are involved in the procedures of the computation and verification of timestamps should be protected and closely synchronized.
- Anonymous remote attempts of authentication should be unauthorized. The number of limited attempts of authentication that are made remotely by any mobile user should be limited to a small number (typically equal to 3).
- Selecting, generating, and managing the security parameters used by authentication schemes should take into consideration the need to reduce the probability of successful attacks.
- When a trust relationship is defined between servers, the authentication-related configurations should be reviewed carefully. A hacker can use the trust relationship to gain access to another host from a compromised host.
- When assessing/auditing the security of an authentication scheme, the underlying cryptographic algorithms and digital signature protocols should be checked as highly secure.
- In assessing an authentication scheme, the potential impact of compromise of keying material should be addressed. In particular, the compromise of long-term keys and past connection keys should be considered.

2.7 Authorization and Access Control in Mobile Communication

After a mobile user has been authenticated by a wireless network, an authorization process should be executed to check whether he is authorized to use resources or access a service he has requested. While the authentication is straightforward, authorization and access control may be controlled by a set of rules that define the access/authorization policy.

2.7.1 Access Control

Typically, the wireless access control system includes three logical entities—an *administration point*, a *personal trusted device* (PTD), and an *access controller*. The administration point defines the computer system, from which the authorization and authentication information is granted to mobile users under the form of certificates. The administration point uses a database to store information. The personal trusted device is often a handheld device that the customer carries with him and uses to access the communication system. The access controllers authenticate and authorize the users.

Usually, a common security policy is set for the network and all the authorization access rules applied within the network should comply with it. The authorization access policies may include some personalized rules that apply for personal data or configurations. The setup of an access control policy depends on the characteristics of the communication network, the service to access, and the distribution of that service. The authorization to access resources and services can be done in several ways and requires the submission of information that can be checked using appropriate databases, libraries/lists, or certificates. Indeed, digital certificates can be used to prove the user's identity, to provide the user's rights, and put restriction on the authorization for specific services.

Access control lists: A network administrator may manage for each service and resource a protected list that describes the resources/services and the users who can use them. The list can be stored in an ad hoc database or simply kept in a file. The authorization list may be set up to allow access depending on different types of information. It may be limited for groups of users, or devices connecting from remote nodes in the mobile network. Updating the access control rights might be hard to achieve, particularly when the access list is distributed. The propagation of updates may be needed in a synchronous way.

Single sign-on: Service providers (such as banks, network operators, and merchants) can allow the access of a mobile customer to services after logging in with username and password. Allowing the customers to utilize the same username and password at several different places, to help memorize them, may generate a security problem. A malicious service provider, for example, can access the customer's information on another service, using the same login and password (or try similar passwords). The need for memorization and use of multiple logins can be made possible using the concept of *sign-on service*, which allows the customer to access several services after a single authentication. For this, an authentication server is responsible to provide the customer with a proof of his identity, which can be passed to the other servers as a replacement for his password. This proof of identity must be encrypted so that it can be passed to the customer without a concern that it can be tampered with.

Kerberos is a well-known system that uses the single sign-on concept (Neuman, 1994). The servers of a Kerberos system trust in Kerberos server with the identities of the other customers. Kerberos uses *tickets* to securely deliver the identity of a customer to a server. The customer gets his initial ticket from the Kerberos server by sending his identity to the server and then decrypting the response with his password. The initial ticket, called ticket-granting ticket, is then used with the ticket-granting server to obtain additional tickets for other servers. The ticket is encrypted with the private key of a server. It contains the *name* of the client, the name of the server, the address of the client, a timestamp, a lifetime and a random session key. One ticket may be used to authenticate the customer to only one server multiple times during the lifetime of the ticket. When using other servers, new tickets need to be obtained from the ticket-granting server.

Certificate authorization: Certificates were originally designed as digitally signed bindings between a subject and a public key. Certificates can be used for different purposes including authorization. They may have different forms. The most used ones are the X.509 and PGP certificates. Certificates that bind a subject and a key are called *ID certificates*. They prove that according to their issuer, the subject of the certificate is holding the private key related to a public key in the certificate. The public key can be used to encrypt confidential information directed to the subject of a certificate. However, sometimes the identity of the owner of a public key is not sufficient. Information about whether a subject is authorized to access is needed.

Simple Public Key Infrastructure (SPKI) addresses this issue. Since an attribute certificate binds the authorization, the subject together with an ID certificate binds the subject and the key together, attribute certificates can be combined with the ID certificates to complete a binding between authorization and the key. If these certificates are controlled by different issuers, both of them must be trusted with the authorization decision. In case of an authorization certificate the permission is mapped directly to a key, which is used as an ID for an individual.

2.7.2 Certificate-Based Authorization

Simple Public Key Infrastructure (Ellison, 1999), represents an alternative to the public key infrastructure that eliminates the notion of global name and reduces the complexity due to X509 standard. It also presents a good framework for authorization provision. SPKIs incorporate the notion of local name. The digital certificates addressed in this subsection try to give a global view of certificate that unifies the concepts of X509 and SPKI certificates by extending them and reducing their limitations. The digital certificates incorporate some added features to the validation, delegation, and authorization fields that well apply for mobile communications. For the sake of uniformity, we consider that there are three types of digital certificates: (a) naming certificates, which provide for (local) names; (b) authorization certificates, which grant a specific authorization from the issuer to a mobile customer. It includes also specific fields for the delegation of authorization; and (c) revocation list certificates, which report on invalid certificates belonging to the previous two types of certificates (Boudriga, 2004).

Let us consider that for a given mobile entity there is a (home) name space, and that the meaning of home name may vary from one customer to another. We define a fully qualified name as a compound expression of the form:

name *k n*1 ... *ns*

where *k* is a public key and *n*1 ... *ns* are (home) names (Ellison, 1999). Naming, authorization, and revocation certificates have the following forms of signed messages:

<cert (issuer (name k n)), (subject p), Val >,
<cert (issuer (name k n)), (subject p), Act, Del, Val >, and
*<crl (issuer (name k n)), (revoked c*1, ..., *cs), Val >*.

where *k* is a public key representing the issuer, who should have signed the certificate using the private key associated with public key *k*, *n* is a local name of the issuer, *p* is a fully-qualified name of a mobile user, *Val* is a field describing validity constraints on the certificates (along the address where to check for revocation, if needed), *Act* stems for a set of actions that the issuer authorizes the mobile user to perform (such an access to a resource or the execution of a procedure), *Del* is a delegation field, which indicates the conditions under which the subject is authorized to propagate the authority of performing actions in *Act* to others. Finally, *cj* (= *(kj,hj)*, *j < s*), represents a pair containing the public key and hash value of a certificate declared by the issuer as invalid.

An action description is a pair:

<action (act-name, in-parameters),π >,

where *act-name* is the identifier of the action, *in-parameters* are the input parameters that are needed to execute the action, and π is a predicate defining the requirements to satisfy before authorizing the execution of the action. The verification of an authorization certificate involves the local environment that is in charge of executing the action and the parameters defined in *in-parameters*.

The delegation *Del* is nothing but a statement to be satisfied before the mobile user (having an authorization certificate) can delegate his/her authorization to another mobile user. The delegation field may vary from a static statement reduced to a bit (this represents the traditional approach for SPKI), which controls the transfer of the authority of delegation, to a time dependent predicate involving different attributes and delegation criteria based on the certificate use, time, and environment. The validation field *Val* contains the period validity, described as an interval of time during which the certificate is valid. It may include different constraints, depending on the purpose of the certificate. Field *Val* can contain a predicate called *now(t)*, which is used to represent authorizations to be performed only once within a short period of time starting from time *t*. This means that, as soon as the execution of the requested action terminates, the certificate is invalidated. It is clear that the use and semantics of predicate *now(t)* should be tightly controlled by the system that allows it.

Various operations can be performed on the aforementioned naming and authorization certificates. Such operations include certificate generation, certificate revocation, and on-line checks. Some other operations can be defined based on the need of the service to access. Among these operations, one can mention the on-line checks, including the validity, authorization, and status checks. The generation process is initiated by the user and managed by the certificate signer, who should be in charge of the appropriate verification, especially when some decisions or/and information are to be included in the authorization, delegation, and validation fields of the requested certificate. The generation process can be decomposed into three sub-processes:

- *User interfacing process*: This is an interactive tool that aims at collecting the needed information from the mobile user, delivering securely the generated certificates to the mobile user, and verifying the delivery acknowledgment.
- *Check realization process*: This is responsible for the verification and validation of the field's content of a certificate. It verifies the compliance with the statements present in the parent certificates and the local policy controlling the access.
- *Propagation process*: This process is responsible for updating all archives, directories, and services about the generation of any certificate. The revocation process of certificates is an operation that can be achieved when triggered by the certificate owner or by the certificate signer.

The revocation process can be decomposed into simpler sub-processes based on the certification practices used in the system where the certificates are utilized. The following can be one approach to decompose it into three sub-processes: a user process, which interfaces the system; a verification process, which checks whether the revocation is authorized; and a publication process, which propagates the CRLs where it is needed.

The on-line checks expressions contain information about the protocol or service used to perform the verification, and to authenticate the identity of the resource that should be consulted using the protocol. All on-line checks can have the following format:

<Req (Checker check-type) (k p Act Del Val) opt>

where *Checker* stems for the specification of one or more uniform resource identifiers that can be contacted to request the on-line check, and *check-type* defines the type of the check. This may be a validity check, which verifies that all conditions appearing in the validation field are satisfied; an authorization check, which verifies that all actions occurring in authorization field can be authorized to be performed; a delegation check, which verifies the validity of the constraints described in the delegation field; or status check, which reports on the state of a resource. Finally, the *Opt* field contains parameters values to be used in the on-line checks. Of course, the chain (*k p Act Del Val*) represents the content of a certificate and can be reduced to (*k p*).

The definition of public key certificate that is provided in this subsection authorizes various insertions in order to satisfy the requirements that mobile users and service providers may need. In addition, it is obvious that the X.509 and SPKI schemes can be integrated within the model based on the home names, the 3-type certificates, and the caching functions. Furthermore, the on-line checks specification developed here can be extended to include on-line requests for operations such as certificate generate and revocation, and security statuses, as well.

2.8 Key Distribution and Management

We discuss, in this subsection, some protocols for establishing a secret key between two entities in mobile communication system. These protocols provide a link-level security for mobile users in a machine-to-machine manner, compared with end-to-end security. The protocols form also the basis for establishing a secret key among a group of mobile users. We analyze these protocols from the vulnerabilities point of view. We also analyze the advantages and limits of each protocol. Three protocols will be considered.

2.8.1 Beller-Yacobi Protocol

The Beller-Yacobi protocol uses a challenge to address the replay attacks. Two reasons explain why this protocol is largely accepted in mobile systems: (a) the digital signature is used on the challenge at the mobile device and (b) the mobile can do most of the signature work when it is in an idle status (Boyd, 1998a). The protocol runs four steps: First, the base station B sends its public key K_B^+ to the mobile M. In the second step, the mobile M generates and encrypts a session key, say k_{BM}, using K_B^+ and sends it to B. Upon receiving the encrypted message, B decrypts the session key. In the third step, B sends to M a large random value N_B encrypted by the session key. In the last step, the mobile user M signs N_B using its private key and sends it back to B along with its identity ID_M public key K_M^+, and certificate *Cert(M)*.

The four operations can be formally described as follows.

1. $B \rightarrow M$: K_B^+, $Cert_B$
2. $M \rightarrow B$: $E_{k_B^+}(k_{BM})$
3. $B \rightarrow M$: $E_{k_{BM}}(N_B)$
4. $M \rightarrow B$: $E_{k_{BM}}(ID_M, K_M^+, Cert_B, E_{K_M^-}(N_B))$

Finally, B decrypts the received message using k_{BM}, decrypts N_B using K_M^+, and checks whether N_B has the expected value.

The protocol is resistant to replay attacks because, if an attacker replays the message 2, he will receive a different random value that he cannot get because he cannot know the needed private key, even if he knows an old session key previously established between B and the mobile user M. Therefore, the attacker will not be able to send the right message 4. However, if an attacker, say A, is a registered user of B, then he can succeed in spoofing as M, provided that he can communicate with M as a base station. The attack that can be launched by A operates as follows:

1. $B \rightarrow A$: K_B^+, $Cert_B$
2. $A \rightarrow B$: $E_{k_B^+}(k_{BA})$
3. $B \rightarrow A$: $E_{k_{BA}}(N_B)$
4. $A \rightarrow M$: ID_A, K_A^+, $Cert_A$
5. $M \rightarrow A$: $E_{k_A^+}(k_{AM})$
6. $A \rightarrow M$: $E_{k_{AM}}(N_B)$
7. $M \rightarrow A$: $E_{k_{AM}}(ID_M, K_M^+, Cert_M, E_{K_M^-}(N_B))$
8. $A \rightarrow B$: $E_{k_{BA}}(ID_M, K_M^+, Cert_M, E_{K_M^-}(N_B))$

During this attack, A starts a session with B and gets the random nonce for the session to be established. Then A starts the setup of a connection with M pretending to be a base station. Thus, A gets the message $(ID_M, K_M^+, Cert_M, E_{K_M^-}(N_B))$ from M and forwards it to B after encrypting it using the session key k_{AM}. This attack cannot be detected by B.

To protect the Beller-Yacobi protocol against this attack, one can require the signature of M on the session key and the identities of the mobile itself and the base station. Formally, the improved protocol operates as follows:

1. $B \rightarrow M$: B, K_B^+, $Cert_M$, N_B
2. $M \rightarrow B$: $E_{k_B^+}(k_{BM})$, $E_{k_{BM}}(ID_M, K_M^+, Cert_M, E_{K_M^-}(N_B))$, $E_{K_M^-}(h(ID_M, ID_B, N_B, k_{BM}))$
3. $B \rightarrow M$: B, $E_{k_{BM}}(N_B)$

Thus, the attacker standing between M and B cannot succeed by just forwarding the message containing M's signature that is acquired in another session with M.

2.8.2 Aziz-Diffie's Protocol

The Aziz-Diffie protocol uses both public-key and secret-key cryptography techniques. The public-key cryptography provides the means for session key setup and authentication (Aziz, 1994). Secret-key cryptography is used to provide privacy for bulk data transmission. The protocol works as follows: First, M sends B its certificate, a challenge, and a list of algorithms. Using the corresponding CA's public key, B can decrypt and get the public key of M. B responds with its certificate and the session key contribution component, say X_B, encrypted by K_M^+ and the preferred algorithm. To avoid the man-in-the-middle attack, a digest of vulnerable items is calculated and appended to the message. Similarly, M responds to B with its contribution component, say X_M, for the session key. With the knowledge of both contribution components, both sides can calculate the session key. The formal description of the protocol is given by

1. $M \rightarrow B$: $Cert(M)$, N_M, $AList$
2. $B \rightarrow M$: $Cert_B$, $E_{k_M^+}(X_B)$, $Balg$, $E_{k_B^-}(h(E_{k_M^+}(X_B), Sel, N_B, AList))$
3. $M \rightarrow B$: $E_{k_B^+}(X_M)$, $E_{k_M^-}(E_{k_B^+}(h(X_M), E_{k_M^+}(X_B)))$

where $AList$ is a list of flags representing secret-key algorithms provided by the mobile M and Sel represents the flag representing the particular algorithm selected by the base station.

One can notice the following facts for Aziz-Diffie protocol: (a) the certificate used in the protocol binds the identity of the certificate owner with its public key, (b) N_M allows avoiding replay attacks; (c) the mobile has to perform two computationally expensive operations during key establishment: one decryption to get in step 2 and one encryption to do the digital signature in step 3. Unfortunately, the protocol has shown some limitations. In fact the following attack can be launched by an attacker, say A, who is also a registered user of base station B (Meadows, 1995).

1. $M \rightarrow B$: $Cert_M$, N_M, $Alist$
2. $A \rightarrow B$: $Cert_A$, N_M, $Alist$
3. $B \rightarrow A$: $Cert_B$, $E_{k_M^+}(X_{BA})$, $Alist$, $E_{k_B^-}(h(E_{k_A^+}(X_{BA}), Sel, N_M, AList))$
4. $A \rightarrow M$: $Cert_B$, $E_{k_M^+}(X_{BA})$, $Alist$, $E_{k_B^-}(h(E_{k_M^+}(X_{BA}), Sel, N_M, AList))$
5. $B \rightarrow M$: $Cert_B$, $E_{k_M^+}(X_{BM})$, $Alist$, $E_{k_B^-}(h(E_{k_M^+}(X_{BM}), Sel, N_M, AList))$
   (The message is intercepted and deleted by attacker A)
6. $M \rightarrow B$: $E_{k_B^+}(X_M)$, $E_{k_M^-}(E_{k_B^+}(h(X_M), E_{k_M^+}(X_B)))$

Just after M starts a session with B, A initiates another session with B by replaying M's challenge. Then A forwards to M the B's contribution component for the session key between A and B (steps 3, 4 in Aziz-Diffie protocol) while discarding the B's contribution component for the session key between M and B. Thus, the mobile calculates the session key with X_{BA} and X_B while the base station calculates it with X_{BM} and X_B. This means that the mobile and the base station agree on the session key with different values, and cannot perform the following encryption and decryption properly.

2.8.3 ASPeCT Protocol

The Advanced Security for Personal Communication Technologies (ASPeCT) is a protocol used within the third generation mobile communications system, also known as UMTS, for secure billing between a mobile user and a value-added service provider (Boyd, 1998b). The protocol is built based on two processes: the Authentication and Initialization Protocol and the Payment Protocol. The first protocol performs authentication between the user and the service provider, establishes the session key, and initializes the payment protocol. On the other hand, the payment protocol is responsible for making payments for a value-added service.

Since the payment is the subject of another chapter, we discuss here the first protocol, which uses asymmetric cryptography. The protocol utilizes three functions denoted by h_1, h_2, and h_3 that are realized using hash functions. In addition, a trusted certificate authority is involved. The following are the three major steps of the protocol:

1. $M \to B$: CA, g^{r_M}, $E_k(ID_M)$
2. $B \to M$: $E_k(r_B, h_2(k_{MB}, r_B, B))$, $Cdata$, $Time$, $Cert_B)$
3. $M \to B$: $E_{k_{MB}}(E_{k_M^-}(h_3(g^{r_M}, g^b, r_B, BCdata, Pdata)), Pdata)$

where CA, B, and M represent the identities of the certification authority, the base station, and the mobile user, respectively, k is a temporarily used key computable by CA, g^b the public key k_B^- contained in $Cert_B$, $Cdata$ is the charging information, $Pdata$ is the data needed to initialize the Payment Protocol, and $Time$ is a time stamp. An operation is voluntarily omitted between step 1 and step 2. It allows B to consult CA to get the public key of M, the value of the secret key k, and a timestamp T in the response message. During the second step, B sends the charging information to M. In step 3, M signs its payment and sends it back to B. The hashing operation avoids message compromise during transmission without detection. The timestamp added in the message aims at preventing replay attacks while the signature prevents repudiation.

The session key k_{MB} is calculated by M as:

$$k_{MB} = h_1(r_B, (k_B^+)^{r_B}) = h_1(r_B, (g^b)^{r_M})$$

and by B as:

$$k_{MB} = h_1(r_B, (k_B^+)^{r_B}) = h_1(r_B, (g^{r_M})^b)$$

References

A. Aziz and W. Diffie, Privacy and Authentication for Wireless Local Area Networks, *IEEE Personal Communications*, 1(1994): 25–31.

D. Boneh, Twenty Years of Attacks on the RSA Cryptosystem, *Notices of the AMS*, 46(2) (1999): 203–213.

C. Boyd and A. Mathuria, Key Establishment Protocols for Secure Mobile Communications: A Selective Survey. *Information Security and Privacy*, LNCS 1438, Springer-Verlag, 1998a, pp. 344–355.

C. Boyd and D.-G. Park, Public Key Protocol for Wireless Communications, in *1st International Conference on Information Security and Cryptology (ICISC'98)*, pp. 47–57, 1998b.

W. Diffie and M. Hellman. New Directions in Cryptography. *IEEE Transactions on Information Theory*, 22(6) (1976): 644–654.

Electronic Frontier Foundation, Cracking DES: Secret of Encryption Research, Wiretap Politics & Chip Design. Oreilly and Associates, May 1998.

T. ElGamal, A Public Key Cryptosystem and Signature Scheme Based on Discrete Algorithms, *IEEE Trans. On Information Theory*, 31(4) (1985): 469–472.

C. M. Ellison, B. Franz, B. Lampson, R. L. Rivest, B. M. Thomas, and T. Ylonen, SPKI Certificate Theory, *Intl. Eng. Task Force*, rfc 2693, Sept. 1999.

U. Feige, A. Fiat, and A. Shamir, Zero-Knowledge Proofs of Identity, *Journal of Cryptology*, 1(2) (1988): 77–94.

S. Garfinkel and G. Spafford, *Web Security, Privacy & Commerce*. O'Reilly & Associates Inc., 2001.

S. Guemara-ElFatmi, N. Boudriga, and M. S. Obaidat, Relational-Based Calculus for Trust Management in Networked Services, *Computer Communications*, 27(12) (2004): 1206–1219.

D. Hankerson, A. Minezes, and S. Vanstone, *Guide to Elliptic Curves Cryptography*, Springer, 1st Ed, 2003.

C. Meadows. Formal Verification of Cryptographic Protocols: A Survey, in Advances in Cryptology—ASIACRYPT'94 (J. Pieprzyk and R. Safavi-Naini, eds.), vol. 917 of *Lecture Notes in Computer Science*, pp. 135–150, Springer-Verlag, 1995.

A. J. Menezes, P. C. van Oorschot, and S. A. Vanstone. *Handbook of Applied Cryptography*. CRC Press, 1996. http://www.cacr.math. uwaterloo.ca/hac.

C. Neuman and T. Tso (1994). Kerberos: An Authentication Service for Computer Networks, *IEEE Communications*, 39(1984): 33–38.

NIST, Advanced Encryption Standard, FIPS-197 (available at http://csrc.nist.gov/publications/fips/fips197/fips-197.pdf.

NIST, Digital Signature Standard (DSS). Technical Report FIPS PUB 186-2, January 27, 2000.

R. Rivest, A. Shamir, and L. Adleman, A Method for Obtaining Digital Signatures and Public-Key Cryptosystems. *Communications of the ACM*, 21(2) (1978): 120–126.

B. Schneier, *Applied Cryptography: Protocols, Algorithms, and Source Code in C*. John Wiley & Sons, 1995.

P. F. Syverson, A Taxonomy of Replay Attacks, in *Proc. of the Computer Security Foundations Workshop VII*, CSFW 1994: pp. 187–191.

W. G. Tzeng and C. M. Hu, Interprotocol Interleaving Attacks on Some Authentication and Key Distribution Protocols, in *Information Processing Letters*, 69(1999): 297–302.

M. J. Weiner, Cryptanalysis of Short RSA Secret Exponents, *IEEE Trans. Inform. Theory*, 36(1990): 553–558.

Chapter 3

Common Techniques for Mobile Communications Security

3.1 Introduction

Mobility is an omnipresent quality currently expected by voice and data cellular subscribers. Seamless mobility and services have been successfully integrated into a wide range of mobile communications including cellular networks over the recent years. Nowadays, an increasing demand for advanced services that is generating new challenges to the cellular technologies can be observed. This demand has forced the convergence between cellular communications and IP networks as cellular services improve by utilizing the capacity of the IP backbone to deliver rich user data.

The challenges presented by the convergence are being addressed to facilitate the successful implementation of new mobile technologies. Each network has been enhanced to provide related services, voice or data. Web data services have been integrated into 2G cellular networks with the introduction of the General Packet Radio Service, GPRS, and voice communications are now transported over IP networks using Voice-Over-IP encapsulation. These adapted data and voice services build the initial bridge between cellular and IP networks and deploy new technology layers for each respective network. IP and cellular networks differ fundamentally in infrastructure, communication protocols, and properties and their convergence has become a technical research challenge. Designing hybrid cellular and IP access in integrated 3G and IP networks is the focus of this chapter.

3.2 Securing Network Protocols

The security of network protocols has attracted a lot of interest in cellular networks, offering IP-based services, and ad hoc networks. One among the large set of network protocols, one special protocol, namely the IP protocol, has been the subject of an important effort. For the lack of space, we will discuss, in the following, the major solution provided to secure IP.

IP security protocol (or IPsec) is a suite of protocols that seamlessly integrate security into the protocol IP and provide security services such as packet source authentication, packet integrity, confidentiality, and protection against replay attacks. In addition, the IPsec provides data privacy, access control, and traffic tunneling. Common applications of the IPsec include the following:

■ *Enabling secure communication across public networks*: this is the initial application of IPsec, since it has been created to provide a solution to IPv4 and be native in IPv6.

■ *Secure intranet and extranet connectivity*: IPsec can be used in conjunction with other security mechanisms to establish secure connections between communicating entities needing peer authentication and packet encryption (at the network layer).

■ *Secure enterprise's connectivity*: Virtual private networks can be built for the needs of enterprises based on IPsec. They present various advantages, since they can save on communication costs and enable companies to build network equivalent to private networks using the IPsec.

■ *Secure remote access*: Using IPsec, the end-users (that can be mobile) can make a local call to their ISP and get securely into his enterprise network. This is able to free companies from communication charges for remote employees.

It stands that a virtual private network (VPN) is a way to use a public communication infrastructure to provide remote sites or individual users with secure access to their enterprise's private network. In order to extend that concept to wireless environments, solutions such as IPsec and SSL/TLS have emerged involving the use of wireless access, either by using a Wireless LAN (WLAN) provided by a Wireless Internet Service Provider (WISP) or a cellular network such as the GPRS. Such a VPN is called wireless VPN. The secure connection can consist of two types of end points, either an individual computer or a LAN with a security gateway. Figure 3.1 depicts an example of wireless VPN. Traditionally the LAN-to-LAN connection, where a security gateway at each end point with known IP addresses serves as the interface between the secure connection and the private LAN, was the most used.

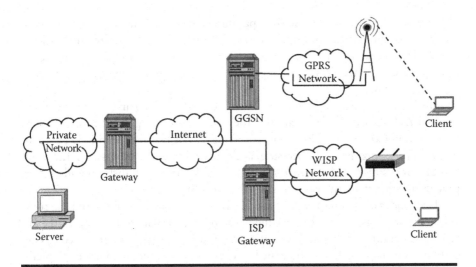

Figure 3.1 Wireless VPN.

Today, when telecommuting using mobile devices such as a laptop are common, the end entities can be involved in the VPN scheme.

3.2.1 IPsec Processing

IPsec uses the Authentication Header (AH) and the Encapsulating Security Payload (ESP) protocols to apply security to IP packets. The AH provides integrity and authentication and non-repudiation, if the appropriate choice of cryptographic algorithms is made. And the ESP provides confidentiality, along with optional (but strongly recommended) authentication and integrity protection. The cryptographic and hash algorithms specified for the use of IPsec include HMAC-SHA1 for integrity protection, and 3DES-CBC and AES-CBC for confidentiality. IPsec uses the concept of a security association as the basis for building security functions into IP. A security association is simply the bundle of algorithms and parameters (such as keys) that is being used to encrypt and authenticate a particular flow in one direction. In order to decide what protection is to be provided for an outgoing packet, IPsec uses the Security Parameter Index (SPI), an index to the security association database (SADB), along with the destination IP address in a packet header, which together identify a SA for that packet. A similar procedure is performed for an incoming packet, where IPsec gets the decryption and verification keys from the security association database to process the received packet.

The SA provides security services by using either the AH or ESP protocol, but not both (for this, two SAs are used if the traffic stream uses both AH and ESP). For typical IP traffic, two SAs are needed: one in each direction that traffic flows (one each for source and destination host). Three things uniquely identify an outgoing SA: a security parameter index (SPI), the IP destination address, and the security protocol (AH or ESP) identifier. Typically, the destination address may be a unicast address, an IP broadcast address, or a multicast group address. However, the management mechanisms currently set up utilize only unicast SAs.

The IPsec processing is mainly classified into outbound versus inbound processing and AH versus ESP applications. The packet processing in IPsec is different between input and output. Protocol processing can be classified into SPD processing, SA processing, header processing, and packet transform processing. The SPD and SA processing are the same for both AH and ESP. The transform and header processing are differently realized with AH and ESP.

Two modes of operation can be used in IPsec: the transport mode and the tunnel mode. In transport mode, only the payload (i.e., the data to transfer) of the IP packet is encrypted and/or authenticated. The routing function is not altered by the extra process made by IPsec, since the IP header is neither modified nor encrypted; however, when the authentication header is used, the IP addresses cannot be translated, as this will invalidate the hash value. The transport and application layers are always secured by hash, so they cannot be modified in any way. Transport mode is used for host-to-host communications. On the other hand, the entire IP packet (data and the message headers), in the tunnel mode, is encrypted and/or authenticated. It must then be encapsulated into a new IP packet for routing to work with a new header.

3.2.1.1 Outgoing Traffic Processing

The first step in the IPsec processing is to query the database of policies, namely SPD, to find the policy to apply on the outgoing packet. The selector is constructed from the traffic information found in the packet, such as the source and destination IP address, the transport protocol, and the source and destination ports. The policy could specify the action to perform on the packet. If the packet must be discarded then, the action is performed and the IPsec processing ends. If the packet must be processed (i.e., IPsec applied), then either a SA exists for the given traffic, and so the SA is retrieved from the database SAD of the SAs, or the SA does not exist, and thus a new SA has to be created for the traffic.

If the SA is retrieved, the system gets the mode to be applied. If the tunnel mode is provided, then a new packet is created. The original packet becomes the payload of the new packet. In this case, the information of the original packet is left unmodified except for the TTL field of the IP header. Therefore, the checksum of the original packet must be recomputed. The header of the new IP packet is constructed from the original header by copying or computing parameters based on the SA content. Once the new packet is created, it may be processed by AH or ESP according to the SA. The next header field should be filled with the identifier of AH or ESP. After the AH or ESP processing, the packet could be reprocessed

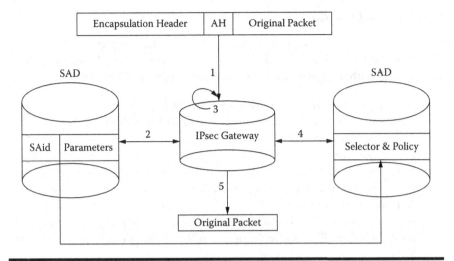

Figure 3.2 Outgoing traffic processing.

again by IPsec, if a bundle of SA is applied, or relayed to the lower communication layer.

It is worth it to notice that fragmentation may occur after the IPsec processing. Fragmentation is needed because the IP datagram can become larger than the maximum transport unit supported by the underlying layer. This operation reduces the size of the IP packet by splitting it into parts with smaller size. Figure 3.2 describes the five steps of the IPsec processing. These steps summarize the above discussion: (1) receiving packet; (2) querying the SPD to discover the policy to apply; (3) querying SAD to retrieve the appropriate SA; (4) processing packet by applying AH or ESP using the SA; and (5) relaying the produced packet. Steps b, c, and d are looped until there are no more policies to apply (Jourez, 2000).

3.2.1.1.1 AH Outbound Processing

AH is applied to an outbound packet through the following steps:

1. Insertion of the AH header in the IP packet to process.
2. Generation of the sequence number. This number is incremented and copied with each AH processing into the corresponding field of AH. It is set to 0 at the generation of the SA.
3. MAC computation. The MAC algorithm specifed by the SA is used to generate a MAC of the packet.
4. Padding. If required, the authentication data field is padded to align it to the IPv4 or IPv6 format.
5. Fragmentation. The IP fragmentation can be applied to the packet after the AH processing.

3.2.1.1.2 ESP Outbound Processing

ESP is applied to an outbound packet only after an IPsec implementation determines that the packet is associated with an SA established for ESP processing. The steps for ESP outbound processing are the following (Kent, 1998; 1998a):

1. Insertion of the ESP fields to the processed packet.
2. Addition of the Padding if required. The fields Pad length and Next header are set to their values.
3. Encryption. The ESP payload is encrypted using the algorithm and parameters specified by the SA.
4. Generation of the sequence number. This is done in the same way as for AH processing.
5. MAC computation. This is done if authentication is required. It is computed on the whole ESP packet with the exception of the last field containing the MAC.
6. Fragmentation. If required, fragmentation is applied to the produced packet after the ESP processing.

3.2.1.2 Incoming Traffic Management

Upon receiving an incoming IP packet, the packet is reassembled. Then, it is processed if, and only if, its Next Protocol field specifies a value related to AH or ESP. Otherwise, it is only checked using the SPD to verify if it matches the incoming policy. If the packet belongs to a traffic that does not have to be IPsec protected, then the packet is relayed; otherwise, it is discarded.

For the other cases, the IPsec processing proceeds as follows: In the first step, the IP destination address, the IPsec protocol, and SPI are used to query the SAD in order to retrieve the SA used (by the sender) to protect the packet. In the second step, the destination checks the selectors to check whether that they are defined by the SA, and if it is not the case, the packet is discarded. In the third step, an entry policy checks the packet selector to find out if this policy has been enforced. In the final step, the original packet is routed to the outgoing interface. During IPsec processing, AH or ESP may be applied.

3.2.1.2.1 AH Inbound Processing

The different steps of the AH incoming traffic processing are described as follows:

- Sequence number validation: If the retrieved SA specifies anti-replay protection, the sequence number is checked. If it is already encountered, the packet is discarded, otherwise it is accepted.
- MAC validation: The MAC value is verified by recomputing its value based on the SA parameters. If the received MAC value and the computed MAC

value are equal, then the packet is accepted. Then the AH is header removed, and the anti-replay is adjusted if necessary.

A sliding window is used to detect the duplication of the sequence number. It maintains the received sequence numbers between the lower and the upper bound of the window. When an incoming sequence number is lower than the lower bound of the window, then the packet is discarded. When the incoming number is greater than the higher bound, the window is slid upward and the sequence number is kept. A minimum window size of 32 bit must be supported. If the received sequence number of the packet falls within the window, then the receiver will proceed to ICV verification. If the ICV validation fails, the receiver has to discard the received IP datagram as invalid. If the ICV verification succeeds, the received window is updated. The audit log entry for this event should include the SPI value, date/time, Source Address, Destination Address, the Sequence Number, and the Flow ID (in the case of IPv6).

3.2.1.2.2 ESP Inbound Processing

The following three steps constitute the main operation to perform by the ESP process on incoming packets:

1. Sequence number validation.
2. MAC validation: If authentication is required, the MAC value is recomputed and checked. If the two codes are not equal, the packet is discarded; otherwise the inbound processing continues.
3. Original packet reconstruction: This is done using a series of three operations: (1) decrypting the ESP Payload data, padding, Pad Length, and Next Header fields using the secret key, the encryption algorithm, the algorithm mode, and the cryptographic synchronization data specified by the SA; (2) adding any padding as specified in the encryption algorithm specification; and (3) reconstructing the original IP datagram from original IP header and the upper layer protocol information in the ESP Payload field for transport mode, or from tunnel IP header and the entire IP datagram in the ESP Payload field for the tunnel mode.

3.2.2 IPsec Limitations

The limitations experienced by the IPsec techniques and implementations can be classified into four classes (Arkko, 2003): (a) the limitations of expressive power in policy specifications; (b) the limitations of application control over policies; (c) the limitation of support mechanisms for authorization and the inability to link authorization decisions with security processing; and (d) limitations of the SAD and SPD protection. In the following, we give some details of these four issues.

3.2.2.1 Limitations of Expressive Power

It appears that, in some situations, the expressiveness of the security policy entries needs to be increased to cover a wider range of objects than the traditional objects (such as IP addresses, upper layer protocol identifier, and port identifiers). Given, for example, the general use of dynamic addressing, the growing use of mobility, and roaming, one can state that an IP address is not sufficient to uniquely identify the host. Thus, the fact that the IPsec SA parameters are closely linked with IP addresses would reduce the autonomy of mobile network nodes in choosing the addresses they use to communicate. In particular, it would be useful to be able to use multiple addresses instead of just one.

3.2.2.2 Limitations of Application Control

The application control of the policies used in IPsec may be required in multiple situations, particularly those involving mobile applications. Examples of applications necessitating the control of the security policy include, for instance, the need to configure the security policy by an application in an environment where the protocols under use are largely deployed without security being turned on; the need to consider dynamic address and port number; and the need of applications even require that the application to be aware of the underlying security mechanisms (such as knowing whether a specific security is in use or not). In addition, some standard specifications of applications even require that the application is aware of the underlying security mechanisms, or at least whether security is turned on or off.

To address this issue, one needs to know that the security decisions require information typically coming from the application layer and that the current IPsec architecture expects that all security processing is performed at the IPsec layer. More crucial, the applications often do not know whether the IPsec was applied.

3.2.2.3 Limitations of the Authorization Procedures

Networks implementing IPsec do not use local access control to provide authorization mechanism. This will reduce the ability of a networked service to provide personalized rights based on parameters involving service and users features. A node can, for example, utilize local access control lists, make use of specific fields in digital certificates, or create separate digital certificate infrastructure for each application it hosts. Unfortunately the IPsec implementations do not cope with all such needs.

On the other hand, the key exchange protocols used in IPsec, such as IKE, do not take into consideration the authorization information that can be extracted from certificates in accepting a specific request for a new SA. This generates some problems in the SA creation phase.

3.2.2.4 Limitations of the SAD and SPD Protection

The current standards, implementations, and practices used in the deployment of IPsec do not show a particular mechanism has been made available to protect the most sensitive components of IPsec suite. An illegal access to the SPD would allow the intruder to modify any policy. An unauthorized intrusion to the SAD would allow the attacker to retrieve all security materials included in stored SAs.

Two approaches can be used to overcome these limitations: The first approach reduces the role of IPsec as a protocol to build exclusively VPNs, and provide appropriate solutions for securing applications. The second approach makes IPsec cooperative with the applications needing advanced security. In the following we discuss the required improvements related to these approaches, particularly when mobility is involved.

3.2.2.5 Application-Specific Security Solutions

These solutions assume that no extra requirements are imposed on the IPsec implementations and that adapted security solutions are provided for the applications. They also assume that it may be possible to provide tools under the form of generic security object formats and a library to support them. The library should allow developers to integrate the following tools to the related applications:

- Tools for the retrieval, verification, processing, and proof of digital credentials (such as digital certificates).
- Procedures for testing major properties including liveness, denial of service detection, and address validity.
- Tools for the relay of signed and encrypted packets; transferring information about the entity the application is communicating with; and responding to authorization requests issued by the application.

Limited libraries exist today. They are typically well suited for traditional applications and lack specific mechanisms (such as address tests) required in the control protocols for mobile systems. Three improvements can be useful for the large deployment of IPsec in mobile systems and mobile applications:

1. *Provide mechanisms for applications to control security policies*: An application can be requested to automatically provide a default configuration for IPsec. This can be provided through an application programming interface (API) and would help guarantee that a security service has not stopped without the knowledge of the user. However, this approach presents a drawback; such default configurations may not always be sufficient or appropriate. For instance, the default configuration may not be compatible with some requirements for protecting the traffic coming from a specific mobile node.

2. *Allow applications to make authorization decisions.* One approach to allow applications to control authorization is to create an application programming interface (API) between IPsec, IKE, and the controlling applications. A standardized API would also make it possible for applications to rely on IPsec and IKE to receive security information by simply copying the IPsec security database to the application layer. However, this is not enough since the application layer policy information has to be completely involved. To make use of the API, applications need tools to deal with authorization issues, including performing all types of verifications related to the use of certificates and using effectively the extensions that make it possible to represent authorization information in an easy way.

3. *Reduce the reliance on IP addresses.* Reducing reliance on the security mechanisms provided by IPsec on IP addresses is necessary to allow for address dynamicity. This reduction should apply to IPsec security associations, in policy entries, and at application layer policies. It should also comply with requirements of the roaming and handover procedures.

3.3 Attacks on IPsec

Several attacks have targeted the IPsec suite. Among these attacks one can consider Bit Flipping attack on CBC mode, attack based on Destination Address Rewriting, attack based on IP Options Processing, attack based on Protocol Field Manipulation, and Packet decryption with Padding Oracle attacks.

3.3.1 Destination Address Rewriting-Based Attack

To explain how this attack is performed, let us consider an attacker located at address AD_{Hck}, two gateways communicating using ESP in tunnel mode without authentication. We also assume that blocks have 64 bits (the attack, however, is feasible when the block size has 128 bits). Finally we suppose that the attacker knows the destination address AD_{Dst} of the destination of the inner packet. The attack takes advantage of the following weakness called flipping bit.

Let $C = <C_0, C_1, ..., C_k>$ be an encrypted packet containing k blocks. Flipping the packet consists of transforming one of the blocks, say C_i, using a chosen mask with same length,

$$C_i' = C_i \oplus Mask$$

The decryption of the modified packet, $C' = <C_0, ..., C_{i-1}, C_i', C_{i+1}, ..., C_k>$, gives the following packet, where $P = <P_0, P_1, ..., P_k>$ is the initial plaintext, D_K is the decryption function used by ESP, and the ciphertext is obtained following Figure 3.3, which depicts the relation $P_j = C_{j-1} \oplus D_K(C_j)$, for all j. We have:

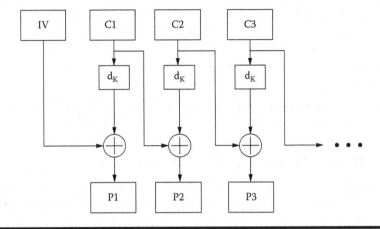

Figure 3.3 CBC decryption.

$$P' = \left\langle P_0, \ldots, P_{i-1}, P_i', P_{i+1}', P_{i+2}, \ldots, P_k \right\rangle$$

$$P_{i+1}' = C_i' \oplus D_K\left(C_{i+1}\right) = C_i \oplus Mask \oplus D_K\left(E_K\left(C_i \oplus P_{i+1}\right)\right)$$

$$= C_i \oplus C_i \oplus P_{i+1} \oplus Mask = P_{i+1} \oplus Mask$$

$$P_i' = C_{i-1} \oplus D_K\left(C_i'\right)$$

This shows that the bits block P_{i+1} is flipped in a similar way to C_i. However, the block P_i is modified in a random way.

The destination address rewriting-based attack is performed in two steps:

1. The attacker captures the encrypted packet, $C = <C_0, C_1, \ldots, C_k>$, written into k blocks of 64 bits. Then he changes block C_2 by applying the following mask $M = ADD_{Dst} \oplus ADD_{Hck}$ on its first 32 bits where the destination IP address is contained. Then if he injects the new packet into the tunnel, the gateway will decrypt the packet, see the destination address in block P_3 and send it in clear to the attacker (except for block P_2). Since modifying C_2, the attacker also disturbs the block P_2. This block contains a part of the IP header. Thus, if some values are no longer valid (such as the checksum), the gateway will drop the packet. To overcome this drawback, the attacker may attempt to modify the last 32 bits of C_2 in a random way, obtaining C_2'' and inject the new packet into the tunnel. The attacker repeats this attempt until the gateway accepts the packet. It has been shown that, after 2^{17} attempts, the probability of attack success is about 60% (Paterson, 2005).

2. Then the attacker intercepts the new encrypted packets from the tunnel, denoted $<\delta_0, \delta_1, \delta_2, .., \delta_k>$, changing the four first blocks by the blocks used in the first step (i.e., C_0, C_1, C_2'', C_1). This gives a new valid header containing the address of the attacker as a destination source. The packet is injected in the tunnel and then the gateway will send it in clear to the attacker.

We can also assume that the attacker does not know the address ADD_{Dst}. In this case, he must be able to capture all the traffic leaving the gateway.

3.3.2 Attacks Based on IP Options Processing

An attack of this type uses the same workflow as in the previous attack. Thus, it performs two steps. The major difference occurs in the first phase, where the mobile attacker modifies randomly the last 32 bits of C_2, and, thus, alters a part of the IP header containing the source address. The attacker also modifies the block C_0 in such manner to have a bigger value of the IHL field in the IP header. As a result, the gateway sees that the header has invalid values and sends an ICMP packet in clear to the random inner IP address containing the modified packet. If the attacker is able to listen to the outgoing packets, he can record this ICMP packet containing the header in clear and a part of the payload. In addition, if the random modifications in C_2 alter the checksum value, the packet will be dropped if it is declared invalid. Therefore, the attacker has to iterate the first step until an ICMP packet is sent. It has been shown that 2^{16} iterations will guarantee that the probability of success of the attack will exceed 55% (Paterson, 2005).

The second phase of the attack is essential to the second step of the first attack. The attacker can reuse the C_0 and C_2 blocks to make new packet, which will generate ICMP packets with the IP header and a part of the payload in clear.

3.3.3 Attacks Based on Protocol Field Manipulation

This attack aims at manipulating the protocol field in the IP header of captured encrypted packet. This attack is efficient when the block size of the encrypted packet is equal to 128 and the encryption algorithm is AES. In this version the protocol field (in the captured packet) lies in the block P_1 of the plaintext. Therefore, by simply flipping the bits in C_0, the field will indicate an upper layer protocol that is not supported by the end host. On receiving the modified encrypted packet, the host will send an ICMP packet named "*port unreachable.*" The attacker still needs to modify the source address but since he modifies only one bit in C_0 until it receives an answer; this may require about 2^{15} iterations.

3.3.4 Proposal Attack

Suppose that the initiator of a communication sends a list of many proposals in order of preference during the SA negotiation, and assume that the least preferred proposal only provides marginal security. The attacker can now modify the responder's SA to select this weak mode, and let the rest of the exchange finish as usual. The initiator will now start using the newly negotiated SA keys, which is considerably weaker than it should be. When the initiator begins to use the weak keys, the attacker can do a brute-force search for the keys. Once found, the attacker has recovered the ISAKMP SA keys and can now negotiate full-strength IPsec SA with the initiator while pretending to be the responder. This is a clear violation of the intention of the protocol. Another refinement is that changing the responder's SA might change the mode being used. An attacker can thus have the responder performing the protocol in one mode, and the initiator the protocol in another mode.

3.3.5 Oracle ESP Padding Attack

The ESP protocol adds some padding, at the end of the IP packet, to have a length equal to a multiple of the block size. It also adds two bytes, the Pad Length (PL) and the Next Header (NH), after this padding (see Figure 3.4 for a description of the IP header). The NH byte takes the value 4 in a tunnel mode. Each IPsec gateway receiving an encrypted packet should check whether the padding of the packets has the described structure. If that is not the case, the gateway will drop the packet. Now let us consider an IPsec tunnel between two gateways using ESP without authentication. Mobile attackers are capable of listening to the traffic and inject packets inside the tunnel. An attacker can then capture an encrypted packet. He then randomly modifies the part of the packet that contains the IHL value or the Protocol field (in a way similar to the second attack). The aim of the attack is to perform a modification to have an ICMP message staying inside the tunnel. So, the attacker does not have to modify the source or destination addresses. For this, he needs to keep trying until the packet generates an ICMP message. Since the ICMP message is still encrypted, we assume that the attacker is able to recognize it.

Bits 0-3	4-7	8-15	16-18	19-31
Version	IHL	Tos	Total Length	
Identification			Flags	Fragment Offset
TTL		Protocol	Checksum	
Source Address				
Destination Address				
Options				
Data				

Figure 3.4 IP packet structure.

Let us now write the received ICMP message under the form $C = <C_0, C_1, ..., C_k>$. The Oracle Padding attack can be launched (Vaudenay, 2002).

Assume that the attacker wants to decrypt a newly received message block C_i'. He starts by sending the message $<C_0, C_1, ..., C_k, R, C_i'>$, where R is a random block and C_is are the blocks of the ICMP ciphertext that the attacker managed to obtain in the preparation phase. On receiving this message, the gateway will interpret the last bytes of C_i' as the padding and will drop the packet if it is invalid. An ICMP message is generated if the last two bytes of $R \oplus D_K(C_i') = R + P_i$ are equal to 0 and 4, respectively, which corresponds to a valid padding of length 0. This means that (a) it takes at most 2^{16} iterations to obtain the ICMP message and (b) the attacker can change the last two bytes of R until an ICMP message is sent. And when this happens, the attacker is guaranteed that he has found the last two bytes of the plaintext P_i.

To decrypt the previous byte, the attacker sends the packets $<C_0, C_1, ..., C_k, R', C_i'>$, where R' is equal to R except that the 6th byte $R'[6]$ of R' is given by

$$R'[6] = R[6] \oplus 1 .$$

The attacker selects random values of byte $R'[6]$ until an ICMP message is sent. This means that the attacker can construct a valid padding of length 1 and, thus, can find the value of the previous third to last byte of P_i. Therefore, the attacker can decrypt all the bytes of C_i using this method by incrementing the value of the padding. It takes at most 256 iterations for each byte.

Finally, let us notice that the most effective countermeasure to avoid these attacks is to use ESP in both encryption and authentication modes since the authentication mode allows the detection of injected packets by an attacker using the message authentication code added at the end of the packet. Another approach to protect against several attacks could be to forbid ICMP message.

3.4 Transport Protocols Security

The aim of this section is to analyze the SSL/TLS and SET protocols in terms of how well they satisfy the security requirements needed by a wireless network. In particular, the wireless version of TLS is studied and its limitations are addressed.

3.4.1 SSL/TLS Features

SSL and TLS are currently the most widely used protocols for providing security for the client/merchant Internet link. SSL is layered on top of an existing reliable protocol suite, namely the TCP/IP. To provide its services, SSL is divided into two layers: the handshake protocol and the record layer. The handshake protocol

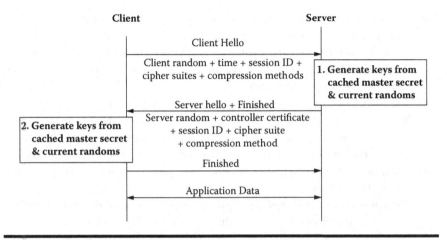

Figure 3.5 Message flow in SSL session.

allows the communicating parties to optionally authenticate each other and then exchange session keys. Upon the termination of handshake procedure, the communicating parties share a secret that can be utilized to construct a secure channel. SSL is an asymmetric protocol that applies the client server paradigm. A typical use of SSL makes a session use the RSA key exchange algorithm with only the server authenticated. Figure 3.5 depicts the message flow to set up a session. Typically, SSL requires seven steps for this (Freier, 1996):

1. The client (or its browser) initiates the communication by sending a *ClientHello* message to the server. The message includes information such as the SSL version, the data compression method to use, a session ID, and a random number that is used in the handshake to prevent replay attacks.
2. In response to the Hello message, the server replies with a *ServerHello* message. This message contains a random number and a session ID attribute that can be used by the client to identify a session with the server. The message is accompanied by a certificate, which contains the server public key, along with the information (optional) to verify the certificate.
3. The client verifies the certificate, by checking its signature. Then, he copies the server's public key of the server, if the certificate is valid. The client then generates a pre-master secret, encrypts it with the server's public key, and sends the ciphertext to the server in a *ClientKeyExchange* message.
4. The server decrypts the *ClientKeyExchange* message using its private key and gets a copy of the pre-master secret chosen by the client. Both the server and the client will use a predefined algorithm to deduce a master secret from the pre-master secret and the random numbers generated by the client and the server.

5. The master secret key is used to generate symmetric keys for encryption and message authentication. The master secret is generally referred to as the state of the session between the client and the server. The session is identified by the two random numbers. The session state is cached by the client and the server for a short period of time.

6. The client sends its certificate (when he has one), if needed by the server. Then he/she sends a *ClientKeyExchange* message containing the key information that will be used to generate a master secret key and keys that will be subsequently used for encryption (on the other direction). The client also sends a *CertificateVerify* message to prove that he/she has the corresponding private key in the certificate.

7. The client sends a *ChangeCipherSpec* message to indicate the starting point of a protected channel. Then he/she sends a *ClientFinish* message containing a hash of the handshake messages exchanged. The message is encrypted and authenticated. The server sends back a *ChangeCipherSpec* message and a *ServerFinish* message.

The re-establishment of an SSL session using a cached state is relatively simple to perform compared to the initialization step. The client may simply specify the session ID of the session (old or current) that it desires to reuse when sending the Hello message. The server checks its cache; if the state is still alive, the old master secret is used to create the secret keys for the client and the server.

It appears clearly from the description of SSL function that:

▪ The SSL protects transaction confidentiality by using symmetric encryption. It protects the confidentiality of transmitted data against interception attacks and provides integrity protection for the transferred data.

▪ The SSL uses the server certificate as the basis for server authentication. To this end, the client can check the server authentication by verifying its ability to decrypt information encrypted using the server's public key. In addition, SSL can provide client authentication, if the client has a public key signed and transported using a certificate that can be checked by the server. SSL provides protection against third party replay attacks (on sessions) by using a random number during handshake.

▪ The SSL does not provide non-repudiation services. In fact, the client and the server do not have any cryptographic evidence to show to a third party that a transaction has taken place.

The transport layer security (TLS) protocol (as introduced in 1995 by the IETF) works in a similar way to SSL, but it presents some differences that we describe as follows (Allen, 1997):

1. For message authentication, TLS relies on the computation of message authentication codes.
2. For certificate verification, TLS assumes that the signed party information includes only the exchanged handshake messages. Conversely, the information in SSL consists of a two round hash of the handshake messages (the master secret and the padding).
3. For secret keys generation, TLS employs a pseudorandom function to generate the keys using a master secret, a label in which the name of the key is specified, and a seed as initial inputs. In the contrary, SSL uses a complex scheme to generate the material.

3.4.2 Security Limitations of SSL/TLS

Despite their large use, SSL and TLS have witnessed some drawbacks that can be summarized as follows:

1. Transaction information is protected against interception attacks only while it is being transmitted. Therefore, sensitive information such as client's account information is available to the merchant. Hence, the clients need to trust the merchant and have to rely on the security of the merchant's Web server. If the merchant server is penetrated, a large number of user account details could be compromised.
2. The SSL/TLS solution provides integrity protection for the transferred data over SSL/TLS sessions; however, it offers no protection against modification of the transaction information by corrupted merchants or clients.
3. SSL/TLS protocol uses the server certificate as the basis of server authentication. Nevertheless, there remain some risks of server masquerading. Man-in-the-middle attacks can be introduced easily by using a sniffing application to intercept the communications between two entities during the initialization step. If an SSL/TLS connection is in use, the attacker can simply establish two secure connections, one with the client and the other with the server. Thereby, the attacker can read and modify the information sent between the two parties and can convince client and server that they are communicating together through a secure channel.
4. The client authentication scheme in SSL/TLS generates a serious threat allowing unauthorized people to perform attacks. Indeed, anyone having access to the client's computer and knowing (or being able to know) the corresponding PIN or password to decrypt the private/secret keys may be able to perform a transaction on behalf of the client. This is particularly important when the merchant uses the client identity to access records containing client private information (e.g., account details and address).

5. The SSL-based approach to build a VPN assumes that the clients can access only to Web server applications. In contrast, an IPsec VPN would provide access to all type of applications.
6. The SSL/TLS solution simply provides a secure means of communication between clients and servers, but does not provide long-term evidence regarding transactions. In fact, session states and secret keys are not stored for short periods.

3.4.3 WTLS

The Wireless Transport Layer Security, WTLS, operates between the transport and the transaction layer. Like SSL, it is responsible for the security of the connection between client and server. The technology behind WTLS is based on TLS. WTLS took the principles of TLS and made adjustments to the wireless environment, taking into consideration the limitation of it resources. The purpose of WTLS is to be a lightweight version of TLS and to provide the following four tasks: (a) to provide privacy, data integrity, and authentication. For this, it supports a number of cryptographic algorithms in order to establish and maintain a secure connection; (b) to provide datagram support when using packet-switched services; (c) to implement an optimized handshake that saves time and bandwidth; and (d) to ensure a dynamic key refreshing (Christinat, 2000).

By allowing the change of keys for encryption and authentication during active connections, WTLS makes it very hard, for an eavesdropper, to decrypt the messages flowing through the connection since the keys are not the same throughout the entire session. How often the keys are changed is decided in the handshake procedure. The WTLS architecture is divided into five parts (as depicted in Figure 3.6). It integrates a Record Protocol and four client protocols that are used in conjunction with the Record Protocol. The major features of the WTLS are described as follows.

3.4.3.1 Record Protocol

The record protocol is divided into four different protocol clients: the Alert, the Application, the Change Cipher, and the Handshake protocols. It takes data that

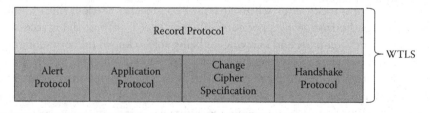

Figure 3.6 WTLS architecture.

has to be transferred to another entity, over the wireless network, and applies different operations on it such as compression, MAC application, encryption, and transmission. When the Record Protocol receives data, it is responsible for decrypting it, verifying it, decompressing it, and sending it to the next layer. However, the compression, authentication, and encryption operations are made optional and their use is decided in the handshake phase. Unlike TLS, the Record Protocol does not allow fragmentation, leaving this task to the transport layer.

In order to protect the payload, explicit sequence numbering can be used. If the datagram transport protocol is used, then this is mandatory. When using sequence numbering, various issues may occur related to duplicate and lost records. To address these issues, a sliding window is used to keep track of the received messages. The sequence numbers always starts with the value zero and ends when the value is $2^{16}-1$. When the upper limit is reached, the secure connection must be closed and a *ChangeCipherSpec* message is sent to reinitialize the sequence numbering.

3.4.3.2 Change Cipher Spec Protocol

If a client or server wants to change the cipher suite negotiated for a session, it sends a Change Cipher Spec message. On the receipt of the message, the receiver gets into a pending state. The receiver then waits until the arrival of a message to start the connection. Then, the client and the server will start a connection after establishing a new cipher suite. When the senders gets the confirmation of the set up of the new suite, the two entities go into the current state and processing can be restarted again.

3.4.3.3 Alert Protocol

This protocol is responsible for the generation and transmission of different alerts between the client and the server. An alert can be a message to close the connection between a client and a server or an error message. A message to close informs the other side of the desire of the issuer to terminate a session. The error alert contains information about the problem encountered and how severe it is. There are three kinds of error alerts: fatal, critical, and warning.

If a fatal alert is sent to a client or a server, the client and the server should immediately terminate the secure connection they have established, since it could be seriously compromised. Other connections using the same session may continue but the session identifier should be invalidated so that a failed connection cannot be used to set up new secure connections. The critical alert causes the connection between a server and a client to terminate. Other connections can continue to use the secure session without invalidating its identifier. This implies that new connections can be established using the secure session, despite the alert. Finally, the warning alert is sent only to inform that the MAC of the receive object is invalid. The connections are not terminated; instead, the packet with corrupt MAC is discarded.

3.4.3.4 Handshake Protocol

All the security related parameters must be negotiated during handshake. These parameters include useful information about the protocol version, the cryptographic algorithm to be used, authentication techniques, and public key techniques to generate a shared secret. The handshake procedure begins with a Hello message (as shown in Figure 3.7). The client sends a Client Hello message to the server. The server replies to the message with a Server Hello message. In the two hello messages, communicating parties agree on the session capabilities.

After the client has sent the Hello message, it starts receiving messages until the Server Hello Done message is received. The server sends a Server Certificate message if authentication is required on behalf of the server. In addition, the server may require the client to authenticate himself. The Server Key Exchange is used to provide the client with the public key, which can be used to conduct or exchange the pre-master secret value.

After receiving the Server Hello Done, the client continues its part of the handshake. At request, the client sends a Client Certificate message where it authenticates itself. Then he sends a Client Key Exchange message containing either a pre-master secret encrypted with the server's public key or the information that

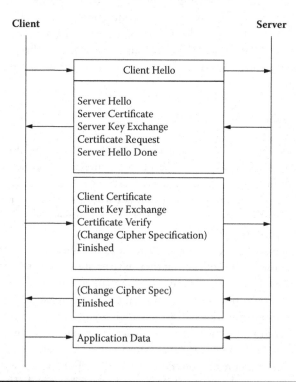

Figure 3.7 Full handshake flow.

both parties can complete the key exchange. Finally, he sends a Finished message, which contains verification of all the previous data including the calculated security related information.

The server responds with the Finished message where it also verifies the exchanged and the calculated information. Besides, both parties must send a Change Cipher Spec message to start to use the negotiated session parameters. If the client and the server decide to resume a previously negotiated session, the handshake may be started by sending a Client Hello message where the Session Identifier is initialized with the identifier of the previous session.

The WTLS also defines an *abbreviated handshake,* where only the Hello and the Finished messages are sent. In this handshake, both parties must have the shared secret that is used as a pre-master secret. Another variation is the optimized full handshake where the server can retrieve the client's certificate using a trusted third party, based on the information provided by the client in the Client Hello message. Using the information provided by the certificates, both parties are able to complete the shared secret values using the Diffie-Hellman key exchange method. The server has to send the Server Hello, Certificate, and Finished messages to the client in order to complete the handshake on the server's behalf. The client responds with the Client Finished message.

3.4.4 Security Features of WTLS

WTLS provides various services. Among these services, one can mention the authentication, the integrity, the privacy, and key exchange.

3.4.4.1 Authentication

The authentication, in the WTLS, is ensured by the digital certificates. Authentication can be mutual, if the client and the server present certificates during handshake, or it can apply only to the identification of the server. Currently, three types of certificates can be used with WTLS; namely, the X.509v3, X9.68, and WTLS certificates are supported. The WTLS certificate is optimized for size. The authentication procedure immediately takes place after the client and server hello messages. When the authentication is used, the server sends a Server Certificate message to the client.

To achieve authentication, the receiving end may receive a chain of certificates, where the first one is the server's own certificate. Each of the next certificates certifies the one preceding it. The verification of all certificates, in the chain, is required to authenticate the sending party. An explicit verification is carried out by each entity, to protect the Certificate messages sent or received. The entity concatenates all the messages received from the server or created by it and calculates a hash value to be signed. This signature is sent to the other entity, which can ensure that authentication is well performed so far.

3.4.4.2 Data Integrity

Data integrity is guaranteed by the use of message authentication codes. The used MAC algorithm is decided at the same time as the encryption algorithm. The decision is made based on a list sent by the client of supported MAC algorithms, where the algorithms are listed with respect to the preference the client has. The server returns the selected algorithm in the Server Hello message. The WTLS supports the most common MAC algorithms, including the SHA and the MD5. It also allows different versions of these algorithms and sizes.

A special MAC algorithm can be used by WTLS, namely the SHA_XOR_40, which is a 5-byte checksum. The algorithm is intended for devices with limited CPU resources. It operates as follows: First the input data is divided into the 5-byte blocks. Then all blocks are XORed one after another. It is required that the XOR MAC must be encrypted and is only used for CBC mode block ciphers. The MAC is generated over the compressed WTLS data. The following values are used to calculate the MAC:

$$(MAC\_Secret, seg\_num \oplus WTLS\_Compressed\_data.record\_type \oplus$$

$$WTLS\_Compressed\_data.data\_length \oplus WTLS\_Compressed\_data.fragment)$$

3.4.4.2.1 Key Exchange

To establish a secure communication channel, initial values to calculate keys and encryption keys are exchanged in a secure manner as described below. The Server Key Exchange message can be used to provide complementary data, when needed, for key computation. The key exchange mechanism of the WTLS also provides an anonymous way to exchange keys. In this procedure, the server sends a Server Key Exchange message, which contains the public key of the server. The key exchange algorithm may be RSA, Diffie-Hellman, or the elliptic curve Diffie-Hellman.

When RSA or the anonymous RSA is used, the client encrypts the pre-master secret with the server's public key and sends it back to the server in a Client Key Exchange message. When the Diffie-Hellman based algorithms are utilized, the client and the server calculate the pre-master secret based on their private key and the counterpart's public key.

3.4.4.2.2 Privacy

Privacy in the WTLS is implemented by means of encrypting the communication channel. The used encryption methods and all the necessary values for calculating the shared secret are exchanged securely during the handshake. The master secret is a 20-byte sequence, which is calculated with the following formula:

$$Master\_secret = PRF(pre\_master\_secret,\ \text{``master secret,''}$$

$$ClientHello.random \oplus ServerHello.random),$$

where PRF is a pseudo-random function that takes as input a secret, a seed, and an identifying label and produces an output of arbitrary length.

The encryption keys are conducted based on a key block, which is computed using the initial values transmitted during the handshake procedure. The key block is given by:

$$key\_block = PRF(master\_secret \oplus expansion\_label \oplus$$

$$seq\_num \oplus server\_random \oplus client\_random).$$

The key block expression uses a sequence number that makes the key block variable. The key block is recalculated after certain intervals based on the key refresh frequency, which is negotiated in the Client hello and the Server hello messages. The expansion label stands just a string expression for calculation. The client uses string "client expansion" and the server "server expansion." The encryption key, the initial vector, and the MAC secret are conducted from the key block based on the key lengths required by the chosen algorithms.

3.4.5 SSH

The SSH protocol allows two hosts (a client and the server) to construct a secure channel for data communication using DSA and Diffie-Hellman key exchange, which provides a shared secret key that cannot be determined by either party alone. The shared secret key is used as a session key. Once an encrypted tunnel is created using this key, the context for the negotiated compression algorithms and the encryption algorithm are initialized. There are three main parts of the SSH protocol: algorithm negotiation, authentication, and data encryption (Barrett, 2001).

The negotiation of algorithms is mainly performed to determine the encryption algorithms, the compression algorithms, and the authentication methods supported and to be used between the client and the server. The negotiation is then followed by the authentication, which is done by a 2-step process: the key exchange and the client authentication. The objectives of the key exchange are to attempt to authenticate the server to the client and to establish a shared key that is used as a session key to encrypt all the data being transferred between the two entities. The session key encrypts the payload and a hash generated for integrity checking of the payload using the private key of the server. The client verifies the server's public key and the server signature received, and then continues with user authentication.

User authentication methods that are supported include, but are not limited to, passwords, public key, OpenPGP certificates, and X509v3 certificates. Once the authentication is successful, one of the negotiated encryption algorithms is used to encrypt the data transferred between the two machines. The key exchange produces two values: a shared secret K, and an exchange hash H. For this, the client generates a random number x where $(1 < x < q)$ and the server generates a random number y $(0 < y < q)$, where q is a prime integer. The management of the key pairs is done as follows:

■ The user creates a public/private key pair, if he intends to use "public key authentication" on any client machine. The public key needs to be added in to the database of the server, before authentication can proceed.
■ Similarly, the server maintains private and public key pairs created by the root. Typically there is a key pair based on RSA and another key pair based on DSA.
■ The user account on the client machine maintains a database of all the public keys of the SSH servers to which a user logged.
■ If the client does not have a matching public key of a server, he can configure the security on his machine so that he accepts the public key provided by the remote server.

Given this brief description of how the public keys are managed, it is easy to observe that the client blindly trusts the server and accepts its public key during an initial connection. An attacker can intercept such exchange scenario and render the SSH channel to be insecure.

Various attacks can be launched on connection using SSH. A first attack is the well-known man in the middle. Suppose that a mobile user A wants to establish a connection with a server S, and that a malicious user M wants to launch the man-in-the-middle attack on them. The attack is conducted as follows:

1. Mobile user A initiates a connection with S, who sends his public key to A, which intruder M intercepts.
2. The attacker M sends his own public key to A, who accepts the new public key and stores it in its database. If it was the "first time authentication," A would blindly add the public key provided by M thinking that it is public key of B.
3. User A sends his username and password to S that is again intercepted by M, who decrypts A's username and password using the session key and his private key.
4. Attacker M then encrypts A's credentials using the public key provided by S and forwards the new packet to S. Then S authenticates M thinking that it has authenticated A. The attacker M can now generate serious damages to A.

A second attack on SSH is a spoofing. It allows an attacker to claim to be mobile user A and to establish a secure channel with mobile C, who thinks that it has a secure connection with A. The attacker hides his real identity and forges a false identification. User spoofing is possible when a mobile user B on a client machine attempts to establish a connection with a remote host C. A malicious server S intercepts the channel when it is in the initial phase, fakes to be remote host C, and replies back with its own public key.

If the SSH client configuration on B is set to non-strict host key checking (which is the default configuration), it would ask B to overwrite the previous key stored in its database for host C and proceed with establishing a connection. If it was the "first time" authentication, B would simply accept the server host key. When the user performs password authentication to the remote server, the malicious remote server accepts the credentials provided by B and then outputs the error message stating that an invalid password was provided. The attacker can know B's credentials retrieved from malicious server and claim to be B.

3.5 Attacks against Transport Security Services

Cryptographic algorithms form a set of primitives that can be used as building blocks to construct security mechanisms that target specific objectives. Network security protocols, such as SSH, SSL/TLS, and WTLS, combine these blocks to provide authentication between communicating entities, and guarantee the integrity and confidentiality of communicated data. However, these security services only specify what functions should be performed, irrespective of how these functions are implemented. In particular, the specification of a security protocol is often independent of the way the encryption algorithms are implemented in software running on a typical processor, or being embedded in a hardware unit, and whether the memory used to store the data computed during the executions of these algorithms is shared by other applications.

Therefore, these security mechanisms are far from being complete security solutions. Actually, cryptographic algorithms are always implemented in software or hardware on physical devices that interact with their environments. These interactions can be monitored by attackers and may result in information useful in breaking the security service. This type of information is called side-channel information (SCA). We will consider in this section some among the multiple SCAs targeting SSL/TLS and WTLS. The SCA attacks have been proven to be several orders of magnitude more effective than the other attacks including mathematical analysis-based attacks.

For the lack of space, we omit to discuss the attacks targeting SSH.

3.5.1 Attacks against SSL and TLS

It has been shown that SSL is not the efficient tool for securing data; rather, it is a mean of negotiating security tools. This means that security of SSL not only depends on the actual SSL specification or implementation, but also on the encryption and authentication algorithms in use. Several attacks are proposed in the literature (Wagner, 2004), some of which I want to introduce here. Most of them are pretty theoretical and do not succeed. (Do they prove that SSL itself is secure?) Other attacks succeed, in theory. They have not been implemented. One of them seems to be implementable, but it reveals a threat against privacy, opposed to confidentiality and authentication, which are SSL's primary aims.

3.5.1.1 Cipher Suite Rollback Attack

Cipher Suite Rollback is a man-in-the-middle attack that aims at leading the attacked parties to use ciphers providing low security, despite what they could possibly agree on. A man-in-the-middle attack intercepts the initial messages from the client to the server during the handshake phase and changes the containing list of preferred ciphers. Since the server will choose the best cipher it supports from this list for later use, the attacker can manipulate to induce the decision he wants. This was rather easy to do in SSL 2.0, since the initial handshake messages were just sent unencrypted and unauthenticated. We note, however, that the attacked entities would still use a cipher, which is not the most protecting. For example, if the induced cipher is the 40-bit DES, then the confidentiality is relatively hard to break online, when acting by man in the middle. To counteract against this attack, SSL 3.0 imposed the authentication of the handshake messages. Nevertheless, it is still possible to attack the handshake protocol.

3.5.1.2 Dropping Change Cipher Specification Messages

As previously stated, when finishing the handshake, the SSL client and server exchange change cipher specification messages to switch their security options. Both sides will, after that, only communicate using the security features agreed on. To force both sides to begin confidential communication without changing to the new agreed-on cipher specification, a man-in-the middle attack can simply attempt to send a Finished message just before cipher specification messages can be sent. Two reasons justify the success of such attack: First, there is no operation in SSL that is responsible for checking whether a change cipher specification has been sent before a Finished message. Second, the change cipher specification message is not authenticated like all the other handshake data.

3.5.1.3 Key-Exchange Algorithm Rollback Attack

This attack aims to force the client and the server to use two different key-exchange algorithms that the attacker can specify. It assumes that the attacker can launch a cipher suite rollback attack. The attack works as follows: The attacker forces the server and the client to use the Diffie-Hellman key exchange and the RSA key exchange, respectively. This attack leads to a particular situation, where the client interprets Diffie-Hellman parameters as exponents and modulus for RSA encryption. In addition, and because of the data structures used, the RSA modulus received by the client will be a prime number, which is the Diffie-Hellman prime modulus.

After performing the ciphersuite rollback, the attacker has to intercept the Diffie-Hellman parameters, say a number g and prime number p (the DF modulus) that the server sends to the client in the server's authentication phase. Then the attacker has to wait for the client's pre_master_secret, which will be encoded with RSA using these parameters. Then, all he has to do is to take the g-th root of this value. He now knows the pre_master_secret and can derive everything he needs to decrypt future SSL communication between the client and server.

3.5.2 Attacks against WTLS

A large number of attacks have targeted the WTLS showing its limitations to provide a robust level of protection. In the following we describe some of these attacks (Saarinen, 1999).

3.5.2.1 Predictable IVs

WTLS uses a linear computation of the initialization vector, even for reliable transports. When a block cipher is used in CBC mode, the IV is computed as follows:

$$IV_s = IV_0 \oplus (s,s,s,s)$$

where s is the (16-bit) sequence number of the packet and IV_0 is the original initialization vector derived during key generation. The plaintext blocks $P_{s,0}, P_{s,1}, ..., P_{s,n}$ in the packet P_s are encrypted into blocks $C_{s,0}, CP_{s,1}, ..., C_{s,n}$ as follows:

$$C_{s,0} = E_K \left(IV_s \oplus P_{s,0} \right)$$

$$C_{s,j} = E_K \left(C_{s,j-1} \oplus P_{s,j} \right), \ j > 0$$

Assume now that a terminal application used by user A allows each keypress typed to be sent as an individual packet. When user A enters his password into the application, the attacker captures the related packets. The attacker now has blocks of type

$$C_{s,0} = E_K \left(IV_s \oplus P_{s,0} \oplus (s,s,s,s) \right)$$

where $P_{s,0}$ contains an unknown letter of A's password. We notice, in addition, that the sequence number s is known to the attacker. Now imagine that the attacker guesses the unknown letter of the password, say m. Then he sends the following packet through A's channel:

$$P_{t,0} = m \oplus (t,t,t,t)$$

where t is the sequence number of this packet. One can see that because (t,t,t,t) cancels out in the CBC computation, a correct guess $m = P_{s,0}$ leads to matching ciphertexts $C_{t,0} = C_{s,0}$. In other words, this is an oracle attack that tells whether the password letter is correctly guessed. Thus, the entire password can be brute forced, letter by letter, with a few tests using this oracle attack.

3.5.2.2 35-Bit DES Encryption

The 40-bit DES encryption method used in WTLS is defined to use five bytes of key built for the encryption service. Because of the parity bits contained in each byte of the DES key, there are only 35 effective key bits in five bytes. This fact reduces by a factor of 32 the size keyspace, in the case of brute-force attack to retrieve the useful part of the encryption key.

3.5.2.3 Unauthenticated Alert Messages

Some of the alert messages used in the WTLS protocol are sent in cleartext and are not properly authenticated. Most of these messages are warnings and do not cause the session to be terminated. Using the fact that an alert message should have a sequence number in the WTLS protocol, an active attacker may attempt to replace an encrypted datagram with an unauthenticated plaintext alert message having the same sequence number and without being detected. This attack can be classified as a *truncation attack* since it allows arbitrary packets to be dropped from their data stream. The countermeasure against this attack requires simply that all messages affecting the protocol state should be appropriately authenticated.

3.6 Public Key Infrastructure

The Public Key Infrastructure (PKI) applies a public key cryptographic method to transmit user's public key and user's identity in a reliable and secure manner. The users of public key cryptography can transmit their public keys to the other users, and should keep the private key corresponding to the public key protected.

3.6.1 PKI Components

The PKI has been used in many business applications and various security service platforms such as user authentication, digital signature, and non-repudiation. It uses two major objects: the digital certificate, as described by the X.509 v3 certificate format and the specification of Certificate revocation list (RFC 2832). The PKIX model defines the elements that a PKI comprises. The PKIX model components integrate four major components: the end-entity, public key certificate, certification authority, and repository. Figure 3.8 illustrates the PKIX model including mobile entities.

3.6.1.1 End Entity

End entities can be considered as the users of the PKI-related services. The term *end-entity* is a generic term that denotes subscribers, network devices (such as servers and routers), processes, or any other entity that has applied and received a digital certificate for use in supporting the security and trust in transactions to be undertaken. An end-entity can also be a third party (an individual or an organization),

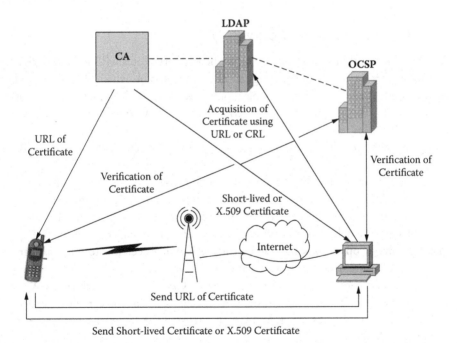

Figure 3.8 Wireless PKI model.

who does not hold necessarily a certificate, but may be the recipient of a certificate (during the execution of a transaction) and who therefore acts on reliance of the certificate and/or digital signature to be verified using that certificate.

3.6.1.2 Public Key Certificate (PKC, or just Certificate)

A PKC acts like an official ID card. It provides a means of identifying end-entities (or their identities) to their public keys. PKCs can be distributed, publicly published, or copied without restriction. They do not contain any confidential information. A PKC is a digital document and a data structure containing a public key, relevant details about the key owner, and optionally some other information, all digitally signed by a trusted third party, usually called certificate authority, which certifies that the enclosed public key belongs to the entity listed in the subject field of the certificate. The advantage of a certificate is characterized by the fact that it is considered impossible to alter any field of the certificate without an easy detection of the alteration. An example of a PKC is the X.509 v3 certificate. It is a widely used certificate format. It is being utilized in the major PKI-enabled applications available in the market place, such as the SSL and the privacy enhanced mail (PEM).

3.6.1.3 Certification Authority (CA)

A certification authority is the issuer of public key certificates within a given PKI. Public key certificates are digitally signed by the issuing CA, which effectively (and legally) binds the subject name to subject public key and the CA's public key that is used to verify the signature on the issued certificates. CAs are also responsible for issuing certificates revocation lists (CRLs), which report on invalidated certificates, unless this has been delegated to a separate entity, called certificate revocation list issuer.

A CA should be involved in a number of administrative and technical tasks such as end-users registration, end-user's information verification, certificate management, and certificate publication. However, some of the administrative functions may be delegated to optional actors, called registration authority (RA). The CA's major operations include certificate issuance, certificate renewal, certificate revocation, and certificate verification. The verification of an end-entity certificate may involve a list of CAs, denoted by CA_j, $1 < j < n$, such that CA_1 is the issuer of the end-entity certificate, CA_{k+1} is the issuer of the certificate issued to CA_k to sign certificates, $k > 1$, and CA_n is a trusted CA (from the verifier point of view).

Therefore, the end-entity certificate represents the starting point to validate a given certification path, which represents a list of certificates signed by CAs and delivered for CAs.

3.6.1.4 Certificate Repository (CR)

A certificate repository is a component (or system) used to store and retrieve certificate-related information such as the PKCs issued for end-entities and the CRLs that report on revoked certificates. A repository can be an X.500-based directory with public access facilities via the Lightweight Directory Access Protocol (LDAP) or the File Transfer Protocol (FTP) so that the certificates can be retrieved by any end-entity for various needs.

It is possible to offload certain verification functions from the end-entity system to a trusted third party, who will act on its behalf. For example, a specific protocol can be set up at the end-entity site to ask a trusted third party about the revocation status of certificates that a mobile user wishes to rely on. Arguably, the trusted third party could be viewed as a virtual repository since the revocation status and the output verification are derived and returned to the end-entity system in response to a specific request.

3.6.1.5 Certificate Revocation List (CRL) Issuer

A CRL is a data structure used to notify someone who wishes to check the status of a certificate he wants to rely on. Typically, a CRL is a signed document that contains reference to certificates, which are decided to be no longer valid. The CRL issuer may be a third party to which a CA delegates the verification of information related to revocation, issuance, and the publication of CRLs. Usually, the CA that issues a certificate is also responsible for issuing revocation information associated with this certificate, if any.

It also happens that a CA transfers the whole revocation function to another CA. CRLs that are issued by the other CA are referred to as indirect CRLs. Therefore, and for the sake of efficiency and coherence, a certificate should include a field indicating the address of the location where CRLs that might include this certificate are published when it would be revoked.

3.6.1.6 Registration Authority (RA)

A RA is an administrative component to which a CA delegates certain management functions related to the registration of users. The RA is often associated with the end-entity registration process. However, it can be responsible for a number of other functions including the following tasks:

- Ensuring the eligibility of applicants to be issued with certificates, while verifying the accuracy and integrity of the required information provided by the applicants.
- Verifying that the end-entity requesting the issuance of a certificate has possession of the private key associated with the public key being provided.

- Generation of key pairs, archiving key pairs and secret keys, and delivering keys to end-entities.
- Conducting the needed interactions with the delegating CA on behalf of the end-entity, in case of key compromise notifications and key recovery requests.

The RAs, however, are not allowed to issue certificates or CRLs. Deployment of an RA can provide two major advantages. First, the RA can help to reduce overall certification costs. This is especially true in large, geographically distributed companies that require their users to be physically present before specific PKI-related activities are permitted. Second, offloading the administrative functions from a CA allows an organization to operate their CA off-line, which reduces the opportunity that an adversary launches attacks against that CA.

Typically, a PKI service scenario involving a mobile user is described as follows:

1. The CA performs user identification through direct contact.
2. CA provides the user with identity and password.
3. A mobile phone generates a key pair and certificate request message.
4. The mobile phone signs certificate request message and digital signature verification key with digital signature generation key.
5. The mobile phone sends them to CA.
6. CA confirms the ownership of the digital signature generation key.
7. CA generates a certificate.
8. CA publishes the generated certificate on a directory.
9. CA sends the certificate information to user.
10. The mobile phone obtains the certificate and can exchange the messages with digital signatures using the public key to another entity.

3.6.2 PKI Functions

The model identifies a number of major functions that support the management process of digital certificates. These functions include registration, initialization, certificate generation, certificate update, certificate revocation, key management, and cross-certification. We describe in the following the main features and requirements of these functions.

3.6.2.1 Registration

End entities must register with the PKI before they can hold a certificate and take advantage of the PKI-enabled services. This step is usually associated with the initial verification of the end-entity's identity and the information the entity provides. The level of assurance associated with the registration process can vary based on the target environment, the intended use of the certificate, the communication system, and the enforced security policy. The registration process can be accomplished directly with the CA or through an intermediate RA. It may also be accomplished on-line or off-line depending on the trust level of the required certificate and the security practices implemented by the issuing authority.

Once the required information is provided by the end-entity and verified in compliance with the applicable policies, the end-entity is typically issued one or more shared secrets and other identifying information that will be used for subsequent authentication as the enrollment process continues. The distribution of the shared secrets is typically performed following specific ways and may be based on pre-existing shared secrets.

3.6.2.2 Initialization

The registration process is followed by the initialization process. This involves initializing the associated trust anchor (or trust point) with the end-entity. In addition, this step is associated with providing the end-entity with its associated key pairs. Key pair generation involves the creation of the private/public key pair associated with an end-entity. Key pair generation can be made prior to the enrollment process or it can be performed in response to it. Key pairs can be generated by the end-entity client system, RA, CA, or some other PKI components such as a hardware security module. However, in the case where the end-entity generates the key pair, the registration process should include the verification that the public key provided by the end-entity is connected to the private key held by the end-entity.

The location of the key pair generation is driven by operational constraints and applicable policies. Moreover, the intended use of the keying material may have an important role in determining where the key pairs should be generated. It is possible that tasks composing the initialization process may occur at different moments and places. However, the task performed by the end-user should not be realized before an explicit certificate request is generated.

3.6.2.3 Certificate Generation

This process occurs after the termination of the initialization process. It involves the issuance of the entity public key certificate by the certification authority. Typically, the generation process organizes the necessary information (including the CA's identity and the revocation address) in a data structure following the X.509 standard and digitally signs it. If the key pair related to the certificate is generated externally to the CA, the public key component must be delivered to the CA in a secure manner. Once generated, the certificate is returned to the end entity and/or published to a certificate repository (Housley, 2002).

3.6.2.4 Certificate Update

Certificates are issued with fixed lifetimes (referred to as the validity period of the certificate). The duration of the pre-fixed lifetimes can be of one year or two years (or even longer). On certificate expiration, the key pair used with the certificate may also be required by the end-entity for different reasons. As a result, the certificate is updated (or renewed) and if lifetime is re-fixed. However, it is preferable that a certificate renewal involves the generation of a new key pair and the issuance of a different public key certificate, since it contains a new public key.

Key pair update can occur in advance of the pair's expiration. This will help to ensure that the end-entity is always in possession of a valid certificate. The key pair update may induce a certificate renewal before the associated public key actually expires. It also provides a period of time where the certificate associated with the initial key pair remains unrevoked, meaning that this certificate can be used for a short window of time to verify digital signatures that were created with this key pair. This will help to minimize inappropriate alert messages that would otherwise be generated to the end-entity.

3.6.2.5 Revocation

Public key certificates are issued with fairly large lifetimes. Nevertheless, the circumstances that existed when the certificate was issued can change to an unacceptable state before the certificate can come to expire normally. Reasons for unacceptability may include private key compromise or change of the information related to the subscriber (e.g., affiliation and name change). Therefore, it may become necessary to revoke the certificate before its expiration date. The revocation request allows an end-entity (or the RA that has initiated the enrolment process) to request revocation of the certificate. Certificate revocation information must be made available by the CA that issued that certificate or by the CRL issuer, to which the CA delegates this function.

X.509 defines a method for publishing the above information via certificate revocation lists (CRLs). The frequency of publication and the type of CRLs used are functions of local policy. Finally, one can notice that end-entities, or third trusted parties operating on behalf, must check the revocation status of all certificates it wishes to rely on. This will be addressed in the sequel.

3.6.2.6 Key Pair Management

Since key pairs can be used to support the digital signature creation, data encryption, and message decryption, an end-entity may need to rely on the CA for the creation of a management of key pairs. When a key pair is used for encryption/decryption, it is important to provide a mechanism to recover the necessary decryption keys when normal access to the keying material is no longer possible, otherwise it will be impossible to recover the encrypted data. Key pair recovery allows end-entities to restore their encryption/decryption key pair from an authorized key backup facility, provided by the CA.

It is also possible that an end-entity's association with an organization can change (e.g., employee resignation, firing, or new appointment), and the organization has a legitimate need to recover data that has been encrypted by that end-entity. It is also possible that access to the keying material may be required in association with legitimate law enforcement needs. Moreover, a CA can provide certification services where key pairs need to be managed at the PKI level. Key pair management includes all functions needed during the key life cycle.

3.6.2.7 Cross-Certification

Cross-certification is the action performed by one CA when it issues a certificate to another CA. The basic purpose of a cross-certification is to establish a trust relationship between two CAs, with which the first CA validates the certificates issued by the second CA for a period of time. Cross-certification is provided to establish the proof of certificate paths for one or more applications by allowing the interoperability between two distinct PKI domains or between CAs working within the same PKI domain. While the former is referred to as inter-domain cross-certification, the latter is referred to as intra-domain cross-certification.

Cross-certification may be unilateral or mutual. In the case of mutual cross-certification, a reciprocal relationship is established between the CAs: one CA cross-certifying the other, and vice versa. Unilateral cross-certification simply means that the first CA generates a cross-certificate to the second CA, but the second does not generate a cross-certificate to the first. Typically, a unilateral cross-certificate applies within a strict hierarchy where a higher level CA issues a certificate to a subordinate CA. However, cross-certification adds an important complexity to the process of validating path certificate.

3.6.3 Wireless PKI

The limitations of the devices and wireless nature of communication system must be taken into consideration when implementing a PKI in a wireless network (wireless PKI or WPKI). In particular, many communication issues have to be solved and make it very difficult to apply wired PKI system to a wireless network. These issues include the optimization of limited resource use, the latency of communication, and the insecurity of connections and devices. A mobile terminal lacks computing capabilities of multiple PKI services such as the key generation, the signature generation, the verification and validation of certificates, the certificate revocation, the revocation verification, and memory size of storing certificate.

3.6.3.1 WPKI Requirements

In order to apply wireless PKI to mobile terminals attached to a mobile communication system and allow providing security at an equal level as that of the wired communication, the following four requirements must be satisfied:

1. *Use optimal digital signature algorithm to be calculated in mobile terminals*: RSA based public key cryptographic algorithm has been selected for digital signature algorithm of PKI for a long time. However, a public key pair generation based on the RSA algorithm in a mobile phone might be time consuming or be impossible due to the lack of memory and CPU performance. Therefore, an alternative public key algorithm to make the key generation possible in the mobile phone may be required. In addition, the time needed to perform a digital signature operation must be acceptable to the users.

2. *Minimize data size to be stored in mobile phone and to be transmitted through wireless bandwidth*: Generally, a certificate used in PKI is ITU X.509 certificate defined by ITU (Housley, 2002). This X.509 certificate has basic fields for certificate verification and many extension fields that are required for certificate path validation. These extension fields increase the size of the certificate and make procedures of certificate path validation complex. Thus, the optimization of certificate profile is required without side effect for the certificate verification and path validation.

   On the other hand, the validation of an X.509 certificate requires CRL verification. To realize this, a mobile device needs to download CRL from CA and check whether a certificate is revoked. This procedure costs the mobile device and wireless transmission considerable overhead. Thus, an efficient and reliable method is required to validate X.509 certificates without direct verification of the CRL mobile device.

3. *Optimize the certificate management protocol (CMP)*: Current wired CMP is based on the SSL protocol, while the certificate requests are issued by the devices and sent using WTLS. Since security based on WTLS does not support end-to-end security, information necessary for the certificate request could not be securely transferred to CA. Therefore, a new wireless CMP (or WCMP) needs to be built and should not be based on SSL, neither on WTLS. However, the WCMP must guarantee the same functions as the wired CMP and be lighter than it and be optimized for processing in mobile devices and transmission over the wireless link.

4. *Optimize certificate validation scheme*: To validate X.509 certificate, certificate chain and CRL must be acquired and verified in the mobile device. If the verification of certificate needs the validation of long chains of certificates, such tasks may appear hard to process. Efficient and reliable methods for certificate path validation suitable for mobile devices are needed.

Solutions to comply with the aforementioned requirements may use the concept of delta CRL, which reduces the size of the CRL to download by simply sending the modifications made to the CRL since the last request. Another approach can delegate all complex verification to entities located on fixed node (in the mobile network) or to trusted third parties having more resources.

3.6.3.2 An Example of WPKI Architecture

Many WPKI architectures have been proposed in the literature. We describe here an example of architecture derived from the one presented by for a WPKI model that satisfies the requirements mentioned before and examine the proposed PKI model, detailed technologies, and its characteristics. Figure 3.8 shows the proposed WPKI model. We assume the following:

- We consider communication between mobile phone and server as content provider, and exclude communication between mobile phones.
- This model has one CA. It is a two-level hierarchical architecture.
- End entity such as a mobile phone or server has only one public key pair and one certificate for one purpose.
- A mobile phone and server have one unique name.
- We consider the possibility that mobile phone receives the wired X.509 certificate owned by a server that was designed for wired Internet.

In this model, we apply X.509 certificate as certificate of mobile phone. Because X.509 certificate owned by mobile phone is verified by server, verification of the certificate is not difficult in the server with enough performance. Even storing of a certificate is burdening to the mobile phone and mobile phone just sends it to other party without any operation for certificate. In this model, CA issues a certificate, publishes its directory, and sends only URL of the certificate to the mobile phone. When a mobile phone communicates with a server, the mobile phone sends an URL of the certificate to the server, not the certificate itself. The server can easily access the directory and acquire the certificate. As a result, the mobile can save memory space for another use.

For server, we use X.509 and a short-lived certificate (Housley, 2002). If a server sends an X.509 certificate to mobile phone, an efficient and lightweight certificate validation scheme might be required in the mobile phone. Sometimes the mobile phone may validate X.509 certificate because it may try to connect a server that was designed for serving only a wired terminal and has only X.509 certificate. We introduce Online Certificate Status Protocol (OCSP), and the mobile phone delegates OCSP to validate certificates rather than validation in the mobile phone by itself. In this case, the mobile phone could avoid the complicated procedure of certificate validation and acquire a result from the trusted OCSP server (Myers, 1999).

A WTLS-certificate can be defined as a short-lived certificate for WTLS connection. A short-lived certificate does not have extensions that are used for certificate path validation and only has a valid period for a short time. It is verified only if signature of CA and valid period for certificate validation are valid. Therefore, mobile phone can avoid burden of CRL download and certificate path validation. We explain the detailed components for WPKI architecture.

3.6.3.2.1 Digital Signature Algorithm

Because a mobile phone has much smaller memory and slower CPU performance than a server, it is hard for mobile phone to run complex public key calculation. We consider an optimal digital signature algorithm for mobile phone. First, generation of public key pair is required for digital signature. The time that it takes to mount a brute-force attack on encipher of the data is directly proportional to the key size used to encipher the data. Although the time depends on the hardware being used, it was estimated that a brute-force attack on a key size of 128 bits for DES algorithm, using multi-trillion dollar specialized hardware, would still take 1,011 years in 1995. We decide that a key size of at least 128 bits would be sufficient to protect the confidentiality of the data. Thus, we choose a RSA 1024-bit key size that is at the same security level as that from 128 bits.

A X.509 certificate (as illustrated in Table 3.1) consists of basic field and extension field. Generation implies that a certificate has to include the specified field, and processing implies that if the specified field is present in the certificate, the field must be examined when the certificate is verified. In basic field, subject unique identifier and issuer unique identifiers are present in the certificate to handle the possibility of reuse of subject and/or issuer names over time. A good profile defines that names should not be reused for different entities and CAs conforming to this profile should not generate certificates with unique identifiers.

Authority key identifier and subject key identifier are used to identify the public key where an issuer and/or subject have multiple signing keys. In the previous section, we assumed that all entities have only one signing key. Thus, we define that these extensions could be processed optionally. The private key usage period extension allows the certificate issuer to specify a different validity period for the private key than validity period of the certificate. We also assume that the private key usage period is the same as the validity period of the certificate and do not use this extension. Because the policy mapping extension is used in CA certificates, we do not define this extension for end-entity.

The subject directory attributes extension is used to convey identification attributes (e.g., nationality) of the subject; this extension is not defined for the end-entity with a unique identifier.

The extended key usage indicates one or more purposes for which the certified public key may be used, in addition to or in place of the basic purposes indicated in the key usage extension. If the extension is present, then it must be examined. Since

Table 3.1 Wireless X.509 Certificate Profile for Mobile Phone

	Generation	Process
Basic Field		
Version	m	m
Serial number	m	m
Signature	m	m
Issuer	m	m
Validity	m	m
Subject	m	m
Subject public key info	m	m
Issuer unique identifier	x	x
Subject unique identifier	x	x
Extension Field		
Authority key identifier	m	o
Subject key identifier	m	o
Key usage	m	m
Private key usage period	x	x
Certificate policy	m	m
Policy mapping	—	—
Subject alternative names	m	m
Issuer alternative names	o	m
Subject directory attributes	x	x
Basic constraints	x	x
Name constraints	—	—
Policy constraints	—	—
Extended key usage	o	m
CRL distribution points	m	o
Domain information	o	o
Authority info access	m	o

Note: m = mandatory, o = optional, x = not recommended, — = not defined.

applying OCSP for certificate validation in this model, we use domain information and authority information access extension for specifying how to access OCSP server (Myers, 1999). For server, CRL distribution points could be used to obtain CRL information. The authority information access and the CRL distribution points extensions must be present in the certificate for certificate validation, but the verifier can choose the certificate validation method, either CRL or OCSP.

3.6.3.2.2 Wireless Certificate Request and Management Protocol

We consider how a mobile phone securely requests a certificate to CA and CA issues it to the mobile phone. The following are requirements of the certificate request protocol.

- The certificate request message is constructed at the mobile phone. This value should include a public key, and the end-entity's reference number (such as an ID) and password. We assume that other requested certificate fields, and additional control information related to the registration process, are made in out-of-band.
- A POP (Proof of Possession) of the private key corresponding to the public key for which a certificate is being requested value is included in certificate request message.
- Method that the certificate request message is securely communicated to a CA.

To satisfy these requirements, a wireless certificate management protocol can be developed on mobile phone. A password could be transferred to a CA by hash value; confidentiality of the password could be guaranteed. We use the public key as one time information for prevention of replay attack.

3.6.3.2.3 Certificate Validation Scheme

As mentioned before, a mobile phone delegates validation authority (VA) to validate certificate in this model. The mobile phone can avoid the burden of CRL download and storage as well as the complicated procedure to acquire and verify certificate chain. For a short-lived certificate, the mobile phone validates the certificate through verifying only signature and valid period in the certificate.

A delta CRL that lists the certificates whose revocation status has changed since the issuance of a referenced complete CRL may be used for CRL verification in a mobile phone. But conformation procedure of complete CRL from delta CRL is not easy for the mobile phone and requires additional module. Also the mobile phone should store the base CRL, finally complete CRL. Thus, we exclude the delta CRL-based CRL verification from our model. Figure 3.9 shows the certificate validation procedure. The server acquires a certificate from directory using the URL of the certificate received from the mobile phone, and validates it using CRL or VA.

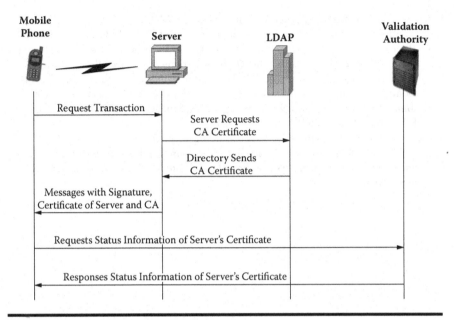

Figure 3.9 OCSP procedure in mobile phone.

Inversely, the server sends to the mobile phone its certificate with CA's certificate and ARL (Authority Revocation List) together. Consequently, the mobile phone need not acquire the CA's certificate and ARL from directory as shown in Figure 3.9. It reduces the number of wireless connections between mobile phone and directory.

References

C. Allen and T. Dierks, The TLS Protocol Version 1.0, IETF, Nov. 1997, available at ftp://ietf.org/internet-drafts/draft-ietf-tls-protocol-05.txt.

J. Arkko and P. Nikander, Limitations of IPsec Policy Mechanisms, *Security Protocols Workshop*, 2003: 241–251.

D. J. Barrett and R. E. Silverman, *SSH: The Secure Shell, The Definitive Guide*, O'REILLY & Associates, 2001.

M. Christinat and M. Isler, WTLS—The Security Layer in the WAP Stack, Colloquium on Information Security, June 22, 2000. Available at: http://www.keyon.ch/en/publications/infosec_wtls_keyon.pdf.

A. O. Freier, P. Karlton, and P. C. Kocher, The SSL Protocol, Transport Layer Security Working Group, INTERNET-DRAFT, Available at: http://wp.netscape.com/eng/ssl3/draft302.txt.

R. Housley, W. Polk, W. Ford, and D. Solo, Internet X.509 Public Key Infrastructure Certificate and Certificate Revocation List (CRL) Profile, IETF RFC3280, IETF Network Working Group, April 2002.

A. Jourez, *An IPSec Primer*, Université Libre de Bruxelles, France, 2000.

S. Kent and R. Atkinson, IP Authentication Header, RFC 2402, 1998.

S. Kent and R. Atkinson, IP Encapsulating Security Payload (ESP), RFC 2406, 1998.

K. G. Paterson and A. K. L. Yau, Cryptography in Theory and Practice: The Case of Encryption in IPsec. 2005. Available at http://eprint.iacr.org/2005/416.pdf.

J. O. Saarinen, Attacks Against the WAP WTLS Protocol, in Secure Information Networks: Communications and Multimedia Security, *IFIP Joint Working Conf. on Comm. and Multimedia Security (CMS '99)*, B. Preneel (Ed.), Sep. 20–21, 1999, Leuven, Belgium. IFIP Conf. Proc., 152(1999), pp. 209–215.

S. Vaudenay, Security Flaws Induced by CBC Padding—Applications to SSL, IPSEC, WTLS, 2002. Available at http://www.mics.org/getDoc.php?docid=196& docnum=1.

D. Wagner and B. Schneier. Analysis of the SSL 3.0 Protocol, February 2004, Available at: http://www.citeseer.net.

Chapter 4

Smart Card Security: The SIM/USIM Case

4.1 Introduction

Smart cards are tamper-resistant modules that are capable of securely saving secret cryptographic materials and executing undetected (or autonomous) executions of cryptographic algorithms. A smart card can be considered as a safe container to store data, in the sense that it is highly protected against all unauthorized or unforeseen access. Smart cards can help users in various sensitive activities. Nowadays, smart cards are typically used in an *application-specific* way, in the sense that they act as the application's security modules. In particular, they are ideally appropriate to operate as personal security modules in mobile systems.

Their usage spans over several application domains including banking, telecommunications, and identity. An example of smart card is given by the SIM module, used in any GSM phone, that implements a special application, which is defined by the GSM standard, to protect the data in the card and to allow access control to the GSM networks. However, it is possible for a GSM subscriber to have more than one SIM in the same mobile terminal (for different network providers) or to use the same SIM in different equipments. Thus, smart cards are trusted personal devices designed to store and process confidential data, and to act as secure tokens for providing access to applications and services.

Open smart card-based platforms used by mobile systems are new generation trusted personal devices with enhanced flexibility in terms of connectivity and interoperability. Smart cards can host several applications and allow new applications to be added after their issuance. Such flexibility adds more concerns about the

127

possibility of logical attacks that can be launched to affect the regular operation of a mobile equipment operation or the work of a large number of devices, and requires the development of special security techniques and tools that can be used to increase the reliability of platforms and applications for trusted personal devices.

An example of use of a smart card is an application allowing a mobile user to have remote access to a service, to which he has a subscription for a digital newspaper. The access allows the user to request the download of the newspaper from a predefined server through a public mobile network. The information provided by the user is automatically used to search for the subscription download server and check whether the user is authorized to download. To perform the download, the mobile user and the server can establish a protected channel; then, they can establish a secured and efficient procedure to build the required security elements using the smart card. A challenge/response procedure can be set up between the user and the server based on sensitive information stored in the smart card. In particular, the mobile user's subscription can be checked using an *authorization certificate* submitted by the user. The smart card safely keeps the authorization certificate and performs the response of the server's challenge.

Integrating smart cards in mobile equipment requires that the following issues be addressed:

■ The definition of a personal security environment, based on smart card, in a mobile communication system.
■ The definition of ad hoc optimized methods to help smart cards be efficiently integrated into the mobile environment, knowing that they have limited resources in terms of computational power and communication bandwidth.
■ The identification of the security features that have to be met by the mobile terminals used to host and/or interact with the personal smart card.
■ The determination of the level of security that can be achieved by the acceptable smart card-based solutions.

On the other hand, since the mobile service usage in business will typically require some form of robust *access control, authorization*, and accounting *mechanism* that allows the service providers to control the use of advanced services and provide correct and protected billing, an important number of security related issues need to be addressed, including the following:

■ *Authentication*: This issue considers the study of the techniques used to involve a smart card to authenticate the cardholder, when he wants to start a connection.
■ *Authorization*: This issue considers the study of ways the smart card can be involved to authorize a user access to a service offered on the mobile network and how authorization control can be set up.

- *Non-repudiation*: This issue considers the techniques the smart card can integrate to create legally-binding evidence of the cardholder's participation in a connection with other parties.
- *Privacy management*: This issue develops the models and techniques the users can utilize to specify their privacy preferences and allow their smart cards to interact with the offered service.
- *Forensic investigation*: This issue develops the investigation techniques that are needed to retrieve and understand which evidence can be obtained from a smart card-based system and how smart cards can be organized to provide evidences such as attacks traces.

Implementing the aforementioned ad hoc security mechanisms was one of the most important issues to address when enterprises started to think about a world-wide mobile telephone system. The SIM cards appeared as a result of the user requirements, provider responsibility, and technological limits.

4.2 Basic on Smart Cards

Basically, a smart card can be seen as small electronic device for digital information processing that is implemented under the form of a tiny computer in the shape of a very small hardware module. Among the most important features of a smart card, one can mention the possibility to protect data stored on the card against unauthorized access and manipulation. A smart card is equipped with an operating system that controls the interface for transferring data between the smart card and a connected reader. Since a smart card can utilize cryptographic algorithms and security protocols, confidential data can be stored on the card in a way that prevents it from being read from the outside by unauthorized individuals or applications.

Three smart card categories can be distinguished. They mainly differ by the functionality they perform and the communication techniques they use to communicate with outside environment:

- *Memory cards*: These cards present surface contacts implementing a memory-only integrated circuit chip. Usually the memory card includes an EEPROM where implemented applications can store their data. An access control to these data is handled by a security logic. Most memory cards have a functionality that is usually optimized for a specific application. A typical application area for memory cards is prepaid telephone system.
- *Microprocessor cards*: These cards present surface contacts implementing a microprocessor-only integrated circuit chip. A microprocessor card integrates an operating system onto the card. This authorizes several applications to be loaded onto a single card. The application specific part of the operating

system is loaded into the EEPROM after the microprocessor card has been produced. Recently developed microprocessor cards allow the card owner to load the application's programs onto the card. They also allow high processing and large memory capacities. This enables these cards to run complex cryptographic algorithms.

■ *Contactless cards*: These cards aim at overcoming drawbacks that physical contacts can generate with the reader. In fact, the most common problems caused by contacts include contamination and electrostatic discharge. Since contactless cards do not necessarily need to be inserted into a reader, several new attractive applications can be offered and the access control system significantly gains comfort. Today, the coverage of a standard contactless card is about 1 meter.

In the following we will focus on the features and use of the second category of smart cards, since the majority of cards used by the mobile communications systems belong to that category (including the SIM and USIM cards).

4.2.1 Components of a Smart Card

Considering a smart card as a computer system distinguishes three components assembled in a single integrated circuit chip. They are the CPU, the memory system, and the Input/Output system. Figure 4.1 depicts the major modules the smart card can contain. The main functions performed by these modules are described as follows.

■ *The CPU*: The CPU is a programmable central processing unit. Current CPUs range from 8-bit micro controllers with a few MHz of clock frequency to highly equipped controllers running at higher rates, passing by a 32-bit RISC processor. Typically the instruction set used in the CPU is based on well-known architectures (e.g., the Motorola 6805 or the Intel 8051). The CPU characterizes the most important difference between a memory card and a smart card. Typically, it is equipped with an operating system that conducts, independently of the system integrating the smart card, all the sensitive operations that can be executed by the smart card.

■ *The memory system*: The memory available on a smart card is limited. It contains three types of memory.

  – *ROM*: The ROM component is a non-volatile persistent memory that is usually easy to use, implement, and communicate with. It contains the card operating system and the needed procedures to perform its (protecting) operations. The typical capacity of the ROM component ranges from 6 kB to 128 kB. Some modern cards replace the ROM component by a *FlashROM*, whose content can be updated by special loading protocols,

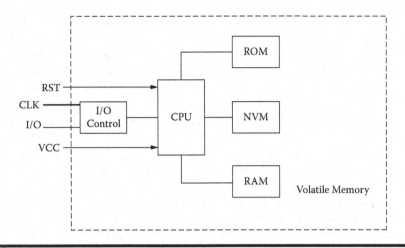

Figure 4.1   Typical smart architecture.

e.g., to test new versions of a card operating system. The ROM also contains the keying materials (such as a secret key) needed to perform the smart card functions.

- *RAM*: Since RAM is rather expensive to manufacture and consumes a rather large portion of silicon, typical sizes of modern smart cards range from 128 bytes to 4 kB. Access to RAM is usually a magnitude faster than access to ROM or EEPROM. Therefore, RAM is used for the runtime control and data stack. Since RAM only ensures volatile storage, it cannot be used for storing persistent data.

- *Non-volatile memory*: In contrast to RAM, the non-volatile memory is persistent, meaning that when the power is turned off, the data stays there for a long time (about 10 years). However, it depreciates after about 100,000 writing operations. Its content can be changed during operation. It is mostly used for persistent but alterable data and dynamically installed applications. Typical sizes range from 4 kB to 64 kB. Writing the data to non-volatile memory takes more time and consumes more power than reading. A common usage of RAM is to store data that can be changed by an application (e.g., PIN codes and account numbers).

■ *The I/O-system*: The I/O-Block offers serial communication with the outside world (I/O) and contains other lines for external clock (CLK), card reset (RST), and the power supply (VCC). The protocol used for the communication between the smart card and outside is based on a master (the card reader) and slave (the smart card) relationship. The card reader sends commands to the smart card and waits for a response that the smart card computes autonomously. The smart cards never initiate data transfer with the card reader.

The file system in a smart card is subject to the non-volatile memory. The SC operating system takes care of the file management. Files are stored hierarchically and the structure consists of three basic elements: Master File (MF), the Dedicated File (DF), and the Elementary Files (EF). Only one MF exists on a SC and it is the root of the file system. A two-byte long file identifier is used to identify each file. A DF usually forms a subdirectory in the file hierarchy, which is rooted by the MF. Finally, the EFs form the leaves of the file hierarchy and they are the files where the data is stored.

Therefore, one can say that basically the smart cards operate as a server in a traditional client/server system. They mainly perform three processes:

1. **Requesting:** A request containing a command to be executed is received by the I/O manager via the serial interface. Error correction due to transmission failures are usually directly handled by the I/O manager.
2. **Processing:** The card interprets and subsequently executes the received command. State transitions may occur during computation. A messaging manager is usually responsible for appropriate de- and encoding of messages. A command interpreter decodes the commands and triggers appropriate actions to perform the interpretation. The return code manager takes the result of the interpreter's computation and generates a corresponding return code.
3. **Responding:** After the card has processed the command, the return code and the computed data are returned to the outside client via the I/O manager. Smart card computations only occur synchronously after an appropriate request has been issued to a card. Hence, smart cards are *reactive* devices that are not able to *proactively* initiate external activities on their own.

4.2.2 Smart Card Applications

Smart cards have been proposed as portable and secure data storage devices for a wide range of applications. In addition, their computing capabilities make them suitable as private key storage devices for asymmetric cryptosystems. Encryption and decryption of data can be performed on request by the smart card operating system in a way that keeps the user's private key secure and cannot be eavesdropped. Smart cards are now everywhere: in GSM phones, in new generation credit cards, and in pay-TV. They are also used for credit cards and prepaid phone cards to provide

- secure access to a network, secure identification, law-strong digital signature;
- secure cellular phones from fraud; and
- secure the piracy of set-top boxes on televisions

Smart cards are particularly useful as crypto devices. A major motivation for this is that they are capable of generating and protecting a private signing key, which must never leave the card. Thus, it is extremely hard for external attackers to gain knowledge of the private key. This is a feature that could otherwise occur through a compromise of the host computer system, for example. This has obvious and immediate advantages compared to protocols and applications providing authentication, authorization, privacy, integrity, and non-repudiation, such as the *PKI*. On the other hand, placing the private key of a user on a smart card never puts the crucial secret of the smart card in a situation where it can be compromised.

Moreover, if the private key is stored in a file, on a hard drive, it is typically protected by a password, and this file can be attacked using dictionary attacks, where commonly used passwords are attempted in a brute force manner until the private key is obtained. On the other hand, a smart card will typically lock itself when a predefined low number of consecutive bad PIN attempts has been done (typical value for the SIM card is equal to 3). Thus, the dictionary attack cannot be considered as a feasible way to access the private key that has been securely stored on a smart card. In addition, smart cards can bring multiple technologies together by storing multiple certificates and passwords on the same card. This solves one of the biggest problems related to password vulnerability. If a smart card is used to store a set of user's multiple passwords, they need only to remember the PIN to submit to the smart card (and, of course, to hold the device) to access each of these passwords. The cardholder does not need even to know the passwords and, therefore, does not have to write them down or share them with other users.

Different standards have been used to integrate smart cards into networked applications to provide security related services. They all follow certain principles in order to be useful and gain acceptance: they are (a) multi-platform, since they should be applicable to the whole wide variety of operating systems and computer architectures; (b) interoperable, meaning that they should be interoperable with other leading standards and protocols; (c) functional, since they should apply to real world problems and markets and satisfactorily address their requirements; and (d) extensible, since any standard should facilitate the development to new applications, protocols, and smart card capabilities that weren't yet around when the standard was created.

The following are some among the emerging standards involved in the integration of smart cards into network security applications:

- **PKCS#11: Cryptographic Token Interface Standard:** This standard specifies an Application Programming Interface (API), called Cryptoki, to devices that hold cryptographic information and perform cryptographic functions.
- **OpenCard:** This is a standard framework announced by International Business Machines Corporation, Netscape, NCI, and Sun Microsystems that provides for interoperable smart card solutions across many hardware and software platforms. The OpenCard Framework is an open standard

providing an architecture and a set of APIs that enable application developers and service providers to develop smart card aware solutions.

■ **JavaCard:** This is a specification that enables the Write Once, Run Anywhere capabilities of Java on smart cards and other devices with limited memory.

Typical applications based on smart cards include the following:

■ *Workstation logon*: Logon credentials can be securely stored on a smart card. The normal login mechanism of the workstation, which usually prompts for a username and password, can be replaced with one that communicates with the smart card.

■ *Dialup access*: Many of the common remote access dial-up protocols use passwords as their security mechanism. Smart cards enhance the security of passwords. Also, as many of these protocols evolve to support public key based systems, smart cards can be used to increase the security and portability of the private key and certificate.

■ *Secure electronic transaction (SET)*: The SET protocol allows for credit card data to be transferred securely between customer, merchant, and issuer. Since SET relies on public key technology, smart cards are a good choice for storage of the certificate and private key.

■ *Law-strong digital signatures*: New digital signature laws are being written by many states that make it the end user's responsibility to protect their private key. If the private key can never leave an automatically PIN disabling smart card, then the end user can find it easier to meet these responsibilities. Certificate authorities can help in this area by supporting certificate extensions that specify the private key was generated securely and has never left the confines of a smart card.

■ *Digital cash*: Smart cards can implement protocols for the management of digital cash and electronic payment (e-payment). In these systems, the underlying keys that secure the architecture never leave the security perimeter of the hardware devices used for transactions.

■ *Networking framework for smart cards*: This application allows a smart card to spontaneously integrate itself into a local environment after insertion into a suitably equipped card terminal. The application might use a mobile code as an enabling technology to complement the card-resident resources with off-card resources in a dynamic way.

■ *Web smart card*: This application implements the idea of a wireless smart card reader in the form of a mobile phone. The smart cards can made reachable from the Internet by means of an architecture that allows a smart card (such as the GSM SIM card) in a mobile phone to appear as a Web server in the Internet. Other Internet nodes can connect to the smart card using the HTTP protocol, anytime and anywhere the user is located.

4.2.3 Security of Smart Cards

Data stored in smart cards are protected and kept secret. Typically, four major components are in charge of the security of a SC: The card body that holds the different components, the chip hardware, the operating system, and the applications. The first component is responsible for the physical security, while the three remaining components contribute to protect the programs and data in the smart card. The smart card operating system and the applications on the smart card need to be able to handle a variety of cryptographic elements including cryptosystem, digital signature, and key management.

Physical security guarantees the temper-resistance character of the smart card. It is ensured by packaging the integrated circuit card (ICC) and its connections into a module that is made in an epoxy resin. Physical attacks typically leave an obvious track on such a package (Figure 4.2).

The key management main objective for the smart cards is to minimize the consequences for the system and the smart card application if one or more secret keys get compromised by an unauthorized entity. Smart cards can easily be taken away and it is therefore more likely that they are exposed to the most severe attacks. Even if somebody breaks the security mechanisms and reads the contents of the card, found keys are just those derived from a master key. Derived keys are unique and they are usually created out of the card specific features, information stored, and a master key using a cryptographic algorithm.

Usually different keys are used for every cryptographic operation in the smart card to reduce the damage in the case that somebody breaks a key. For each type of key, a separate master key must exist to generate the needed derived keys. In practice, more than one key generation is stored in the smart card in order to change the version of the keys in case one or more keys get compromised. However, a change of the key generation is not always due to an intrusion of an unauthorized person. Normally, a version change takes place at fixed or variable intervals of time.

Dynamic keys represent another well-known security practice used by the smart cards to communication with outside environment. For instance, the so-called

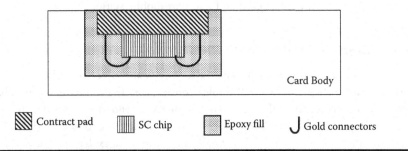

Figure 4.2   The physical security of a smart card.

session keys are generated and used for data exchange. These keys are generally created from a random number that is sent to the other party. To be able to use the keys stored in the smart card, a special key number is required. The operating system responsibility is to guarantee that a specific key can only be used for the purpose it was created for. To differentiate between the versions of a key, a version number is used. The following list gives the typical key parameters stored on a smart card:

Key number: Key reference number, unique within the key file
Version number: Version number of the key
Application purpose: Identifies the cryptographic algorithms and the procedures with which the key may be used
Disable: Allows the key to be temporarily or permanently disabled
Retry counter: This counter keeps track of non-successful attempts to use the key with a cryptosystem
Max retry counter: If the retry count reaches a maximum count, the key is blocked
Key length: Length of the key (i.e., the actual key)

4.3 Smart Card and Communication

Smart cards basically offer a combination of the following objects in a mobile environment: (a) a *secure* and *tamper-resistant* storage for a set of objects including cryptographic keys and (b) an implementation of algorithms (such as cryptographic algorithms). This combination, associated with mobility, makes smart cards a very attractive device for off-line and on-line communicating systems. In other words, a smart card allows storing a secret, implementing a number of cryptographic algorithms, performing computations with that secret without revealing it, and participating in communication sessions. Two types of communication are of utmost interest: the communication to and from a smart card to the terminal it is attached to take place via the serial interface; and the communication between the mobile station integrating the smart card and a similar mobile station.

4.3.1 Communication with the Terminal

Data transfer (defined in the ISO/IEC 7816 specification) between the card reader and the card takes place on a half-duplex connection. The standard specifies basic electronic characteristics of integrated circuit cards and power needed for data transfer. In addition, it specifies the structure of the answer-to-reset and describes the data transmission protocol. Informally speaking, data exchange is always initiated by the host (or the card reader) and never by the client (i.e., the smart card).

The smart card receives the command from the reader, executes it, and sends back its response to the host.

The ISO/IEC 7816-3 standard specifies a total number of 16 protocols. The protocols are named 'T=*' (for *= 0,...,15). In practice, only four protocols are of interest. The T=0 and the T=1 are predominating in international use. T=2 is still in the design process and a standard will be available in a few years, while T=14 is a protocol for national use. The T=0 protocol was the first internationally standardized smart card transmission protocol. It is the most widely used protocol, since it is the protocol used in GSM cards.

The internationally standardized data unit for the data exchange between the card reader and the smart card is called APDU (Application Protocol Data Unit). The APDUs can be understood as boxes that either contain a command sent from the card reader to the card or a response from the card to the card reader. A distinction is made for different purposes of APDUs. An APDU used in the transmission protocol layer is called TPDU (Transmission Protocol Data Unit). The protocol T=0 is byte-oriented and each TPDU implements a simple error detection mechanism (based on the parity check at the end of each byte). On the detection of an error, the byte is resent. The TPDUs are subdivided at the application protocol layer into two types of APDUs, namely the command APDUs (C-APDUs) and the response APDUs (R-APDUs). A command APDU has two elements, a header and a body. The length of the header is fixed to 4 bytes, and the length of the body varies, depending on the amount of the included data. A response APDU is composed of a body and a trailer. The body is optional and the trailer mandatory.

Several applications can be stored in a multi application smart card. Logical channels make it possible to address up to four applications at the same time. Physically there is still the single serial interface, but on a logical level it is possible to have four connections. However, one major limitation for the communication on the logical channel can occur: The external process and the application on the smart card must be mutually synchronized and interference with a communication in progress is not allowed, since the R-APDU does not include any information about the logical channel and that it is not possible to recognize to which C-APDU the received R-APDU belongs.

One possible application for the logical channels could be the following scenario. A GSM mobile phone is equipped with a multi application smart card allowing the cardholder to simultaneously perform a call, on a first logical channel, and execute the calendar application stored on the smart card. Using a second logical channel he searches in his calendar for the desired day and is now able to provide the information about confirming the appointment. Every logical channel can be seen as a separate smart card. However, a large memory may be necessary to store all the information needed by the applications. In addition, the logical channels produce several management requirements for the smart card operating system.

4.3.2 Subscriber Identity Module

The smart cards used in mobile communication systems are better known under the name SIM, for the GSM network, and USIM for the UMTS network. On the other hand, the wireless identity module (WIM, or WAP Identity Module) is a security module that can be used to establish a secure communication using the WAP protocol (Chapter 7 will give the details and limitations of the WAP). It stores the necessary private keys for authentication. It is also used to store and process information needed for user identification. Sensitive data can be stored in the WIM and all operations involving the data (such as computing a signature, verifying a signature, encrypting, decrypting, or deriving a key) can be performed in the WIM.

The SIM has marked a tremendous pioneering work in memory management and protections of sensitive data on smart cards. For that reason, we discuss in the following various features provided by the smart cards, to the mobile communications, by simply discussing the case of the SIM. The primary function of the SIM in a GSM network is to authenticate the validity of a MS when accessing the network (Chapter 5 will detail this process). In addition, the SIM provides a way to authenticate the user and may also store other subscriber-related information or applications. The SIM can be incorporated into a multi application smart card and thus may contain non-GSM functions. It has two main purposes. First, it ensures access to a particular GSM network, and second it associates the use of the network with an AAA function. In addition to this basic functionality, a SIM is completely involved in (a) managing services and supplementary applications; (b) the storing of data related to dialing, short messaging, subscriber information, and mobile phone setting; and (c) the subscriber administration (Rankl, 2004).

The GSM 11.11 specification defines twenty-two APDU commands that can be executed on the SIM card. The commands can be classified into four categories: (a) the security commands, which provide for the management of the PIN (see Chapter 5 for the details) and the execution of the authentication algorithm; (b) the file operation commands, which are used to manage files and read objects; (c) the SIM application toolkit (SAT) commands, which allow applications, existing in the SIM, to interact and operate with any mobile device that supports the specific mechanism(s) required by the application; and (d) the miscellaneous commands. The following commands are among the most important commands:

Security commands

Change CHV:	Change the PIN
Unblock CHV:	Reset the PIN retry counter
Verify CHV:	Verify the PIN

File operation commands

Status:	Read various data from the currently selected file
Read Binary:	Read from a file with a transparent structure
Read Record:	Read from a file with a record-orientated structure

SAT commands

Envelope: Pass data to a value added service of the SIM
Fetch: Retrieve a SAT command from the SIM in the mobile device
Terminal Profile: List all functions of the ME with respect to the SAT

Miscellaneous commands

Get Response: Command specific to T=0 protocol to request data from the
                  smart card
Sleep: Command for putting the smart card into a low power state

4.3.3 The Smart Card in UMTS

Similarly to the GSM networks, where the SIM card has proven to be a useful security component, the security in UMTS is mainly based on a subscriber-related smart card (called the USIM) which has to be present in the terminal to provide essential security to the UMTS service. Authentication of the mobile user to the network will be carried out by secret keys and cryptographic algorithms stored on the USIM. Of course, such a key never leaves the card in plain format, and thus has to be operated in the USIM card. However, opposed to GSM, the USIM is responsible for the authentication of the serving network using a similar procedure. On the other hand, similar to the GSM, real-time encryption does not take place on the USIM card due to resource limitations. Therefore, the USIM secures the generation of the session keys after negotiation between the USIM and the serving network. Then, this key is handed over to the mobile terminal.

The USIM files system consists of three first-level dedicated files, the DF_GSM, DF_TELECOM, and DF_UMTS. DF_UMTS contains the UMTS authentication application. Elementary files located in the directory DF_UMTS include authentication of relevant data that are irrespective of the used protocol. EF_CHV contains the card holder verification and administrative attribute values. The cipher key is stored in EF_CK, the integrity key in EF_IK. EF_IMUI contains the IMSI. EF_SPID stores the service provider ID. The EF_SSD references the security mechanisms that are implemented in the USIM.

The following GSM commands are used by the GSM application as well as by the UMTS application: CHANGE CHV, GET RESPONSE, READ BINARY, SELECT, UNBLOCK CHV, UPDATE BINARY and VERIFY CHV. On the other hand, the UMTS authentication application includes further commands.

■ The MANAGE SECURITY ENVIRONMENT command is used to refer to control data elements for a security environment. The data objects will be referenced in the event of a command accessing this object.
■ The SECURE READ BINARY function reads a string of bytes from the current EF and encrypts the bytes using a symmetric encryption function and a secret that is referenced by a preceding MANAGE SECURITY ENVIRONMENT command.

- The INTERNAL AUTHENTICATE command initiates the computation of authentication data and the computation and storage of a cipher key and integrity key by the smart card using the challenge data sent from the terminal and a relevant secret key stored in the card.
- The GENERATE PUBLIC KEY PAIR command initiates the generation and storing of a temporary Diffie-Hellman public key pair in the USIM. The public key is delivered to the terminal as a random challenge.
- The MUTUAL AUTHENTICATE command allows the authentication of the network by the card, the authentication of the card by the network, and the establishment of a session key between the USIM and the network.
- VERIFY CERTIFICATE command allows the verification of a certificate in the smart card. The certificate content is delivered to the card in the data field. The card retrieves a public key from the certificate, which can be used for the verification of authentication data in a subsequent MUTUAL AUTHENTICATION command.

4.4 Attacks against Smart Cards

Smart cards can protect the stored data they contain against unauthorized access. However, the strength of the mechanisms built for protection seems to be frequently overestimated. Analyzing the security of a smart card-based system is useful to determine the security level of this system. It means checking whether the mechanisms properly protect the information stored inside the smart card and estimating the cost, measured in terms of time, money, and effort, that an attacker has to pay to launch a successful attack.

Two classes of attacks will be presented, in particular, in the following. They are the invasive attacks and the non-invasive attacks. An attack on a smart card is called *invasive* if it involves a tampering of the device that is clearly visible for anyone. On the other hand, *non-invasive* attacks can usually be performed by making use of a smart card device for some small amount of time; invasive attacks can require hours of work in specialized labs and are therefore available only to highly skilled and funded attackers.

In this section, we discuss some among the known and well-documented attacks against smart card-based systems. A particular interest will be given to the attacks against the smart card itself, its interaction with the system, and the API and OS it uses. We will not consider attacks targeting issues associated with digital signature, authentication, and non-repudiation schemes.

4.4.1 Invasive Attack Techniques

Most of the invasive attacks presented in this subsection require an utter damage of the card hardware. In addition, there is a small probability that these attacks could be performed without knowledge of the user (who will realize that he no longer owns his card, for instance). The examples of attacks considered in the following include removing the chip from the card, reverse engineering the chipset, and microprobing.

4.4.1.1 Removing the Chip from the Card Attacks

Often, the removing attacks aim at taking away the chip package. The typical chip module consists of a thin plastic basis plate of about a square centimeter with conductive contact areas on both sides. One side is visible on the final card and makes contact with the card reader; the silicon die is glued to the other side, and connected using thin wires. The chip side of the plastic plate is then covered with epoxy resin. The resulting chip module is finally glued into the card. Removing the chip is therefore physically easy. It can be done using a sharp knife, by heating the card plastic until it becomes flexible, or by covering the chip with hot fuming nitric acid and waiting until the acid dissolves the epoxy resin.

4.4.1.2 Reverse Engineering Attacks on the Chipset

The smart card modules can be reverse engineered. To do so, the first step is to create a map of a new processor. It could be done by using an optical microscope with a camera to produce high-resolution photographs of the chip surface. Lower layers can only be recognized in a second series of photographs after the metal layers have been removed. This can be achieved, for example, by submerging it for a few seconds in hydrofluoric acid. More sophisticated tools such as focused ion beam (FIB) workstations can be used. Details on how to examine circuits, tools, and methods are present in literature (Daniel, 1998).

Using laser interferometer stages, a FIB operator can navigate blindly on a chip surface with 0.15 μm precision, even if the chip has no recognizable surface structures. Chips can also be polished from the back side down to a thickness of just a few tens of micrometers. Using laser-interferometer navigation or infrared laser imaging, it is then possible to locate individual transistors and contact them through the silicon substrate. This rear-access technique is about to become very common.

To counteract these attacks, a number of copy trap features can be incorporated into the chip designs and to introduce complexity into the chip layout and to use non-standard cell libraries. Many of the copier traps can be based on holes made to isolate layers. However, new sophisticated techniques have been developed to detect such traps. In addition, complexity reduction measures should be taken to allow non-standard cells being reconstructed properly.

4.4.1.3 Microprobing Attack

This attack aims at removing the chip from the card and interacting directly with its components. Microprobing needs to remove at least part of the passivation layer (which is a layer of silicon nitride or oxide that is responsible of protecting the chip from the environmental influences and ion migration) before probes can establish contact. This can be done by the use of microprobing needles that remove the passivation just below the probe contact point using ultrasonic vibration. The resulting hole in the passivation layer can be made so small that only a single bus line is exposed and accidental contacts with neighboring lines are prevented. In addition, microprobing attacks can be combined with the usage of a FIB. This can be used to explore the silicon and modify the chip structure by creating new interconnect lines and even new transistors.

In particular, using microprobing with two needles allows an attacker to set or reset any given bit in an EEPROM and modify any given bit in the ROM. An interesting example of microprobing attack is presented in (Zanero, 2001). It uses a laser cutter to destroy the last bit of the carry register that feeds the output of a round as the input of the next. An effect of this operation is that the least significant bit of the output of the round function is set to zero. Information about the round keys of previous rounds are then deduced using differential cryptanalysis. In addition, a non-negligible part of the information related to the secret key can be recovered to make key search easy.

A chip modified in this way will have the property that encryption and decryption are no longer inverses. So, to counteract this attack, a simple self-test procedure can be added to the smart card that takes an arbitrary input, encrypts and decrypts under an arbitrary key, and compares the result with the original block. Another solution involves disconnecting almost all of the CPU from the bus, leaving only the EEPROM and a CPU component that can generate read accesses.

4.4.1.4 Semi-Invasive Attacks

This category shares a part of characteristics of the invasive attacks. Semi-invasive attacks require depackaging the chip in order to get access to the chip surface, but it does not need the passivation operation. Moreover, it does require creating contacts to the internal lines. Semi-invasive attacks could be performed using tools such as the UV light, the X-rays, or other sources of ionizing radiation. A semi-invasive attack can be conducted simply by illumination (Zanero, 2001). The illumination of a target transistor causes it to conduct, thereby inducing a transient fault. Such attacks are finely grained and powerful: it has been demonstrated that it is possible to change any individual bit of an SRAM array in the smart card.

4.4.2 Non-Invasive Attacks

A non-invasive attack on a smart card has limited effects in different ways; it must occur while a card is still operating in a black box manner, any manipulation must be performed on the bytes getting in or going out of the smartcard, or on the environmental conditions. A non-invasive attack has, however, the advantage of leaving the device completely physically undamaged, and thus it can be difficult to detect. However, in general, a non-invasive attack requires that the software and hardware of the smart card are known to the attacker. Four major classes of non-invasive attacks can be distinguished.

1. *Timing attacks*: In this type of attacks, various byte patterns are sent to the smart card to be processed (e.g., signed by the private key, encrypted using a secret key, or used to compute a specific function). Information such as the time required to perform the operation and the number of 0s (and 1s) in the input bytes are used to eventually obtain sensitive information stored in the smart card (e.g., the private key or the secret key) (Kocher, 1996).

   There are logical countermeasures to this attack but not all the smart card manufacturers have implemented the countermeasures. In addition, this attack requires that the attacker knows the PIN required to get to the card, or can mislead the user into signing the byte patterns of his choosing (often, this attack is referred to as a *chosen-plaintext attack*).

2. *Software attacks*: Some of attacks can be launched using procedures or programs. For example, a Trojan horse application could be used to transport an attack. The Trojan horse can wait until the user submits a valid PIN from a trusted application, thus enabling usage of the private key, and then asks the smart card to digitally sign some rogue data (for example, a legally binding contract or strong digital signature). The operation completes but the user never knows that his private key was just used against his will.

   A countermeasure to prevent this attack is to use a *unique-access device driver architecture*. With this architecture, the operating system guarantees that only one application can have access to the smart card at any given time. This prevents the attack but also diminishes the ease of use of the smart card because multiple applications cannot use the services of the smart card at the same time. Another way to prevent the attack is by using a smart card that enforces a "one private key usage per PIN entry" policy model. In this model, the user must enter his PIN every single time the private key is to be used and therefore the Trojan horse would not have access to the key. This is also rarely convenient for the end-user experience.

3. *Power and electromagnetic analysis attacks*: A power analysis attack aims at measuring the fluctuations in the energy consumed by the smart card. The various instructions cause different levels of activity in the instruction decoder and arithmetic units; they can often be properly distinguished, and parts of algorithms can be reconstructed. These techniques fall into the category of information monitoring. They are of large interest because they can be applied to a large number of vulnerable products on the market today. The attacks are easy to implement. They can be automated (and so can be used by low-skill attackers) and have a very low cost per device.

Simple Power Analysis (SPA) involves directly observing a system's power consumption to obtain information on the executed sequence of instructions. If the attacker has access to just one transaction, a limited amount of information can be leaked. Attackers with access to multiple transactions knowing the internal mechanisms of the particular chipset under use can be more and more challenging. On the other hand, differential power analysis (DPA) is based on the phenomenon that storing a 1-bit in a flip-flop consumes typically more power than a 0-bit. Also, state changes typically cause extra power consumption. In addition to large-scale power variations due to the instruction sequence, there are effects correlated to data values being manipulated. These variations tend to be smaller and are sometimes overshadowed by measurement errors and other noise, but there are effective mechanisms for treating such problems (Messerges, 1999). In these cases, it is still possible to compromise the system using statistical functions tailored to the target algorithm.

An attack with strong similarities with DPA is called the *Electro Magnetic Analysis* (EMA). The idea of this attack is to measure the field radiated by the processor and correlate it to the activities of the processor. An interesting work depicted in Quisquater (2001) shows that the electromagnetic attack obtains at least the same result as power consumption and consequently must carefully be taken into account.

A work presented in Kocher (1999), along with a complete description of this technique, explains how DES can be broken with it, if poorly implemented in a smart card, and how this attack can be easily turned away by avoiding that key material is used to choose between two branches of a jump. The various instructions cause different levels of activity in the instruction decoder and arithmetic units and can often be quite clearly distinguished, such that parts of algorithms can be reconstructed. Various units of the processor have their switching transients at different times relative to the clock edges and can be separated in high-frequency measurements.

The DPA can be used to break the implementations of almost all symmetric or asymmetric algorithms. For example, a 128-bit Twofish secret key, which is believed to be safe, was recovered from a smart card after observing 100 independent encryptions (Chari, 1999). Public key algorithms can be, particularly, analyzed using the DPA by correlating candidate values for computation intermediates with power consumption measurements. For modular exponentiation operations, it is possible to test exponent bit guesses by testing whether predicted intermediate values are correlated to the actual computation. In addition, it is possible to reverse-engineer even unknown algorithms and protocols.

There are techniques for preventing DPA and related attacks. These techniques fall roughly into three categories. Firstly, we can reduce signal size, such as by using constant execution path code, choosing operations that leak less information in their power consumption or adding extra gates to compensate for the power consumption. Unfortunately, such signal size reduction cannot reduce the signal size to zero and an attacker with an infinite number of samples will still be able to perform DPA on the signal. Secondly, we may introduce noise into power consumption measurements but like in the previous case, an infinite number of samples will still enable statistical analysis. In addition, execution timing and order can be randomized. Designers and reviewers must approach temporal obfuscation with great caution because many techniques can be used to bypass or compensate for these effects. Another technique involves the use of non-linear key update procedures.

4. *Fault generation attacks*: These attacks rely on stressing a smart card processor in order to make it perform illegal operations or give faulty results. There is a wild variety of forms these attacks can assume. Under-voltage and over-voltage attacks can be used to disable protection circuits or force processors to perform wrong operations. Power and clock transients can also be used to affect the decoding and execution of individual instructions. By varying the parameters, the CPU can be made to execute a number of completely different wrong instructions. Sometimes it can be fairly simple to conduct a systematic search. For example, low voltage can facilitate other attacks too: one card has at least an on-board analogue random number generator, used to manufacture cryptographic keys and random numbers, which will produce an output of almost all 1's when the supply voltage is lowered slightly.

Every transistor and its connection paths act like an RC element with a characteristic time delay; the maximum usable clock frequency of a processor is determined by the maximum delay among its elements. If an attacker applies a clock pulse shorter than normal or a rapid transient in supply voltage, this can affect only some transistors in the chip. By varying the parameters, the CPU can be made to execute a number of completely different wrong instructions, sometimes including instructions that are not even supported by the microcode. Thus, this technique allows corrupting data values as they are transferred between the registers and the memory.

A possible countermeasure would be to remove completely the clock, transforming the smart card processors in self-timed asynchronous circuits. Then, the external clock will be used as a reference only for communication. In this case, clock variations will just cause data corruption. Another solution assumes that the security processors have sensors that cause a reset when voltage or other environmental conditions go out of predefined range. Unfortunately, the generation of the environmental alarms will cause some degradation in terms of robustness.

4.4.2.1 Differential Fault Analysis (DFA)

The differential fault analysis is a powerful attack on cryptosystems that can implemented in smart cards, provided that the device can be made to deliver erroneous output under stress (such as heat, vibration, pressure, and radiation). Then a cryptanalyst comparing correct and erroneous outputs has a dangerous entry point. An example of DFA attacks targeted DES and broke it with 200 cyphertexts in which one-bit errors has been introduced in the literature.

If a smart card computes an RSA signature, say S, on a message m, modulo $n = p \times q$ by computing it modulo p and q separately and then combining them using the Chinese Remainder Theorem, and if an error can be induced in either of the former computations (by variably reducing the clock period, for example), then one can factor n at once. If e is the public exponent and the signature $S = M^d$ (*mod n*) is correct modulo p but incorrect modulo q, then we should have

$$p = gcd(n; S^e - M).$$

The obvious defense against DFA is to add error-checking to the cryptographic device. If it never delivers erroneous output, DFA becomes impossible. If erroneous output is made rare enough, DFA becomes impractical. There are a number of techniques proposing error checking. A first protection would be to perform the encryption several times, compare the results, and reject them if they were not equal. The goal is to find methods more efficient than that. Two other mechanisms have been proposed; they are the parity checks and modular arithmetic checks. Parity-checking can protect any cryptographic operation that amounts to a permutation of bits. This covers several operations that are subsets of full bitwise permutations. Parity-checking, then, can protect any cryptographic operation that consists of concatenating bits, or the inverse, splitting a large object up into parts. Parity-checking can also protect the XOR operation that is so common in cryptography since the bit parity of the result of an XOR is the XOR of the parities of the inputs.

On the other hand, any arithmetic operation on integers can be checked by performing the same operation modulo any convenient base. Using a larger base or checking against more than one base improves the odds of catching an error.

4.4.2.2 Data Remanence

Smart cards suffer from data remanence problems. Recent results have shown various remanence problems occurring with the use of EEPROM and Flash memories. Remanence can be observed, for example, with a flash RAM that leaves copies of sensitive data around in mapped-out memory to avoid unnecessary and time consuming block-erase commands. To prevent attacks based on data remanence, several rules should be observed including (a) the cryptographic variables should not be stored in RAM for long periods of time; (b) the EEPROM/Flash cells should be "cycled" several times with random data before writing anything sensitive to them; and (c) the increasing density of semi-conductor memory and the special techniques such as multilevel storage should be utilized.

4.5 Security of Log Files in Smart Cards

Smart cards are becoming an essential security instrument in mobile communications in the light of the explosive increase of electronic services. Since they are engaged in the security sensitive applications, a number of protection mechanisms are required. These mechanisms can be provided at the hardware level (as part of the physical security of the chip), at the software level (as part of operating system), and at the communication architecture level (as part of communication protocol). It is then important that some mechanisms be available to recover the smart card application to a safe state, logging certain events in order to provide as whether the security of the smart card or an application has been compromised, or identify when an application has used the functionality of another application involved with the smart card. Therefore, smart cards need to log securely the relevant events internally in dedicated log files in the smart card.

Log files can be used for different purposes including the following tasks that are of utter importance to mobile communications:

■ Store securely critical events to enable mobile users to be held accountable for their security related operations.
■ Detect security breaches after they have occurred.
■ Recover the system involving the smart card to a safe state after failure occurrence.

The management of smart card log files has received little attention due to two facts. First, the smart cards were confined mainly to single-application architectures. Second, the technology used in the traditional smart cards was limited. However, the recent advances in smart card systems suggest that log file mechanisms could be set up to enhance the overall smart card security, particularly when smart cards are getting increased processing power, larger memory, and advance operating system features. In all cases, the access to the smart card log files should be controlled to make sure that they are not overused or that the log file information is not disclosed to unauthorized entities.

Four types of event files (EF) are possible in a smart card's filesystem:

1. *Cyclic EF*: these files implement a circular buffer where the atomic unit of manipulation is the record. A record is a group of bytes that have a known coding: every record of the same file represents the same kind of information.
2. *Linear-fixed EF*: the atomic unit for these files is the record, instead of the byte. In a linear-fixed EF, all the records have the same length.
3. *Linear-variable EF*: these files are similar to the linear-fixed EFs, but the length of a linear-variable EF may vary from one record to another.
4. *Transparent EF*: these files are organized as a sequence of bytes. The file system allows reading all or only a subset of their contents by specifying an interval.

However, SIM cards do not allow linear-variable EFs, implementing only transparent, linear-fixed, and cyclic EFs. On the other hand, smart card APIs allow client applications to communicate with the smart card resident applications. Until several years ago, there were no card reader independent APIs and a developer, needing to communicate with a smart card application via a smart card reader, had to obtain the driver specific to the smart card reader and integrate it in the client application.

Recently, the personal computer smart card (PC/SC) standards (and other systems) were developed to allow interoperability of smart card reader and cards in a PC environment and serve a mechanism that permits multiple applications to share access to a single smart card. The major component of the PC/SC architecture is the smart card resource manager (Markantonakis, 1999). It is responsible for controlling all accesses to the smart card relevant resources in the system and tracking them. It is typically provided as a component of the operating system. A second important device is the so-called interface device component (IFD), which is the actual interface between the smart card and the outside environment.

As part of an operating system, the resource manager has a good protection; however, this component should hold a database for every IFD and smart card information related to their availability statuses. This makes the resource manager subject to different attacks such as the denial of service, since the databases can be modified. To add on security, the notion of exclusive access to a smart card is provided in the PC/SC. This property is important since it forces the uninterrupted execution of a sequence of smart card operations, provided that certain conditions have been defined.

4.5.1 Modeling the Log File Manager in Smart Cards

Three models can be distinguished to handle the behavior and security of log files for events related to the smart card activities, namely the threat model, the event model, and the entity model.

4.5.1.1 The Threat Model

This model describes the assumptions made about the security threats targeting the smart card and its operational environment. Threats considered by this model can affect the integrity and confidentiality of the log file content and operation, but do not consider the operational environment of the smart card. The assumptions include the following list (Markantonakis, 1999a):

- The smart card holder can attempt to modify or delete the log file information. He may want prevent the logging of certain events.
- The smart card's data can be attacked while stored in the smart card or while they are transmitted (for verification purposes to the verifier, for example).

- The smart card is a physically secure device with appropriate tamper-resistance mechanisms.
- A smart card application may represent a threat to the log files. Rogue applications, for example, can perform attacks to modify data or the code of card applications.
- The verifier is assumed to be trusted. It is in charge of ensuring that the log files are properly received. It also has to protect the integrity, confidentiality, and availability of the log files.

4.5.1.2 The Event Model

The entities involved with the log file manager and relationships between them are depicted by Figure 4.3, where C represents the smart card, U is the smart card holder, AP is an application running on the smart card, SP is a service provider (via applications), M is an attacker, the ALSS is audit log storage server. This entity serves as a verifier of the log file stored in the smart card, LF is a log file, LFM is the log file manager (the smart card operating system), A is the verifier (or dispute resolver), and PD is the smart card reader/writer used for the smart card.

The selection of events to log in the smart card is particularly difficult to make. This selection takes into consideration different factors including the security policy of the system involving the smart card and the space allocated to the logging. Certain events would be stored in the ROM of the card while some others are stored in the EEPROM.

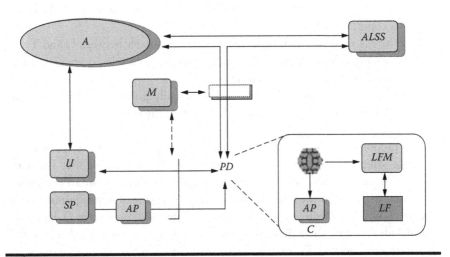

Figure 4.3 **Entities involved in the log file management of a smart card.**

4.5.1.3 The Entity Model

The events are objects logged for the need of applications running on a smart card. Typically, three types of events can be stored in a smart card (Trane, 1996): (a) the value event, which represents the change of a variable value from an old value to a new one. An example of a value event is an event that reports on the variation on a counter (the number of a phone number, for example); (b) the transition event, which stores the difference computed between the old value of a variable and its new value along with the name of the transaction performed on the variable. An example of transaction is the removal of a certain amount of money from a prepaid account for mobile calls; and finally (c) the action event, which contains the names of actions performed on some variables, and eventually their arguments and their results. An example of an action event is an event reporting on the change of a PIN.

4.5.2 Secure Logs in Untrusted Environment

Let us discuss the problem of log file use in a smart card environment. Consider the problem on an untrusted device (say U, for untrusted) generating and maintaining a log file, and a trusted machine (say T) frequently accessing the log file. Device U can be the smart card and T could be a server in a secure location on the mobile network. An approach has been proposed to solve this problem by Shneier (1998). The approach assumes the following facts:

- No security measure can protect the entries added to the log file after an attacker has gained control of device U. A measure can be provided to refuse to an attacker the ability to read, alter, or delete log entries made before U was compromised.
- If there is a reliable, high bandwidth channel available between U and T that is continuously available, then U simply encrypts each log entry as it is created and sends it to T using the channel.
- No encryption method can be used to actually prevent the deletion of log entries and the storage capacity of U is sufficiently large.

The proposed architecture proceeds with four steps. In the first step, U and T exchange some information in order to enable U to create the log file. The main information exchanged is an authentication key A_j that is to generate an encryption key K_j that will encrypt the jth entry of the log file. The second step requires that the jth entry is encrypted with key K_j. The third requires that a message D_j is written that includes the data and a time stamp along a normal close message. The A_j and K_j are irretrievably deleted. Finally, the fourth step considers the verification of the log file. The verification can be made by having T receiving the log file and validating it since it knows all the encryption keys.

The approach proposed by Schneier (1998) cannot apply perfectly to smart cards involved in mobile communication, for different reasons. First, it assumes that the log file is sufficiently large. Second, the owner of U is assumed to be legitimate user. In that case, the owner of U knows the A_0 and it is able to deduce all the keys A_j and K_j and can create a whole false log file.

A similar solution has been provided by Bellare (1997). The method involves generating message authentication codes in such a way that, even when the MAC key is compromised, the attacker cannot forge past log files entries. However, the attacker can delete log entries, but cannot modify stored entries without being detected. The method guarantees the following facts: assume that the time is slotted into periods $E_j = [T_j, T_{j+1}]$, and that the system has been compromised during the jth period E_j, then the attacker cannot forge log entries that have been made before time T_j. This implies that no guarantees can be provided for the log entries produced after T_j. This feature is achieved thanks to the variability of the MAC key from one period of time to another. Indeed, the MAC key K_j used during period j is obtained using a hash function of the key K_{j-1}. Moreover, the key K_{j-1} is immediately deleted after the derivation of K_j. The initial key K_0 can be used by the checker to verify the MAC of all log entries. It can be stated that, when the periods are short, the method avoids remote logging, and log replication.

A third method more appropriate to smart card in mobile communication has been proposed in Markantonakis (1999a). The main entity provided in this method is the log file manager (LFM) since it is the only entity authorized to access the smart card log file. The LFM performs three tasks: it creates and updates the log files in the smart card; it takes control of the log file download procedure; and browses the log files while they are stored. The components of LFM are depicted by Figure 4.4. They are:

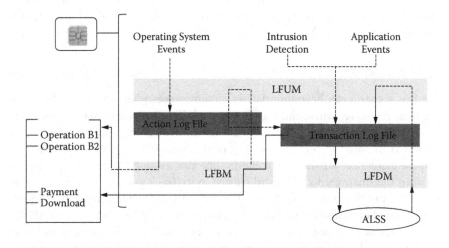

Figure 4.4   Entities authorized to access the log file in a smart card.

- The log file update manager (LFUM): It is responsible for identifying the events to be logged both at the application and the operating system level. It identifies security critical events and updates the log files.
- The log file download manager (LFDM): It is responsible for securely downloading the log files from the smart cards to an audit log storage server, if needed. The LFDM also informs the LFUM about the result of the log file download
- The log file browse manager (LFBM): It is responsible for providing the cardholder with browse functionality to the log files.

A set of general requirements have been established to provide secure transfer of log files. Four major requirements can be selected (Markantonakis, 1999a):

- No log file should be lost during transmission. This imposes that the LFDM and the ALSS should receive the appropriate acknowledgments, before accepting the log files.
- The system should be secure in the sense that it must be very hard for participants, such as the cardholder and the verifier, that are involved to deny the existence and origin of the log files. The privacy and integrity is achieved by the use of encryption mechanisms and message authentication codes.
- The log file chains must be verifiable. Changes to the log entries must, after the log file has been transmitted, be identified.

Various scenarios can be proposed to provide secure transfer of the log files. Three methods can be distinguished depending on the place where the ALSS is implemented. The ALSS can be placed in the cardholder-controlled device; it can be placed in a remote zone, where confidentiality is not a problem; or it can be placed in a remote zone where the confidentiality should be provided (meaning that the ALSS cannot read it).

4.5.2.1 First Scenario

The first scenario can be appropriate in circumstances where the cardholder does not trust the server where the log file is stored. The cardholder may also want to attack the log files by deleting certain entries in them. In such situations, the only interest is the need to provide integrity of the files after they are transferred. To propose a solution for the first scenario, one can assume that a key K is shared between the smart card and a third trusted party (such as entity A in Figure 4.3). The key K can be stored in the card during its issuance or later transmitted securely to the card. Let us also assume that the smart card is able to manage a number N that cannot be modified by external events. The solution can assume that, before transmitting a log file m, the card first authenticates the cardholder (by requesting the PIN, for example). Then, the log file manager can proceed to the transfer of the log file by sending the following object to the server:

$$\langle M_N, MAC_k(M_N) \rangle$$

where $M_N = \langle m, N, E_k(h(M_{N-1})), h(m)) \rangle$, N is the current value of the internal number (taken as a message sequence number), ID_K is the key identifier, E_K is the encryption function using K, and h is a known hash function. Then, the log file manager securely protects the hash value $h(M_N)$ and deletes the stored value $h(M_{N-1})$.

Message M_N creates in its second component a hash chain (assuming that M_0 is known; Haber, 1969). On receiving M_N, the server can acknowledge by sending $h(m)$. On receiving the acknowledgment, the log file manager can perform different verifications including the verification of the hash value correctness. However, some problems can occur. First, the manager can receive an invalid hash value. In that case, it can request the receiver to resend the acknowledgment. It can assume the occurrence of communication problems if invalid replies persist. Second, an intruder can attack the server by gaining access to the log files. Fortunately, he cannot modify it because it is protected.

4.5.2.2 Second Scenario

In the second scenario, the log file manager transmits the log files to a physically secure server. A confidentiality service has to be provided on the connection between the server and the smart card. A solution for this can prevent an attacker monitoring the traffic flowing on that connection from getting the content of the transmitted log files by simply using a secret or a public cryptosystem. A solution based on the use of public keys performs the following two tasks: registration and secure transfer.

During registration, the smart card and the server authenticate each other. Authentication of the server (and actually its selection) can be made using digital certificates and verifying the public key it contains. The authenticated server can then acquire a copy of the card's public verification key, which can be used to verify the signature sent by the log file manager. To achieve authentication, information such as the card's identification, identification of the card's public verification, and the identification of the server need to be exchanged.

The secure transfer involves the encryption of the information exchanged between the log file manager and the server. Assuming that a public key cryptosystem is used, the transfer will be conducted by having the log file manager sending the following message, containing the log file m,

$$\langle E_{K_S^+}(M_N), Sg_{K_C^-}(M_N) \rangle$$

where

$$M_N = \langle m, N, CS_N \rangle, \ CS_N = h\Big(h\big(CS_{N-1}\big), N, m\Big)$$

where $E_{K_S^+}(X)$ is the result of encrypting X using the public key K_S^+ of the server, $Sg_{K_C^-}(X)$ is the signature made on X using the private K_C^- in the card, CS_0 is a predefined string including the card's identification and the server identification, N is the sequence number of the message (internally stored), and h is a hash function. On receiving the aforementioned message, the server can decrypt its first component; checks whether it is a replay and whether it is the destination; and uses the card's public verification key to verify the signature on the second component of the received message. Upon successful verification, the server sends as an acknowledgment back to the log file manager the following message

$$\Big\langle E_{K_C^\pm}\big(X_N\big), Sg_{K_S^-}\big(S_N\big)\Big\rangle$$

where $X_N = \langle N, ID_C, h(CS_N, N)\rangle$. Subsequently, the log file manager can check whether the log file m has been transmitted properly and securely.

4.5.2.3 Third Scenario

A solution for this scenario can be a combination of the solutions provided in the previous scenarios since the server does not have access to the content of the log files and the information exchanged between the card and the server is vulnerable and should be encrypted.

4.5.3 Partitioning Attack

We now show how partitioning attacks can be formulated on any implementation of any algorithm on a smart card in which countermeasures against differential side-channel analysis have not been properly applied. Clearly, the partitioning attack would be highly dependent on the algorithm being implemented and the architecture being used in the smart card. Obviously, it would also require some imaginative work from the attacker as to the types of software countermeasures being used. Therefore, we only describe these at an abstract level.

Assume that the implementation of a particular algorithm verifies the following: the relevant bits or their values thereof in some intermediate cycle, of the algorithm, are not *statistically independent* of the input, output and sensitive information. In that case, we say that the implementation violates the Cardinal Principle (Rao, 2002). An example of implementations that do not satisfy the cardinal principle is the COMP128 occurring in some GSM SIM cards (see Chapter 5). Without loss of generality, assume that the values of the relevant bits at that intermediate cycle depend on some parts of the input and the sensitive information. The statistical distribution of the side channel signal for each of the intermediate cycles for any input can be estimated with reasonable accuracy by a sampling technique. This can be done by repeatedly invoking the operation on the same input and extracting the resulting signals. By performing these estimates on several inputs, the intermediate cycles where the statistical distribution is input-dependent can be localized. However, the estimation of the statistical distribution is likely to be unnecessary in practice; in fact, an estimation of a parameter of the distribution (such as the mean signal) should be sufficient.

Once the input dependent intermediate cycles are identified, a particular attention can be put in the first one among them; the others can be revisited subsequently to the analysis of the first identified cycle. Using the information describing the algorithm being implemented, the characteristics of the device, some reasonable assumptions about the implementation, and the experiments performed with different inputs, it should be possible to isolate a small-valued function of the input that affects this intermediate cycle. It happens that, in some cases, the function could generate a word derived from the input and possibly some secret information during the initial processing step of the algorithm.

A heuristic that can be used to isolate the small valued function would be to vary a few bits of the input while keeping the rest of the bits the same, to see whether the distribution is affected; or it would be to compute correlations between the input bits and the signal to identify how far these input bits are directly manipulated in the computation and whether these bits are correlated at this cycle. If there is no input correlation at this cycle, then one can still estimate how far this cycle is from the other cycles that manipulated inputs. A third heuristic would compute correlations with the bits resulting from the mixing operation rather than the input bits, and the algorithm shows that it performs some form of input mixing operation before performing sensitive operations.

After the small-valued function has been defined, the next step is to perform statistical characterization of the signal distributions for all possible values of the function, using known inputs that affect the function. For example, if the function returns the byte located at a specific place in the input, a statistical characterization can be performed with all possible values for this byte (with other bytes kept unchanged). Then, it should be possible to divide the various values of the function into different partitions based on statistical similarities of the side-channel signals created by these values.

In concluding a general picture of the partitioning attack, one can say that knowing the partitions and the values that fall into each partition should provide useful information such as the processing that has occurred. If this processing involves secret information, then knowing the specification of an algorithm and some features about its implementation, one can derive a hypothesis for the observed partitioning behavior. This hypothesis, together with the actual values that fall into each partition, puts constraints on the sensitive information that was involved in the computation and may imply a leakage of sensitive information.

After being similarly analyzed, each subsequent cycle where a different partitioning behavior is observed would provide an opportunity for information leakage about the sensitive information employed in the algorithm. If enough such cycles can be found and exploited, then little work remains to be performed to disclose completely the sensitive information. Even if this is not the case, it is quite possible that the entropy of the sensitive information could be so reduced that exhaustive search-based attacks becomes feasible (Rao, 2002).

4.6 Forensics and SIM/USIM Cards

In the face of the increase and sophistication of the security incidents targeting communication systems, security specialists have shown a significant interest in a new emerging field of networked computers security, the forensic investigation of security incidents. Defined by the literature as the preservation, identification, extraction, documentation, and interpretation of computer data as evidences (Kruse, 2001), the digital forensic investigation aims to meet a lot of objectives while performing post-incident analysis. The objectives include: evidence collection to carry out digital postmortems, rebuilding of the potential conducted attack scenarios, and proof of hacker malice starting from the collected evidence. However, the complexity of the conducted attacks makes investigation a challenging problem, as attackers try to remove, hide, or alter any sign of suspicious action.

The smart card in the investigation process is receiving a lot of interest. Proven methods for the collection, validation, identification, analysis, and interpretation of smart-based information have been built for the purpose of facilitating the reconstruction of events found to be criminal. The smart card can be accessed by mounting the card in a standard smart-card reader. To access the card, a software implementing the smart card access mechanism is needed. The content of the smart card is organized as a series of files containing binary data that can be downloaded once the user has authenticated himself with a PIN, in general. The best forensic procedure would be to image the entire contents by dumping the entire memory of the smart card and compute a hash value of this memory. There is currently no tool available to do this.

There are, however, tools available to download binary contents of individual files in the SIM cards and store them as individual files. Examples of such tools include the *Sim Manager Pro* (SIMMAN) and SIM-Scan (SIMSCAN). There are also available administrative tools that are able to synchronize data such as text messages between a SIM card and a computer.

4.6.1 SIM Card Investigation

Let first explain why SIM card investigation is valuable and what pieces of information might be extracted from a SIM. For this, one can notice the following important facts:

■ The subscriber of a mobile telephony system essentially wants a means to communicate: this implies an exchange of information (including data) that might be useful for investigation.
■ Every GSM telephone system traces the position of mobile equipments to exchange information over the wireless link. In most cases, the univocal relationship between the user and his mobile equipment is very interesting from an investigator's point of view. Such a relationship highlights a clear difference between the fixed telephone networks, where a terminal identifies only a geographical location (e.g., home and business) but not the users of that terminal, and the GSM.

The SIM stores several types of information, including information about the subscriber, information about acquaintances of the subscriber (by maintaining a list of the numbers they call or they are called from), information about SMS traffic (by storing SMS messages sent and received by the subscriber), *information about subscriber's location* (by storing the last location where the subscriber has been registered by the network), information about calls (by storing the last numbers dialed are stored in a file in the SIM filesystem), information about the provider, and information about the SIM (e.g., the unique ID of the SIM).

Moreover, the evidences that can be derived from the SIM card are stored in the following files:

Phase:	Phase ID 1 byte
SST:	SIM Service table 5 bytes
ICCID:	Serial Number 10 bytes
LP:	Preferred languages variable
SPN:	Service Provider name 17 bytes
MSISDN:	Subscriber phone number variable
AND:	Short Dial Number variable
FDN:	Fixed Numbers variable
LND:	Last Dialed numbers variable
EXT1:	Dialing Extension 1 variable
EXT2:	Dialing Extension 2 variable
GID1:	Groups 1 variable
GID2:	Groups 2 variable
SMS:	Text Messages n * 176 bytes

SMSP:	Text Message parameters variable
SMSS:	Text message status variable
CBMI:	Preferred network messages variable
PUCT:	Charges per unit 5 bytes
ACM:	Charge counter 3 bytes
ACMmax:	Charge limit 3 bytes
HPLMNSP:	HPLMN search period variable
PLMNsel:	PLMN selector variable
FPLMN:	Forbidden PLMNs 12 bytes
CCP:	Capability configuration parameter 14 bytes

All of the stored data can potentially have evidentiary value. However, most of the files refer to network internals that the user never sees, and therefore does not represent evidence on the usage of the mobile telephone as such. We therefore limit the discussion here to the files that typically represent relevant evidence for a particular use. In addition, we focus the reader's interest on the forensic process that can be built on the SIM. This will better show the investigation that can be built on a mobile communication system such as the GSM network.

- **Location information, IMSI, MSISDN:** The LOCI-file byte 5-9 contains among other information the Location Area Identifier (LAI) where the mobile is currently located. This value will be retained in the SIM card when the mobile is shut off. Thus, it is possible for an investigator to determine in which Location Area the mobile was located when it last was operating. The network operator can assist the investigator in identifying which area the identifier corresponds to. It should be noted that a location area can contain hundreds or even thousands of cells. Which cell the mobile was last camping in is not stored in the SIM card.
- **Serial number, IMSI, and MSISDN:** These numbers provide a unique identification of the customer and his equipment. The serial number identifies the SIM itself. The IMSI is the customer identification, whereas the MSISDN is the phone number to the mobile.
- **SMS and SMSP:** The SMS service allows the user to insert a short text message on the phone and send it to another user via a central Short Message Service Centre (SMSC). The SIM provides a storage space for text messages. Typically, a SIM card has 12 slots for text messages. In addition, most of the GSM phones allow the user to store text messages in the mobile equipment. Each SMS slot on the SIM contains a status byte and a TPDU. The TPDU consists of the following elements: the ISDN number of the service center, the ISDN number of the sender (or recipient, depending on status) of the message, the date and time (in seconds) the message was received by the service center, referring to the clock on the service center, the phonebook number, and the message itself.

■ **Short Dial Numbers:** To aid the user in remembering numbers, most phones have an ability to store commonly dialed phone numbers. Most SIM-cards have around 100 slots for storing short dial numbers. On GSM phones older than around 1999 this was the only mechanism for storing numbers. On most modern phones, the phone also has its own memory and the user can choose to use one of the two memories or both.

In the SIM, short numbers are stored in a binary encoded format, containing a name and a number in each slot. Special programs such as Cards4Labs and Sim-Surf Profi are capable of decoding the format. When a short-number is deleted, the information in the slot is overwritten with hex value FF. Thus, it is not possible (or at least not feasible) to recover deleted short dial numbers. The slots will normally be allocated in sequence, so identifying empty slots between used slots will normally indicate that a stored number has been deleted.

■ **Last Numbers Dialed:** The SIM also has the ability to store the numbers last dialed. Most cards have only 5 slots for this. The numbers are stored in a binary encoded format that can be interpreted by programs. Most phones do not use this feature, however, and store a calling log on phone memory instead. Investigators should therefore also investigate the phone for calling logs.

4.6.2 SIM Card Files Investigation

Every file in the SIM card is clearly identified by its *ID*, which acts as the name of the file. No two files in the whole file system can have the same ID. The operations allowed on the file system are coded into a set of commands that an interface between the smart card and mobile equipment can deliver to the smart card, and then waits for responses. The interface acts as the *master* and the SIM card as the *slave*. This is different in so-called *proactive smart cards*, which are capable of issuing commands to the interface. The aforementioned commands, by means of which it is possible to interact with a SIM card's file system, are the following:

SELECT: this command selects a file for use and makes the header of that file available to the IFD;

STATUS: this command has the meaning of a SELECT with MF as argument;

READ BINARY: this command reads a string of bytes from the current EF;

UPDATE BINARY: this command updates a string of bytes in the current EF;

READ RECORD: this command reads one complete record in a record-formatted file;

UPDATE RECORD: this command updates one complete record in a record-formatted file;

SEEK: this command searches the records of a record-formatted file for the first record that starts with the given pattern;

Table 4.1 Access Conditions and Level Coding for SIM Cards

Level	Access condition
0	ALWays
1	CHV1
2	CHV2
3	Reserved for GSM future use
4–14	ADM
15	NEVer

INCREASE: this command adds the value passed as a parameter by the IFD to the last increased/updated record of the current cyclic EF and stores the result in the oldest increased/updated record. It is used for incrementing time or charge information;

GET RESPONSE: this command is when some data is needed to be communicated from the smart card to the card reader.

It is worthy to note that there is no command to remove or create files in the SIM card. In addition, no command to quickly browse the file system is made available. The SIM cards implement many security systems to protect their data. One such security system is the *access conditions*. Access conditions are constraints to the execution of commands that aim at filtering every execution attempt to make only authorized entities to be served, and only for the duration of their authorization. Sixteen access conditions can be distinguished (as depicted in Table 4.1). Table 4.1 shows that every file in the file system has its own specific access conditions for each command applicable on it. Access conditions are organized in levels, but this organization is not hierarchical, in the sense that an authorization applicable for higher levels does not imply immediately for lower levels.

The meaning of these access conditions is summarized as follows:

ALW: the command is always executable on the file;

CHVi: the command is executable on the file only if one among *Card Holder Verification i (CHVi)* code or *Unblock Card Holder Verification i (UNBLOCK CHVi)* code has been successful, for *i=1, 2*;

ADM: allocation of these levels is a responsibility of the administrative authority which has issued the card: the card provider or the telephony provider that gives the card to its subscribers.

NEV: the command is never executable on the file.

References

M. Bellare and B. Yee, Forward Integrity for Secure Audit Logs, Tech. Rep. CSE Dep. Univ. California at San Diego, 1997.

F. Casadei, A. Savoldi, and P. Gubian, Forensics and SIM Cards: An Overview, *Int. J. of Digital Evidence*, 5(1)(2006): 1–21.

S. Chari, C. Jutla, J. R. Rao, and P. Rohatgi, A Cautionary Note Regarding Evaluation of AES Candidates on Smart-Cards, *AES Second Candidate Conference*, Rome, Italy, 22-23.3.1999, available at http://csrc.nist.gov/encryption/aes/round1/conf2/papers/chari.pdf.

J. H. Daniel, D. F. Moore, and J. F. Walker, Focused Ion Beams for Microfabrication, *Engineering Science and Education Journal*, pp. 53–56, April 1998.

S. Haber and W. S. Stornetta, How to Timestamp a Digital Document, *Journal of Cryptology*, 3(2) (1996): 99–111.

ISO 7816 *Identification Cards—Integrated Circuit(s) Cards with Contacts*, parts 1–3 available at http://www.geocities.com/ResearchTriangle/Lab/1578/iso7816.txt, and part 4 available at: http://www.geocities.com/ResearchTriangle/Lab/1578/iso78164.htm.

P. Kocher, Timing Attacks on Implementations of Diffie-Hellman, RSA, DSS, and Other Systems, in *Proc. of Cyrpto 96*, LNCS 1109, Springer, 1996, pp. 104–113.

P. Kocher, J. Jaffe, and B. Jun, Differential Power Analysis, *CRYPTO '99 Proceedings*, Springer–Verlag, pp. 388–397, 1999, http://www.cryptography.com/resources/whitepapers/DPA.pdf.

W. G. Kruse and J. G. Heiser, *Computer Forensics: Incident Response Essentials*, Addison-Wesley Professional, 2001.

C. Markantonakis, Interfacing with Smart Card Applications: The PC/PS and OpenCard Framework, *Elsevier Information Security Technical Report*, 3(2)(1999): 82–89.

C. Markantonakis, Secure Logging Mechanisms for Smart Cards, University of London, Dec. 1999.

T. S. Messerges, E. A. Dabbish, and R. H. Sloan, Investigations of Power Analysis Attacks on Smartcards, USENIX Workshop on Smartcard Technology, Chicago, Illinois, USA, May 10–11, 1999.

J. J. Quisquater and D. Samyde, ElectroMagnetic Analysis (EMA): Measures and Couter-Measures for Smard Cards. In *Smart Card Programming and Security* (E-smart 2001), Cannes, France, LNCS 2140, pp. 200–210, September 2001.

W. Rankl and W. Effing, *Smart Card Handbook*, Wiley, 3rd edition, 2004. Information (available online at: http://www.wrankl.de/ SCH/SCH.html).

J. R. Rao, P. Rohatgi, and H. Scherzer, Partitioning Attacks: Or How to Rapidly Clone Some GSM Cards, in the *Proc. of the 2002 IEEE Symp. On Security and Privacy (S&P'02)*.

B. Shneier and J. Kelsey, Cryptographic Support for Secure Logs on Untrusted Machines, in *Proc. of 7th USENIX Security Symposium*, Usenix Press, Jan. 1998, pp. 53–62.

Sim-Manager Pro, Software package, Commercial product available at: http://www.txsystems.com/.

Sim-Scan, Software package, Freeware, available at: http://users.net.yu/~dejan/.

P. Trane and S. Lecomte, Failure Recovery Using Action Logs for Smart Cards Transaction Based Systems, in *3rd IEEE Int. On-Line Testing Workshop*, 1997.

S. Zanero, SMART CARD CONTENT SECURITY: Defining Tamperproof for Portable Smart Media, Dip. Elettronica e Informazione, Politecnico di Milano.

II

ATTACK AND PROTECTION TECHNIQUES IN MOBILE COMMUNICATION NETWORKS

Chapter 5

Security of GSM Networks

5.1 Introduction

The Global System for Mobile Telecommunication (GSM) has become one of the most popular systems for mobile communication. GSM provides terminal mobility and allows users to roam seamlessly from one GSM network to another. It is characterized by a special feature, the separation of the user identity from the terminal phone equipment. In fact, the subscriber identity is inserted in a *Subscriber Identity Module* (SIM) that can be added to any GSM mobile terminal. The SIM carries sensitive data that are utilized to authenticate the subscriber and provide confidentiality of the exchanged messages. In 1989, the responsibility for GSM development was transferred to the European Telecommunications Standards Institute (ETSI), and the *Phase 1* of the GSM specification was published in 1990. The first commercial GSM service was deployed in 1991 and in 1995, the *Phase 2* of the GSM specification was completed (ETSI, 1977). The first GSM services in the United States started the same year.

Security in GSM is an important issue because the mobile users are likely to transmit sensitive data over a network infrastructure that is not truly secure. The security weaknesses of GSM stem from some trust-related hypotheses made by the developers, including the lack of node authentication and some design flaws in the security protocols. The most important threat for the GSM is, however, linked to the fact that the subscriber may believe that the entire structure is secure and may erroneously trust it to exchange confidential information. Nonetheless, since all wireless networks suffer from multiple exposures posed by the wireless environment, the security and confidentiality in GSM were some of the reasons for which this standard was considered superior to other mobile communication systems. The GSM success has later inspired other systems such as the Digital Enhanced Cordless Telecommunications (DECT) and Code Division Multiple Access (CDMA).

However, some security problems have occurred with the GSM operation. These problems include, but are not limited to, the following: (a) security is obscure, meaning that none of the security algorithms used by GSM is available to the public; (b) the GSM provides only access security, which means that all communications between the user's mobile terminal and the base transceiver are encrypted. However, all communications and signaling messages are generally transmitted in clear text in the GSM network; (c) the cryptographic mechanisms are difficult to upgrade; and (d) the mobile subscriber visibility is missing.

5.2 GSM Mobility Scheme

The Global System for Mobile communication is a set of ETSI standards specifying the infrastructure for digital cellular services. The standard is adopted and deployed by more than 90 countries in multiple locations in Europe, Asia, and Australia.

5.2.1 GSM Infrastructure

A GSM network (as depicted by Figure 5.1) involves nine major components: the *Mobile Station* (MS), the *Subscriber Identity Module* (SIM), the *Base Station Subsystem* (BSS), the *Base Station Controller* (BSC), the *Transcoding Rate and Adaptation Unit* (TRAU), the *Mobile Services Switching Center* (MSC), the *Home Location Register* (HLR), the *Visitor Location Register* (VLR), and the *Equipment Identity Register* (EIR). Together, all these components constitute a Public Land Mobile Network (PLMN). The major functions of these components are described as follows.

5.2.1.1 The Mobile Station (MS)

The MS is carried by the subscriber. It is constituted by the mobile equipment (ME) and a smart card referred to as the Subscriber Identity Module (SIM). The typical ME is the mobile phone. Inserted into the ME, the SIM card allows the subscriber to receive calls at the ME and make calls from that ME. The SIM stores sensitive data that are protected by the subscriber's personal identity number (PIN). Chapter 4 studies in details the structure and usage of the SIM card. Particularly, the SIM card contains the following subscriber related information (ETSI, 509):

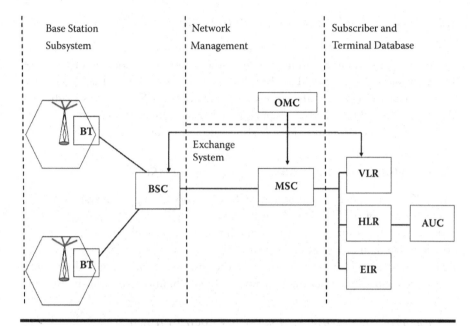

Figure 5.1   The GSM architecture.

- The *International Mobile Subscriber Identity* (IMSI): This number uniquely identifies a subscriber. Its provision is necessary to access the GSM services. IMSI is used by the network for purposes including universal identification and roaming.
- The cryptographic algorithms *A3* and *A8* and a secret *subscriber authentication key* K_u: They provide security functions for authenticating the mobile user via his SIM card, and generating the session keys for confidentiality needs, respectively.
- The temporary network related data: The temporary data mainly include the *Temporary Mobile Subscriber Identity* (TMSI), which is an identifier assigned to the subscriber for a limited interval of time, the *Location Area Identifier* (LAI), and the forbidden *Public Land Mobile Networks* (PLMN).
- The *Card Holder Verification Information* (CHVI): The information authenticates the user to the card and provides protection against the use of stolen cards.

The Personal Identification Number is used to protect the information stored in the SIM card. The management of the PIN is made in a way that would block the use of the SIM after a predefined number of false PIN values have been introduced (typically, this number is equal to 3). A Personal Unblocking Key (PUK), known only by the owner and the network, is made available for the SIM owner to unlock a blocked SIM.

5.2.1.2 The Base Station Subsystem (BSS)

The BSS controls the radio related tasks and provides connectivity between the network and the mobile stations via the radio interface. It consists of the Base Transceiver Station (BTS) and the Base Station Controller (BSC). The BTS sets up the radio transceivers that handle the radio link with the MS and covers a radio cell identified by the BTS. The BSC manages the radio communication and takes care of all the needed control functions. It also controls a set of BTSs.

5.2.1.3 Mobile Services Switching Center (MSC)

The MSC is the main component of the GSM network management system. It controls a large number of BSCs and acts like a switching node. It also provides all the management functions for terminal mobility including registration, authentication, location, handover, and call routing. Similar to a digital telephone exchange, a router, or a switch, it is responsible for the routing of incoming and outgoing calls. In addition, it handles the assignment of user channels on the air-interface.

5.2.1.4 The Operation and Support System (OMC)

The OMC is connected to all equipments in the switching system and to the BSC, as well. The purpose of the OMC is to offer the customer cost-effective support for centralized, regional, and local operational and maintenance activities that are required for a GSM network. It also provides a network overview and supports the operational maintenance activities.

5.2.1.5 Home Location Register (HLR)

The HLR is a database that stores and manages the mobile subscriber specific parameters (or administrative information) of a large number of registered subscribers along with their current location. The parameter values stored for a subscriber are permanent. The most important parameter of a subscriber stored in the HLR is the shared authentication key Ku and the IMSI. Every user is assigned to a unique HLR. A unique HLR is usually assigned to a PLMN. The HLR plays an important role in various tasks such as the roaming of mobiles to foreign networks.

5.2.1.6 Visitor Location Register (VLR)

The VLR component is a database designed to off-load the HLR of user database related functions. Like the HLR, the VLR contains subscriber information, with the difference that it relates only to the subscribers who roam in the area assigned to the VLR. When a subscriber roams away from his/her own network, information is forwarded from the subscriber's HLR to the VLR of the serving network, in order to perform the authentication process. Typically, when a subscriber moves out of a VLR area, the HLR takes care of the relocation of the subscriber information from the old to the new VLR. Notice that a VLR may be associated with several MSCs; but, a MSC is always assigned to only one VLR.

5.2.1.7 Authentication Center (AuC)

The AuC contains a database that is used to store the identification and authentication information related to each subscriber. Typically, the AuC is an important part of the HLR. The attributes in this database include the subscriber's IMSI, secret key K, LAI, and TMSI. The AuC is responsible for generating triplets of values consisting of a random field, called RAND, an assigned response (denoted by and SRES), and session key K, which are stored in the HLR for each subscriber and a call made by the subscriber.

5.2.1.8 Equipment Identity Register (EIR)

Since the subscriber identity and the mobile equipment (ME) are processed independently by the GSM system, it is possible to operate any GSM ME with any valid SIM card. This makes cellular terminal theft an attractive task for hackers. To protect against thefts, the Equipment Identity Register (EIR) was introduced in the GSM system. Every GSM terminal has an internationally unique identifier, called the International Mobile Station Equipment Identity (IMEI), which cannot be altered without destroying the terminal. IMEI contains a serial number and a type identifier. The EIR is a repository that maintains three lists: the white list, black list, and grey list. The white list contains all number series of equipment identities that are permitted for communication. The Black list contains all equipment identities that need to be disqualified. Mobile equipments appearing in the grey list are not disqualified (unless they are on the black list or out of the white list), but are tracked by the network for specific purposes.

5.2.2 Mobility Management

The mobility management is concerned with the functions of tracking the location of roaming mobile subscribers, registering the location information at the appropriate components, and handling connection handoffs for mobile users during communication. Typically, the handoff is defined as the transfer of a cellular phone transmission from one radio frequency within a cell to another radio frequency in the same cell or an adjacent cell. Handoffs occur when a GSM terminal passes out of the range that the serving cell can handle. In that case the signal is passed from one base station to the next. The handoff is transparent to the user and typically will not result in a noticeable loss of service. The transition process and the process required to make the move are both referred to as the *handoff*.

The major features of these functions are detailed in the following:

■ *Connection handoffs*: A handoff can be performed between two channels in the same cell, between channels in different cells that are located under the coverage of the same BSC, or between cells that are covered by different BSCs (these handoffs are called external handoffs). While the BSC can manage the handoffs under its coverage, the MSC is necessarily involved in managing external handoffs. When a BSS indicates that an external handoff is required, the decision of when and whether the handoff should occur is then taken by the MSC, which uses the signal quality measurement information reported by the mobile terminal to make the decision. The information is pre-processed at the BSS. The original MSC handling the call keeps control of this call in an external handoff to a different MSC. Moreover, when a BSS performs an internal handoff, it informs the MSC about the completion of the handover. The need for a connection handoff is indicated by the mobile station or by the BSS as it keeps monitoring the quality of the signal received. The MSC may also have to initiate a connection handoff for traffic reasons or availability of resources in an attempt to balance out the traffic load in the network.

■ *Location management*: Location information is maintained and used by the network to locate the mobile user for call routing purposes. The network registers the location of a mobile user in the HLR to which the user is subscribed. Each BSS keeps broadcasting on a periodic manner the cell identities on the broadcast control channels of the cells under its coverage. The mobile users within each cell keep monitoring such information. As changes in location are detected, with respect to the last information they have stored, they report the new location to the BSS, which sends it to the VLR of the MSC to which it is connected. The VLR, in turn, sends the location information to the user's HLR, where it is also stored. In the same period of time, the HLR directs the old VLR to delete the old location of the mobile from its database and transmits a copy of the user's service profile to the new VLR. Location updating is performed by the mobility management protocol.

■ *Call routing management*: A call may be initiated by a mobile user to another mobile or fixed-network user. To route a call to a mobile user, the network signaling needs to locate the mobile. The HLR is questioned for the routing information required to extend the call to the visiting MSC. The visiting MSC (or more specifically the VLR within the new MSC) is identified in the mobile's HLR by the Mobile Station Roaming Number (MSRN), which is defined as the telephone number used to route telephone calls in a mobile network from a GMSC (Gateway Mobile Switching Centre) to the target MSC. The MSRN is sent to the HLR on location updating or call initiation. The VLR then initiates the paging procedure and the MSC pages the called mobile station with a paging broadcasted to all BSSs of the location area, as the BTS of the mobile may not be known. The VLR may also assign the mobile a new TMSI for the session. This allows the identification of the exact BSS resulting in connection establishment and user authentication. The BSS may then assign a traffic channel to that session.

5.2.3 Protocol Architecture

The GSM protocol architecture is depicted in Figure 5.2. The protocol layering consists of three layers: the physical layer, the data link layer, and the routing layer. The message layer (or Layer 3) protocols are used for resource management, mobility, and code related management messages between the entities involved.

5.2.3.1 Physical Layer

All the techniques and mechanisms used to make communications possible on the mobile radio channel with some measure of reliability between a mobile and its base station represent the physical layer or the Layer 1 procedures. These mechanisms consist of modulation, coding, timing, power control, and other details that control the establishment and maintenance of the channel. Basically, the physical layer contains the following mechanisms (Redl, 1995):

■ *The frequency-division multiple access and time-division multiple access*: The GSM uses Time-division multiple access (TDMA) on top of Frequency-division multiple access (FDMA) to allow users to access the radio resources in a GSM cell. With FDMA, the users are allocated a channel among a limited

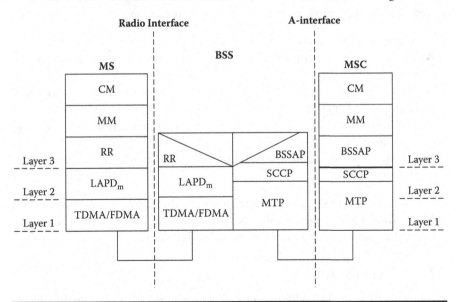

Figure 5.2 The GSM protocol architecture.

number of channels ordered in the frequency domain. In TDMA, users share a physical channel where they are assigned time slots. GSM subscribers are placed onto a physical channel with respect to simple FDMA techniques. Then, the channel is divided up (in time) into frames, during which eight different users share the channel. A GSM time slot lasts 577 µs and each user is allowed to utilize the channel for a slot every 4.615 ms (meaning that 577 µs × 8 slots = 4.615 ms).

■ *Radio channel*: The radio channel is represented by a pair of radio frequencies that can be utilized by two entities to communicate with each other. Channel coding is applied to the channel in order to minimize some damaging effects on the quality of the transmitted signal. The frequencies used by the GSM network are defined in the FDMA part of the physical layer. GSM is allocated three different frequency bands: the 900 MHz, 1800 MHz, and 1900 MHz bands. The frequency bands employed within each of the three ranges are similarly organized. Thus, we only give some details for the frequency usage in the 900 MHz band. In that range, two 25-MHz frequency bands are used. The mobile station transmits in the 890- to 915-MHz range and the base station transmits in the 935- to 960-MHz band.

5.2.3.2 Data Link Layer

The data-link layer is responsible for the correct and complete transfer of information blocks between the Layer 3 entities over the GSM radio interface. Layer 2 constructs the envelopes that are responsible for encapsulating the data to be transmitted. The protocol implements the following basic functions:

■ The organization of the information issued by Layer 3 into cells (or frames).
■ The peer-to-peer transformation of signaling data using well-defined frame formats.
■ The establishment, the supervision, and the termination of one or more data links on the signaling channels.
■ The acknowledgment of transmission and reception of numbered information frames.
■ The non-acknowledgment of unnumbered information frame transmission and reception.

The BTS passes the signaling messages between the mobile station and the BSC or MSC. The BTS infrequently takes part in the conversations except when it has to respond to commands for adjustments in its operations.

5.2.3.3 Message Layer

The network layer in the GSM architecture, also referred to as the signaling layer, uses a protocol that contains all the functions necessary to establish, maintain, and terminate mobile connections for the services offered within a GSM PLMN. The network layer also provides the control functions to support additional services such as the short message services. There are well-defined procedures and data structures for the Layer 3 protocol, which contains three sub-layers defined as follows:

■ Radio resource management (RR): The RR sub-layer is responsible for the management of the frequency spectrum, the GSM system's reaction to the changing radio environment, and every task related to maintaining a clear channel between the GSM network and the mobile station. The RR responsibilities include channel assignment, power-level control, time alignment, and handover from one radio cell to another. The RR sub-layer handles all the procedures necessary to establish, maintain, and release dedicated radio connections.

■ Mobility management (MM): The MM sub-layer is in charge of coping with the tasks of handling mobile users that are not directly related to radio functions. These tasks include all the operations that a fixed network would perform to connect a user to it, while taking into consideration the fact that the mobile user may roam to another place during connection. They include (a) the location update procedure, which is in charge of storing and tracking the mobile user location; (b) the mobile identification procedure, or more precisely the mobile authentication; (c) the IMSI attach procedure, which is nothing but the location update procedure after power on; (d) the IMSI detach procedure, which tells the GSM network, just after mobile powers off, that the mobile station is no longer in service; and (e) the TMSI reallocation procedure.

■ Connection management (CM): The CM sub-layer manages all the functions essential for circuit-switched call control in the GSM. While these functions are provided by the call control entity within the CM sub-layer, other entities within the CM sub-layer are responsible for the provision of supplementary services such as the SMS. The call control responsibilities are almost identical to those provided in a fixed communication network. Thus, the CM sub-layer is virtually unaware of the mobility of the users. Specific procedures defined for this purpose include but are not limited to (a) the call establishment procedures for *mobile-originated calls*; (b) the call termination procedures for *mobile-terminated calls*; (c) the procedure handling the changes of transmission mode during ongoing calls; and (d) the call reestablishment after interruption.

5.3 GSM Security Model

The GSM provides authentication of users and encryption of traffic across the A-interface. These services are provided in response to several requirements established among the major objectives of the GSM systems. In this section, we first discuss some of these requirements. Then we explain how GSM provides a solution that attempts to satisfy the requirements.

5.3.1 Security Requirements

GSM, like many other cellular networks that serve a large number of users, contains many valuable assets that may constitute serious vulnerability sources and need protection against misuse and malicious attacks. Two classes of requirements are valuable for GSM: the requirements for mobile user's privacy and the requirements for data integrity protection. A subscriber to a GSM network necessitates protection in the following activities: call setup, voice-based services protection, privacy of location, privacy of calling patterns, privacy of user identity, and protection of data (Vijaya, 865). In the following, we explain how these requirements occur.

5.3.1.1 Protection of Call-Setup Information and Communication Services

During the call-setup process, the mobile terminal transmits important call-setup information to the GSM network. This information contains the calling party number, the calling card number, and the service type requested. This information must be protected and secured against eavesdroppers. In addition, all communication services (including spoken communication) must be properly encrypted by the cryptographic system, when requested, so that it cannot be intercepted by any malicious user listening to the radio interface or other interfaces of the system.

5.3.1.2 Privacy of User-Location, Calling Patterns, and User-Data

Any out-leaking of signaling information on the GSM network may enable a hacker to approximately locate the position of a subscriber and reduce the subscriber's privacy. Information related to traffic generated by a particular user and his/her calling patterns (such as the caller-id) should not be made available to attackers. Therefore, measures should be taken to protect the mobile subscriber from attacks against his/her privacy of location, to keep calling patterns inaccessible to eavesdroppers, and to protect subscriber identification information against hackers. In addition to securing the transmitted data, there must be a provision in the network and the terminal to check whether the data it receives has been altered. This property is traditionally called *Data Integrity*.

On the other hand, theft of services and equipments is a significant problem in GSM networks. The network subsystem does not worry whether a call has been originated from a legitimate or a stolen mobile terminal as long as its bills get paid from the correct account. There are two kinds of theft that could be performed, the theft of personal equipment and the theft of the services offered by the service provider. Cryptographic protection measures must be designed to make the reuse of stolen terminals as difficult as possible. Further, it should block theft of services made possible by techniques such as cloning.

5.3.1.3 Replication and Clone Resistant Design

Replication is a damaging attack in GSM communication systems. Cloning refers to the ability of an intruder to determine information about a personal mobile terminal and duplicate it, meaning that the intruder creates a duplicate copy of it using the collected information. This kind of fraud can be easily accomplished by legitimate network users, since they have all the information they need to clone their own personal information stored in the SIM in the terminal. Doing so, multiple users can use one account by cloning personal equipment. It also could be done by an external user who desires to get benefit of services on the expense of legitimate users. The cryptographic protection for the GSM network must include some tools for clone-resistance. Security must be provided for the radio-interface, the network databases, and the network interconnections so that personal equipment information can be kept secure. Since the mobile terminal can be utilized by anyone, it is necessary to identify the right person for billing purposes.

5.3.1.4 Equipment Identifiers

In GSM systems, where the account information is logically and physically separated from the terminal, stealing personal equipments could be an attractive and lucrative business for attackers. To avoid such a threat, personal equipments must have unique (worldwide) identification information that reduces the potential of stolen equipment to be re-used. The identifiers may take the form of tamper-resistant values that are permanently integrated into the mobile terminals.

5.3.2 GSM Security Model

Authentication and confidentiality are achieved by giving the mobile user and the network a shared secret, which is a 128-bit value K_u, stored in the SIM card. This key is not directly accessible to the user. Each time the user connects to the network, the network authenticates the user by sending a random number RANDG to the MS. The SIM card then uses the random number and the secret key to run an authentication algorithm, provide a value SRES, and create a session key K_s. The user is authenticated if the provided SRES matches the value of SRES that is computed separately by the GSM network using the same parameters and algorithms. The encryption is performed using key K_s. The mechanisms used in GSM networks to provide anonymity, authentication, and confidentiality to the mobile users are detailed in the following.

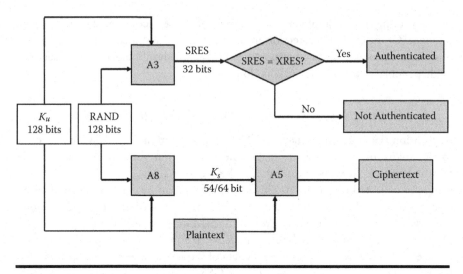

Figure 5.3 GSM security model.

The security model aims at providing a solution that satisfies, to a certain extent, the aforementioned security requirements. Security services provided by GSM include anonymity, user authentication, and confidentiality. Unfortunately, this model does not provide any security features such as network authentication. The attack discussed in the following sections show to what extent some security services are missing. Figure 5.3 depicts the security model used by GSM.

5.3.2.1 Anonymity

Limited anonymity is provided in GSM by the use of temporary identifiers. When a mobile user powers on his/her mobile terminal, the real identity (IMSI) is used to identify the MS to the network and then a temporary identifier Temporary Mobile Subscriber Identity (TMSI) is allocated as a temporary local identifier of the MS to the network in future sessions. The TMSI has significance only within a given location area. According to the ETSI specification, the network always encrypts TMSI before sending it to the MS.

A location update request is generated in the mobile station receiving a TMSI. Outside the location area where the user is served, it has to be combined with the local area identifier to provide for a clear identity. Usually the TMSI reallocation is performed at least at each change of a location area, as the location update request is issued by the MS to the network. From that moment until the mobile powers off or moves to another location, the temporary identifier is used. Nevertheless, it appears that it is possible to determine the temporary identifier being used by a mobile user by tracking the user (Brookson, 1994).

5.3.2.2 *Authentication*

Since any malicious user is able to overhear a communication on the radio medium, authentication, which aims to prove that users are who they claim to be, is an essential element of a mobile network. Authentication involves two functional components, the SIM card in the mobile and the Authentication Center (AuC). One of the most important security functions of the SIM is to authenticate the subscriber to the network. This process guarantees that the MS requesting service is a legitimate subscriber and not an intruder or a cloned user. The GSM network verifies the identity of the subscriber through a *challenge-response* process. When a MS requests a service, the network randomly generates the real number RAND and challenges the MS by sending it RAND. The MS should answer correctly the challenge before being granted access by sending back the expected value of SRES.

The challenge RAND sent by the GSM network to the MS consists of a 128-bit number that is randomly generated in an arbitrary manner, so that it has a quasi-null chance of being repeated, since otherwise, an attacker can easily build a file of (RAND, SRES) pairs and use the collected information to gain illegal access to services. When the MS receives RAND, it passes it to the SIM for processing. The SIM takes the RAND value and the secret 128-bit key K_u and produces a 32-bit response (or SRES). The response is transferred out of the SIM into the terminal, where it is relayed to the network. This is the MS's response to the network's challenge. Meanwhile, the network (or precisely the AuC) has performed the same set of operations (Figure 5.3) and sent the appropriate SRES to the BSS covering the MS. Using the same value of RAND and an identical copy of K_u, the GSM network has computed its own response to the challenge.

When the network receives SRES from the MS, it compares it to its own SRES. If the two values are identical, the network can suppose that the MS is legitimate. Then it can decide to allow the provision of the requested service. If the two values are not equal, the GSM network assumes that the SIM does not have the appropriate secret key K_u. Hence, it decides to refuse the provision of service to the MS. Consequently, an attacker collecting and storing the SRES responses will not be able to successfully reuse any of them because the RAND value changes with every access attempt (or almost every access). Even if a particular RAND challenge happens to be reused (allowing an attacker to impersonate a legitimate subscriber of the GSM network), the connection to the service cannot last a long time since the GSM network has the ability to do again the authentication process of the MS (using another value for RAND) and reiterate it as often as it wants. Thus, the next challenge, received by the MS (or by the SIM it contains) from the GSM network, would be a new RAND value for which the attacker would not probably have the right SRES. In addition, it should be noticed that a key feature of the GSM security solution is based on the fact that a subscriber's key K_u is kept secret and is never transmitted over the GSM network. It is kept stored in the SIM card and the AuC. To calculate SRES, an algorithm called A3 is used. It utilizes a hash function built using a function called COMP128, which is also used in the generation of a session key needed for confidentiality of calls and exchanged data. COMP128 algorithm was designed to be a reference model for GSM implementation, but for a variety of reasons it has been adopted by almost all the GSM providers. Unfortunately, some weaknesses have been discovered for COMP128. After it was cracked (in April 1998), new stronger versions, called *COMP128-2* and *COMP128-3*, have been developed. Nonetheless, the breakage did not convince the GSM operators to adopt one of the new versions; instead, they kept the old failing version of COMP218. One among the reasons invoked by the operators is the large amount of cost involved in replacing COMP128.

5.3.2.3 Confidentiality

In addition to the information needed for the authentication of subscribers, the SIM card also provides the information needed to encrypt the radio connection between the MS and the covering BTS. More specifically, an algorithm called A8 is used to generate a session key K_s with each accepted connection. Key K_s is utilized for voice and data encryption before transmission on the radio link. The algorithm used for computing the 64-bit K_s is invoked according to Figure 5.3.

On the other hand, since the GSM network uses the time division technique to share the radio channel with up to eight other users, each user takes its turn using the common radio channel, sending and receiving information only during one of the eight available time slots in every frame. A GSM conversation uses two frames, one going from the base station to the MS (i.e., on the downlink) and the other going from the MS back to the base station (uplink). Each of these frames contains 114 bits of user information, which is often digitized and compressed speech. Therefore, every 4.615 milliseconds the MS receives 114 bits of information from the base station and transmits another 114 bits to the base station. These 228 bits require encryption to protect it from hackers.

Using RAND and secret key K_u, the SIM runs the A8 algorithm to produce the 64-bit long session key called K_s. K_s is transferred out of the SIM and into the MS, where it is used by a third algorithm called A5. Algorithm A5 uses K_s and the current publicly known frame number to produce a key stream of 228 bits, decomposed into two halves. While the first half encrypts the downlink frame (*dl*), the second half is used to encrypt the uplink frame (*ul*). For each new frame to be transferred, a new 228-bit key stream is produced by the algorithm A5 to encrypt (and decrypt) the frame. The algorithm A5 lives in the hardware part of the terminal, and never in the SIM card. It has to operate quickly and continuously to generate a fresh set of 228 bits every 4.615 milliseconds. In addition, because GSM terminals are designed to operate in different networks, the A5 algorithm should be common to all GSM networks. Presently, there are at least two known versions for algorithm A5. The first version, called *A5/1*, provides the strongest level of encryption through the radio link. Although officially using 64 bit keys, the keys are no more than 54 bits long, since the last ten bits are forced to be equal to 0. The second algorithm, denoted by *A5/2*, is considered to be much weaker than *A5/1*. It is designed for areas where there is a benefit in achieving an uncomplicated cracking of encrypted conversations.

5.4 Basic Attacks on GSM

Various flaws have been noticed in GSM networks, making possible the construction of damaging attacks. This section describes the major flaws and discusses some among the well-known attacks that can be performed on GSM systems targeting security services such as the anonymity, authentication, and confidentiality of the GSM.

5.4.1 GSM Security Flaws

Among the major flaws, one can describe the following (Quirke, 2004):

■ No authentication of the network is provided to the user: The authentication procedure described in the previous sections does not require the network to prove its knowledge of the user key K_u. Thus, it is possible for an attacker to set up a false base station (or RBS, for rogue BS) with the same mobile network code as the subscriber's network. Since the authentication procedure initiation is under the control of BS, the RBS may choose to simply send the RAND and ignore the response. It does not have to activate ciphering, either. The attacker can set the cell reselection parameters of his false base station to values that will highly encourage the MS victims to camp on it. The subscriber can then unknowingly be making calls or sending text messages that could be intercepted.

■ Common implementation of A3/A8 is flawed: The most common implementations of the A3 and A8 algorithms use the procedure COMP128, which generates the 64-bit Ks and the 32-bit SRES from the 128-bit RAND and the K_u input. This algorithm is seriously flawed, in the sense that some chosen values for the input RAND will provide enough information to determine the key K_u in significantly less than the large number of attempts required by a brute force (of the order of 2^{128}).

   Another flaw noticed with the procedure COMP128 is characterized by the fact that, when the 64-bit key session is generated, the algorithm sets its least significant 10 bits to 0. This reduces the strength of the data encryption algorithm. Unfortunately, this flaw is also present with the COMP128-2. In addition, the earlier attacks based could typically crack a SIM in approximately 2^{17} values for RAND.

■ Vulnerabilities in the subscriber identity confidentiality mechanism: To avoid mobile terminals being addressed or identifying themselves in plaintext by their IMSI and prevent an eavesdropper listening in on the initial plaintext stage of the radio communication learning that a particular subscriber is in the area, the GSM system has provided the TMSI and maintains a database in the VLR mapping TMSIs to IMSIs. If the network somehow loses track of a particular TMSI, it must ask the subscriber to submit its IMSI over the radio link, using a special mechanism for identity request. Thus, the IMSI is sent in plaintext. Combined with the aforementioned flaw (stating that the network does not authenticate itself to a user), an attacker can use this to map a TMSI to its IMSI.

■ Over the air cracking of K_u: Combined together, the aforementioned flaws can result in a serious attack. The main steps of this attack can be performed as follows:

  – The attacker can imitate a valid base station with the same mobile network code as the customer's network.
  – The attacker attempts to establish a radio connection to a mobile station MS using its TMSI.
  – Once the connection is established, the attacker gets the MS's IMSI by sending to the MS an identity request, to which the terminal must respond.
  – The attacker keeps choosing various values for RAND and submitting them to the MS via the AUTHENTICATION REQUEST messages. The MS simply returns the expected SRESs.
  – The attacker collects the (RAND, SRES) pairs until he gains enough information to derive the key K_u.

5.4.2 Impersonation Attacks

In impersonation attacks, the attacker is willing to impersonate the network with respect to the MS, impersonating the MS with respect to the network, or combining both operations to perform a *man-in-the-middle* attack. A malicious adversary impersonating one of the two entities is able to perform a large spectrum of illegitimate actions, including (a) listening to private traffic; (b) modifying, deleting, re-ordering, or replaying messages; and (c) spoofing and behaving as a repeater relaying signaling and user data between the two communicating parties. The required equipments to achieve a man-in-the-middle attack are made of a modified BTS in conjunction with a modified MS. The modified BTS impersonates the network to the MS, while the modified MS impersonates the MS to the network. The term rogue base station (RBTS) will be used to refer to the modified, while BTS will mainly refer to a legitimate base station.

A legitimate BS continuously broadcasts replica bursts on the base channel to make it possible for MSes to find the serving radio channel. When the channel is found, all the information needed about the network to request services will be gained from this channel. Recall that this information includes cell identity, network identity, control channel structure, list of channels in use, and details of the access protocol. An attacker equipped with a RBTS placed between a MS and a legitimate BTS, providing higher power levels than the BTS, is able to make the MS use the RBTS's channel and have total control over which system information the MS gets and which messages will reach the BS from this MS. In this case, the MS is referred to as captured MS. The captured MS identity will then be used to provide fabricated messages on behalf of the MS. Therefore, capturing a MS not only gives the attacker the ability to impersonate the MS owner, but also induces a denial of service on the MS. The captured MSes have no contact with the network and are therefore unable to get services. The attack is easy to perform; it is also easy to detect. A denial of service attack does not, however, require that the attacker use a RBTS; the attacker only needs to jam the radio signals.

5.4.3 Attacks against Anonymity

In addition to the attack used to impersonate subscribers using the identity request message, one can build attacks against user anonymity. An attacker may want to track some subscriber's movements and/or calling patterns and thus needs to know the IMSI or the TMSI(s) of the MS. This information, if compromised, may also be used to launch attacks and impersonate individuals. If the attacker can get the IMSI of a subscriber or associate a TMSI currently being used in the cell with a specific IMSI, then the anonymity of the user of the system owning that IMSI is compromised. The compromised anonymity leaves the door open for the attacker to perform traffic analysis, i.e., to observe the time, rate, length, sources, or destinations of messages on the radio interface or other system interfaces in the network. Attacks on the anonymity of mobile users of a GSM network can be made through passive or active monitoring, as described below.

5.4.3.1 Passive Monitoring

Every time a mobile station is turned on, an IMSI attach operation is performed by the MS to indicate that the IMSI is active in the GSM network. The IMSI attack is realized using the location updating request. Since the MS's IMSI is not registered in the network (except in the HLR), it is not associated with a session key K_s when the attachment is performed, and encryption cannot be applied. Therefore, the IMSI has to be transmitted in a clear text. An attacker listening to the traffic is able to extract the IMSI and can then register that user in the current area. Another situation, where the IMSI is transmitted in clear, was discussed in the previous subsection. Before a location updating procedure is terminated, the subscriber is assigned a TMSI to be used in the near future when communicating with the network.

According to the GSM specifications, the TMSI should be encrypted prior to its transmission and has to be used in future communication sessions with the network. An operator following the ETSI specifications, assigning TMSI to the subscriber after IMSI attach has been performed, will make it very difficult for an attacker lacking decryption capabilities to track the MS. Trying to passively track GSM users and eavesdropping on the users' permanent identity is easy to perform if the network operator does not encrypt the TMSI before transmission on the radio link, even though the passive nature of the attack limits the possibilities. Passive monitoring is, however, inefficient and time-consuming because the attacker needs to wait either for MSs to perform IMSI attach or for a database failure to occur in the network, which probably does not happen so often.

5.4.3.2 Active Monitoring

As passive monitoring is considered inefficient for tracking GSM users, an alternative solution for tracking is realized if the attacker is able to communicate with the MSs to track. Active monitoring requires, however, that the attacker has more advanced equipments than those needed to perform passive monitoring. While scanning for GSM radio frequencies is sufficient to perform passive monitoring, the attacker in the active monitoring is in need of base station functions that provide the ability to fabricate messages and initiate special procedures. To track a MS, the attacker can make use of the identification procedure, which is initiated by the transfer of an identity request message to the mobile station, thereby asking it to transmit a specified identification parameter that can be specified using the IDENTITY TYPE information. Since GSM networks do not use message authentication to check message origin on the radio link, an attacker with sufficient base station functions is able to use these messages, using an active attack, to retrieve the information that a legitimate base station can get.

The attacker starts by capturing the MSs as described in the previous subsection. Every close MS in the area will request a dedicated channel and initiate a location update procedure with the RBTS. This is a man-in-the-middle attack, and the messages transmitted by the MS are relayed by the attacker to the legitimate network, if necessary. Relaying the whole session between the MS and BTS may be needed if the channel used between the MS and the RBTS is not the same as the one used between the RBTS and the BTS. The attacker should first ensure that he gets the identity information of the MS. When the attacker has the IMSI of the MS and is able to uniquely identify it, the next step is to capture the TMSI that the network allocates to the MS, so that the attacker would be able to associate the IMSI and the TMSI. This will enable the attacker to track the MS's movements and the type of traffic/services that the subscriber utilizes, since the TMSI is used in communications when it has been issued. However, the TMSI is encrypted before transmission on the radio link; therefore, the attacker needs to suppress the encryption somehow. This can be done in several ways. For example, the attacker can create a situation

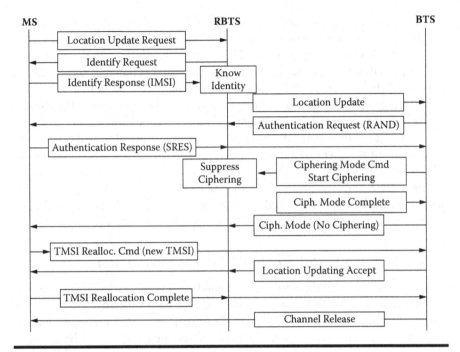

Figure 5.4 Capturing GSM user identities.

where the two legitimate communicating entities believe that they have incompatible encryption capabilities.

Figure 5.4 depicts the process of message exchange between the entities performing this attack. As it can be seen, the attacker inserts his/her own messages (i.e., Identity Request), discards, and fabricates responses (Ciphering Mode Command and Ciphering Mode complete, respectively). The attacker can even relay modified messages. The attack is made feasible because of the lack of message integrity assurance and the lack of network-authentication.

5.4.4 Attacks on the Authentication Algorithm

The difficulty in starting using an algorithm different than COMP128 remains in the fact that the algorithm resides inside the SIM, meaning that the subscribers having active subscriptions (or SIM cards) algorithm are forced to keep using their SIM cards with the old algorithm on the introduction of a different algorithm even though it is stronger. It is, however, possible to include better secure versions of the procedure COMP128 in the new SIM cards that are handed to new subscribers. This unfortunately poses a crucial problem for the efficiency of the authentication process. Another fact adds serious consideration: despite the decision of keeping the design of the procedure COMP128 non-available to public, one can find that it has been reverse engineered and crypt-analyzed. In fact, since the GSM specification for SIM cards is widely available, all that is needed to clone a SIM card is the 128-bit COMP128 secret key and the IMSI, which is embedded in the SIM card.

The procedure COMP128 is basically a keyed hash function that takes a 128-bit key and 128 bits of data RAND and uses them to produce a 96 bit hash value. The first 32 bits of the hash constitute the response SRES to the challenge. The remaining 64 bits are used as the session key K_s for voice encryption using some version of the A5 algorithm. By copying K_s and the IMSI into an empty SIM card (that is easy to acquire) the attacker can authenticate himself to the network as the legitimate subscriber, and thus can call any other MS for no cost. The attacker can also use the captured key K_s for decrypting all the calls from and to the subscriber. The attacks performing such actions are called *Cloning attacks*. Cloning can be done either by having physical access to the SIM card to be cloned or over the air.

5.4.4.1 Cloning Based on Physical Access to the SIM

If the attacker has *physical access* to the SIM module, several attacks can be launched to clone it. Some of these attacks focus on using the flows generated with the cryptographic algorithm located in the SIM, while other approaches use vulnerabilities in the SIM card itself related to COMP128. An example is given by the so-called *chosen-challenge attack*. The attack uses flows in the hashing function to deduce the secret key. The attacker constructs a number of specially-chosen challenges and submits them to the SIM card, which applies the procedure COMP128 to its secret key and the chosen challenge, and returns a response. By analyzing the responses, the attacker is able to determine the value of the secret key. As a result, the attacker is able to gain access to the secret key of the MS. The attack exploits a lack of diffusion, meaning that some parts of the output hash depend only on some parts of the input made to the algorithm.

Launching the chosen-challenge attack would require having physical access to the victim SIM card, a smartcard reader, and a computer to conduct the operation, and would allow up to 150,000 queries to be sent to the SIM card. Assuming that the average SIM reader can issue 6.25 queries per second, the whole attack would therefore take approximately 8 hours. This processing time can be reduced using appropriate measures such as overclocking the SIM card or having a higher frequency oscillator on the SIM card reader. The chosen-challenge attack is the most common attack on the GSM SIM modules. A major countermeasure against this attack is simply the replacement of the hash function used for authentication by a stronger one. It should be noticed that COMP128-2 has remedied the aforementioned limitations. COMP128-2 would require more sophisticated methods to gain the secret key K from newly issued SIMs. On the other hand, any discovery of a common vulnerability in smart cards immediately affects the security of the information stored in the SIM because it is implemented on a smart card. A relatively new class of attacks on smart cards called *optical fault induction* can be used to break the security of GSM SIM modules (Skorobogatov, 2002). According to this attack, the illumination of a target transistor causes it to conduct, thereby inducing a transient fault. The authors of the paper were able to expose the circuit of the SIM card to the light by deleting most of the protective coating from the surface of the micropro-cessor circuit embedded in the card. They also could focus the light on individual transistors within the chip by beaming the light through a microscope, and then by sequentially changing the values within the transistors used to store data, they were able to reverse engineer the memory address map. This operation allowed the authors to extract the secret data from the smart card. Another significant weak-ness was lately discovered on a smart card launching a new class of side channel attacks called *partitioning attacks*, which exploits vulnerabilities in the execution of COMP128 table lookups (Rao, 2002). The attack results reveal that the secret key can be recovered from a SIM card with less than 1,000 challenges with random inputs. The number of challenges can strongly be reduced to a few challenges that are chosen adaptively. This means that, if the attacker enters in control of a SIM card for only one minute, the secret key can easily be extracted from it. Therefore, partitioning attacks seem to be more efficient than the attacks that attempt to break the procedure COMP128 as well as optical fault induction attacks. However, the hardware currently required to perform the partitioning attacks is difficult to pro-vide and generally can only be found in R&D laboratories.

5.4.4.2 Cloning Attacks over the Air

The attacker launching an attack over the air needs to use a rogue base station and has to know a target IMSI or TMSI. When these resources are accessible, the attacker starts by capturing one or more MSs. The captured MS will imme-diately make a location update request. After the channel allocation is completed, the attacker initiates an authentication process. Immediately after the attacker has a challenge-response pair, he initiates a new authentication procedure. The MS is required to respond to every challenge made by the GSM network. This process can continue until the attacker gets the required number of pairs to initiate a cloning procedure. The number of frames exchanged between the network and the MS, for one authentication process, can be estimated to 66 frames. Since the duration of one TDMA frame is 4.615 ms, the duration t of the whole signaling sequence is equal to

$$t = 4.615 \times (66 \text{ frames}) = 0.30459 \text{ s}.$$

The time needed to get the needed number of challenge-response pairs for the aforementioned attack can be estimated. For this, let us assume that the crypto-graphic attack requires approximately 150,000 challenge-response pairs, the time required by the attack is approximately equal to 13 hours (about 45 689 seconds). This means that the MS has to be accessible to the attacker over the air for the whole time it takes to collect the information needed. This shows that the attack is unrealistic, because it is not feasible that the MS remains accessible for that period of time. To overcome this limitation, the attack can be performed in several parts; instead of performing a 13-hour attack, the attacker could interrogate the MS for a short period every day, for a certain number of days.

Finally, let us notice that this attack can be launched simultaneously against as many MSes in the radio range as the RBS can have channels. In addition, if the attacker is capable of getting the information gathered by cloning the SIM and can intercept the RAND value over the air during the call establishment procedure, then all the requirements for determining the confidentiality key are satisfied. The protection against cloning over the air attacks is to limit the number of times a SIM card can be authenticated to a value considerably smaller than 150,000. However, a drawback can be observed with this protection: A new SIM card has to be issued and distributed to the mobile subscriber every time this limiting value is reached. This would require unacceptable costs for both the subscriber and the GSM provider (Brookson, 2002).

5.5 GSM Encryption Algorithms

To provide confidentiality, the GSM network requests the MS to start encrypting using encryption algorithm A5 each time the information is transmitted on the radio link. The purpose of the A5-algorithm is to provide encryption of the radio connection between the MS and the BTS. To explain this, let us recall that a GSM conversation is divided into frames of 4.615 ms. Each frame contains 114 bits of information. Therefore, every 4.615 ms the MS sends to and receives from the BTS a 114-bit frame. The two frames are encrypted in both the BTS and the MS using the A5 algorithm with the same input values.

The algorithm A5 takes as input the session key K_s and the (publicly known) frame number, and produces a key-stream of 228 bits. The first half (114 bits) is used to encrypt the information sent from the MS to the BTS and the other half is used to encrypt the information sent from the BTS to the MS. The key-stream is then XORed with the 114 bits of information to be sent and the result is the 114 bits of encrypted information. On the other end, the original information is taken back by XORing with the same keystream. The algorithm A5 is a pseudo-random bitstring generator that uses a number of linear feedback shift registers (or LFSR). The LFSRs are pseudo-random bit generators. In each LFSR, some bits in specific places are XORed and put at the end of the register, thereby shifting the bits one place to the side and generating an output-bit. There are several versions of the A5-algorithm. They are numbered A5/*. Algorithm A5/0 provides no encryption. We will discuss in the following two algorithms, A5/1 and A5/2.

5.5.1 Algorithm A5/2

The A5/2 stream cipher uses four LFSRs, say R1, R2, R3, and R4. The registers have the lengths 19, 22, 23, and 17 bits, respectively (Figure 5.5). In each LFSR some bits in specific places, called *taps*, are XORed and put in the end of the register after shifting the bits one place to the side, thereby generating an output-bit (the one extracted from the register after shift). The sequence of operations constitutes the clocking of the LSFR. Figure 5.5 shows how the clocking of R1, R2, and R3 is handled. It shows that:

- Register R1 has taps at R1[18], R1[17], R1[16], and R1[13]
- Register R2 has taps at R2[16], R2[13], and R2[8]
- Register R3 has taps at R3[17], R3[15], and R3[13]
- Register R4 has taps at R4[16] and R4[15]

When a register is clocked among the aforementioned registers, four operations are performed. They are the following:

- The bits in the register *taps* are XORed;
- The bits in the entire register are shifted one place to the left;

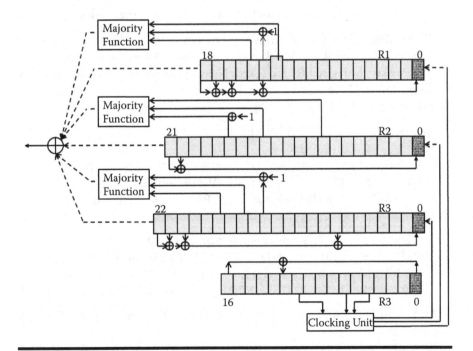

Figure 5.5 Structure of A5/2.

■ The bit is taken out of the register after the shift is used in the function for the output stream; and
■ The result of the XOR-operation of the taps is put in rightmost place in the register (i.e., R*[0], for *= 1,..,4).

The computation of the output bit issued from R1, R2, and R3 was performed using a major function *maj* defined by $maj(a,b,c) = a.b \oplus b.c \oplus c.a$. It is realized as follows:

■ In each LFSR, the majority function is computed. This function computes a bit from the content of three places in the register R1 by taking

$$a = R1[15], b = R[14] \oplus 1, \text{ and } c = R[12].$$

■ The result of the majority-functions of R1, R2, and R3 are XORed with the leftmost bits of R1, R2, and R3 to produce the bit for the output stream.
■ The register R4 and the clocking unit controls the clocking of R1, R2, and R3 as follows:
  – The bits at R4[3], R4[7], and R4[10] are inputs to the clocking unit)
  – The clocking unit performs a majority-function, which decides which of the LFSRs R1, R2, and R3 are to be clocked in the following way:
    • R1 is clocked if and only if R4[10] agrees with the result of *maj*.
    • R2 is clocked if and only if R4[3] agrees with the result of *maj*.
    • R3 is clocked if and only if R4[7] agrees with the result of *maj*.

Performing the clocking unit this way, one can say that at least two among the registers R1, R2, and R3 are clocked synchronously. When the two registers are clocked, we say the register R4 is clocked. Now, the ingredients needed to describe the algorithm A5/2 are covered using the session key K_s and the frame number *f*. The algorithm performs the following:

```
Set R1, R2, R3, and R4 to 0
For i = 0 to 63 do
  Clock all LFSRs (while ignoring the outputs);
  R1[0] := R1[0] ⊕ Ks[i]; R2[0] := ← R2[0] ⊕ Ks [i];
    R3[0] := R3[0] ⊕ Ks [i]; R4[0] := R4[0] ⊕ Ks [i]

For i = 0 to 21 do
  Clock all LFSRs (while ignoring outputs).
  R1[0] := R1[0] ⊕ f[i]; R2[0] := R2[0] ⊕ f[i];
    R3[0] := R3[0] ⊕ f[i]; R4[0] := R4[0] ⊕ f[i]
  Force the bits R1[15], R2[16], R3[18] and R4[10] to be 1

Run the preceding two steps for 99 clocks and ignore
  outputs
Run the preceding two steps for 228 clocks and use the
  output as the key stream
```

5.5.2 Algorithm A5/1

The algorithm A5/1 is built in a simpler way, compared to A5/2, from three linear feedback shift registers (LFSR) of lengths 19, 22, and 23 bits. We denote these FLSR by $R1$, $R2$, and $R3$, respectively (as depicted by Figure 5.6). The rightmost bit in each register is labeled as bit zero. The taps of $R1$, $R2$, and $R3$ are placed at bit positions 13, 16, 17, 18; at bit positions 20, 21; and at bit positions 7, 20, 21, 22, respectively. When a register is clocked, its taps are XORed together and the result is stored in the bit zero of the left-shifted register.

The registers are clocked in a stop/go procedure using a majority rule as follows. Each register has a single clocking tap (bit 8 for $R1$, bit 10 for $R2$, and bit 10 for $R3$). With each clock cycle, the majority function of the clocking taps is computed and only those registers whose clocking taps agree with the majority bit are actually clocked. Thus, at each step, at least two registers are clocked. The process of generating the keystream from K_S and the frame counter f is carried as follows:

```
Set all LFSRs to 0
For i = 0 to 63 do
  Clock all LFSRs (ignoring outputs).
  R1[0] := R1[0] ⊕ KS [i]; R2[0] :=R2[0] ⊕ KS [i]; R3[0]
    :=R3[0] ⊕ KS [i]

For i := 0 to 21 do
  Clock all LFSRs (ignoring outputs).
  R1[0] := R1[0] ⊕+ f [i]; R2[0] := R2[0] ⊕ f [i]; R3[0]
    := R3[0] ⊕ f [i]
```

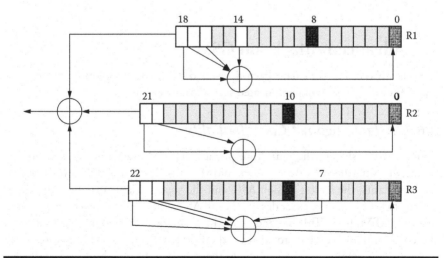

Figure 5.6 Structure of A5/1.

Run the preceding two steps for 100 clocks and ignore the outputs
Run the two steps for 228 clocks and use the outputs as the key stream

Many differences can be observed between the algorithms A5/2 and A5/1. These differences include the following:

■ The algorithm A5/2 uses register R4 that is in charge of controlling a clock unit, which is initialized at each cycle.
■ The algorithm A5/2 builds on the registers R1, R2, and R3 majority functions to carry out the process of bitstream generation.
■ The algorithm A5/2 discards 99 bits of outputs while A5/1 discards 100 bits of output.
■ A5/2 imposes that one bit in each register is equal to 1 after initialization.
■ Despite the fact that the clocking mechanism is the same, the input bits to the clocking mechanism are taken from R4, in the case of A5/2, while they are collected from R1, R2, and R3, in the case A5/1.

In addition, the algorithm A5/1 output is based on the summed output of 3 LFSRs whose clock inputs are controlled by a unique majority function of certain bits in each LFSR. This scheme seems an interesting solution. However, it has been shown that the algorithm A5/1 could be cracked in a very short time (of the order of the second) on a typical PC (Biryukov, 2000). The attack exploits flaws in the algorithm when storing tables as well as exploiting the poor single-bit taps used to control the LFSR clocks.

On the other hand, A5/2 has been deliberately weakened and has been demonstrated to be also flawed. In fact, A5/2 can be cracked very easily and thus is even weaker than A5/1.

5.6 Advanced Attacks on GSM

Various advanced attacks have been developed for the GSM communication systems. They all implement complicated scenarios of actions.

5.6.1 Attacks against Confidentiality

As it has been shown, the radio link privacy of GSM telephone conversations is protected using the A5 stream cipher, and this algorithm was shown to have two major variants: the strong version A5/1 and the weak variant A5/2. It is, however, worthy to know that the exact design of both A5/1 and A5/2 was reverse engineered in 1999 (GSM, 929). This has encouraged the emergence of many attacks against the confidentiality service provided by the GSM networks. In the following, we discuss some among the major attacks that we classify into brute-force attacks, cryptanalysis-based attacks, and non-cryptanalysis attacks.

Some of the attacks use the weaknesses in the GSM architecture and/or flaws in the protocols used in the communication between the GSM networks and the subscribers. The following facts will be used to describe some among these attacks:

- It is assumed that most of the GSM mobile phones can communicate with different base stations and networks. This is possible because all the manufacturers follow the standards developed by the ETSI.
- The same secret key K is used for the different encryption algorithms A5/1, A5/2, and A5/3. This means that breaking one of these algorithms and retrieving the session key will threaten the confidentiality of the conversation even when the stronger version of the algorithm is used later.
- A base station does not need to authenticate itself to the MS it is communicating with. The messages it exchanges with the MS are not authenticated and their integrity is not guaranteed.
- It is mandatory for A5/1, A5/2, and A5/0 to be implemented on every MS. The reason is to allow an easy roaming between different technologies and providers that potentially use different encryption algorithms.

Recall that the confidentiality of a conversation in the GSM depends on the secrecy of the session key K. An attacker that discovers the key is able to decrypt to the data exchanged during the conversation. Assuming that A5/2 is a weak cipher, it is more practical to try to derive the session key K by attacking A5/2. When the session key is retrieved, the attacker is able to decrypt and listen to the conversation even if it is encrypted using A5/1 or A5/3. The cryptanalysis of A5/2 will be addressed in the following subsection.

5.6.1.1 Brute-Force Attacks

The confidentiality of GSM is protected by the secrecy of the symmetric key Ks, which is used to encrypt communication over the air for an established session. A brute-force attack aims at utilizing all possible keys with a given plaintext until the appropriate key is found. As mentioned earlier, Ks is 64 bits, but since its last 10 bits are set to zero, the key space is reduced and the cryptanalysis effort would cost 1024 times less. It is known that A5/2 can be broken in real time, whereas A5/1, the stronger variant, is susceptible to brute-force attacks. In fact, an easy computation would show that, using a traditional computer, it is easy to have a set of 30.000 parallel A5/1 implementations clocked at a speed higher than 3.2 GHz, so that each A5/1 implementation would generate one output bit for each clock cycle and that the system would try approximately more than 10^6 keys per second for each A5/1 implementation. A brute-force attack would therefore require about less than 18 hours, using all the parallel implementations. Further optimization can be achieved to reduce this period of time. For example, the system can stop trying a specific key after the first invalid key stream.

Finally, let us notice that brute-force attacks are not appropriate for real-time operation since it is generally possible to find a key given a few hours. However, a brute-force attack may easily be used offline to find the key utilized in a specific conversation, meaning that the attacker intercepts and stores the conversation and decrypts it at a later time.

5.6.1.2 Cryptanalysis-Based Attacks on A5/1

Recently, cryptographic attacks on the algorithms protecting the confidentiality of GSM calls have been developed. Some of these attacks have been presented (Barkan, 2003; Ekdahl, 2001). In particular, a real-time cryptanalysis attack of A5/1 has been studied. This attack is called the *Biased Birthday attack*. To perform the attack, two minutes of data collection and a few seconds of processing time are required. The basic idea used in the first attack is to find out the initial internal state of the algorithm, assuming that the attacker has complete knowledge of the outputs of the A5/1 algorithm during some initial period of the conversation. Knowing that the GSM mobile station sends a new frame every 4.615 milliseconds, each second of conversation contains about 2^8 frames. When the attacker has found the initial state of any frame, running the algorithm in the reverse direction to derive the session key Kc can be done, knowing the (publicly known) frame number.

During pre-computation, the attack pre-computes a large set A of states and stores them along with their output prefixes. The algorithm $A5/1$ has a relatively small number of internal states, since it contains only n = 2^{64} states defined (as the shift registers R1, R2, and R3 have 19 + 22 + 23 = 64 places) and a large set B of states through which the algorithm progresses during the actual generation of output bits (the output prefixes). Any intersection between A and B will make it possible to identify an actual state of the algorithm from stored information. The output prefix is the first $log(n)$ bits in a state's output sequence. The pairs (prefix, state) are sorted into increasing prefix values to allow the use of the prefix as an index. Searching for common states in A and B can then be efficiently done by probing the sorted data A on the hard disk with prefix queries from B.

Since the attack requires that the attacker knows some pseudorandom bits generated by the $A5/1$ algorithm in some frames, the problem of how to get access to these bits should be solved to achieve the global objective. Depending on the amount of known plaintext needed, the attacker can try to guess the bits that are needed. This would, however, increase the period time complexity of the attack, making it impractical. The attacker may be able to derive this amount of key stream if he can mount a man-in-the-middle attack by asking the MS, after encryption is enabled, to respond to certain signaling requests that yield responses with content that the attacker can guess with high probability of success. Assuming that the attacker succeeds in deriving the needed amount of key stream, the next step in the biased birthday attack can be performed, that is, calculating the session key itself. This step takes several minutes in the best case. However, this is a too long period of time to comply with real-time requirements, since it is very unrealistic that several minutes of the key stream would be available to the attacker. Thus, the problem with the Biased Birthday attack is the amount of known plaintext required. Another attack against A5/1 algorithm can be found in Ekdahl (2001). It brings some ideas from the correlation attacks and exploits the bad key initialization procedure in A5/1. This weakness is linked to the fact that the key and the frame counter are initialized in linear manner. This fact allows the hacker to launch an attack that makes it possible to separate the session key from the frame number in binary linear expressions. The proposed attack is, however, linear in the length of the shift registers and its implementation requires the knowledge of the 40 first bits from two (possible non-consecutive) frames. The complexity of the attack is low and requires very little pre-computation time and memory. In the presence of the required known plaintext, the attack takes about five minutes on a simple PC. However, this makes it hard to use the attack in real-time, but it can be executed after the conversation has been stored.

5.6.2 Cryptanalysis Attacks on A5/2

Several attacks on A5/2 to eavesdrop on GSM user's conversations have been presented.

5.6.2.1 Using A5/2 Cryptanalysis to Target the Confidentiality of A5/1 and A5/3

An attack targeting A5/1 and A5/3 can be prepared on A5/2 as follows: Once a victim MS is captured under a RBTS using the fact that the MS listens and responds to the strongest signal it receives, the attacker waits until the MS requests a service. Then the RBTS impersonates the network to the calling subscriber, and the subscriber to the network until the session key is found. We know that, in the standard case, before a conversation starts the network requires the caller to be authenticated. The network sends RAND to the attacker. The attacker forwards RAND to the subscriber, who computes SRES and returns it to the attacker believing that the attacker is the network. Recall that the attacker is not able to compute SRES even if RAND is known, because the attacker may not know the secret key *K*. The attacker relays the signaling messages between the network and the MS until SRES is received from the MS. Instead of forwarding SRES to the network, the attacker transmits the ciphering mode command message to the caller MS asking it to encrypt using A5/2.

The captured MS, believing that it is authenticated by the network, starts encryption using A5/2 and responds with an encrypted acknowledgment of the ciphered mode. The attacker can use cryptanalysis of A5/2 to retrieve the encryption key *K* using the data he has received from the subscriber. When the key is retrieved, the attacker sends the authentication information (SRES) to the waiting network. The attacker should succeed in finding the session key before the assigned timer expires (which is used to ensure that the network will not wait forever for an answer), because if the timer expires the network may abort the transaction. Now, when the network receives the SRES, it continues with the rest of the call establishment procedure, until the call is established.

The attacker now has the session key that can be used to generate the key stream. He examines the message exchange between the network and the MS to determine which encryption algorithm the network is asking the MS to use, in order to use the right encryption algorithm when relaying messages between the communicating parties. Even if the network now tells the MS to use *A5/1* the attacker is able to decrypt it, since the same session key will be used by that algorithm, together with the publically frame number f of each transferred frame. Finally, the attacker receives the *A5/2*-encrypted frames from the MS. Having decrypted these frames, he saves them.

In practice, the described attack may not be successful because of the known plaintext requirement. It is, however, possible to obtain the required amount of known plaintext if the attacker can make the MS send signaling messages with content that is known or almost known to the attacker. Another reason for the attack failure may also be the processing time needed by the cryptanalysis, which may result in the expiring of the assigned timer, which will result in the network denying service to the legitimate MS.

A second attack can use the attacker's SIM. It is based on the attacker's capability of impersonating the MS and a BTS. The attacker uses his own SIM when communicating with the network on behalf of the subscriber. The attacker simply uses a compromised secret K together with the related IMSI and ensures that the actual owner of the communication parameters is not active in the same region to avoid the network getting suspicious.

The attacker (or his RBTS acting as a legitimate customer with a valid SIM) attempts to capture MS, whose IMSI/TMSI is known. He ensures that all messages/service requests that are exchanged between the MS and the network are relayed by the RBTS. Once the attacker detects that the captured MS is demanding a service, the attack starts. In this case, the captured MS demands that a call be set, and the RBTS makes exactly the same request to the network using the stolen IMSI/TMSI. The attacker allocates the MS a channel and starts an authentication with it, exactly the same way the network would do. The SRES provided by the MS is useless for the attacker and thus it is discarded. Then, the attacker asks the MS to use no encryption (*A5/0*) and the MS sends to the RBTS the call-setup message with the number it wants to call. The attacker obtains the number f and can initiate a call-setup procedure with the network on his/her behalf to the number that the legitimate subscriber wishes to call, using the *attacker's* identity.

The network suspects nothing and uses the key of the attacker during authentication and session key generation. This way the attacker will obtain the session key and will be able to decrypt the call. The BTS commands the RBTS (which is acting on behalf of the MS) to encrypt the communication. Finally, the call is transmitted unencrypted from the MS to the RBTS. The attacker, who operates the RBTS, eavesdrops on the call, encrypts it with the session key obtained from the RAND received from the BTS, and sends it to the BTS. The unencrypted call set up between the MS to the RBTS is connected to the encrypted call between the RBTS to the BTS. Therefore, although the call between the RBTS and the BTS is an encrypted real call, the network does not suspect that something is working wrong.

In conclusion, one can say that with this attack: (a) the call is made on the attacker's long term registration and not with the MS; (b) the attack can be detected later if the related bill is checked; and (c) the MS receiving the call will see the number of the attacker's subscription and not that of the legitimate calling party when the call is established. These facts ensure that a digital investigation would reconstruct the attack and identify the attackers. This limitation can be easily overcome by making certain that the phone number of the attacker is kept secret.

5.6.2.2 The Known Plaintext Attack

We describe here how a known plaintext attack can be performed against A5/2, assuming that the attacker is aware of the message that is encrypted and the resulting cipher-text. Knowing the way the LFSRs are initialized, we will first claim that it is possible to find the confidentiality key K_c from the initial state of the LFSRs. Then, we state that it is possible to find the initial state of all the LFSRs given a sequence of frames (each frame has 114 bits of cipher-text). Therefore, combining these two facts, the attack can be conducted to break the cipher.

1. Getting K_c from the initial state: Consider the way the LFSR are initialized. At the begining all the LFSRs are all zeros; each LFSR has a certain length. One can say that since a feedback function is attached to each register, each bit in the K_c can be expressed as an XOR of bits from the register. All the feedback functions have different irreduceable polynomials from where the feedback functions are constructed. Let $K_c[0]$, …, and $K_c[63]$ be the variables corresponding the bits in K_c. The attacker is able to construct a system of 64 linearly independent equations that express each bit in a register in terms of $K_c[i]$'s. This system can be solved for the $K_c[i]$'s using Gauss elimination and backwards substitution, for example.

   For *A5/2*, the set of equations defines a 64x64 matrix that is invertible. So, given the internal state of the registers after initialization with K_c, we can go back and get K_c.

2. Getting the initial state: Given a sequence of successive frames, we describe how the attacker can get the initial state of all the registers R1, R2, and R3, assuming that he knows the content of R4$_j$, where j is the number of the first frame in the sequence. For this, let's define for each frame in the sequence a set of variables V$_j$ containing the following variables:

- One variable for each bit in the registers R1, R2, and R3, which are not forced to 1. Call the bits in R1, R2, and R3 by x0, ..., x18; y0, ..., y21; and z0, ..., z22, respectively. In total, we have 61 variables.
- The following quadratic variables have a value of 594, when j is at maximum:
  - $xi \cdot xk$, where $0 < i < j$ and $1 < k < 18$
  - $yi \cdot yk$, where $0 < i < j$ and $1 < k < 21$
  - $zi \cdot zk$, where $0 < i < j$ and $1 < k < 22$
- The constant 1, which describes the bits in each register that are forced to 1. This makes a total of 656 variables in V_j.

Observe now that the variables in any V_{j+i} can be represented as a linear combination of the ones in V_j, for any integer i. Considering the initial states of two different frames with numbers j and j+ i, for some i, we can calculate the XOR-difference between the initial states of the frames. In order to solve a linear equation system of 656 variables we need 656 linearly independent equations. However, the attacker is in fact only interested in 61 of these 656 variables, namely the ones describing the initial state of the registers. This problem has been addressed in the literature.

Now, we can get the initial state and derive the key K_c knowing the content of $R4_j$, except for the three positions in the registers that were forced to be 1. As there are two possibilities for each position, this gives us eight possible equation systems. We end up with eight possible values for K_c, from which the wrong values can be easily excluded. Therefore, one can say that given a cipher-text and the corresponding plaintext for a few successive frames, one can find K_c, and break the cipher by trying all 2^{16} possible values of the register R4.

5.6.3 Denial of Service Attacks

Several DoS attacks have been developed against GSM networks during the recent years. The DoS attacks can be performed by different means including physically disturbing the radio signals or by logical procedures. We discuss in this subsection some attacks belonging to these two categories.

5.6.3.1 Physical Intervention-Based Denial of Service Attacks

The physical attacks are the most straightforward attacks. In performing physical DoS attacks, the attacker prevents the mobile user traffic or a signaling traffic from being transmitted on any network link, whether the link is wired or wireless. The attack is performed using some specific physical means. Examples of physical intervention on a wired link include wire cutting, while examples of physical intervention on a wireless interface include jamming the connection. The equipment that jams the GSM radio signals can be placed in the area where a traffic to be interrupted is transmitted. As a consequence, the mobile device placed within the equipment's range will not function properly. Since jamming targets radio frequencies, one can notice that the frequency jumping makes the jamming more difficult to achieve than one can think about. There are common examples of large jamming causing real problems for GSM operators. Radio jamming attacks can target more than one mobile station or a given set of frequencies. Some GSM operators have suffered heavily from jamming activities and had serious problems stopping this type of DoS attack.

Another attack aiming at making the MSs in a local area unreachable can be performed based on the following steps: An attacker operating a modified base station (or RBTS), transmitting the base channel with a higher signal strength, will force the MSs under its radio coverage to camp on the radio channels of the false base station. Then, the RBTS can stop transmitting and can replay this several times. This physical DoS attack makes the mobile stations unreachable for serving network, during short periods of times.

5.6.3.2 Logical Intervention-Based Denial of Service Attacks

An attacker can perform DoS attacks on the GSM systems by logical procedures including de-registration, location updates, and replaying attacks. In the sequel, we recall the main ideas behind these attacks:

- *De-registration*: The attacker spoofs a de-registration request (i.e., IMSI-detach message) that a MS sends to the network. The network de-registers the subscriber from the visited location area and informs the related HLR to do the perform action. The user is consequently made unreachable for other mobile subscribers.
- *Location updates*: The attacker spoofs a location update request sent by a MS in a location area different from the area in which the subscriber is roaming. The network registers the subscriber in the new location area and the target user will be paged in that new area. The user is subsequently unreachable for mobile terminated services.
- *Replay attacks*: Various signaling messages sent or received by a MS can be eavesdropped by an attacker and replayed in the same radio area or in a different area. This can cause a service requested by the MS to be stooped, the MS to be unreachable, or the resources offered by GSM network to be overused. Therefore, replay attacks can be considered as damaging DoS attacks.

5.7 Improving GSM Security

Many revisions to the GSM standards have been made, since its first deployment to add technologies such as GPRS and EDGE and new security solutions. The standards are nowadays evolving toward 3G technologies such as UMTS. Indeed, recent versions of the standard, particularly UMTS, have benefited from the security flaws discussed in this chapter. They are described below and introduced as important to the solution provided by the GSM system. Improvements include newer A3/A8 implementations, improvements to the algorithms A/5, and GPRS encryption algorithm.

5.7.1 Improving Authentication Protection

Some proposals were made available recently to provide new user authentication approaches for GSM. With the approach presented in Aydemir (2005), the calling mobile user, say *UserA*, makes use of a password instead of the embedded key. Doing so would reduce the dependency of the authentication process on the SIM card. In addition, *UserA* would be able to reach his account via any cellular phone, Internet, or a special network, without the help of a SIM card. The users can reach their address books, get their personal information, or redirect their calls. In fact, the authentication of *UserA* to the HLR (via VLR) uses several items: (a) a password, which is set by the user; (b) a random nonce RAND, which is generated by VLR to provide freshness guarantees for the session; and (c) three other random nonces n_1, n_2 and c, which are generated by *UserA*. The nonce n_1 is used to prove the correct decryption of an important message exchanged with the HLR, n_2 masks the session key K, and c protects the authentication request against replay by a malicious adversary.

The authentication protocol starts with the *UserA*'s authentication request, which consists in sending a first message containing the user identity (IMSI) to VLR. The VLR responds by sending a message containing a random number *RAND*. Upon receiving *RAND*, *UserA* generates the three random nonces n_1, n_2, and c. Then, he encrypts *RAND* with his password and encrypts n_1, n_2, c, and the result obtained for *RAND* using the HLR's public key and sends the computed value to the VLR with a random challenge r_A. The message sent to the VLR has the following form:

$$<E_{KHLR}(\ n_1,\ n_2,\ c,\ H_{Pass}(RAND)),\ r_A>.$$

The VLR takes the received message and encrypts its first part along with RAND using the symmetric key (K_V) it has with the HLR. The result is then sent to the HLR. Knowing the VLR's symmetric key, the HLR can decrypt the received message, decrypt the message encapsulated from *UserA*, and check whether RAND is well-signed by *UserA*. Finally, the HLR generates a session key k_s to be used between the VLR and *UserA*, encrypts it with the VLR's symmetric key, and sends the following message to the VLR,

$$<E_{KV}(k_s),\ n_1,\ n_2+k_s>.$$

The VLR then decrypts the first component of the received message and sends to UserA the message:

$$<E_{ks}(r_A),\ n_1,\ n_2+k_s,\ r_B>,$$

where r_B is a new challenge provided to *UserA,* who responds by sending back $E_{ks}(r_B)$. A security analysis of the authentication scheme shows that:

- The correct value of n_1 in the fifth message indicates that the HLR that has decrypted the fifth message, has identified the user and sent the message.
- The nonce n_2 protects HLR's response against dictionary attacks that can be made by an attacker who gets to know the session key to get knowledge of n_1 and n_2.
- Random c protects the first message against regeneration by VLR.

Let us now consider the case where a mobile phone operating with this protocol is lost. Let us notice, however, that the current systems do not provide automatic key disabling after the loss of terminal device. He has to call the service provider's operator and prove his identity and reveal his private information to disable his SIM card. Protecting against loss can be performed within the authentication scheme as follows: During the initialization of the mobile client's account, the mobile client generates two random numbers t and t_2, computes the hash value $u = H(t)$, and creates a *key disabling ticket* τ as equal to $E_{Khlr}(H(u),\ t_2)$. Then, the random t_2 and the hash u are deleted and τ and t are taken out of the phone to a storing medium. On the occurrence of a loss or capture, the user immediately sends t and τ to the HLR. The HLR decrypts τ gets u, and checks whether u is equal to $H(t)$. If they are equal, then the account of the user is locked and further access is not possible. Let us now notice that the random t_2 occurring in the ticket τ is used for the protection of the password against dictionary attacks on τ. If t_2 is not available, a malicious adversary seeing τ can generate u, encrypt it with a candidate password and the public key of the server, and checks whether it is equal to τ. If that is the case, the selected password is correct.

5.7.2 Other Improvements

Several improvements have been made available to overcome the drawbacks of the GSM networks related to security. These improvements address newer implementations of the major algorithms and the intrusion detection.

5.7.2.1 GSM—Newer A3/A8 Implementation

As it has been shown in a previous section, newer versions of the algorithms A3 and A8 have been introduced based on new implementation of the procedure

COMP128, called COMP128-2 and COMP128-3. These algorithms have shown a reasonable behavior. However, COMP128-2 is still suffering from the 10-bit weakening problem of the ciphering K_i. COMP128-3 uses the same basic algorithm, but with a truly 64-bit K_i. The COMP128-2 and COMP128-3 algorithms have stopped (so far) SIM cloning and have made the K_u extraction over the air unfeasible.

In addition, to overcome the aforementioned limitations, the 3GPP project has defined a completely new and open set of authentication algorithms for use with the UMTS system. These algorithms will be discussed in Chapter 6.

5.7.2.2 GSM—A5/3 Ciphering

It is known that GSM standard supports up to 7 different algorithms for A5. Until recently, only the A5/1 and A5/2 algorithms were used. In 2002, the GSM standard added a stronger algorithm, called A5/3, which is based on the Kasumi core (the core encryption algorithm for UMTS; Blunden, 2002). Only few networks and handsets support this algorithm currently, however. Similarly to the A5/3 algorithm, a new GPRS encryption algorithm, called GEA 3, based on the Kasumi core has been added to the GPRS system.

In UMTS, the possibility of an attacker imitating the network has been removed by means of a two-way authentication procedure. The procedure for which the mobile authenticates itself to the network is largely the one performed by GSM. The procedure adds the following: the network sends an Authentication Token (AUTN) along with the RAND. The AUTN consists of a sequence number (SQN) encrypted using the RAND and a key called the root key (K) and the MAC code, which works much like the GSM SRES but in the opposite direction. Therefore, if the XMAC does not match the MAC calculated by the SIM, the MS sends an authentication reject message to the network and the connection is dropped. The model used for authentication and key generation is quite similar to the GSM model, where all authentication and key generation functions are performed in the SIM card and the network's AuC. However, the algorithms typically have stronger input parameters and operate on a hidden 128-bit root key, known only by the SIM and AuC. Like GSM, these algorithms can be operator specific. Finally, to stop an attacker from performing a replay attack by resending the legitimate network's authentication request, the SIM card is required to keep track of the sequence numbers used. A comparative study will be made in Chapter 6 (with GSM) to show how UMTS solves some of the limitations of GSM. The study also shows the attacks that are still being performed on UMTS. The study aims at proving that limitations of the GSM systems can be truly reduced, but at high price (on the MS).

5.7.2.3 Protecting against Distributed DoS Attacks

Availability is among the most critical security services that are targeted by distributed denial of service (DDoS) attacks in wireless cellular networks. These attacks are difficult to prevent, detect, or defend against. A few architectures have been proposed to provide a defensive and reactive architecture that uses diverse algorithms to cope with diverse types of DDoS attacks.

To detect DDoS attacks in cellular networks, two main techniques need to be implemented. The first technique is *Radio Intrusion Detection,* which needs to be implemented in the BTS to detect attacks that aim to consume radio resources and exploit vulnerabilities related to the openness and scarcity of the radio channels. This detection technique is based on analyzing the characteristics of the detected signals and the information they contain. The second technique is the distributed intrusion detection. It is crucial for detecting DDoS attack occurrences, since the analysis of the behavior of only one user does not suffice to prove that he is participating in a DDoS attack. Detection agents need to be distributed through backbone and access network to the cellular. They are intended to perform a continuous monitoring of the users' behaviors and send alerts to other nodes whose role is to correlate different alerts and decide on the existence of an attack. The architecture, denoted by CODERA and discussed in Rekhis (2007), consists of the following:

- Radio intrusion detection systems (RIDS): These IDSs, which are implemented in the BTS level in the case of GSM, perform a real-time monitoring of the radio interface. They are able to analyze all the traffic carried in signaling and traffic channels. A RIDS can operate either as an autonomous system or as a participant in a distributed defense system. In the autonomous mode, it detects attacks and responds to them without communication with any other entity. In the distributed mode, the RIDS considers received attack alerts from other participants. This mode is able to monitor many parameters like signal characteristics (e.g., power, frequency) and calling behaviors (e.g., number of calls, rate of incomplete calls). The RIDS has also the ability to execute a multitude of algorithms to detect abnormal behavior.
- Analysis servers: These nodes are deployed in the BSC. They have two major roles: first, the monitoring of all traffic that pass through the BSC and the supervision of many parameters linked to data transmission such as bit rate and jitter; second, the aggregation of alerts that are collected from others RIDSs. Analysis servers are the core of detection operation in Distributed IDSs. They receive alerts from detection nodes in order to decide on the existence of the attack.
- Rate limiters: These nodes represent reactive components that are able to perform selective rate-limiting and reduce the aggressiveness of packets that pass through them. Rate limiters are useful when countering attacks based on overwhelming the victim by useless packets.
- Classifier nodes: These nodes are used to differentiate between legitimate and malicious packets, dedicate their available bandwidth to legitimate traffic, and cooperate with other defense nodes to ensure good service for legitimate users. Note that the classifier functionality encompasses rate limiter functionality.

References

O. Aydemir, GUAP—A Strong User Authentication Protocol for GSM, Master Thesis, Bilkent Univ., Ankara, Jan. 2005.

E. Barkan, E. Biham, and N. Keller, Instant Cipher-Text-Only Cryptanalysis of GSM Encrypted Communications, 2003 (available at http://www.cs.technion.ac.il/~biham/publications.html).

A. Biryukov, A. Shamir, and D. Wagner, Real Time Cryptanalysis of A5/1 on a PC, in Fast Software Encryption, vol. 1978, *Lect. Notes in Comp. Sci.* pp. 1–18, 2000.

M. Blunden and A. Escott, Related Key Attacks on Reduced Round KASUMI, in the *Proc. of the Fast Software Encryption (FSE 2001)*, Matsui (Ed), Lecture notes in computer science, vol. 2355, pp. 277–285, Springer 2002.

C. Brookson, Security and Cryptography Applications to Radio Systems, in *IEE Colloquium on GSM Security: A Description of the Reasons for Security and the Techniques*, 1994.

C. Brookson, Can You Clone a Smart Card (SIM)?, Technical report, 2002 (available at http://www.brookson.com/gsm/clone.pdf).

P. Ekdahl and T. Johansson, Another Attack on A5/1, Abstract, *Proceedings of International Symposium on Information Theory (ISIT)*, Washington, 2001 (available at http://www.it.lth.se/patrik/publications.html).

European Telecommunications Standards Institute (ETSI), European Digital Cellular Telecommunications System (Phase 2+) (GSM), *General description of a GSM Public Land Mobile Network* (PLMN), 1997, http://www.etsi.org.

GSM 02.17 (ETS 300 509): European Telecommunications Standards Institute (ETSI), European Digital Cellular Telecommunication System (Phase 2); *Subscriber identity modules (SIM), Functional Characteristics*, (available at http://www.etsi.org).

GSM 04.07 (TS 100 929), European Telecommunications Standards Institute (ETSI), European Digital Cellular Telecommunication System (Phase 2+), Security Related Network Functions (available at http://www.etsi.org).

J. Quirke, Security in the GSM System, *Ausmobile* 2004 (http://www.ausmobile.com).

J. R. Rao, P. Rohatgi, H. Scherzer, S. Tinguely, Partitioning Attacks: Or How to Rapidly Clone Some GSM Cards, in the *Proceedings. 2002 IEEE Symposium Security and Privacy*, 2002, pp. 31–41.

S. M. Redl, M. K. Weber, and M. W. Oliphant, *An Introduction to GSM*, Artech House, 1995.

S. Rekhis, A. Chouchane, and N. Boudriga, Detection and Reaction Against DDoS Attacks in Cellular Networks, The International Conference on Information & Communication Technologies: From Theory to Applications (SCS-ICTTA'08) April 7–11, 2008, Damascus, Syria.

P. Skorobogatov and R. Anderson, Optical Fault Induction Attacks, in *CHES 2002*, Springer-Verlag, pp. 2–12, 2002.

C. Vijaya, *Security, Authentication and Access Control for Mobile Communications*, http://www.ittc.ku.edu/rvc/documents/865/865_securityreport.pdf.

Chapter 6

Security of 3G Networks

6.1 Introduction

The third generation (3G) proposal for cellular communications aimed at providing global roaming for mobile users, high transmission bandwidths, and protection to sophisticated services such as the global positioning systems and multimedia on the demand of mobile users. In the mid-1980s, the International Telecommunications Union (ITU) challenged the communication community to produce a single and worldwide standard capable of offering high speed communication, better QoS support, and enhanced security compared to 2G cellular networks. Ten years later, the concept of International Mobile Telecommunications 2000 was established and a series of technical specifications has been approved under the identification IMT-2000 (ITU, 1997). This specification was meant to be a unifying description that is able to attract various technologies covering many frequency bands, channel bandwidths, modulation formats, and network operators.

The following is a (non-exhaustive) list presenting the major objectives related to the communication features that IMT-2000 intends to provide:

- To make it easy to offer mobile subscribers a wide range of services, regardless of their location; and to offer the best possible quality of service by providing a large radio coverage and transport higher bandwidth.
- To make larger the number of services that can be offered, regardless of limiting constraints such as radio transmission, spectrum efficiency, and system economics.

- To offer high speed packet data rates such as:
  - 2Mbps provided for fixed environments;
  - 384 Kbps for pedestrian;
  - 144 Kbps for vehicular traffic.
- To maintain user mobility based on the registration of mobile terminals and the provision of the mobile subscribers with individual cards (such as the subscriber identity module cards used in GSM).
- To enhance on the security of the second generation systems, by addressing and correcting real and perceived weaknesses in GSM and other 2G networks.
- To permit and support international operation and roaming of mobile subscribers.

The specifications made publicly available for the 3G networks define a set of security features and fix operational security objectives that are needed to set up trustable environments and provide secure services. Such environments can be characterized by three major features: (a) a variety of mobile payment schemes (e.g., the prepaid payment, autonomous payment systems, and pay-as-you-go) is available; (b) an increased control over the mobile user's service profile is allowed for the subscribers; and (c) the mobile terminals can be used as a platform for sophisticated applications such as m-payment, m-commerce, and multimedia delivery.

Among the security objectives of 3G networks, one can distinguish the following non-exhaustive list:

- To accomplish the required mobile user authentication scheme based on unique user identification, unique user numbering, and unique equipment identification.
- To adopt the challenge and response authentication concept based on a symmetric secret key shared between the SIM card and the authentication center in the home environment, in use in the GSM networks.
- To ensure that a message generated by a user is adequately protected against misuse or mis-appropriation and minimize the likelihood of attacks by restricting access to services presenting vulnerabilities.
- To protect mobile users against misuse and theft of mobile stations by maintaining a list of stolen mobile stations identities and monitoring traffic for their use.
- To ensure that resources and services provided by 3G networks and home environments are appropriately protected.
- To support emergency services by providing useful information for the emergency calls. Such information includes user identity, location information, and any other information that might be needed for local authorities.

The frequency spectrum band allocated to IMT-2000 by ITU has two parts. They are the [1885 MHz, 2025 MHz] sub-band and the [2110 MHz, 2200MHz] sub-band. In 1999, the ITU attached significance to the fact that the following radio interface technology proposals should fulfill the IMT-2000's requirements, and accepted them as compatible: (a) the IMT Direct Spread (IMT-DS) and (b) the IMT Multicarrier (IMT-MC). The *3rd Generation Partnership Project Agreement* (3GPP,-) was formalized and signed. This agreement marked the creation of a standardization organization involving several telecommunications standards bodies and organizational partners. This organization, which is called 3GPP, seeks to build specifications for UMTS, a third generation system dealing with IMT-DS and deployed on two technologies: the extended GSM/GPRS network and UTRA radio interface. Besides, the specification of the second interface (i.e., IMT-MC) and its use is promoted by the 3GPP2 standardization organization (3GPP2,-). While 3GPP specifies a new radio interface, 3GPP2 specifies an interface compatible with the IS-95 systems. This alternative 3G proposal is known as CDMA2000.

6.1.1 Security Challenges

The main security challenges desired by ITU can be stated in very simple terms: The 3G standard should at least meet the following two requirements: (a) 3G security must be equivalent to the fixed network security; and (b) user privacy must be maintained while roaming. The first requirement implies that the same level of security should be provided despite the differences between security of the actual wireless networks and the security provided in fixed networks. These differences have been justified mainly by the following reason: The fixed networks benefit from the fact that to intercept a transmission the attacker needs physical access to the fixed network, whereas he only needs to be in a radio range of the wireless network to launch an attack targeting it. Four basic differences can be distinguished with respect to launching security attacks: the available bandwidth; the allowable error rates; the communication latency and variability; and the mobile station power limits. The second requirement implies that when the user is roaming, no secure connection between the home network and the roaming user is provided. The user's data might have to be sent over an unsecured connection. This might be used by an intruder to violate the users' privacy.

To meet the aforementioned requirements, the security models of 3GPP and 3GPP2 have been built and enhanced to cover the following objectives (3GPP, 205; Blumenthal, 2002):

- *Improve the 2G security architecture*: The enhancements attempted to address the subscriber authentication, the subscriber identity confidentiality, the radio interface encryption, the use of (removable) subscriber identity modules, and the creation of a secure application layer between the mobile phone and its home network.
- *Guarantee an adequate level of the protection offered*: An adequate level of protection should be provided to mobile subscribers, to all the information

generated and sent by users over the network, and to all the resources and services offered by the serving networks.

■ *Make certain the existence of specific features*: Specific security features should be available on 3G networks including at least a few encryption algorithms that might be used on a worldwide basis, an acceptable standardization of security properties, and the ability to extend security mechanisms by adding some features to them.

6.1.2 Security Threats

The past worldwide experience in using communication networks has led to a large range of threats that should be addressed by the 3G framework. The major attack categories experienced with 3G networks include (but are not limited to) the following:

■ *Misusing network services*: In this category, an intruder attacks network services with the objective of driving the service to a denial of service or reducing its availability. An example of attack in 3G is given as follows: An attacker can flood the call forwarding service with call forwarding requests and can cause a denial of service.

■ *Eavesdropping transmission*: In this attack category, an attacker manages to intercept a transmission. This can be done during voice transfer, signaling activity, and authentication process. Eavesdropping can cause privacy problems. The data collected by eavesdropping might also be used to perform attacks on 3G networks. For example, an attacker can view the call forward number and track the location of the victim.

■ *Manipulation attacks against messages*: In this class of attacks, an intruder manages to manipulate a transmission between two parties in order to modify flowing messages, corrupt transactions between the two parties, or simply change the exchanged packets.

■ *Man-in-the-middle attacks*: In this case, an intruder places himself between two parties involved in a transmission. Neither party is aware of the intruder's presence and both think that they are actually communicating with each other, while the intruder talks with each of them (3GPP, 200; Blumenthal, 2002). As these threats are the basis for which the 3G security architecture is developed, they will be used in this paper as a basis for a comparison with algorithms and protocols.

■ *Unauthorized access to networked services*: In this case, an intruder manages to get unauthorized access due to some masquerading or misuse of the access rights. An example of these would be a rogue shell attack, which allows an intruder to open a session on a remote system.

6.1.3 Retention of 2G Robust Features

Several security mechanisms have been shown to be robust and useful in 2G communication systems. 3GPP standards have to build on these mechanisms and retain their advantages. These mechanisms rely on four major issues: (a) the SIM-based authentication; (b) the confidentiality of user traffic on the air interface; (c) the radio interface encryption; and (d) the confidentiality of user identity on the radio interface.

- *SIM-based authentication*: The essential feature in 2G network security is the use of the SIM card as a removable security module, which is issued by and managed with the Home Environment operator. It is independent of the terminal where it is integrated. It was shown that the concept of SIM card had been the most significant in maintaining the security of GSM and that the user practice of security is highly similar to what the user is used to apply in a banking transaction, which can be described by: the client takes care of his digital card, reports its loss immediately, and does not release the PIN to other individuals.

  The 3G systems have to apply the challenge and response authentication mechanism based on a symmetric secret key shared between the SIM and the Authentication Centre in the Home Environment. The authentication key and algorithm for the challenge/response mechanism are not required by the Serving Network, which helps keep the level of trust placed in many possible serving networks to a minimum. This method also allows the algorithm to be made specific to the home domain, meaning that the impact of any compromise can be confined to the user base of just one operator rather than to the entire user community worldwide. However, problems with insufficient algorithms and conditions regarding configuration of authentication should be addressed. The strength of the encryption scheme should be maintained at a higher level in the 3G networks. And, a method for the negotiation of cryptographic algorithm should be included.

- *Confidentiality of user traffic on the radio interface*: Air interface encryption is being considered not only for the profit of the user. In fact, confidentiality is actually essential for the network operator, since it allows him to ensure that the validity of the authentication at the start of the call is preserved throughout the whole call duration and prevents a session from being hijacked. However, confidentiality of ubiquitous user traffic is a chimera that can hardly be fulfilled, essentially due to concern about restrictions on the use of encryption in several countries. On the other hand, the utilization of integrity protection in the signaling activity can be seen as an alternative means of providing authentication.

■ *Confidentiality of the user identity on the radio link*: It has been shown that the decision of allowing the network access to a mobile subscriber requires that the subscriber identifies himself by an identity known to the system, just like a user ID when accessing a computer system. In GSM, once this initial identification is sent over the air, then a temporary international mobile subscriber identity (TIMSI) is assigned. As shown in Chapter 5, the temporary identity is local to the area where the mobile is located and might be reassigned to another subscriber as soon as the first subscriber moves out of that area. The major objectives of the TIMSI are to reduce the exposure of the real user identity on the air interface and to prevent flow analysis by ensuring that information on the subscriber's use of a service or mobility cannot be collected. To benefit from these advantages, the concept of TIMSI is retained by the 3G standards.

6.2 The 3G Networks

Wireless network evolution can be characterized by an increase in functionality and its support for a growing number of services. However, this also means that the 3G networks are becoming increasingly complex in terms of architecture. 3G networks are likely to share the same major components within their communication architecture. For the sake of clarity, we describe in the following the UMTS architecture and several basic associated security functions.

6.2.1 Network Architecture

UMTS network presents an implementation of 3G-mobile systems, which is compatible in some way with the Global System for Mobile communication and the General Packet Radio Services networks. The fundamental difference between GSM/GPRS and UMTS is that the latter supports higher access rates. This is achieved through a Wideband Code Division Multiple Access radio interface for the land-based communication system, named UMTS Terrestrial Radio Access Network (UTRAN). Recent versions introduce new concepts and advanced features including the shift to an all-IP network architecture and the integration of an open service architecture, which aims at allowing network operators to offer third party access to their UMTS service architecture.

A UMTS network is logically divided into two parts, which are referred to as the *Core Network* (CN) and the *Generic Radio Access Network* (GRAN). The core network reutilizes several elements already present in GPRS and GSM networks (3GPP, 900). It consists of two overlapping domains: the *Circuit-Switched* (CS) domain and the *Packet-Switched* (PS) domain. The CS domain is made up of entities that allocate dedicated resources to the user traffic, control the signals when the connections are established, and release them when the sessions terminate. Often,

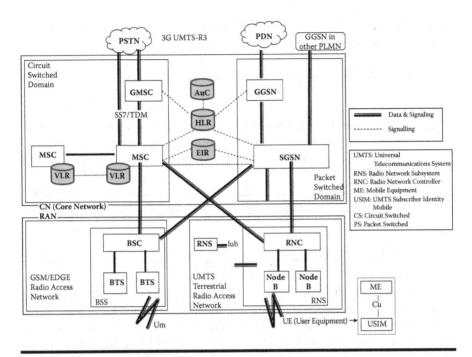

Figure 6.1 Architecture of a UMTS network.

voice calls are handled by the functions developed within the CS domain. The entities in the PS domain are responsible for transporting the user data in the form of autonomous packets, which are routed independently of each other. This attempts to overcome the limitations of 2G networks to transmit data efficiently. The user can set up a connection to and from external packet data networks and other wireless networks.

Let us now identify the key components of the UMTS architecture and explain the underlying procedures and interfaces. The basic architecture of an UMTS network is divided into three components as depicted in Figure 6.1. These components are: the mobile station (MS), the access network, and the core network (CN). The access network handles all the functions related to the radio resources and air interface management, while the core network performs switching functions and interfaces to external networks.

6.2.1.1 The Mobile Station (MS)

Similar to GSM, a MS is defined as a device allowing a user access to network services and the Universal Subscriber Identity Module (USIM). It is involved in any major UMTS procedures, call setup and management, handoff procedures, and mobility management. The USIM contains the functions and data needed to identify and authenticate users, as well as a copy of the user's service profile and the security elements needed for confidentiality and integrity services. UMTS mobile stations can operate using one of the three modes:

1. The circuit switching mode of operation, which allows the MS to be only attached to the CS domain and which can only operate services of the CS domain;
2. The packet switching mode of operation, which allows the MS to be only attached to the PS domain and which may only operate services of the PS domain, while not preventing CS-like services to be offered over the PS domain; and
3. The PS/CS mode of operation, where the MS is attached to the PS and CS domains and is capable of simultaneously operating PS services and CS services.

The USIM is an application stored in a removable smart card, which interoperates with the mobile equipment to provide access to 3G services. Similar to the SIM card, USIM has the following features: it unambiguously identifies a unique mobile subscriber; it stores subscription related information; it authenticates itself to the network and vice-versa (mutual authentication); it provides security functions; and finally, it stores information elements such as the preferred language, international mobile subscriber identity (IMSI), and cipher key.

6.2.1.2 The Access Network (UTRAN)

The UTRAN manages all the functions related to the radio resources and air interface management. As depicted in Figure 6.1, the UTRAN consists of two types of components, the *Node-Bs* and the *radio network controllers*, which play roughly equivalent roles to those performed in GSM by the base transceiver stations and the base station controller, respectively.

1. *Node B*: This is the physical unit for radio transmission/reception with mobile stations located within their radio cells. The base transceiver station of the UTRAN serves one or more radio cells. The main tasks of Node B are the air interface transmission/reception and CDMA physical channel coding. It also measures the quality and strength of the connections and determines the Frame Error Rate. It transmits these data to the RNC as a measurement report for handoff and macro diversity. Some of its functions are error detection on transport channels and indication to higher layers, modulation/demodulation of the physical channels, radio measurements and notification to higher layers, and power weighting. In addition, Node B also participates in the power control as it enables the MS to adjust its power.
2. *Radio Network Controller (RNC)*: This component manages the radio resources of each of the Node Bs that are under its control. The RNC connects Node B

to the transport network. It is responsible for handoff decisions that require signaling to the MS. The Node B resources are controlled from the RNC. The typical functions of RNC are the radio resource control, the admission control, the channel allocation, the power control settings, the handoff control, the macro diversity, and the encryption. Figure 6.1 shows that the RNC is connected to the core network's CS domain through the IuCS interface, and to the PS domain through the IuPS interface. The RNC is a part of the path to/from the core network for the services under use by the user equipment. Some other tasks performed by the RNC include the processing of voice and data traffic, the handoff between cells, and the call establishment and termination.

6.2.1.3 The Core Network (CN)

The CN is the structure responsible for transporting the user's data to its destination. It involves the use of a number of switching entities and gateways (such as the MSC, the Gateway MSC, the SGSN, and the GGSN) to the external networks (such as the Internet). It also maintains information regarding the user's access authorizations (involving the AuC and the EIR). Therefore, the CN also includes databases that store user profiles, and mobility management information (e.g., HLR and VLR).

Now, let us give a brief description of the role of the main elements that are specific to the core network's CS domain:

■ *Mobile Switching Center (MSC)*: This is the main component of the network's CS domain. It is the interface between the cellular network and the external fixed circuit-switched telephone networks such as PSTN. The MSC performs the routing of calls from the external network to any individual mobile station and all the switching and signaling functions needed by the mobile stations located in a geographical area defined as the MSC area. The additional functions performed by the MSC include (a) carrying out the procedures required for location registration and handover; (b) collection of data for charging purposes; and (c) encryption parameter management. Additional MSCs might coexist within the same cellular network if the traffic handled requires more exchange capacity than the one provided by a network using only one MSC. The IuCS interface links the MSC with the RNC in the UTRAN and some interfaces interconnect the MSC to the PS domain, the PSTN, the other MSCs, and the registration components in the network.

■ *Home Location Register (HLR)*: The HLR module, in UMTS, like the HLR in the GSM, stores data relevant to every mobile subscriber of the services provided by the mobile network. The data is collected when the user registers with the network. There are two types of information in an HLR register entry: the permanent and the temporary identities. The permanent data does not change except when a subscription parameter is required to be modified. The temporary data change continuously. It changes from controlling MSC to another and may even change from one radio cell to another and from a call to another. The permanent data relevant includes the IMSI and an authentication key. A mobile network can integrate several HLRs if it has a large size or when it covers a large area.

■ *Visitor Location Register (VLR)*: The VLR is generally implemented in connection with a MSC. It holds information related to every mobile station that roams into the area it covers via an associated MSC. Therefore, the VLR contains information about the active subscribers in its network. As the subscriber registers with different networks, the information in his HLR is copied to the VLR in every network visited, and discarded when the subscriber leaves that network. The information stored by the VLR is quite the same as that stored by the HLR. However, this may not be true if the mobile is roaming under the VLR.

■ *Authentication Center (AuC)*: The AuC is physically located with an HLR. It is responsible for the storage, for each subscriber, an authentication key K, as well as the corresponding IMSI. The AuC plays a crucial role in the network's security architecture, as will be discussed in the sequel since it is in charge of the generation of important data used in the authentication and the encryption procedures. Its role will be discussed in details in the sequel. The functions it performs to support roaming will also be addressed (mainly in Chapter 10).

The components of the PS domain in a UMTS network are nothing but upgraded versions of those defined for the GPRS networks. They are described below:

■ *Serving GPRS Support Node (SGSN)*: This component is responsible for the management of mobility and the handling of IP packet sessions. It routes user packet traffic from the radio access network to the ad hoc Gateway GPRS Support Node, which in turn provides access to external packet data networks. In addition, it generates the records to be used by other modules for charging purposes. The SGSN helps controlling access to network resources, preventing unauthorized access to the network and other specific services and applications. The IuPS interface links the SGSN, the main component of the PS domain, with the RNC in the UTRAN.

■ *Gateway GPRS Support Node (GGSN)*: The GGSN is the gateway between the cellular network and the packet data networks such as the Internet and corporate intranets. Similar to the SGSN, the GGSN also collects the billing information, which is forwarded to the Charging Gateway Function (CGF).

6.2.2 UMTS Security Architecture

The security generic architecture of UMTS networks is built on a set of security characteristics and protection mechanisms. A security characteristic is a service capability that complies with one or more security requirements. A security mechanism is a process that is used to carry out a security function. Figure 6.2 depicts the way security functions are organized together in five classes. Each class is facing a specific threat and achieving specific security objectives. In the following, we give a description of these classes:

- **Network access security** (Class I): The functions of this class provide secure access to 3G services and protect against attacks on the radio link.
- **Network domain security** (Class II): This class contains functions that allow the nodes in the operator's network to securely exchange signaling messages, and protects against attacks on the wired network targeting UMTS.
- **User domain security** (Class III): This class of functions aims at securing the access of mobile stations to the UMTS network and services.
- **Application domain security** (Class IV): The functions belonging to this class aim at enabling applications implemented at the user domain and the provider domain to securely exchange messages.
- **Visibility and configurability of security** (Class V): This class of functions allows the user to get information about the security functions that are in operation for him. The class also allows the user to check whether the provision of a service depends on the activation of some security features.

In addition, the network access security functions (Class I) can be further classified into three categories: entity authentication functions, confidentiality

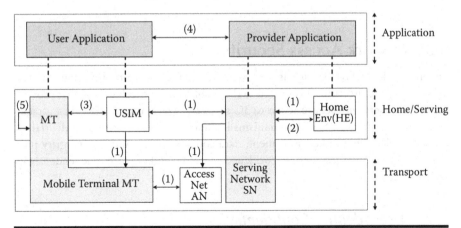

Figure 6.2 UMTS security architecture.

functions, and data integrity functions. The following is a list of the major elements in the first category:

- *Subscriber authentication*: the serving network confirms the identity of the subscriber.
- *Network authentication*: the subscriber confirms that he is connected to a serving network, which is authorized by the subscriber's home network to provide him with services; this includes the guarantee that this authorization is recent. It should be noted here that the concept of authentication presents many understated aspects.
- *Signaling data integrity and origin authentication*: the following security features are provided with respect to the integrity of data on the network access link:
- *Integrity algorithm agreement*: the mobile station (MS) and the serving network (SN) can securely negotiate the integrity algorithm that they use.
- *Integrity key agreement*: the MS and the SN agree on an integrity key that they may use subsequently; this is realized as part of the protocol that also provides entity authentication.
- *Data integrity and origin authentication of signaling data*: the receiving entity (MS or SN) is able to verify that signaling data has not been modified in an unauthorized way since it was sent by the sending entity (SN or MS), and that the data origin of the signaling data received is indeed the one claimed. The use of the integrity feature for signaling data is mandatory. This security feature has no equivalent counterpart in GSM standard. It provides protection against the attacks based on rogue BS, as the origin of signaling messages required to set up a communication with a mobile can now be authenticated by the mobile.

6.3 Network Access Security

The network access security is an essential class of security functions in the 3G-security architecture. It is related to the set of security mechanisms that provide the mobile users with secure access to 3G services, as well as protect against attacks on the radio interface. Such mechanisms include: the user identity confidentiality, the authentication and key agreement, data confidentiality, and the integrity protection of signaling messages. Network access security takes place independently in each service domain.

6.3.1 User Identity Confidentiality

User identity confidentiality allows the identification of a user on the radio access link by means of the Temporary Mobile Subscriber Identity (TMSI). This implies that the confidentiality of the user identity is protected, almost always, against passive eavesdroppers. The initial registration is an exceptional case where a temporary identity cannot be used, since the network does not yet know the permanent identity of the user. The allocated temporary identity is transferred to the user once the encryption is turned on. A TMSI in the CS domain (or P-TMSI in PS domain) has a local significance only in the location area or the routing area, in which the user is registered. The association between the permanent and temporary user identities is stored in the Visited Location Register or the Serving GPRS Support Node (VLR/SGSN). On the other hand, if the mobile user arrives into a new area, then the association between the permanent and the temporary identities can be obtained from the old location or routing area. If the address of the old area is not known or the connection cannot be established, then the permanent identity must be requested from the mobile user.

To avoid user traceability, which may lead to the compromise of the user identity confidentiality and the user location tracking, the user should not be associated for a long period of time with the same temporary identity. Moreover, the signaling or user data that might be used to reveal the user's identity must be encrypted on the radio access link. The following security features related to user identity confidentiality are guaranteed:

■ Confidentiality of the user identity: the permanent identity (IMSI) of a mobile user to whom a service is provided is protected against eavesdropping on the radio access link.
■ Confidentiality of the user location: the presence in an area or the arrival of a user in a certain area is protected against eavesdropping on the radio access link.
■ Untraceability of users: an intruder cannot deduce, by eavesdropping on the radio access link, whether specific services are delivered to the same user.

To achieve the aforementioned objectives, the user is normally identified on the radio access link by a temporary identity by which he is known at the serving network. To avoid user traceability, which may lead to the compromise of the mobile user identity confidentiality, the user should not be identified for a long period by means of the same temporary identity. In addition, it is required that any signaling message or user packet that might reveal the user's identity is ciphered on the radio access link. These features are identical to those provided by the GSM standard. They protect against passive attacks, but not against active attacks.

■ Identification of the mobile equipment: This feature is the same as in GSM. In certain cases, the serving network may request the mobile station to send its international mobile equipment identity (IMEI). However, neither GSM nor UMTS provide a method for authenticating the mobile equipment identity. This is mainly due to the complexity of designing and implementing a robust system. This means that any network features that are based on the

IMEI, such as the barring of stolen phones, relies on the terminal providing the legitimate IMEI to the UMTS network. The UMTS standards therefore impose requirements on terminals to protect the integrity of the IMEI so that it cannot be tampered with or modified.

■ Authentication of the user to the USIM: This feature is the same as in the GSM networks. It provides the property that access to the USIM is restricted until the USIM has authenticated the user. Thereby, it is ensured that access to the USIM can be restricted to an authorized user or to a number of authorized users. To realize this feature, the user and the USIM must share a secret (the personal identity number) that is stored securely in the USIM. The user gets access to the USIM only if he proves knowledge of the secret.

■ Authentication of the USIM–terminal link: This feature ensures that the access to a mobile terminal or another user's equipment can be restricted to an authorized USIM. To this end, the USIM network and the terminal must share a secret that is stored securely in the USIM and the mobile equipment. If a USIM node fails to prove its knowledge of the secret, it will be denied access to the mobile equipment.

6.3.2 Authentication and Key Agreement

The authentication and key agreement mechanism lead to the mutual authentication between the mobile user and the SN showing knowledge of the shared secret key. They also allow the derivation of the encryption and integrity keys for the session setup. The authentication method is composed of a challenge/response protocol (as depicted by Figure 6.3), and is chosen so as to achieve maximum compatibility with the GSM/GPRS security architecture. This would help to facilitate the migration from GSM/GPRS to UMTS. Furthermore, the USIM and the home equipment keep track of two counters, denoted by SQNMS and SQNHE, to support the network authentication. The sequence number SQNHE is an individual counter for each user, while the SQNMS denotes the highest sequence number that the USIM has accepted.

The UMTS authentication and key agreement (UMTS AKA) is a security mechanism to achieve the authentication and provide all the key agreement features described. This mechanism is based on the challenge/response authentication protocol similar to GSM's subscriber authentication and key establishment protocol, which makes it easier the transition from a GSM network to a UMTS network. The challenge/response protocol is a security measure set up for an entity to verify the identity of another entity without revealing the secret password shared by the two entities. The key concept is that one entity must prove to the other that it knows the password without actually revealing or transmitting it. Figure 6.3 depicts the AKA in 3G networks.

The UMTS AKA process described in this subsection is invoked by a serving network after the first registration of a user, a service request, a location update

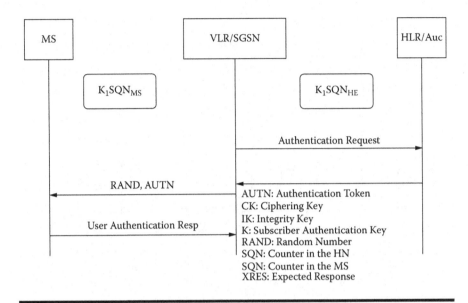

Figure 6.3 3G authentication and key agreement.

request, an attach request, and a detach request or connection re-establishment request. In addition, the relevant information about the user must be transferred from the user's home network to the serving network in order to complete the requesting process. The home network's HLR/AuC provides serving network's VLR/SGSN with Authentication Vectors (AVs), each one holding the information fields described as follows: on the receipt of a request from the VLR/SGSN, the home network authentication center (HN/AuC) forwards an ordered array of authentication vectors (AV) to the VLR/SGSN. Each AV consists of a random number RAND, an expected response XRES, a cipher key CK, an integrity key IK, and an authentication token AUTN.

Figure 6.4 shows the AV generation process by the HE/AuC. The HE/AuC starts with generating a fresh sequence number *SQN*, which proves to the user that the generated AV has not been used before, and an unpredictable challenge RAND. Then, using the secret key *K*, it computes the following components using specific functions denoted by *f1, f2, f3, f4,* and *f5*:

■ The first component is the message authentication code (MAC). It is given by:

$$MAC = f1(K, SQN, RAND, AMF),$$

where *f1* is a message authentication function, and the Authentication and key Management Field *AMF* is used to fine tune the performance or bring a new authentication key stored in the USIM into use.

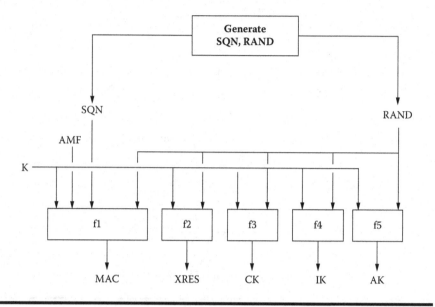

Figure 6.4 Generation of authentication vectors.

■ The second component is the expected response $XRES$ (given by $f2(K,RAND)$), where $f2$ is a (possibly truncated) message authentication function.
■ The following three components are the cipher key CK, the integrity key IK, and the anonymity key AK. They are given respectively by:

$$CK = f3(K, RAND), IK = f4(K, RAND), \text{ and } AK = f5(K, RAND),$$

where $f3, f4$, and $f5$ are specific key generating functions.
■ The last component is the authentication token AUTN. It is computed by the home equipment using the following expression:

$$AUTN = \langle SQN \oplus AK, AMF, MAC \rangle$$

It is worthy to notice that the design of the generation functions $f1, f2, f3, f4$, and $f5$ is based on the same basic algorithm and that they differ from each other in an essential way interdicting any deduction of any information about the output of one function from the output of the others. Because the five functions are used by the AuC and in the USIM, which are controlled by the home operator, the selection of the algorithms implementation is in general operator specific. However, an example of an algorithm set has been proposed called MILENAGE (3GPP, 205).

When the VLR/SGSN initiates an authentication and key agreement procedure, it selects the next AV from the vector of AVs, and forwards the parameters RAND

and *AUTN* to the user. Based on the same secret key *K*, the USIM computes the key *AK* using *f5(K, RAND)* and retrieves the *SQN* by applying the formula:

$$SQN = (SQN+AK) + AK.$$

Then, it computes *XMAC = f1(K, SQN, RAND, AMF)* and checks whether the received *AUTN* and the retrieved *SQN* values were indeed generated in AuC. If this is the case, the USIM computes the *RES = f2(k, RAND)* and activates the MS to send back a user authentication response. Finally, the USIM computes the keys *CK, IK* using:

$$CK = f3(k,RAND) \text{ and } IK = f4(k,RAND).$$

The VLR/SGSN compares the received *RES* with the *XRES* field of the *AV*. If they match, it considers that the authentication and key agreement exchange has been successfully completed. Finally, the USIM and the VLR/SGSN transfer the established encryption and integrity protection keys (*CK* and *IK*) to the mobile equipment and to the radio network controller, respectively, to allow them to use encryption and integrity functions.

6.3.3 Data Confidentiality and Integrity Protection of Signaling Messages

Once the user and the network have authenticated each other, they start securing the communication they have established between them. The user packets and signaling messages sent over the radio interface are subject to encryption using the function *f8*: The encryption/decryption process takes place in the MS and the RNC on the network side. The *f8* is a symmetric synchronous stream cipher algorithm that is used to encrypt frames of variable length. The main input to the *f8* is a 128-bit secret cipher key *CK* (3GPP, 201). Additional inputs are used to ensure that two frames are encrypted using different keystreams; they are formed by a 32-bit value called COUNT, a 5-bit value denoted by BEARER, and a 1-bit value called DIRECTION. The key stream and encryption is given by the following formulas:

$$keystreamBlock = f8(CK, BEARER, DIRECTION, length)$$

$$Cipher\text{-}text = keystreamBlock \oplus plaintextBlock$$

The output is a sequence of bits (or the *keystream*) of the same length as the frame. The frame is encrypted by XORing the data with the keystream. An implementation of the function *f8* is based on the Kasumi algorithm (3GPP, 908). The following security features are guaranteed by UMTS and are related to confidentiality of data on the network access link:

■ *Encryption algorithm agreement*: the mobile station and the UMTS network can securely negotiate the encryption algorithm that they want to use in the communication set up.

■ *Encryption key agreement*: the MS and the UMTS network should agree on the encryption key that they may use to protect the communication they set up.

■ *Confidentiality of user data and signaling messages*: neither the user packets nor the sensitive signaling messages can be overheard on the radio access interface.

The first two security features are also provided by the GSM network; however, the entities between which protection is provided are different. In UMTS, the protection extends to the RNC, so that microwave links between the base stations and the RNC are also covered. On the other hand, the radio interface in the 3G-mobile systems has also been designed to support integrity protection on the signaling channels. This enables the receiving entity to verify that the signaling messages have not been modified in an unauthorized way since they were sent. Furthermore, it ensures that the origin of the received signaling data is indeed the one claimed.

The integrity protection mechanism is not applied for the user plane due to performance reasons. A function, referred to as $f9$, is used to authenticate the integrity and the origin of signaling messages between the MS and the RNC in the UMTS (3GPP, 908). It computes a 32-bit message authentication code *MAC*, which is appended to the frame and is checked by the receiver. The main inputs to the algorithm are an 128-bit secret integrity key *IK* and the variable length frame, denoted by *MESSAGE*. Additional inputs are added to ensure that the MACs related to two frames with identical contents (i.e., MESSAGE) are different, for example. These inputs include a 32-bit value *COUNT*, a 32-bit value *FRESH*, and a 1-bit value *DIRECTION*. The expression is formally given by:

$$MAC = f9(IK, COUNT\text{-}I, MESSAGE, DIRECTION, FRESH).$$

To provide the functions $f9$, the UMTS uses an implementation based on the Kasumi algorithm. This will be shown in the sequel.

6.4 Network Domain Security

The network domain security (NDS) functions ensure that the exchanged messages for signaling within the UMTS core network are protected. Different protocols and interfaces are used to provide the control plane signaling. Among these protocols, one can mention the mobile application part (MAP) and the GPRS tunneling protocol. These are protected by standard procedures based on the existing cryptographic techniques. Specifically, the IP-based protocols are protected at the network level by means of IPsec (Blumenthal, 2002), while the protection for the SS7-based protocols is realized at the application layer. In the following, the NDS context for IP-based and SS7-based protocols is presented. Furthermore, the employment of traditional security technologies, originally designed for fixed networking, such as the firewalls and the static VPNs, is examined in order to safeguard the UMTS core network from external attacks and to guard the user data when they are flowing over the public Internet.

6.4.1 IP-Based Protocol

The UMTS network domain control plane is sectioned into security domains, which typically coincide with the operator borders. Security gateways (SEGs) are entities at the borders of the IP security domains used for securing the native IP-based protocols. It is noted that NDS does not extend to the user plane, which means that a packet flowing into the user plan will not be protected by the SEGs. The key management functionality is logically separate from the SEG; and key administration centers (KACs) negotiate the IPsec security associations (SAs) by using the Internet Key Exchange (IKE) protocol, on behalf of the network entities (NEs) and the SEGs. The KACs also distribute the SAs parameters to the NEs or the SEGs through standard interfaces.

To secure the IP traffic between two network equipment (NEs), two modes may be applied: a hop-by-hop or an end-to-end scheme. The former scheme requires the originating NE to establish an IPsec tunnel to the appropriate SEG in the same security domain and forwards the data to it. The SEG terminates this tunnel and sends the data through another IPsec tunnel to the receiving network. A second tunnel is terminated by the SEG in the receiving domain, which in turn uses IPsec to pass the data to its final destination. The latter scheme implies that an IPsec SA is established between the two NEs. This scheme can also be applied in the case where the two NEs belong to the same security domain.

Node authentication can be accomplished using either pre-shared symmetric keys or public keys. Using pre-shared symmetric keys means that the KACs or the NEs do not have to perform public key operations. It also means that there is no need for establishing a public key infrastructure. The IPsec can be configured either in transport mode or in tunnel mode. Whenever at least one endpoint is a gateway, then, the tunnel mode suits better. Finally, the IPsec protocol shall always be the ESP (i.e., the encapsulation security payload protocol), given that it can provide confidentiality and integrity protection at the same time.

6.4.2 SS7-Based Protocols

The NDS for SS7-based protocols can mainly be implemented at the application layer. Specifically, when the transport of traffic relies on the SS7 protocol or on a combination of the SS7 and the IP protocol, then the security service should be provided at the application layer. On the other hand, when the transport is based only on the IP protocol, the security may be provided at the network layer, exclusively by using IPsec or in addition to the application layer security. The needed SAs for the signaling protection at the application layer will be network-based and can be negotiated similarly to the IP-based architecture. Moreover, the end-to-end protected signaling messages will be indistinguishable to unprotected signaling traffic to all parties, except for the sending and receiving entities.

It is worth noticing that in the UMTS R4, the unique protocol that is to be protected is the MAP. The complete set of enhancements and extensions that facilitate the MAP security is referred to as the MAPsec (3GPP, 200). The MAPsec covers the security management procedures, as well as the security of the transport protocol including data integrity, data origin authentication, anti-reply protection, and confidentiality. However, the IKE adaptation is required.

6.4.3 Traditional Network Security Features

In addition to the security features that are provided by the 3G security architecture, the mobile network operators can apply the traditional security technologies used in wired networking to protect the UMTS core network and any internetwork communications between the 3G network and the external networks. Typically, the user data in the UMTS backbone network are transmitted in clear-text. This may expose them to different threats. Moreover, the inter-network communication is generally based on the publicly known Internet. This may enable performing IP spoofing by malicious third parties who may get access to the resources in the 3G mobile network. To counteract against these threats, the mobile providers can use three complementary technologies: the intrusion protection systems, the intrusion detection systems, and VPNs.

The intrusion protection systems can be characterized as a set of mechanisms that aim at enforcing a security policy on the data flowing from and to a corporate network. They can be implemented at the borders of the mobile networks (or on some access points in the network) and can allow the analysis, the correlation, and the filtering of data flows originated from (or to) specific foreign networks. For this, several techniques are developed including various types of networked firewalls that protect the UMTS backbone from unauthorized penetration. On the other hand, application firewalls prevent direct access through the use of proxies for services, which analyze application commands, perform authentication, and keep logs protected.

Nevertheless, the intrusion prevention systems do not provide privacy and confidentiality, in general. Solutions based on VPNs have to be set up to complement them to protect data in transit. VPN establishes a secure tunnel between two points, encapsulates and encrypts data, and authenticates and authorizes user access of the corporate resources on the network. They extend the dedicated connections between remote branches, or remote access to mobile users, over a shared infrastructure. Their deployment provides a two-fold benefit for the enterprise: the low cost of initial investment and operation and security level offered to the internal communication. Different mechanisms that provide VPN over wireless and mobile IP networks have been developed.

The border gateway is an element that resides at the border of the UMTS core network and provides the appropriate level of security policy (e.g., firewall). It also maintains static pre-configured security tunnels (e.g., IPsec tunnels) granting VPN services to specific peers. A border gateway may be able to serve as a gateway between the PS domain and the external IP network that is used to provide connectivity with other PS domains located in other core networks.

The application-layer security builds security features into individual applications. They operate independently of any network security measures. Many applications have special security requirements that simply cannot be met by common network security services. However, the security at this level is by far the easiest to deploy, as long as all users are running a homogeneous set of applications on standard platforms. While these methods are effective for solving specific security problems, such solutions are by their nature limited to their specific domains. For example, the transport layer security works fine for simple client-service environments, but in the cases where the service needs a large number of cross references to other servers, the number of required key exchange operations may be overloading for the clients.

On the other hand, the shortcomings of the security mechanisms used in the 3G networks exacerbates the need for new detection techniques that should defend against sophisticated mobile attacks. In the literature, many attempts have been made to fulfill this need. Most of the existing approaches rely on intrinsic signal characteristics to detect intrusion events. The wireless intrusion detection system is a network component aiming at protecting the network by detecting wireless attacks, which target wireless networks having specific features and characteristics. Wireless intrusions can belong to two categories of attacks. The first category targets the fixed part of the wireless network, such as MAC spoofing, IP spoofing, and DoS; and the second category of these attacks targets the radio part of the wireless network, such as the Access Point (AP) rogue, noise flooding, and wireless network sniffing. The latter attacks are more complex because they are hard to detect and to trace-back (Mateli, 2006; Meddeb, 2007).

To detect such complex attacks, the WIDS deploys approaches and techniques provided by intrusion detection systems (IDS) protecting wired networks. The deployment of these approaches in wireless environment requires some modifications. Features and characteristics of wireless environment make the use of traditional approaches of detection very difficult. The major feature is mobility, where information has to be gathered from different mobile sources, which may require a real time traffic analysis. Moreover, there are no clear differences between the "normal" and "abnormal" behaviors in mobile environment.

Therefore, traditional approaches of detection have to be revised. The signature-based approach in wireless networks may require the use of a knowledge base containing the wireless attack signatures while an anomaly-based approach requires the definition of profiles specific to wireless entities (mobile users and AP). The wireless intrusion detection can be done by monitoring the active components of the wireless network, such as the APs. Generally, the WIDS is designed to monitor and report on network activities between communicating devices. To do this, the WIDS has to capture and decode wireless network traffic (Hall, 2005). Some WIDSs can only capture and store wireless traffic. For example, the system, called WITS (Valli, 2004), retains multiple log files that contain system statistics and sufficient network related data in order to trace back the intruder. Other WIDSs are able to analyze signal fingerprints, which can be useful in detecting attacks.

6.5 User, Application, and Visibility Domain Security

User domain security ensures secure access to the MS. It is based on a physical device called UMTS Integrated Circuit Card, which can be easily inserted and removed from terminal equipment, containing security applications such as the USIM. The USIM represents and identifies a user and his association to the home equipment. It is responsible for performing subscriber and network authentication, as well as key agreement, when 3G services are accessed. It may also contain a copy of the user's profile.

The USIM access is restricted to an authorized user (or to a number of authorized users). To accomplish this feature, the user and the USIM must share a secret (similarly to the GSM). In addition, access to a terminal or to other user equipment can be restricted by the user security domain to an authorized USIM. To this end, the USIM and the terminal must also share a secret. If a USIM fails to prove its knowledge of the secret, then the access to the equipment is denied.

6.5.1 Application Domain Security

The application domain security deals with secure messaging between the MS and the SN or the SP over the network with the level of security chosen by the network operator or the application provider. A remote application should authenticate a user before allowing him to utilize the application services, and it could also provide for application-level data confidentiality. Application-level security mechanisms are needed because the lower layers' functionality may not guarantee end-to-end security provision. Lack of end-to-end security could be envisioned when, for instance, the remote party is accessible through the Internet.

USIM Application Toolkit provides the capability for operators or third party providers to create applications that are resident on the USIM. To assure secure transactions between the MS and the SN or the SP, a number of basic security mechanisms such as entity authentication, message authentication, replay detection, sequence integrity, confidentiality assurance, and proof of receipt have been specified and integrated in the USIM Application Toolkit.

The Wireless Application Protocol (WAP) is a suite of standards for delivery and presentation of Internet services on wireless terminals, taking into account the limited bandwidth of mobile networks, as well as the limited processing capabilities of mobile devices. To connect the wireless domain to the Internet, a WAP gateway is needed to translate the protocols used in WAP segment to the protocols used in the public Internet. To secure data transmission in the WAP architecture, the Wireless Transport Layer Security (WTLS) protocol is employed. WTLS has been optimized for use over narrow-band communication channels providing also datagram support. It ensures data integrity, privacy, authentication, and denial of service protection. The WAP gateway automatically and transparently manages wireless security, and conveys protected data between the WTLS and TLS security channels for Web applications that employ standard Internet security techniques with TLS.

On the other hand the protocol WAP 2.0 has proceeded to the re-design of the WAP architecture by introducing the existing IP stack into the WAP environment. The new architecture allows a range of different gateways, which enables conversion between the two protocol stacks anywhere. A TCP-level gateway allows for two versions of TCP, one for the wired and another for the wireless network, on top of which a secure TLS channel can be established all the way from the mobile device to the server. The availability of a wireless profile of the TLS protocol, which includes cipher suites, certificate formats, signing algorithms, and the use of session resume, enables end-to-end security support at the transport level allowing interoperability for secure transactions.

6.5.2 Security of Visibility and Configurability Domain

Although the security measures provided by the SN should be transparent to the end user, the visibility of the security operations as well as the supported security features should be provided. Such support may include indication of (a) the access network encryption; (b) the network wide encryption; and (c) the level of security provided (particularly, when the user roams from a 3G network to a 2G network).

Configurability enables the mobile user and the HE to configure whether a service provision should depend on the activation of certain security features. A service can only be used when all the relevant security features are in operation. The configurability features that are proposed include the following: (a) enabling/disabling user-USIM authentication for certain services; (b) accepting/rejecting incoming non-ciphered calls; (c) establishing non-encrypted calls; and (d) accepting/rejecting the use of certain encryption algorithms.

6.6 Security Functions

We discuss in the section the main features and constructs of the functions used by USIM to provide integrity and confidentiality.

6.6.1 Integrity and Confidentiality Algorithms

The mechanism that provides the security service for the signaling information transmitted between the mobile station and the network is based on the UMTS Integrity Algorithm (UIA), which is implemented in the mobile station and in the module of the RNC. The UIA of interest in this subsection is the $f9$ algorithm. The procedure of data integrity verification provided by $f9$ works using four steps: in the first step, the $f9$ function set up in the user equipment computes a 32-bit message authentication code (MAC-I) for data integrity based on its input parameters, which include the signaling data (MESSAGE). In the second step, the message authentication code computed is attached to the signaling information and sent over the radio interface to the RNC. In the third step, the RNC computes XMAC-I on the signaling data received in the same way as the mobile station has computed MAC-I. In the fourth step, the integrity of the signaling information is determined by simply comparing the MAC-I and the XMAC-I.

A detailed description of each of the input parameters is out of the scope of this chapter; further details concerning their meaning can be found in the literature. Figure 6.5 shows that the internal structure of the $f9$ algorithm uses the shared integrity key *IK* and is based on a chain of block ciphers implementing the KASUMI algorithm. The outputs of the block ciphers are 64-bit long, but the output of the whole algorithm is 32-bit long.

Contrasting with the integrity algorithm, which only operates on the signaling data, the confidentiality function $f8$ applies on the signaling data and the user data. The algorithm defined to perform the confidentiality tasks operates in the following way: First, the $f8$ algorithm in the subscriber equipment computes an output bit stream, using the ciphering key *CK*, and some other parameters. Second, the computed output bit stream is XORed bit by bit with the data stream, also called plaintext, in order to obtain a ciphered data block or cipher-text. Third, the cipher-text is sent to the network through the radio interface. Fourth, the $f8$ algorithm in the RNC uses the same inputs as the user equipment, including the shared cipher key *CK*, to generate the same output bit stream that was computed in the user equipment. Finally, the output bit stream is XORed with the cipher-text received to

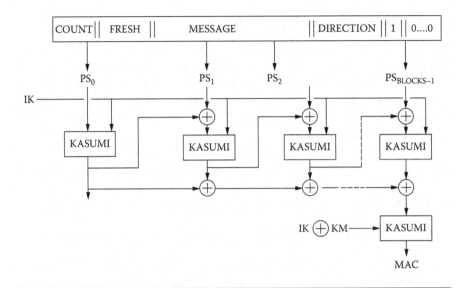

Figure 6.5 The structure of integrity algorithm f9.

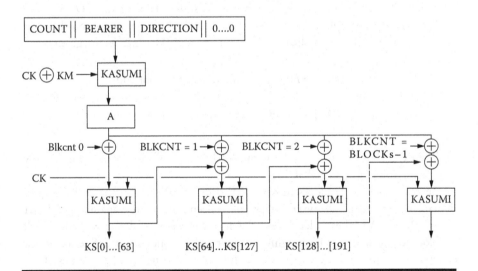

Figure 6.6 The confidentiality algorithm f8.

retrieve the initial information. Figure 6.6 depicts the structure of the *f8* algorithm. Once again, it can be seen that several blocks based on the KASUMI blockcipher are present; this time the blocks are connected in the so-called output-feedback mode. Each block generates 64 bits of the output bit stream and forwards its output to the input of the following block. Figure 6.6 illustrates the *f8* structure.

6.6.2 *The KASUMI Block Cipher*

As can be noticed from the former subsection, the KASUMI block cipher is at the core of the integrity and confidentiality mechanisms in UMTS networks. It operates on 64-bit data blocks controlled by a 128-bit key, say K (3GPP, 202). KASUMI has the following features: (a) it is based on eight rounds of processing, (b) input plaintext is the input to the first round; (c) the Cipher-text is obtained as the last round's output. The encryption key K is used to generate a set of round keys (KL_i, KO_i, KL_i) for each round i. Each round computes a different function, as long as the round keys are different. The same algorithm is used both for encryption and decryption.

Then for each i, $1 < i < 8$, the Kasumi defines

$$I_i = L_i \| R_i, \quad R_i = L_{i-1}, \qquad L_i = R_{i-1} \oplus g_i \left(L_{i-1}, RK_i \right)$$

where g_i denotes the round function, L_i is the left half part of I_i, and RK_i is the ith round key. The output of KASUMI is $I_S = L_S \| R_S$. It can be seen that the functions g_i, $1 < i < 8$, are defined using two sub-functions FL (for $i = 2, 4, 6, 8$) and FO (for $i = 1, 3, 5, 7$). If the round key RK_i comprises the subkey triplet (KL_i, KO_i, KI_i), the FL function has a simple structure and consists of logical operations and shifts on the inputs. FO is more complicated. The structures of the function are given by

$$g_i \left(X, RK_i \right) = FO \left(FL \left(X, KL_i \right), KO_i, KI_i \right), \text{ for odd } i$$

$$g_i \left(X, RK_i \right) = FL \left(FO \left(X, KO_i \right), KL_i, KI_i \right), \text{ for even } i$$

The expression of FO and FL are out of the scope of this chapter. Let us now discuss the robustness of KASUMI.

Several attacks have addressed KASUMI (Biham, 2005; Blunden, 2002). In particular, higher order differential attacks have been constructed against KASUMI (Sugio, 2007). A higher order differential attack is a powerful and versatile attack on block ciphers. It can be roughly summarized as follows: (a) define an attack equation to estimate the key by using the higher order differential properties of the target cipher, (b) derive the key by solving an attack equation. Linearizing attack is an effective method of solving attack equations. It linearizes an attack equation and determines the key by solving a system of linearized equations using approaches such as the Gauss-Jordan method. We enhance the derivation algorithm of the coefficient matrix for linearizing attack to reduce computational cost. In addition, the elimination of most unknown variables in linearized equations has been applied to an attack of the five-round variant of KASUMI and has shown that the attack complexity is equivalent to $2^{28.9}$ chosen plaintexts and $2^{31.2}$ KASUMI encryptions (Sugio, 2007).

6.7 Security Features of 3G Networks

6.7.1 *Mitigating 2G Weaknesses*

The 3G objectives have addressed the mitigation of the following weaknesses observed in the security of 2G networks:

- active attacks using a rogue BTS are launchable in 2G networks;
- the cipher keys and authentication data are transmitted in clear between and within networks;
- the encryption is only performed on the wireless link. This may result in the transmission of user and signaling data in a non-crypted form across micro-wave links (from the BTS to the BSC, in the case of GSM);
- the data integrity is not provided. Data integrity defeats certain rogue BTS attacks and provides protection against channel hijack;
- the IMEI is an unsecured identity;
- there is no HE knowledge or control of how an SN uses authentication parameters for HE subscribers roaming in that SN; and
- the 2G systems do not have the flexibility to upgrade and improve security functionality over time.

Mitigation has mainly addressed the following security issues and attacks (3GPP, 900):

6.7.1.1 *Denial of Service*

DoS attacks launched using request spoofing have been made unfeasible for 3G networks by simply providing integrity and non-replay of signaling requests. Such attacks include the following:

- *User de-registration request spoofing*: This attack requires a modified MS and exploits the weakness that the 2G network cannot authenticate the messages it receives over the radio interface. The intruder may spoof a deregistration request (IMSI detach) to the 2G network. Then the network de-registers the user from the visited location area and instructs the HLR to do the same. The user is subsequently unreachable for mobile terminated services. Integrity protection of critical signaling messages protects against this attack in the 3G networks. More specifically, data authentication and replay inhibition of the de-registration request allows the serving 3G network to verify that any de-registration request is legitimate.
- *Location update request spoofing*: An attack that requires a modified MS and exploits the weakness that the 2G network cannot authenticate the messages it receives over the radio interface. The user spoofs a location update request in a different location area from the one in which the user is roaming. The

network registers in the new location area and the target user will be paged in that new area. The user is subsequently unreachable (where he is actually) for mobile terminated services. The integrity protection of critical signaling messages made the 3G networks protects against this attack. More specifically, data authentication and replay inhibition of the location update request allows the serving network to verify that the location update request is legitimate.

6.7.1.2 Identity Catching

Several attacks can be launched against the user identity confidentiality in 2G networks. The following attacks have been counteracted in 3G networks:

■ *Passive identity catching*: This is a passive attack that requires a modified MS and exploits the weakness that the GSM network may sometimes request the user to send his identity in a clear form. The identity confidentiality mechanism provided in 3G networks counteracts this attack. The use of temporary identities allocated by the serving network makes passive eavesdropping inefficient since the user must wait for a new registration or a mismatch in the serving network database before he can capture the user's permanent identity in plaintext. One can notice, however, that the permanent identity may be protected in the event of new registrations or serving network database in 3G networks.

■ *Active identity catching*: This is an attack that requires a modified BS and exploits the weakness that the 2G network may request the MS to send its permanent user identity in a clear form. An intruder attracts the target user to camp on his false BS and consequently requests the target user to send his permanent identity in cleartext (by forcing a new registration or claiming a temporary identity mismatch due to database failure). The identity confidentiality mechanism provided in UMTS networks counteracts this attack by using an encryption key shared by a group of users to protect the user identity. We notice, however, that the size of the groups chosen should be small, in general, since if it is too large the group encryption key might be vulnerable to attack.

6.7.1.3 Impersonation of the Network Attacks

These attacks aim at impersonating a legitimate network. The ultimate objective of the attacks is to eavesdrop on user data or send to a user information that is subsequently thought to be initiated from an authentic network or a mobile user. Three attacks can be distinguished.

■ *Impersonation of the network by suppressing encryption between the target user and the intruder*: This attack requires a modified BS and exploits the weakness that the MS cannot authenticate messages received over the radio interface in the GSM network. The target user is attracted to camp on the rogue BS. When the intruder or the target user initiates a service, the intruder does not enable encryption by spoofing the cipher mode command. The intruder maintains the call as long as it is required (or his attack remains undetected). The 3G network provides a mandatory cipher mode command with message authentication and replay inhibition to allow the mobile to verify that the encryption has not been hidden by an attacker.

■ *Impersonation of the network by suppressing encryption between the target user and the legitimate network*: This attack requires a modified BS/MS and exploits the weakness that the 2G network cannot authenticate messages received over the radio interface. The target user is attracted to camp on the rogue BS/MS. When a call is set up, the false BS/MS modifies the encryption capabilities of the MS to make it appear to the network that a real incompatibility exists between the network and the MS. The network may then decide to establish an un-enciphered connection. After the decision not to encrypt has been made, the intruder removes the connection with the network and impersonates the network to the target user.

To protect against these attacks, 3G networks set up a mobile station classmark with message authentication and replay inhibition to allow the network to verify that the encryption has not been suppressed by an attacker.

■ *Impersonation of the network by forcing the use of a compromised cipher key*: This is an attack that requires a modified BS and the possession by the intruder of a compromised authentication vector. In 2G networks, this attack exploits the weakness that the user has no control upon the cipher key. The target user is attracted to camp on the rogue BS/MS. When a call is established, the false BS/MS forces the use of a compromised cipher key on the mobile user. Therefore, the intruder can maintain the call as long as the attack remains undetected. The presence of a sequence number in the challenge allows the USIM to verify the freshness of the cipher key to help protect against forced reuse of a compromised authentication vector.

However, one can notice that the 3G architecture does not protect against forced use of the compromised authentication vectors that have not yet been used to authenticate the USIM card. Thus, the 3G networks are still vulnerable to attacks using compromised authentication vectors that have been intercepted between generation in the authentication center and their use. To overcome this, the SN (transitively via the HE) should handle authentication vectors securely.

6.7.1.4 Eavesdropping on User Data

These attacks aim at eavesdropping on user data that is transmitted through the legitimate network to the intended recipient. Three different attacks can be launched in the GSM network.

- *Eavesdropping on user data by suppressing encryption between the target user and the intruder*: Such an attack needs a modified BS/MS and takes advantage of the weakness that the MS cannot authenticate messages received over the radio interface in the 2G network. The target user is attracted to camp on the rogue BS. When the target user or the intruder initiates a call, the network does not enable encryption by spoofing the cipher mode command. The attacker, however, sets up his own connection with the legitimate network using his own subscription. The attacker may then subsequently eavesdrop on the transmitted user data.

  A mandatory cipher mode command with message authentication and replay inhibition allows the mobile to verify, in the 3G networks, that the encryption has not been suppressed by an attacker.

- *Eavesdropping on user data by suppression of encryption between the target user and the legitimate 2G network*: This attack uses a modified BS/MS and utilizes the weakness that the 2G network cannot authenticate messages received over the radio interface. When the target mobile user is attracted to camp on the rogue BS/MS, the target user or the real network sets up a connection and the false BS/MS modifies the ciphering capabilities of the MS to make it appear to the network that a real incompatibility exists between the network and the MS. The network may then decide to establish an un-enciphered connection. After the decision not to cipher has been made, the intruder may eavesdrop on the user data.

  Message authentication and replay inhibition of the mobile's encryption capabilities allows the 3G network to verify that the encryption has not been suppressed by an attacker.

- *Eavesdropping on the user data by forcing the use of a compromised cipher key*: This attack is based on the use of a modified BS/MS. It requires the possession of a compromised authentication vector. It exploits the weakness that the user has no control on the cipher key in the GSM network. The target user is attracted to camp on the false BS/MS. When the target user or the intruder sets up a service, the false BS/MS forces the use of a compromised cipher key on the mobile user while it builds up a connection with the genuine network using its own subscription.

  The presence of a sequence number in the challenge allows the USIM card to verify the freshness of the cipher key to help the protection against forced re-use of a compromised authentication vector. However, similar to the preceding attack, the architecture does not protect against force use of compromised authentication vectors, which have not yet been used to authen-ticate the USIM.

6.7.1.5 Attacks Aiming at the Impersonation of the User

This category of attacks aims at using weaknesses of the mobile network to imper-sonate legitimate mobile users. Three types of such attacks can be distinguished in the GSM network.

■ *Impersonation of the user through the use of a compromised authentication vector*: An attack that requires a modified MS and the possession by the intruder of a compromised authentication vector that is meant to be used by the 2G network to authenticate a legitimate user. The intruder uses that data to impersonate the target mobile user toward the network and a third party. The presence of a sequence number in the challenge means that the authentication vectors cannot be reused to authenticate USIM cards. This helps to reduce the opportunity of using a compromised authentication vector to imperson-ate the mobile users. However, the network is still vulnerable to attacks using compromised authentication vectors, which have been intercepted between their generation in the authentication center and their use and destruction in the serving network.
■ *Impersonation of the user through the use by the network of an eavesdropped authentication response*: This is an attack that requires a modified MS and exploits the weakness that an authentication vector may be used several times. The intruder eavesdrops on the authentication response sent by the user and uses the collected information when the same challenge is sent later on. Subsequently, ciphering has to be avoided between the attacker and the network. The intruder uses the eavesdropped response data to impersonate the target user toward the network and a third party.

    The presence of a sequence number in the challenge means that the authen-tication vectors should be reused to authenticate USIMs.
■ *Hijacking outgoing calls in networks with encryption disabled*: This attack needs a modified BS/MS in order to be executed. While the target mobile user is camping on the false base station, the intruder pages the target user for an incoming call. The user then initiates the call establishment procedure, which allows the intruder to appear between the serving network and the target user, modifying the signaling elements accordingly (in such a way that, for the serving network, it appears as if the target user wants to set up the origi-nated call). The network does not enable encryption. After authentication, the intruder drops the connection with the target user, and uses the connection with the network to make fraudulent calls on the target user's subscription.

    The integrity protection of critical signaling messages counteracts against this attack. Indeed, the data authentication and replay interdiction of the connection setup request allows the serving network to verify that the request is legitimate. In addition, periodic integrity secure messages during a connection help protect against hijacking of non-encrypted connections after the initial connection establishment. However, hijacking the channel between periodic integrity protection messages is still possible, although this may be of limited use to attackers. In general, connections with encryption disabled will always be vulnerable to channel hijacking.

- *Hijacking outgoing calls in networks with encryption enabled*: This attack requires a modified BS/MS. In addition, the intruder has to attempt to suppress encryption by modification of the message in which the MS informs the network of its encryption capabilities. The integrity protection of critical signaling messages defends against this attack. More specifically, the data authentication and replay inhibition of the MS classmark and the connection setup request help prevent the suppression of encryption. It also allows the serving network to verify that the request is legitimate.
- *Hijacking incoming calls in networks with encryption disabled*: This attack requires a modified BS/MS. While the target user camps on the false base station, an associate of the intruder makes a call to the target mobile user, using his IMSI. The intruder acts as a relay between the network and the target user until authentication and call establishment have been performed between the victim and serving network. After authentication and call setup, the intruder releases the target user, and uses the connection to answer the call made by his associate. Therefore, the target mobile user will have to pay for the communication.

Integrity protection of critical signaling messages, in 3G networks, guards partly against this attack. More specifically, the data authentication and replay inhibition of the connection accept message allows the serving network to verify that the request is legitimate. In addition, periodic integrity protected messages during a connection help defend against hijacking of non-enciphered connections after the initial connection establishment. Nevertheless, hijacking the channel between periodic integrity protection messages is still possible, although this may be of limited use to attackers. Typically, connections with ciphering disabled will always be vulnerable to some degree of channel hijacking.

6.7.2 New Security Features and the Security of New Service

The new security service features that will be secured by the 3G networks cannot be exhaustively listed. However, the environments where these features are likely to be developed can be addressed. 3G security will be in charge of the security of these environments. These environments can be characterized by the following aspects:

■ The environments should allow different providers of services to operate. Providers will be able to provide content services, data services, and HLR-based services (or telephony based services);

■ The environments should allow the use of different mobile payment systems. A variety of prepaid and pay-as-you-go services will be accessible. Payment should be made by service, and long term subscription between the user and a unique network operator may not be the main means.

■ The environments should allow an increased control for the user over their service profile, as well as over the capabilities of their terminal. It should be possible to download new services and functions;

■ The environments should be robust to active attacks and should tolerate passive attacks. They should not allow active attacks to use equipment to impersonate parts of the network to actively cause failures in security. In passive attacks, the attacker that is outside the system can listen in, while hoping security lapses occur;

■ The environments should allow the profusion of secured non-voice services, since the latter will be as important as, and even more important than, voice services;

■ The environments should allow the use of mobile terminals as a platform for e-commerce and other e-services. Multi-application smart cards where the USIM is one application among many can be used securely with the mobile terminal.

6.8 Attacks on 3G Networks

Many attacks can be launched against the 3G networks despite the security efforts made to protect these networks. Among these attacks, one can distinguish the following two attacks that are well known in the 2G networks:

1. *Camping on a false BS attack*: This denial of service attack exploits the weakness that a user can be conducted to camp on a false base station. Its effects are comparable to those of the radio jamming, which is very difficult to thwart effectively in any radio system. Once the victim user camps on the radio channels of a false base station, he is out of reach of the paging signals of the serving network to which he has registered. The security features for 3G networks do not counteract this denial of service attack. But fortunately, this attack only persists as long as the attacker is active, unlike other attacks whose effects persist beyond the moment where the involvement of the attacker stops.

2. *Camping on a false BS/MS attack*: This is another denial of service attack. It allows a false BS/MS to first act as a repeater during some period of time and

relay some requests in between the network and the victim user. Second, it allows the attacker to subsequently modify or ignore certain service requests and/or paging messages to be delivered to the victim user. The security architecture of UMTS does not propose any mechanism to prevent against false BS/MS relaying of messages between the network and a victim mobile user, nor does it prevent the action of a false BS/MS ignoring certain service requests and paging requests. However, the integrity protection of critical message may help to prevent some forms of these attacks, which are induced by modifying certain messages. One can also notice that the denial of service in this case only persists for as long as the attacker is active.

Various reasons can be cited to explain what makes many attacks easy to launch. First, the availability of inexpensive off-the-shelf mobile radio test equipments makes it easy to impersonate some parts of the network. The intruder can analyze traffic, eavesdrop, intercept signaling messages, modify signaling messages, and jam the radio interface using these equipments. Second, the integration of the PSTN and the Internet into UMTS networks has opened additional vulnerabilities and provided malicious attackers easy access through cross network servers. The Internet, for example, is open and very easy for malicious attackers to break into. Breaking into an Internet server providing Cross Network Service opens up the opportunity for the attacker to target the 3G networks.

Examples of cross network services include the Call Forwarding Service (CFS), the Client Billing Services (CBS), and the Location Based-Instant Messaging (LB-IM) services. The CFS and CBS Services are triggered when a signaling message arrives at the Subscriber Locator Agent (SLA, within the HLR) that is associated with the mobile user. On the arrival of the signaling message, the SLA sends a database query to its user terminal data source to find the terminals registered for the user. Subsequently, it queries the Location Data Source and finds the Foreign Location Agent (FLA, within the VLR) where the subscriber is registered. Then, the FLA invokes the Routing Agent (in the MSC) to provide a routing number to route the call to the Session Control Agent where the subscriber is currently roaming. The routing number is returned to the Subscriber Locator Agent (HLR). The call is routed to the Session Control Agent (at the MSC). The SCA will invoke the Subscriber Services Support Manager to check if there are any Cross Network Services listed for the call receiver.

If the Cross Network Server is the CBS, after the authentication, the CBS Client Checker Agent checks if a client corresponding to the caller ID exists in Client Phone book data storage. If there is a match, the CBS Forwarding Agent is invoked, which will forward the call to the appropriate number and invoke the CBS Timer Agent. When the call is over the CBS Bill Calculator will calculate the charging amount.

Finally, the LB-IM is triggered when a location track request arrives at the LB-IM Request Manager. The LB-IM Request Manager will check if the requestor belongs in the location and time visibility group by checking the Permissions data store. If the requestor satisfies the constraints, the LB-IM Location Tracking Agent fetches the location of the subscriber by querying the 3G network entities.

Let us now consider how attacks can happen using the aforementioned cross network services. With CFS, attacks can occur at the mail server or CF server.

■ Attacks at the mail server can use vulnerabilities observed with the mail transfer agent (such as a badly configured sendmail daemon). These attacks allow sending emails with a false source identity to the victim; the CF server will assume that it is checking the correct constraints, but in reality the victim may receive unwanted calls and may have an important number of calls sent to victim's voice mail.

■ Attacks at the CF server allow sent spoofed messages to the CFS Subscriber Parameter Manager Agent to request the change of some parameters. They also allow modifying the email cache and subscriber preferences, by gaining privileges, and causing a denial of service, by flooding the CFS Authentication Agent.

With the CBS, attacks can occur at the CB server. The attacks include a denial of service on the CBS authentication agent, or obtaining privileges to modify the timer agents, bill the calculator agents, and the forwarding agents. The attacks targeting the CB server may result in the victims receiving large bills or being unable to receive service on time.

Finally, with the LB-IM service, spoofed messages may be sent to the LB-IM Subscriber Parameter Manager Agent requesting a change in permissions resulting in leakage from the victim's privacy.

6.8.1 Classification of Attacks on 3G Networks

A classification of attacks on the 3G network can be approached using three dimensions. They are (a) the attack categories; (b) the attack means; and (c) the physical access dimension, where attacks are classified based on the level of physical access the attacker has to the 3G wireless telecommunication network.

In the first class, attacks are classified based on their type. Typically, five types of attacks are distinguished:

1. *Interception*: The attacker intercepts information or reads signaling messages on a cable, but does not modify or delete them. Such attacks affect the privacy of the subscriber and the network operator. The attacker may use the data obtained from interception to analyze traffic.

2. *Fabrication/Replay*: In this case the attacker may insert spurious objects into the system. These objects depend on the target means and physical access

type. The attacker may insert false signaling messages, fake service logic, or fake subscriber data into the communication system. The effects could result in the attacker masquerading as an authority, for example.

3. *Modification of Resources*: The attacker causes damage by modifying system resources, meaning that he may modify signaling messages in and out of the cable. He may modify the service logic or modify the subscriber data in the entity.

4. *Denial of Service*: The attacker causes an overload or a disruption in the resources or applications connected to the 3G system, forcing the network to operate in an abnormal manner. The abnormal behavior may include a legitimate subscriber not receiving service, an illegitimate subscriber receiving service, or the entire network to be disabled.

5. *Interruption*: The attacker can cause an interruption of operation by destroying resources. He may delete signaling messages from and to the cable. He may delete a subscriber data in an entity, such as an HLR, and he may stop the delivery of a service to a mobile user.

The second class contains attacks that are classified based on what means are used to cause the attack. The attack major means are as follows:

1. *Data-based attacks*: The attacker targets the data stored in the 3G communication system. The damage can be caused by modifying, inserting, and/or dropping the data stored in the system.

2. *Messages-based attacks*: The attacker launches attacks against the 3G communication system targeting the signaling messages. The attacker may insert, modify, replay, and drop the signaling messages flowing to and from the network.

3. *Service Logic attacks*: The attacker causes important damages by simply attacking the service logic running in the various 3G network entities. An example of damage would be a complete deletion of logic running on the MSC.

Finally, the third class may be further classified into five categories based on the type of physical access that the attacker can launch his attack on a single infrastructure or on a cross infrastructure:

1. *Physical Access attacks* I (the attacker obtains access to the air interface using a physical device): Typically, the attacker has access to an inexpensive off-the-shelf equipment that he uses to impersonate some parts of the network. He may build a part of a rogue base station. Victims camping on the rigue base station are subject to various attacks. Attackers may also use modified mobile stations to broadcast at a high frequency, eavesdrop, and execute man-in-the-middle attacks.

2. *Physical Access attacks* II (the attacker obtains access to the cables connecting the 3G network switches): Typically, only the authorized personnel can access the 3G switches; but, if an attacker has access to cables connecting these switches, they may cause considerable damage by disrupting the normal transmission of signaling messages.

3. *Physical Access attacks* III (the attacker has access to some sensitive components of the 3G network): In this case, the attacker may be a displeased employee who has managed to obtain access to the 3G node switch. The attacker can cause important impairments by editing the service logic or modifying the subscriber data (related to the user's profile, security, and services) stored in the 3G network entity.

4. *Physical Access attacks* IV (the attacker has access to some links connecting the Internet to the 3G network): This is a cross infrastructure cyber attack, where the attacker can cause a certain harm by disrupting the transmission of signaling messages flowing between the link and inserting some signaling messages into the link between the two networks.

5. *Physical Access attacks* V (the attacker has access to Internet servers or cross network servers providing services to mobile subscribers connected to the 3G network): This is a cross infrastructure cyber attack, where the attacker can cause harmful damage by editing the service logic or modifying subscriber data (profile, security, and services) stored in the cross network servers.

6.8.2 Examples of Attacks

Among the attacks that can be launched on a 3G network, we distinguish in the following examples of the so-called cross infrastructure cyber attacks. Such attacks can be organized into simple or complex scenarios of actions. They can be launched in a single step or using multiple stages. In a multiple stage, a cross infrastructure cyber (CIC) attack can use attacks in each step that belong to different attack categories. It also can use different attack means. In addition, such attacks may target different cross network servers from one phase to another. The cross network servers can offer services using multiple phases, and therefore each service phase may be subject to a cross infrastructure cyber attack step. The effects generated by an attack in a step can have direct or indirect effects on the different phases of the service provided by a cross network service. Examples of single stage CIC attacks include the following.

6.8.2.1 Attacks Targeting the Call Forwarding Service

In such attacks, the attacker may view the caller identities of the calls to a victim and provide the collected information to malicious entity. The attacker may also view the call forward number and track the locations of the victim. He may view the authentication messages and use them for a replay attack in a later moment. The attacker may view emails obtained by the CF server from the mail server and obtain a privilege to access to the victim's personal information. The attacker may steal personal information of mobile users stored in the subscriber parameter data store of the CF or mail server.

All the attacks mentioned in the preceding paragraph can be classified as interception attacks. On the other side, performing an insertion attack, the attacker can flood the CF server with multiple call forward requests, and then can cause a denial of service. The attacker can send a large number of authentication requests to the CF and mail servers. He may also request email messages from the mail server with the help of the authentication messages, if the latter has been captured previously. The attacker may insert some subscribers who have not paid for the service into the CF Subscriber Parameter data sources. He can also insert false emails into the email data cache through the CFS mail server agent.

On the other hand, when performing a modification attack, the attacker can modify the parameters in the messages such as the caller identity so that the calls are forwarded incorrectly. The attacker can modify the number of call forward, changing the authentication challenge response to an incorrect value so that the CF and mail servers can never be authenticated. The attacker may also have the capability to modify the call forward numbers for subscribers at the subscriber parameter data store, and change the time stamps of the cached email data.

Using the interruption attacks, the attacker may delete all messages on the link arriving to the CF server or to the mail server, giving a feeling that these servers are not operational. Denial of service attacks may be performed by changing the CF number, since the victim does not gain access to the voice message or to the call itself. Sending two or three call forward numbers to the session control agent (at the MSC) may cause confusion and the call may not be handled properly. The attacker may delete certain target subscriber profiles in the data sources so that they may not receive the CF service. Service logic of certain entities may be completely deleted, such as the CFS filtering agent, so that they become unable to provide any service.

6.8.2.2 Attacks Targeting the Client Billing Service

Physical Access attacks of level IV targeting the CB service include the following major attacks.

- *Interception attacks*: the attacker may view the authentication messages and use them in a replay attack in a later moment.
- *Insertion attacks*: The attacker can flood the CB server with Bill Client Requests and can possibly cause a denial of service. He can send a large number of authentication requests to the CB server and cause a denial of service.

- In the *modification attack*, the attacker can modify parameters in the messages (e.g., caller ID) so that the calls are forwarded incorrectly and change the authentication challenge response to incorrect value.
- In the *interruption attack*, the attacker may delete all messages on the link arriving to the CB Server giving an impression that these servers are not functional.

Physical Access attacks of level V that can be launched against the CB service include the following major attacks.

- *Interception attacks*: the attacker may steal personal information of subscribers stored in the client preferences data source in CB Server.
- *Insertion attacks*: subscribers not paying for the service can be inserted by an attacker into the CB data source. The attacker may also insert false charges into a victim's bill.
- *Modification attacks*: the attacker may also be able to modify the forward numbers for subscribers. He can corrupt the logic in the CBS timer agent and the CBS bill calculator. Attacks against the CBS client checker agent can change the bill amount or the time log of the mobile user. Denial of service may be caused by simply removing the victim's name from the client phone book.
- *Interruption attack*: the attacker can corrupt the phone book and delete victims from the client phone book data source so that they may not be able to get the CF service.

6.8.2.3 Attacks Targeting the Location Based Instant Message System

The attacker can perform an interception attack by viewing the location track requests and reading the location field within the location track responses. The attacker may also view the authentication messages and use them subsequently in a replay attack. The attacker may also steal personal information of mobile users stored in the Permissions of the LB-IM. The attacker can launch an insertion attack by flooding the LB-IM server with location track requests and induce a denial of service. He can send a large number of authentication requests to the LB-IM server and cause a denial of service. He also can insert into the LB-IM permissions some mobile users not paying for the service. Finally, the attacker may insert authentication information to the LB-IM to authenticate fake core network entities.

To launch a modification attack, the attacker can modify parameters in the messages in the location track request and the location where the buddy is located or change the authentication challenge response to incorrect value so the LB-IM server is never authenticated. The attacker may also be able to modify permissions and view the victim's locations. The service logic in the LB-IM request manager may be modified so that the requests are not properly checked.

In the Interruption attack, the attacker may delete all messages on the link arriving to the LB-IM server, giving the impression that this server is not functional. The attacker may delete some mobile users' profiles in the data sources so that they may not receive LB-IM service. In addition, the service logic of some agents may be completely deleted.

References

3GPP TS 35205 (v3.0.0), 3G Security: Specification of the MILENAGE Set: An Example Algorithm Set for the 3GPP Authentication and Key Generation Functions f1, f2, f3, f4, f5, and f5*, Release 99, April 2001.

3GPP TR 33.908 (V3.0), 3G Security: General Report on the Design, Specification, and Evaluation of 3GPP Standards Confidentiality and Integrity Algorithms, Release 99, March 2000.

3GPP TS 33.200 (v4.3.0), 3G Security; Network Domain Security; MAP Application Layer Security, Release 4, March 2002.

3rd Generation Partnership Program, Document 2: KASUMI Specification. Technical Specification, 35.202, Release 5, Version 5.0.0.

3rd Generation Partnership Project: Technical Specification Group SA WG3; A Guide to 3rd Generation Security (3G TR 33.900 version 1.2.0), 2000.

3rd Generation Partnership Program, Document 1: f8 and f9 Specification, Technical Specification 35.201, Release 5.Version 5.0.0.

3rd Generation Partnership Project, 3GPP Specifications Home Page (available at: http://www.3gpp. org/specs/specs.htm).

3rd Generation Partnership Project 2, 3GPP Specifications Home Page (available at: http://www.3gpp2.org/Public_html/specs/index.cfm).

E. Biham, O. Dunkelman, and N. Keller, A Related-Key Rectangle Attack on the Full KASUMI, Advances in Cryptology, Proc. ASIACRYPT '05, LNCS 3788, pp. 443–461, Springer-Verlag, 2005.

U. Blumenthal, M. Marcovici, S. Mizikovsky, S. Patel, G. S. Sundaram, and M. Wong, Wireless Network Security Architecture. *Bell Labs Technical Journal*, Volume: 7, Issue: 2, Pages: 19–36, Dec. 2002.

M. Blunden and A. Escott, Related Key Attacks on Reduced Round KASUMI, *Proc. Fast Software Encryption 8th International Workshop*, LNCS 2355, pp. 277–285, Springer-Verlag, 2002.

J. Hall, M. Barbeau, and E. Kranakis, Using Mobility Profiles for Anomaly-Based Intrusion Detection in Mobile Networks, *The 12th Annual Network and Distributed System Security Symposium*, San Diego, CA, 3–4 February 2005.

International Telecommunications Union, International Mobile Telecommunications-2000 Recommendation ITU-R M.687-2 (02/97).

P. Mateli, Hacking Techniques in Wireless Networks, *Handbook of Information Security*, H. Bidgoli (Ed.), John Wiley & Sons, pp. 83–93. USA, 2006.

A. Meddeb and N. Boudriga, Intrusion and Anomaly Detection in Wireless Networks, *Handbook of Research on Wireless Security*, Y. Zhang, J. Zheng, and M. Ma, 2007.

N. Sugio, H. Aono, S. Hongo, and T. Kaneko, A Study on Higher Order Differential Attack of KASUMI, *IEICE Transactions on Fundamentals of Electronics, Communications and Computer Sciences*, 2007 E90-A(1): 14–2.

C. Valli, WITS: Wireless Intrusion Tracking System, *3rd European Conference on Information Warfare and Security*, pp. 28–29, UK, June 2004.

Chapter 7

Wireless Local Area Network Security

7.1 Introduction

Wireless local area networks are gaining growing popularity and interest compared to traditional cellular telephony thanks to their ease of deployment, cost effectiveness, important throughput, and support of mobility and multimedia applications. In particular, wireless networking with the 802.11 standard is becoming a popular method for interconnecting computers. In fact, IEEE 802.11 standards targeting the Local Area Networks (LANs) environments are currently deployed everywhere; they are used to extend the wired LAN infrastructure and to achieve cross building interconnection while guaranteeing a nomadic access to other networking technologies. A wireless local area network (WLAN) is a network designed as an enhancement to wired LAN using the radio technology. Thus, a WLAN combines data connectivity with user mobility, while offering cost advantages over wired networks. WLAN solutions are widely recognized as a general-purpose connectivity alternative for a broad range of enterprises and are used in a large range of fields. Many WLAN-based solutions have been made available. They include the following cases:

- *Corporate networks*: WLANs can overcome the problems with wire network deployment. An example of corporate network can be deployed between different corporate campuses in different geographic areas, where wireless access might be integrated with an appropriate mobility protocol, so that the user can move from an area to another without a need for manually changing the network setting. However, a corporate network should be able to provide the following features: (a) only authorized users can be allowed to access the network; (b) only trusted system administrators are allowed to take care of user registration; (c) digital credentials are used to authenticate users—they are stored at a local trustful storage; and (d) out-of-band key distribution is performed.

■ *Access to cellular networks*: WLANs represent an interesting access technology to cellular networks including those involved in 2G, 3G, and 4G networks. Combining a wireless LAN technology with a UMTS network, for example, allows the user to benefit from a high-performance network wherever it is feasible to deploy WLANs, and to use UMTS elsewhere. It can be assumed that digital credentials for the security needs are generated for the users of the access network and stored at different locations. However, the key distribution is difficult to organize since a centralized approach is not applicable.

■ *Home networks*: The home environment represents an interesting area of WLAN usage. High-speed access to the Internet from homes, local multimedia applications, and voice over IP are among the category of attractive services in homes. Often, it is useful to assume that the digital credentials used to identify the home entities can be stored at the access point; they are used to control the access to the home environment. However, two major concerns can be noticed with the home networks. First, interference between close home environments can occur. Second, the administration of user registers does not clearly exist.

A plethora of other applications are emerging nowadays. In addition, a new generation of WLAN network technologies is currently under development to meet the networking requirements of tomorrow's applications. These requirements comprise support for quality of service, better security efficiency, shorter handover, and increased throughput.

The lack of efficient security in the existing commercial WLAN products had resulted in the development of security solutions allowing the management of a large number of users and the centralized database of users' digital credentials. The deployment of WLANs in public or home environments requires, however, the definition of novel specific security policies. This is particularly true in the cases of remote access to corporate networks, for Internet service providers, and for cellular third generation cellular networks. However, any solution should not limit the system scalability.

A wireless LAN offers important advantages with respect to wired networks, including the following: First, a wireless LAN allows the mobile terminals to be fully mobile as long as they remain within the radio range. Second, the setting of a WLAN network is an easy and fast process, particularly in the cases where it is not possible to deploy wired infrastructures because of the nature and topology of the covered area. Even when the terminals do not necessarily need mobility management, a WLAN avoids the load of having cables between the mobile terminals. Cases where a WLAN can be easily deployed include battlefield applications and search-and-rescue operations. In addition, it is worthy to notice that, while the direct cost of a small size WLAN may be higher than the cost of a wired local area network, making the network bigger is less expensive. This is due to the fact that there is no need for additional wirelines, no cost for material, and added effort in the network maintenance. In fact, transforming the topology of a WLAN to add, remove, or displace a terminal is a simple task.

On the other hand, some drawbacks need to be considered with the deployment of WLANs. They include the following:

- The power of the radio signal goes weaker with the distance to the WLAN access point (proportionally to the inverse of the square of the distance). Thus, the mobile nodes attached to the access point have a limited radio range and a restricted visibility of the network, when there is more than one access point in the WLAN. This causes the well-known hidden station problem and message collision would occur, since a mobile node may start transmitting when it should not.

- The functioning of the network is highly influenced by the environment it is expected to work in. In fact, radio waves are absorbed differently by objects such as walls, trees, and human bodies. They are differently reflected by other objects (such as pipes, metallic objects, and water). Wireless networks are also subject to interferences with the signals generated by equipments that share the same radio band.

- The data rate is often lower than the rate provided by the wired networks, because of the limitation of the radio range, the possibility of interference occurrences, and the quasi omnipresence of packet collisions. However, it is worthy to notice that some novel standards offer very high data rates for the WLANs.

- A WLAN does not allow transmitting and listening on the same channel and at the same time; this is due to the limitations of the medium. The multi-access control protocols implemented lead to higher chances of message collisions. Therefore, collisions and interferences make message losses more likely in WLANs.

- The mobile terminals attached to a WLAN have limited batteries and computation power. This may generate high communication latency. In fact, the mobile terminals may be turned off most of the time (for power-saving needs). A mobile terminal can turn on its receiver periodically, making necessary that the other terminals wait until they wake to communicate with it.

- Wireless local area networks are inherently less secure as data is transmitted over radio links. In fact, transmissions between any pair of terminals in a WLAN can be eavesdropped by any similar equipment that happens to be in the radio range.

The characteristics of wireless access and mobility call for a security framework. The requirements for an efficient framework include authentication and authorization to protect network resources, data encryption, and user integrity. Privacy regarding the content and user location can also be provided. WLAN systems implement and define different levels and components of the security framework. However, the currently used solutions lack the analysis of an overall framework on the subject of the requirements that would apply for private, public, and virtual private networking taking in WLANs. This is perhaps due to the fact that the environments are different in their nature. In addition, the following three features should be satisfied by the WLANs: (a) the roaming should be made possible between these environments; (b) the usability of the system from the user's perspective should be provided; and (c) the security services should be transparent to applications and end-users in WLANs. They also should be transparent to the transport protocol, except when it is specifically required for given services.

Security mechanisms implemented for WLAN systems are deployed at two layers: layers two and three. The mechanisms implemented at layer two aim at providing the wire equivalent privacy. This means that the wireless physical medium in the protected WLAN should be as secure as an equivalent wire medium appears to be. Several protocols have been developed; they differ sensibly from one WLAN technology to another. The security services provided at layer three include support for secure IP mobility, roaming between different domains, and user's authentication. Authentication at layer three is basically based on user's credentials, knowing that the accounting and authorization data are often independent from the utilized service.

Security mechanisms are required to avoid threats in a cost effective way. When security services are not deployed in a convenient way, they represent system vulnerabilities that an adversary can exploit to attack a WLAN. The most important security services that can be considered for WLAN networks include confidentiality, authentication of users, authentication of access points, data integrity, non-repudiation of origin, non-repudiation of delivery, auditing and logging, denial of service prevention, and traffic flow analysis prevention. This set of services can be complemented by mechanisms for host security, data driven attack prevention, and organizational security policies.

For the sake of clarity, this chapter focuses only on the security flaws relative to IEEE 802.11 standards. In the following, we first present the basics of the 802.11 standards. The Physical and Media Access Control layers, the shared key authentication, and the WEP protocol are discussed. Second, the chapter discusses the security problems that are specific to 802.11. Exploits of the 802.11 MAC layer are presented to exhibit attacks on availability, integrity, and authentication. WEP attacks are used to express attacks on integrity and confidentiality. The effectiveness of attacks targeting the 802.11 networks and using common tools available on the Internet is also discussed.

7.2 Basics on WLANs

WLANs intend to offer the same services provided by the traditional wired Ethernet for mobile users within a small to medium area. They guarantee high rates, are easily scalable, and support traditional mobility management mechanisms such as handover and roaming. Therefore, they may be deployed as an access network to different core networks such as the Worldwide Interoperability for Microwave Access (WiMAX). Various types and versions of WLAN can be distinguished including 802.11a, 802.11b, 802.11g, and HiperLAN2. The early WLANs suffered from several drawbacks: they were slow, expensive, and proprietary in nature.

In 1990, the IEEE 802.11 project was initiated to develop a Medium Access Control (MAC) and Physical Layer (PHY) specification for fixed, portable, and moving device within a local area. The 802.11 standard initially used frequencies in the 2.4 GHz ISM band, and later the 5 GHz band was added to the standard. These frequency bands were chosen because they were available in so many countries, reducing potential interoperability problems and making the devices legal to use in these countries. Transmissions occur among 14 overlapping 22-MHz channels for 802.11b and g, or eight channels for 802.11a. One main reason can be given to justify the channel hopping: the security is increased due to the difficulty and expense of obtaining equipment that can listen to such communications. However, this obstacle is defeated with the recent explosion of 802.11-compatible devices.

802.11 became the de facto standard for wireless networking, with a variety of products using the IEEE 802.11, 802.11a, 802.11b, 802.11g, and 802.11i standards. Some of the extensions add quality of service and base-station roaming. Of uttermost interest to this chapter are the added security enhancements that 802.11i brings to the standard. Table 7.1 describes the major extension provided to the 802.11 standard.

Table 7.1 Extensions to the 802.11 Standard

802.11d	Adds additional regulatory domains for other countries
802.11e	Adds Quality of Services (QoS) enhancements for multimedia and Voice Over IP (VOIP)
802.11f	Inter-Access Point Protocol (IAPP) for roaming between base stations
802.11h	Adds Dynamic Frequency Selection for Europe
802.11i	Adds security enhancements. Mainly replacing WEP with the Temporal Key Integrity Protocol (TKIP). WPA security was taken from this standard and ADDES TO 802.11a, b, and g standards to overcome weaknesses in WEP.
802.11j	Same as 802.11h, but for Japan

7.2.1 The 802.11 MAC Layer

Basically, the WLANs integrate two kinds of devices, the mobile stations (MS) and the access points (AP). A WLAN includes at least one AP, which is in charge of managing the MSs within its coverage area and serving as gateway to the wired network. To cover a larger area, several APs can be deployed. In that case, a different channel will be assigned to each AP to reduce the interferences while each AP may forward the data packets over the wired network to reach a MS under a different AP. Besides, mobile users may roam between the APs while keeping connectivity to the WLAN. Figure 7.1 depicts a generic WLAN.

The MAC layer sits between the PHY layer and the Logical Link Control (LLC) Layer. It is in charge of coordinating access to a shared radio channel and improving communications over the wireless medium. To communicate on the WLAN, the transmitting station should first get control of the radio channel that it shares with other stations. The station gains control by using Carrier Sense Multiple Access with Collision Avoidance (CSMA/CA) to coordinate with the other stations to send frames. When the channel is free, any of the stations is able to transmit; when the channel is not available, all mobile stations should wait until the transmitting station finishes its transmission.

IEEE 802.11 standards introduce two different access mechanisms: the mandatory Distributed Coordination Function (DCF) and the optional Point Coordination Function (PCF). The DCF implements a distributed contention-based channel access through the CSMA/CA (Carrier Sense Multiple Access with Collision Avoidance) technique, while the PCF implements a central control of the channel based on the polling concept. When CSMA/CA is implemented, each station that wishes to transmit a frame should sense the medium before transmitting. If the channel is found idle at least for a DCF inter-frame space time period (DIFS), the station starts its transmission while the other stations should wait until

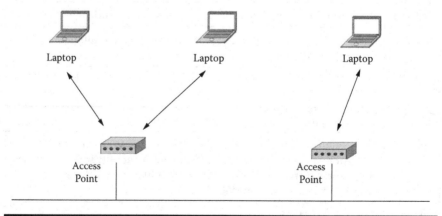

Figure 7.1   Example of WLAN architecture.

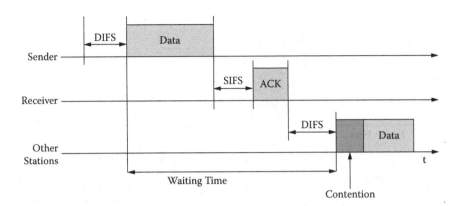

Figure 7.2   DCF basic access mechanism.

the medium is again sensed as idle for at least a DIFS time period. The destination station should acknowledge the frame reception by sending an ACK frame after a short inter-frame space time period (SIFS) as illustrated by Figure 7.2.

Two or more stations may sense the medium as idle and begin their transmissions at the same time, thus leading to collisions. To address this issue, each station has to wait for an additional time if the medium was busy just before that station started waiting the DIFS period or if the medium was sensed busy during the DIFS period. This additional time is a random backoff value measured in time slots that avoid collisions by preventing all stations from transmitting as soon as the medium becomes idle for the DIFS period. The backoff mechanism achieves collision avoidance. After choosing the backoff value, the station wishing to transmit decrements its backoff as the medium is sensed idle at least for a DIFS time period. If during this backoff process the medium is sensed busy, the station will stop its backoff timer. The backoff timer is resumed when the medium is sensed idle for the DIFS period again and the station should only transmit when the backoff timer reaches zero.

The random backoff value is chosen from the interval [0, CW], where CW is a real number called the *contention window*. At the first transmission attempt, CW is set to the minimum contention window size, say *CWmin*. After each unsuccessful transmission, CW is doubled, until it reaches the maximum contention window size, say *CWmax*. When the sender does not receive the ACK frame within a specified timeout (due to unsuccessful transmission, for example), it assumes that a collision occurred and enters into the backoff period again after waiting for the medium to be idle for a DIFS. However, the number of retransmission should not exceed a given limit. The backoff mechanism is also implemented when the sender station wishes to transmit a second frame just after receiving the ACK of the previous frame. This post backoff operation allows the other stations to decrement their backoffs and get access to the medium.

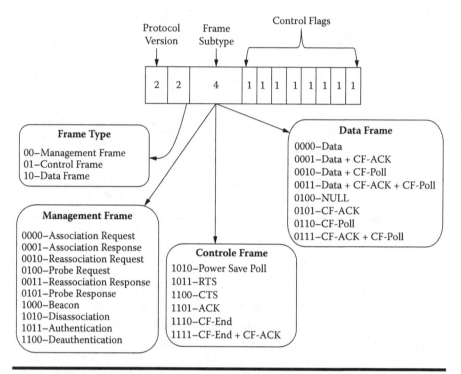

Figure 7.3   Frame control field.

The connections in 802.11 are controlled by exchanging management frames between the AP and the user. Three types of frames can be distinguished: the management frames, the data frames, and the control frames. Management frames are specific packets that are used to set up connections between a station and an AP. Figure 7.3 depicts the significance of the frame control field.

A frame control field determines the management frame subtype. It contains a 4-bit pattern that determines whether the frame is a beacon, probe, authentication request/response, association request/response, or other type of frame. Beacon and Probe Request frames are used for scanning. Two types of scanning can be distinguished: passive and active scanning. Passive scanning allows the wireless station to scan the available channels, looking for beacon frames from APs and sending a response. On the other hand, active scanning allows the wireless station to send out probe requests instead of waiting for a beacon frame from the access point. When the AP receives a probe request, it responds with a probe response frame. Stations also have pre-emptive routing. In addition, the wireless stations periodically check for other APs and change their association to another AP that offers better quality than the serving.

Data frames are in charge of transporting data intended for the higher layers in the network architecture. However, before these data frames can be transmitted, a station must first be authenticated and associated with the AP. Control frames are used to handle handshaking between different stations, solve the "invisible node" problem (where there are stations within range of the access point but not within range of each other), and avoid having collisions between any nodes that cannot see each other.

7.2.2 Basic Authentication and Encryption

After the initial handshake between an AP and the wireless user, the authentication of the wireless station is started. Two authentication mechanisms are defined in the 802.11 standard to authenticate wireless clients. The first is called the *Open system authentication*, and the second is referred to as the *Shared key authentication*. Using the former method, the AP accepts any wireless user without any verification of its identity apart from the client returning its MAC address as a component of the response message. This method is a one-way authentication and does not provide guarantees to the wireless station that it has actually connected to the intended AP. The latter method sets up a two-stage challenge-response between the AP and the wireless station (WS). The AP generates a random challenge and sends it to the WS. The WS then encrypts the message using the RC4 encryption algorithm with a secret key known to both the WS and the AP. The WS then sends back the encrypted response to the AP. The AP then decrypts the message with its copy of the key and compares it to the original message. The AP then authorizes access only if the decrypted response matches the original message.

After authenticating a wireless user, the AP initiates an association with it by picking up a management frame with the association request bit set in the management frame subtype from the user. Each discovered AP is checked for signal strength, supported data rates, and the Service Set Identifier (SSID). If there are several access points within the wireless user range, the WS can prioritize its connections with them based on four parameters: the signal strength between AP and the WS; the users currently associated with the AP; the user's encryption and authentication capabilities; and the user desired data rates.

To transmit data to a wireless network, a wireless station must be authenticated by an AP and associated with it. The authentication and association sequence can be described by a three-state machine (see Figure 7.4). The three states are unauthenticated and unassociated, authenticated and unassociated, and authenticated and associated. The transition from one state to another is controlled by the reception of management frames.

The address frames contain MAC addresses. The two address fields in the management packet are the Source Address (SA) and Destination Address (DA). The Service Set Identifier (or SSID) is contained in another field. Depending on the vendor's implementation, the client can select an available BSS. This is usually done by inspecting the management frame and extracting the network name or SSID.

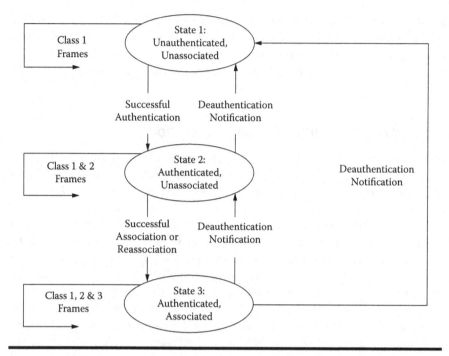

Figure 7.4 Finite state machine depicting authentication and association of users.

The MAC layer provides privacy and encryption by implementing the Wired Equivalency Protocol (WEP). The WEP is an optional feature that encrypts only the body of each frame using a pre-shared key common to the access point and all authorized clients. WEP was designed to prevent disclosure of information sent over the wireless network and to ensure the integrity of messages. Like the aforementioned shared-key algorithm method, WEP uses the RC4 encryption algorithm to create a pseudo-random sequence of bits that is XORed with the data to protect. In the following we discuss the implementation and drawbacks for WEP.

7.3 Defeating the WEP Protocol

The 802.11 standard uses shared key authentication to authenticate wireless clients and WEP for security and privacy. Many papers have been published describing the weaknesses of both of these methods.

7.3.1 WEP Design Issues

The Wired Equivalent Privacy was the security solution adopted by the early versions of IEEE 802.11 standards. As its name indicates, WEP aimed at providing a security level equivalent to the one provided by the wired LAN in terms of protection of network access. In fact, the wireless LANs present two major security vulnerabilities that may be easily exploited. First, wireless signals are broadcasted and may be easily eavesdropped. Second, connecting to a WLAN does not require a physical access; therefore, malicious users may easily connect to the APs and enjoy the provided services. To address those vulnerabilities, WEP protocol encrypts the transmitted messages and authenticates the mobile users before giving them access to the wireless LAN.

A wireless station wishing to connect to the network begins by sending an Authentication Request in order to request the shared key authentication. The AP answers by sending a random authentication challenge generated using the WEP algorithm. After that, the mobile station encrypts the challenge using a locally configured WEP key and then sends the resulting authentication request to the AP. The AP decrypts the received message using the shared WEP key and verifies whether it is the initial random message that it generated. If this is the case, the AP deduces that the mobile station is authenticated, since it is the only one able to encrypt the challenge with that particular shared key. Consequently, the AP sends back an authentication success notification and grants the network access. However, if the authentication process fails, an authentication failure notification is sent back to the mobile station and the latter will not have the right to access the network.

Once the mobile station is authenticated, it will be able to exchange encrypted messages with the AP. Such messages are encrypted by the same WEP key used for authentication. Generally speaking, the sender of a message m should initiate the RC4 encryption algorithm with its secret 40 or 104-bit key. The output of this operation is a long pseudo-random sequence K. Finally, the plaintext is XORed byte by byte with that pseudo-random sequence in order to obtain the encrypted message $c = m \oplus K$. The receiver should initiate the RC4 algorithm with the same secret key in order to obtain the same pseudo-random sequence K, then sums (XOR) the cipher text with that sequence to recover the plaintext

$$m = c \oplus K = (m \oplus K) \oplus K = m \oplus (K \oplus K)$$

Nevertheless, the consecutive messages should not be encrypted using the same pseudo-random sequence because this may result in a very weak encryption. To address this issue, WEP appends a 24-bit random Initialization Vector (IV) to the secret key before initializing the RC4 algorithm and uses a different IV to encrypt each new frame. It is worth noticing that the receiver should know the IV in order to be able to decrypt the message while an attacker is not able to decrypt the transmitted message by only knowing the IV. Therefore, the IV is transmitted in clear to the receiver along with the cipher-text c. To guarantee integrity, the sender also appends an Integrity Check Value (ICV) consisting in a 32-bit CRC checksum to every frame before encrypting it so that a malicious user is not able to compute a new CRC value for a modified message without knowing the secret key:

$$c = \mathrm{RC4(IV},K) \oplus \left(m \| \mathrm{CRC}\right),$$

where $\|$ is the concatenation operator.

7.3.2 WEP Vulnerabilities

The WEP security architecture suffers from multiple flows related to the protocol design. A first problem occurs with the authentication service. The WEP only considers the authentication of mobile stations without requiring the authentication of the AP. Therefore, a malicious user may spoof the AP identity without being detected. Then it can cause serious damages to the WLAN. Besides, the authentication procedure is vulnerable to message injection attacks, thus enabling identity spoofing attacks. More precisely, when the plain text and the related cipher text of a message are known by a malicious user, the latter may sum them to derive the key stream. As an attacker sniffing the authentication process may get the challenge text sent by the AP and the encryption of that challenge sent by the legitimate mobile user, he may simply determine the key stream and then request authentication from the AP using that same key stream.

On the other hand, the WEP does not define how to securely maintain a key base and renew the keys for a better security. Moreover, the same key is used for authentication and confidentiality services, which is definitely inadvisable, since a hacker may exploit the weaknesses of both services to deduce the key. In addition, since the IEEE 802.11 standard requires that each mobile station (in the WLAN) possesses a unique key that is known by the AP only, this makes complicated the key administration at the AP. Meanwhile, the WEP requires the existence of a shared key (known by the AP and each mobile station) that should be used to encrypt the messages broadcasted by the AP. Unfortunately, the majority of the implementations support only this shared key option; therefore, an internal attacker may decrypt all the exchanged messages and impersonate other users.

The integrity of the WEP encrypted messages is also easily compromised. First, the linearity of the CRC with respect to the XOR function enables attackers to modify the encrypted frame and the encrypted CRC without being detected. To illustrate how the attack is performed, let us denote the original encrypted message as $c = K \oplus (m\|\mathrm{CRC}(m))$, where K is the pseudo-random sequence. Let us then assume that the attacker has sniffed the original message m and that he wants to modify it using by adding Δm. To succeed in launching his attack, the malicious user should obtain

$$\left(m \oplus \Delta m\right) \| \mathrm{CRC}\left(m \oplus \Delta m\right) \oplus K$$

from the original message $(m\|\mathrm{CRC}(m) \oplus K)$.

Knowing the equality $CRC(X \oplus Y) = CRC(X) \oplus CRC(Y)$ and that $CRC(\Delta m)$ can be computed without needing the secret key K and the IV, the attacker needs to compute $CRC(\Delta m)$, and then he can compute the sum $(\Delta m)\|CRC(\Delta m)$ with the original encrypted message. Thus, we obtain:

$$c \oplus (\Delta m)\|CRC(\Delta m) = K \oplus \left(m\|CRC(m)\right) \oplus (\Delta m)\|CRC(\Delta m)$$

$$= (m \oplus \Delta m)\|CRC(m) \oplus CRC(\Delta m) \oplus K$$

$$= (m \oplus \Delta m)\|CRC(m) \oplus (\Delta m) \oplus K.$$

Moreover, the WEP architecture does not integrate a mechanism for replayed messages detection. Thus, the malicious users may replay any message that has been already accepted by the AP and cause denial of service.

The confidentiality of WEP encrypted messages can also be easily compromised since it is based on the vulnerable RC4 stream cipher. First, the pseudo-random sequences used for the encryption should be different; otherwise, serious attacks may be conducted by moderately experienced hackers. In fact, an attacker may sniff two consecutive encrypted messages M_1 and M_2 and then apply an XOR on them, thus obtaining $M_1 \oplus M_2$. After that, the attacker may easily break the encryption as M_1 and M_2 are not pseudo-random sequences. Unfortunately, the different pseudo-random sequences used for WEP encryption are generated using only a 24-bit long IV. If we consider that there are approximately 17 million possible IV values and that a mobile user can approximately transmit 500 full length frames in one second and use a different IV for each transmitted frame, we will deduce that the totality of the possible IVs will be used in about 7 hours.

Consequently, the pseudo-random sequences will be reused in 7 hours. Moreover, if n users use the same shared secret key, then the pseudo-random sequences will be repeated after $7/n$ hours. The problem becomes more serious when the IV values are easily predictable. In fact, many WEP implementations initialize the IV with 0 and then increment it for every sent frame. Besides, RC4 cipher may produce a non-random output when used with a weak key; therefore, the attacker can deduce the bits of the seed from the first few bytes produced by the algorithm where the seed denotes the secret key concatenated to the IV. As the IV is transmitted in clear, it is easy to detect when a weak key has been used, then break the full 104-bit secret key by eavesdropping only a few million messages.

7.3.3 Defeating Shared-Key Authentication

The authentication mechanism in 802.11 is not particularly effective at keeping attackers from associating with the network. An attacker can easily defeat the shared-key protocol. An attacker eavesdropping on the network captures the random challenge clearly transmitted and the encrypted response. This gives the attacker the plaintext, cipher-text, and an initialization vector value. Let m, c, and v be these objects, respectively. Using these objects, the attacker can recreate the pseudo-random stream generated by the RC4 algorithm by simply computing

$$RC4(v,K) = c \oplus m = \left(m \oplus RC4(v,K)\right) \oplus m$$

Since the attacker can re-create the keystream $RC4(v, K)$, he can then request authentication, receive the plaintext challenge, encrypt it with the recovered keystream m_2, and authenticate himself to the WLAN by sending

$$c_2 = \left(m_2 \oplus RC4(v,K)\right).$$

This will cause the AP to authenticate the attacker. However, this is only a partial solution. The attacker needs now the WEP key to encrypt the traffic he wants to send and decrypt the traffic he receives. This part of the attack is done by the following attacks.

7.3.3.1 Attacks against Key Distribution

The 802.11 standard does not address key distribution, leaving it to the equipment manufacturers. However, manually entering a key or updating the keys of several is a complicated task. Using a pass phrase mechanism also causes a severe weakness (Newsham, 2001). The weakness comes about by reductions in entropy when the passphrase is converted into the WEP key. A mechanism to brute force all the possible keys in the keyspace against some captured packets can be developed, for particular key generators. A successful key recovery can be detected by testing the key on more than one packet, checking against the ICV, in order to reduce the chance of a false positive. This means that the WEP is vulnerable to dictionary attacks, and tools exist that perform dictionary attacks. Various freeware tools have been available to provide a dictionary attack against captured packets (Blunk, 2002).

7.3.3.2 Passive WEP Decryption

Several attacks have been designed to decrypt WEP encrypted packets without any direct knowledge of the WEP key. Each IV is 24 bits long, leading to 2^{24} possible keystreams per WEP key. Despite the fact that the WEP should not reuse any of the IVs and that the WEP key needs to be changed before all the possible IVs are exhausted, practices have shown that the WEP key is infrequently changed because of the lack of an automated mechanism for this. The limited number of IVs enables an attacker to record all the possible IV values and look for collisions. The amount of packets necessary for a collision is deceptively small and can be calculated by using the Birthday Attack equation that allows to find the least n packets allowing to get at least 50% of a collision (or deduce the IV), knowing that the number N of possible packets is 2^{24}:

$$n \simeq 1.2\sqrt{N}$$

According to this equation, there is a probability higher than ½ to obtain a collision after receiving approximately 4,900 packets. This situation may occur in less than 3 seconds on a busy 11Mbps network. These IV collisions can then be used to decrypt traffic on the wireless network. Four methods of doing so can be found in the literature (Borisov, 2001). Let us now give a brief description of these methods.

The first attack is the *passive attack to decrypt traffic*, where an attacker guesses the content of a packet by statistical analysis. This attack works because when there is an IV collision, the XOR of both encrypted packets is the XOR of both plaintext messages:

$$m_1 = \left(c_1 \oplus RC4(v,K)\right) \text{ and } m_2 = \left(c_2 \oplus RC4(v,K)\right)$$

$$c_2 \oplus c_1 = \left(m_2 \oplus RC4(v,K)\right) \oplus \left(m_1 \oplus RC4(v,K)\right)$$

$$= m_2 \oplus m_1$$

Once the plaintext for any message is known for a particular IV, then all encrypted packets using the same IV can be decrypted.

The second attack considers the fact that it is possible to alter a packet and adjust the CRC so that the packet would be considered valid because the bit order for each packet is fixed, even after encryption. This is known as the *active attack for traffic injection*, or the *bit-flipping attack* (as depicted in Figure 7.5). A bit-flipping attack allows the attacker to change the cipher-text in a particular way in order to result in a predictable change of the plaintext (e.g., flipping bits), although the attacker is not able to learn the plaintext itself. This attack does not target the cipher itself, but against a particular message. Packets forged using this method would be accepted as valid by the access point but would likely be meaningless to the upper network layers. In the extreme case, this could become a denial of service against all messages on a particular channel. The attack is especially dangerous for 802.11 networks since the attacker can turn it into a similar message but one in which some important information is altered. For example, a change in the destination address might alter the message route in a way that will force re-encryption with a weaker cipher, thus possibly making it easier for an attacker to decipher the message. This attack can be achieved because of the following identity:

$$c_1 \oplus m_1 \oplus m_2 = c_2$$

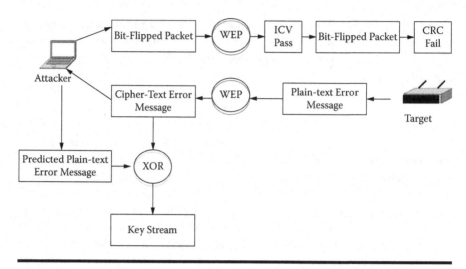

Figure 7.5   Bit-flipping attack.

In the *active attack from both ends*, the destination address of a captured encrypted packet is altered to a machine outside the WLAN. When the modified packet is re-transmitted by the attacker, the access point decrypts the packet and sends the plaintext to another machine most probably controlled by the attacker.

The last example of attacks is called the *table based attack*. It can be successful on networks that generate large traffic. In this attack, the attacker uses some of the techniques described above and monitors network traffic noting IV collisions between known and unknown messages and making a table that would allow the decryption of future traffic using that IV. Since every packet has a maximum size of 1500 bytes, collecting every possible initial vector would require 2^{24}x1500 bytes (approximately 23 Go). Once this data is collected, and if the WEP key remains unchanged, the attacker can record and later read network traffic, when needed.

7.4 Attacks Targeting WLANs

WLAN networks have unique vulnerabilities that make them a perfect target of attack. An attack against a WLAN can take place anywhere. Comprehending the details of various attacks against wireless LANs is critical to determine an appropriate defense strategy. Some attacks are easy to implement but are particularly critical. Other attacks are, however, more difficult to mount but can have devastating consequences. WLAN security presents a serious risk to determine. By knowing the risks involved in the network and making informed decisions about security measures, the WLAN manager has a better chance to protect the assets and the users of the network.

7.4.1 Denial of Service Attacks

Denial of service (DoS) attacks, which aim at preventing access to network resources, can be devastating and difficult to protect against. Typical DoS attacks involve flooding the network and preventing other legitimate users from accessing services on the network. DoS attacks can target different layers. At the application and transport layers, there is nothing fundamentally different between DoS attacks on wireless and wired networks. However, there are essential differences in the interaction between the network, data link, and physical layers that increase the risk of DoS attacks on WLAN.

If a WLAN allows any client to associate, then it is vulnerable to network-level DoS attacks. Since IEEE 802.11 network is a shared medium, a malicious user can flood the network with traffic, denying access to other devices associated to the targeted access point. As an example, an attacker can associate to a victim 802.11b network and send an ICMP flood to the gateway. While the gateway may be able to withstand the amount of traffic, the shared bandwidth of the 802.11b infrastructure is easily saturated. Other clients associated to the same access point will have difficulties sending packets. Given the relatively slow speed of 802.11b networks, a network DoS may happen inadvertently due to large file transfers or bandwidth-intense applications. A few bandwidth-consuming applications on a WLAN can hamper access for the other associated stations. However, these unintentional attacks may become less frequent with the deployment of high-speed WLAN technologies.

At the data-link layer, ubiquitous access to the medium again creates new opportunities for DoS attacks. Even with the wired equivalent privacy (WEP) turned on, an attacker has access to the link layer information and can perform some DoS attacks. Without WEP, the attacker has full access to manipulate associations between the mobile stations and the access points to terminate access to the network. If a client is not using WEP authentication, then he is vulnerable to DoS attacks from spoofed APs. In fact, if the client is configured to associate with any available AP, it will select the AP with the strongest signal regardless of the extended service set identifier. A malicious AP can therefore effectively black-hole traffic from a victim by spoofing the desired AP.

A physical DoS attack against a wired network requires very close proximity to the victim host. However, attackers can launch a physical attack from farther distances. They can use a device that will saturate the WLAN frequency bands with noise. If the attacker can create enough noise to reduce the signal-to-noise ratio (SNR) to an unusable level, then the devices within range of noise will be effectively taken offline. The devices will not be able to pick out the valid network signal from all the random noise being generated and therefore will be unable to communicate. Using a device that produces a lot of noise at 2.4 GHz is a relatively easy and inexpensive operation. For this, common commercial devices can easily take down a WLAN. A cordless phone, for example, can overlap on the frequencies used by 802.11b. There are also DOS attacks that can be launched from other networking protocols. In particular, Bluetooth uses the same band as 802.11b and 802.11g and the modulation used in these technologies is susceptible to interference from the modulation used in Bluetooth networks. As time passes, the 2.4 GHz band will become more crowded, making unintended DoS attacks against 802.11b networks commonplace.

7.4.2 Man-in-the-Middle Attacks

Man-in the-middle attacks in WLAN have two major forms: eavesdropping and manipulation. Eavesdropping occurs when an attacker receives a data communication stream. The eavesdropper can record and analyze the data that he is listening to. On the other hand, a manipulation attack requires the attacker to have the ability to receive the victim's data and be able to retransmit the data after changing it.

7.4.2.1 Eavesdropping

Eavesdropping in WLAN is easy to perform since wireless communications are not easy to confine to a physical area. A nearby attacker can receive the radio waves on the WLAN without any substantial effort or equipment (passive eavesdropping). All frames sent across the wireless medium can be examined online or stored for later examination. Although the transmission distance of WLANs is typically limited to hundreds of meters, this limitation is based upon the use of small antennas built into PC cards. When more sensitive antennas are used, it becomes possible to receive the radio transmission from WLANs situated at a considerable distance. In fact, certain types of antennas with a very high level of directional sensitivity can be used to receive WLAN at distances reaching several miles.

Several layers of encryption can be implemented to obscure the transmitted data in an effort to prevent attackers from collecting useful information from the network traffic. Since the ability of an attacker to eavesdrop on wireless communications is undefeated, the data-link encryption mechanism WEP was developed. If the traffic is not protected at the link layer using WEP, then the higher layer security mechanisms, such as IPsec, SSH, or SSL, must be utilized to protect the data. Unfortunately, WEP experiences several flaws that have been discovered. For example, software such as AirSnort and Network Stumber can be used to reconstruct the WEP key in use, if a sufficient number of frames are captured. In addition to the use of software, it appears that capturing several frames having same IVs can enable a frequency analysis that could result in revealing the content of the encrypted frames to be decrypted. The weaknesses in WEP drastically increase the risk of eavesdropping.

Working on the basis that one cannot decrypt a signal he cannot hear, a valuable countermeasure against eavesdropping is to obscure or hide RF signals from unauthorized third parties. Several approaches can be used to achieve this objective including (a) the antenna positioning and shielding use; (b) the control of the use of a particular antenna, when the WLAN device supports antenna diversity; (c) the control of transmitted signal strength; and (d) the use of directional antennas and shielding for access points.

7.4.2.2 Manipulation

Manipulation takes eavesdropping to more damaging steps. An attacker who can successfully manipulate data on a network can successfully send data masquerading as a victim computer. Furthermore, the attacker can gather sensitive data by introducing a rogue AP into the WLAN coverage area. The rogue AP can be configured to look like a legitimate AP, since many wireless users simply connect to the AP with the best signal strength. Once the user is associated, all communications can be monitored by the attacker through the rogue AP (active eavesdropping). The attacker may, for example, change the content of emails or transactions. He can also choose not to forward packets he receives, effectively denying use of the network from the victim.

The ability of an authorized third party to masquerade as a legitimate user of a wireless LAN can range from very simple to complex undertaking, with the degree of complexity based on security effect. If the victim's WLAN does not employ any security, it becomes a relatively simple process for an authorized third party to determine the SSID in use by an access point and gain access to the victim's network. If the WEP is enabled, gaining access to the victim's network becomes more difficult but not impossible due to the weaknesses of the WEP. Depending on the security used by the WLAN it can be made difficult for an unauthorized third party to masquerade as a legitimate user. However, even if they gain an RF capability to the victim's network, an additional barrier exists through the use of authentication, authorization, and accounting.

7.4.3 Message Modification and Injection

Messages encrypted by WEP can be modified. First, we show that messages may be modified in transit without detection, in violation of the security goals. Attacks aiming to modify cipher-texts use the fact that *the WEP checksum is a linear function of the message*. By this, we mean that

$$CRC(m \oplus m') = CRC(m) \oplus CRC(m')$$

for all m and m'.

One consequence of the aforementioned property is to allow controlled modifications on the transmitted cipher-texts without affecting the checksum. Let us assume that an attacker has intercepted a cipher-text c before it could reach its destination and assume that c is the encrypted form of an unknown message m. This means that

$$c = RC4(v, K) \oplus \langle m \| CRC(m) \rangle$$

for an initial vector v and a key K. One can state that it is possible to find a new cipher-text c' associated with a plaintext m'. Then there is δ such that $m' = m \oplus \delta$ decrypts. The attacker can selected arbitrarily δ and m'. Then, the attacker can replace the original cipher-text by the new cipher-text by spoofing the source. By decrypting c', the recipient will obtain the modified message m', provided that the checksum is valid.

Now let us see how to obtain c' from c so that c' decrypts to m' instead of m. The key observation is to note that stream ciphers, such as RC4, are also linear. Let XOR the quantity $\langle \delta \| CRC(\delta) \rangle$ with the right and left terms of the equation defining c. We have

$$c' = c \oplus \langle \delta \| CRC(\delta) \rangle$$

$$= RC4(v, K) \oplus \langle m \| CRC(m) \rangle \oplus \langle \delta \| CRC(\delta) \rangle$$

$$= RC4(v, K) \oplus \langle m \oplus \delta \| CRC(m) \oplus CRC(\delta) \rangle$$

$$= RC4(v, K) \oplus \langle m \oplus \delta \| CRC(m \oplus \delta) \rangle \qquad \text{(because CRC is linear)}$$

$$= RC4(v, K) \oplus \langle m' \| CRC(m'') \rangle$$

As a result, we have shown how to modify c to obtain a new cipher-text c' that will decrypt to $m \oplus \delta$. Thus, the WEP checksum fails to protect data integrity.

It is worth noting that this attack can be applied without full knowledge of the plaintext m. The attacker only needs to know the original cipher-text c and the desired plaintext difference δ, to calculate $c' = c \oplus \langle \delta \| CRC(\delta) \rangle$. For example, to modify the first bit of a message, the attacker select $\delta = 100 \ldots 0$.

On the other hand, the protocol WEP does not provide secure access control. We use the fact that the WEP checksum is an unkeyed function of the message. Consequently, the checksum field can also be computed by the adversary who knows the message. This property allows the attacker to elude the access control measures. In fact, if an attacker can get hold of an entire plaintext corresponding to some transmitted frame, he can inject arbitrary traffic into the WLAN. Knowing both the plaintext and the cipher-text would reveal the keystream, which can be reused subsequently to create a new packet, using the same initial vector.

Let us, finally, notice finally that it is possible to reuse old IV values without triggering any alarms at the receiver. Therefore, it is not necessary to block the reception of the original message. Once we know an initial vector v along with its corresponding keystream sequence $RC4(v,\ K)$, this property allows us to reuse the keystream indefinitely and elude the WEP access control mechanism.

A natural defense against this attack would be to disallow the reuse of an initial vector in multiple packets, and require that all receivers enforce this interdiction. However, the 802.11 standard does not do this. In addition, the 802.11 standard strongly recommends against IV reuse, but it does not require it to change with every packet. Hence, every receiver must accept repeated initial vectors.

7.4.4 Message Decryption

It has been shown in the previous section that an attacker has the ability to modify encrypted packets without detection. The attacker can also decrypt messages sent over the air. Since WEP uses a stream cipher presumed to be secure (RC4), attacking the cryptography directly is probably hopeless because it may take some unacceptable period of time for the attacker. However, the attacker can involve the AP indirectly in such a purpose. The idea, then, is to mislead the access point into decrypting some cipher-text for the attacker. The ability to modify transmitted packets provides two simple ways to abuse the AP in this way.

7.4.4.1 IP Redirection

An *IP redirection* attack can be used when the WEP access point acts as an IP router with Internet connectivity. This is a common situation for WLANs, since it is fairly common that the WEP is used to provide network access for mobile stations. In this case, the idea is to sniff an encrypted packet off the air and use an attack to modify it so that it has a new destination address that the attacker can control. The AP will therefore decrypt the packet, and send the packet to its (new) destination. The modified packet will flow *from* the WLAN *to* the Internet without being stopped by a firewall. Once it reaches its destination, the attacker can read the packet in the clear.

The easiest way to modify the destination IP address is to figure out what the original destination IP address is, and then modify it to the new address. Discovering the original destination IP address is often easy to perform. On can consider for example that all the incoming traffic will be intended to an IP address on the wireless subnet, which should be easy to determine. Once the incoming traffic is decrypted, the IP addresses of the other ends of the connections will be exposed, and outgoing traffic can then be decrypted in the same way.

However, a condition is required for the success of the IP redirection attack. The attacker also needs to ensure that the IP checksum in the modified packet is still correct—otherwise, the decrypted packet will be dropped by the AP. Since the modified packet differs from the original packet only in its destination IP address, and since both the old and new values for the destination IP address are known, one can compute the required change to the IP checksum caused by the address modification. Various methods can be used to correct the checksum. Three methods can be distinguished:

- *The IP checksum for the original packet is known*: Let in that case x and y be respectively the original and new checksum. A solution would be to modify the IP checksum by XORing in $x \oplus y$, which will change the IP checksum to the correct value of y.
- *The original IP checksum is not known*: If x is not known, the task is harder. Let z be the difference $y - x$, the attacker needs to compute $x \oplus y$ using z and some guesses. Indeed, z does not provide enough information to compute $x \oplus y$. False guesses will be silently ignored by the access point.
- *Organize that $x = y$*: A change in another field can often balance the change in the destination field in order to keep the checksum of the packet unchanged. For this, any header field that is known to the attacker and does not affect packet delivery can be considered (including the source address).

7.4.4.2 Reaction Attacks

Reaction attacks can be performed whenever WEP is used to protect TCP/IP traffic (Bellovin, 1996). However, these attacks do not require connectivity to the Internet, so it may apply even when IP redirection attacks are not achievable. In a reaction attack, the attacker monitors the reaction of a recipient of a TCP packet and uses what he collects to infer information about the unknown plaintext. The attack relies on the fact that a TCP packet is accepted only if the TCP checksum is correct; and when it is accepted, an acknowledgment packet is sent in response. The acknowledgment packets are easily identified by their size, without requiring any effort of decryption. Thus, the reaction of the recipient will disclose whether the TCP checksum was valid when the packet was decrypted.

The attack proceeds as follows: the attacker intercepts a cipher-text c, flowing from user A to B, with unknown decryption P. The attacker flips a few bits in m and adjust the encrypted CRC accordingly to obtain a new cipher-text c' with valid WEP checksum. Then, he transmits c' in a forged packet to the AP. Finally, he watches to see whether the eventual recipient sends back a TCP acknowledgment packet; this will allow the user to tell whether the modified text passed the TCP checksum and was accepted by the recipient.

The presence or absence of an ACK packet is able to reveal one bit of information on the unknown plaintext P. By repeating the attack for many choices of bit position, the attacker can learn almost all of the original plaintext, and then deduce the few remaining unknown bits using classical techniques. Therefore, the attacker has exploited the receiver's willingness to decrypt arbitrary cipher-texts and feed them to another component of the system that leaks little information about its inputs. The recipient's reaction to the forged packets can be viewed as a side channel that allows us to learn information about the unknown plaintext. In addition, one can say that the recipient is used as an oracle to unknowingly decrypt the intercepted cipher-text for the attacker.

Now let us see how to choose new forged packets c' to mislead the recipient into revealing information about the unknown plaintext (Borisov, 2001). Let us first recall that the TCP checksum is the one's-complement addition of the 16-bit words of the message M, meaning that the TCP checksum on a plaintext m is valid only when $m = 0 \pmod{2^{16}-1}$. Let now $c' = c \oplus \delta$ where δ only specifies which bit positions to flip. It is chosen as follows: take arbitrarily a number i, set bit positions i and $i + 16$ of δ to 1, and let the other components of δ set to 0. Then it can be shown that, for all p, the equality $p = p \oplus \delta \pmod{2^{16} - 1}$ holds exactly when $p_i \oplus p_{i+16} = 1$. Since the TCP checksum is valid for the original packet (meaning that $c = 0 \pmod{2^{16} - 1}$), this means that the TCP checksum will be valid for the new packet (meaning that $c \oplus \delta = 0 \pmod{2^{16} - 1}$) just when $c_i \oplus c_{i+16} = 1$. Fortunately, this gives one bit of information on the plaintext.

7.5 WiFi Protected Access

Realizing that a robust security solution should replace the original WEP as soon as possible and that it was not possible to wait for the complete ratification of IEEE 802.11i standard, the WiFi alliance, which is a non-profit international association formed in 1999, used the ready portions of IEEE 802.11i standard to define the WiFi Protected Access or WPA.

7.5.1 WPA Design Issues

The WPA solution tried to overcome the design weaknesses of the WEP architecture while proposing an effective key distribution method (Masica, 2007). In fact, WPA has introduced the Temporal Key Integrity Protocol (TKIP) that relays on RC4 but uses a much longer IV and a per-packet encryption key in order to cure some of the WEP vulnerabilities. TKIP also introduced the message integrity check as it provided a good integrity level without requiring a lot of computing resources.

To guarantee authentication, the WiFi alliance provided two versions for WPA: the WPA per-user based security designed for enterprises (also known as WPA enterprise mode) and the WPA pre-shared key mode designed for consumers referred to as the consumer mode. The WPA enterprise mode authentication is based on the 802.1x and one of the Extensible Authentication Protocol (EAP) types available today (WPA, 2003).

EAP defines the messages to exchange at the data link layer level in order to authenticate users (Aboba, 2004). It can be used for dial-up lines with Point to Point Protocol (PPP), wired LANs, as well as for wireless media (Simpson, 1994; IEEE, 2003). EAP is "extensible" since it may use a plethora of authentication methods such as the TLS protocol, MD5, and the security tokens. The EAP packet's format is made up of four fields: the Code, the Identifier, the Length, and the DATA (Soltwisch, 2004). The code field is one byte long; it indicates the packet's type that can be (a) a Request, (b) a Response, (c) a Success, or (d) a Failure. The Identifier field, which is also one byte long, is used to match a Response packet with the correspondent Request packet. The two bytes long Length field indicates the length of the whole EAP packet including the header and the data. Finally, the Data field can be empty or be several bytes long depending on the EAP packet type.

The authentication protocol involves the following layers: the lower layer, the EAP layer, the EAP peer and authenticator layers, and the EAP method layer (Figure 7.6 depicts the EAP model). The lower layer monitors the transmission and the reception of the data frames in the correct order between the peer and the authenticator. The EAP layer guarantees a reliable transmission of the EAP packets via the lower layer. It implements request retransmission, in the case of response loss, and detects duplicated packets thanks to the Identifier field. It also delivers and receives EAP messages to and from the EAP peer and authenticator layers.

The EAP layer demultiplexes the incoming EAP packets to the EAP peer and authenticator layers using the code field. Received EAP packets with Code field equal to 1, 3, or 4 are respectively Request, Success, and Failure packets. They are delivered by the EAP layer to the EAP peer listener, if present. EAP response packets (Code=2) are delivered to the EAP authenticator listener, if present. Finally, the EAP methods implement the authentication algorithms and receive and transmit EAP messages via the EAP peer and authenticator layers. Since fragmentation support is not provided by EAP itself, this is the responsibility of EAP methods. The EAP methods implement the authentication logic and determine whether the supplicant is a legitimate user.

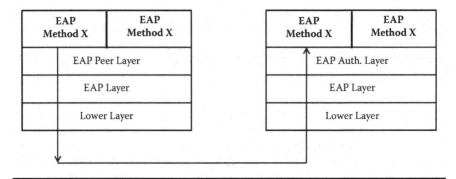

Figure 7.6   EAP multiplexing model.

EAP methods can be removed or added whenever needed; nevertheless, the supplicant and the authenticator should implement at least one sufficient method for both of them to correctly fulfill authentication. The reliability of the authentication process highly depends on the authentication method that will be used. For this reason, the EAP draft requires EAP method specifications to include a "Security Claims" section that indicates the intended use (e.g., physically secure or insecure protocol), the adopted authentication method, the claimed security properties, the key strength, the description of the key hierarchy, and the indication of the vulnerabilities.

Some EAP methods enable the derivation of keying materials on the client and the authenticator sides without carrying such information on the network. However, when an authentication server is added, the authenticator does not need to implement the EAP methods as the authentication decisions are taken by the server. But in this case, the keying material should be transmitted from the authenticating server to the authenticator. We distinguish three special EAP methods that do not implement authentication algorithms. They are used for requesting a peer identity, conveying a displayable message (from the authenticator to the peer), and declining an authentication type proposed by the authenticator.

The mandatory EAP authentication method supported by all EAP implementations is the MD5-Challenge. In fact, the authenticator should send a challenge message to the wireless user and wait for a response from him. If the user's response is valid, the client will be considered as legitimate and the authentication process will succeed. A second authentication method that is supported by EAP is the *One Time Password* (or OTP), which uses a key list of one time valid challenge/password pairs. On the other hand, the EAP-TLS authentication method described in Aboba (1999) based on the Transport TLS aims at guaranteeing the privacy and the integrity between two communicating applications. EAP-TLS takes advantage of protected cipher suite negotiation, mutual authentication, and key management features of the TLS protocol in order to securely authenticate both the client and the authenticator. The EAP Expanded Type, which has the EAP Type 254, may be used for vendor-specific uses of EAP. Finally, a tunneled method is an authentication method having other methods running with it; it is considered as a unique authentication but it is also vulnerable to man-in-the-middle attacks (as it has been shown in Chapter 3).

EAP over LAN (or EAPOL) is an encapsulation technique that is adopted in the LAN environments for exchanging EAP packets between the client and the authenticator. The EAPOL packet's format contains four fields. The Protocol Version field is one byte long that indicates the EAPOL version (currently, this field is equal to 1). The Packet Type field is also one byte long. It determines the packet's type, which can be an EAP packet, an EAPOL-start frame, an EAPOL-Logoff request frame, and an EAPOL-Key frame when the packet field values are 0, 1, 2, and 3, respectively. The third field is the *Packet Body Length* field. It is a 2-byte long field and may take the value "0"; it indicates the length of the packet's body in bytes. Finally, the packet body field may be either an EAP packet (when the Packet Type field contains the value "0") or a key descriptor.

IEEE 802.1x is a port-based network access control standard that was ratified in 2001 and originally designed for the modem connections and the wired LANs (IEEE, 2001). More specifically, IEEE 802.1x authenticates and authorizes devices that are attached to a LAN port having point-to-point connection characteristics, where a LAN port refers to a single point of attachment to the LAN infrastructure (WPA, 2003). 802.1x devices can be a supplicant, an authenticator, a Network Access Server (NAS), and an Authentication Server (AS). The supplicant may be the wireless client, the authenticator may be a 802.1x capable AP, while the authentication server may be simply a Remote Access Dial In User Service (RADIUS) server.

The authenticator plays the role of a security guard as it initially blocks the access to the network resources until the authentication process succeeds. The authentication process is performed as follows: the supplicant supported by the client requests access to the WLAN. The AP hearing the request asks for the wireless node's identity. The wireless node answers by sending its identity using identity hiding so that a hacker cannot spoof it. It is worth noticing that the supplicant and the authenticator use the EAP protocol to communicate and exchange the previously described messages. Then, the authenticator encapsulates the supplicant's identity to the RADIUS format and relays it to the authentication server. The authentication server verifies the identity, sends back a success message to the authenticator, and opens a port to the supplicant if the user is legitimate. Upon an authentication failure, the authentication server sends a failure message to the authenticator. The latter relays the response to the supplicant and prevents that supplicant from accessing the network.

To provide confidentiality and integrity services, the WPA is based on the TKIP protocol, which uses 128-bit keys and implements a key management method. The TKIP will be detailed in the following sub-section; however, it is important to know that once the user credentials are accepted, the authentication server uses 802.1x to create a unique pair-wise key relative to that session. This key is relayed to the client and the AP. A key hierarchy and management system is then set up using the pair-wise key in order to dynamically generate unique keys that serve for encrypting each data packet exchanged during that session. TKIP also adds a Message Integrity Check (MIC, or Michael) to protect the packets' integrity. The MIC provides a strong mathematical function; the receiver and the transmitter each compute and compare the MIC in order to detect any modification in the packet's content. In the case of check failure, the packet can be dropped. Table 7.2 resumes the WEP and WPA differences in terms of the adopted encryption and authentication mechanisms (WPA, 2003).

Controversially, the WPA consumer mode also known as WPA-PSK or WPA-Home does not rely on the relatively costly RADIUS authentication infrastructure; it rather implements a simple shared secret authentication combined with a TKIP key management and a Michael integrity checking. Such shared secret is a

Table 7.2   Comparison between WEP and WPA

	WEP	WPA
Encryption	Flawed, cracked by scientists and hackers	Fixes all WEP flaws
	40-bit keys	128-bit keys
	Static—same key used by everyone on the network	Dynamic session keys per user, per session, per packet keys
	Manual distribution of keys—hand typed into each device	Automatic distribution of keys
Authentication	Flawed, used WEP key itself for authentication	Strong user authentication, utilizing 802.1x and EAP

password configured into the AP and each 802.11 device. Despite the simplicity of the WPA-PSK and its low implementation cost, it has been shown that it is more vulnerable. For instance, impersonation between stations or a station impersonating an AP is possible.

7.5.2 WPA Vulnerabilities

WPA was designed to cure the WEP vulnerabilities; nevertheless, it suffers from some weaknesses especially caused by the PWA-PSK mode and the vulnerabilities of the 802.1x method. Particularly, the PSK version of WPA is vulnerable to the offline dictionary attack since the information required to create and verify a session key is broadcasted (Takahashi, 2004). In fact, a master key, called the PMK, is produced by running a special function on a pre-shared pass phrase and an SSID. Both the host and the AP use this master key, along with the MAC addresses and two nonces, in order to create the a session key (called PTK) and install it on both sides. The PMK is generated by providing the pass phrase, SSID, and the SSID length to a key generating algorithm that produces a value of 256 bits. Since the SSID is easily recoverable, one can notice that only the pass phrase would have to be guessed in order to determine the valid key session PMK. Furthermore, in the generation of the PTK (for cracking needs), the attacker needs only the PMK to be determined since all other fields can be trivially discovered.

Note that the PTK consists of 4 keys: the Key Confirmation Key (KCK), the Key Encryption Key (KEK), the Temporal Key 1, and the Temporal Key 2. In addition a wireless user should generate the PTKs and then run MD5 hash function on the KCK and the EAP packet to be sent. The attacker may begin a dictionary attack by simply extracting the hash value of this packet and then comparing it to the hash of his guessed PTKs. This hash is then added to the EAP packet and sent over the network. Now, the attacker can utilize the hash portion of this packet and match it with the hash result of his guessed PTK and collected EAP packet; the correctly guessed pass phrase produces the same signature. Therefore, the attacker, by passively sniffing two of the EAPOL packets, can perform an offline dictionary attack. Furthermore, the attack can be launched as long as there exits an active session within the wireless network. A well known disassociation attack can be used to trigger a re-association between the host and the AP so that the attacker can gather the necessary packets.

Notice finally that it has been demonstrated that any key generated from a pass phrase that is smaller than 20 characters is highly vulnerable to attacks (Moskowitz, 2003). Tools exploiting such vulnerability are available on the Internet and can be easily used to attack the WPA-PSK mode. Among these tools, one can mention AirCrack and coWPAtty.

It is also valuable to note that the 802.1x authentication procedure may be affected by snooping attacks. In fact, snooping consists in listening to the authentication traffic and collecting valuable information such as identity and passwords. For instance, when the roaming procedure is supported, the AP may rely on the peers' identity to select the best authentication server. Consequently, the identity information may be easily exposed. The key recovery attacks may then follow the snooping as the attacker who collected the information can reproduce the steps taken by the EAP method to derive the keys. If the attacker fails in recovering the keys, he may try to directly attack the cipher suite by a downgrading attack targeting the cipher suite negotiation. The EAP protocol may also become vulnerable to man-in-the-middle attacks when peer authentication is not applied. If that is the case, the attacker may pretend to be the authenticator when it communicates with the wireless user and may pretend to be that user when it communicates with the AP, becoming able to read and modify the complete authentication process.

The rogue authenticator attack is a form of a man-in-the-middle attack. It allows a malicious peer to pretend to be the authenticator and query the client for valuable information; then he may use the collected information for dictionary attacks. Piggybacking may also be performed by gaining access to an authenticated session of a supplicant and then sending data on the behalf of the victim. Session hijacking is an advanced form of piggybacking attack since the attacker will control the whole authenticated session instead of simply sharing it.

The service availability may also be affected when the authenticator (i.e., the AP) is flooded with counterfeit information so that it can no longer start an authentication session with the peer or when packets are injected in order to abort an authentication session. For instance, the injection of EAP Nack messages during the EAP method negotiation or pretending that the supplicant has sent a disassociation request to the AP may lead to serious DoS. Moreover, repeatedly sending a spoofed EAPOL-Logoff (e.g., EAPOL-Logoff frames are not authenticated) may prevent the legitimate user from accessing the network services. Finally, an attack can be launched by simply sending multiple EAPOL-start packets to the AP in an attempt to bring it down.

To guarantee integrity, both WPA and IEEE 802.11i implement the MIC mechanism. The security level of the MIC is generally measured in bits. In fact, if the security level of a MIC is n bits, then an attacker may on average forge a packet in about 2^{n-1} packets (Cam-Winget, 2003). The MIC used by WPA provides only 20 bits of security; therefore, TKIP requires a rekey after detecting a MIC validation error and needs to limit rekeying to once per minute.

7.6 IEEE 802.11i and WiFi Protected Access 2

After the ratification of IEEE 802.11i in 2004, the WiFi alliance introduced an interoperable implementation of the standard that was referred to as WPA2. In addition to TKIP encryption and 802.1x/EAP authentications, WPA2 supports the use of the encryption algorithm Advanced Encryption Standard (AES) that has been adopted as an official government standard by the United States (Takahashi, 2004). The AES will secure the communication between mobile users operating in the ad hoc mode; it is based on a mathematical ciphering algorithm that uses Counter Mode with Cipher Block Chaining Message Authentication Protocol and 128, 192, or 256 bits keys. WPA2 may secure the devices implementing the IEEE 802.11b, IEEE 802.11a and IEEE 802.11g versions; however, a hardware upgrade for implementing the AES algorithm is required for these platforms. WPA2 mixed mode devices will support both WPA and WPA2 architectures contrarily to the WPA mixed mode devices which supported WEP and WPA. Consequently, the transition to WPA2 clients and APs can be gradually achieved without compromising the WLAN security.

7.6.1 IEEE 802.11i Design Issues

The complete specification of the IEEE 802.11i standard was ratified in 2004. It defines two classes of security algorithms, the Robust Security Network Association (RSNA) and the Pre-RSNA. The latter class implements the WEP solution and does not use the 4-way handshake authentication, provided by WPA, while the former implements the TKIP, the Counter-Mode/CBC-MAC protocol (CCMP), the 802.1x authentication, and the 4-way handshake authentication and key management protocols.

IEEE 802.11i is backward compatible with WPA as it implements both TKIP and 802.1x protocols. Nevertheless, the standard also implements AES as encryption algorithm and introduces key caching and pre-authentication to achieve fast and secure roaming. The key caching consists in storing the information related to the wireless station with an AP so that, when the station changes its managing AP and then returns back to that AP, it does not have to provide re-authentication credentials once again. Pre-authentication is achieved when authentication data are sent between APs so that a roaming station does not have to authenticate to each AP. The AES implementation supported by IEEE 802.11i is used in Counter Mode for guaranteeing confidentiality, while it is used in Cipher Block Chaining Message Authentication Code (CBC-MAC) mode to provide authentication and integrity services (Masica, 2007). It is worth noticing that the use of these modes is mandatory in 802.11i while the use of TKIP is optional. The latter was implemented only to guarantee the compatibility of the standard with the existing hardware.

7.6.1.1 Temporal Key Integrity Protocol (TKIP)

The Temporal Key Integrity Protocol (TKIP) aims at curing the multiple WEP vulnerabilities. In fact, the protocol supports a new key generation scheme and a longer initial vector IV to properly utilize RC4 algorithm. Besides, a more robust integrity protection mechanism (known as *Michael*), along with a replay detection mechanism, are used to guarantee integrity services. To guarantee data confidentiality and counteract against the WEP key recovery attacks, TKIP adopts a per-packet encryption-key generation based on a mixing function (Cam-Winget, 2003). The mixing function uses a non-linear substitution function or S-box in order to combine the base key, the transmitter MAC address, and the most significant four bytes of the packet sequence number. The resulting intermediate value is then mixed with the two least significant bytes of the packet sequence number in order to produce the per-packet encryption key. To save the processing resources of the mobile host, the same intermediate value is cached and used to generate the keys of up to 2^{21} packets without compromising the security. In fact, the two least significant bytes of each packet sequence number will differ; therefore, we will obtain a different per-packet key. It is also valuable to mention that using the same base key for all network members is possible since the intermediate value varies according to the MAC address of each host, thus resulting in a different per-packet key at each host.

The TKIP solution is based on two distinct generated keys: a 128-bit encryption key produced by the previously described mixing function and a 64-bit key called the Michael key used for the integrity protection. These keys should be fresh enough (e.g., no relationship should exist between the instantiated keys and the old ones and the lifetime of the key should be kept small) in order to guarantee a good security level. To address this issue, IEEE 802.11i adopts the IEEE 802.1x standard to achieve key management along with the authentication process. In fact, IEEE 802.1x authenticates the mobile user and then generates a fresh master key and distributes it. The mobile station and the AP will use this master key for generating the key pairs while specific WEP key IDs will be used to verify the keys freshness (Cam-Winget, 2003). Special rekey key messages with keying materials are transmitted to the AP and the mobile users in order to ask them to regenerate new keys.

To guarantee data integrity, each mobile user implementing TKIP and wishing to transmit a packet should compute a keyed function of the data and then send

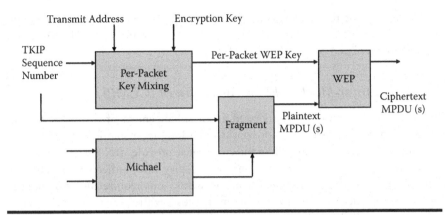

Figure 7.7   The TKIP solution phases.

the resulting value as a tag within that packet. The receiver should re-compute the MIC, then compare the obtained and the received values in order to accept authentic packets and reject non-authentic ones. Since it is better not to implement complicated conventional MIC algorithms such as HMAC-SHA1 and DES-CBC-MAC, the Michael MIC was designed (Ferguson, 2002; IEEE, 2003). The Michael MIC is a keyed hash function that uses a 64-bit Michael key and an arbitrary long message as input in order to produce a 64-bit Michael value. The TKIP solution phases are described in Figure 7.7.

Basically, the Michael function works as follows: the Michael key is divided into two 32-bit keys while the message is partitioned into 32-bit blocks, padded at the end with a single byte having the hexadecimal value of 0x5a, then followed by a number of zero bytes (between 4 and 7) in order to obtain an overall length that is a multiple of 4. Once the function's input is obtained, the Michael function processes multiple operations such as XOR, left rotation, right rotation, addition modulo 2^{32}, and swapping in order to generate the MIC (Cam-Winget, 2003; IEEE, 2003). To achieve swapping, the XSWAP function swaps the position of the two least significant bytes and the two most significant bytes in a word so that XSWAP(ABCD) = BADC where A;B;C;D are bytes. The last step consists in appending the resulting value as a tag to the packet, then encrypting the whole using RC4 and the per-packet encryption key. Note that the per-packet key mixing function and the WEP encryption are applied to the packet fragments (or MPDUs) while the Michael function is applied to the whole packet (or MSDUs).

To protect against replay attacks, it is common to bind a packet Sequence Number (SN) to each packet with a MIC and then reinitialize that sequence space whenever the MIC key is replaced. Since the TKIP was limited by the implementation constraints, it extended the WEP format to use 48-bit sequence number but associated the sequence number to the encryption key instead of the MIC key (Cam-Winget, 2003). After the transmission begins and the TKIP keys are set up, the sender and the receiver initialize the SN to zero. Then, the sender increments the SN with each transmitted packet while the receiver verifies the SN correctness and discards replayed packets (Walker, 2002).

7.6.1.2 Counter-Mode/CBC-MAC Protocol (CCMP)

The TKIP architecture may be viewed as an enhancement of the WEP as it is based on RC4; therefore, it is not considered a long-term solution. To provide a high security level, the adopted architecture must provide strong encryption and special mechanisms for preventing integrity, replay and authentication attacks. Nevertheless, these supported mechanisms should not require high commutating power in order to minimize the hardware costs while preserving the devices' resources. For all these reasons, the IEEE 802.11i implements a security solution based on CCMP, which uses the counter (CTR)-mode with cipher block chaining-message authentication code (MAC)-mode (CCM).

AES is a block cipher that divides the data to encrypt into blocks and then uses a symmetric key having the same length of the block in order to produce the cipher text. AES may implement different operation modes. The Electronic Codebook or ECB divides the messages into 128-bit blocks while padding the last block with zeros in order to obtain the required length; then it encrypts each block individually using the same key. The shortcoming of this mode lays in the fact that the identical plaintext fragments will have the same cipher text, thus enabling an attacker to collect valuable information. Moreover, using ECB implies using a MIC if the transmitted packets need to arrive in the correct order. The CTR, which is the second operation mode, uses an "added counter" encrypted with AES then XORed with the message so that no more padding bits are needed and no more regularities are observed.

The third mode (or CCM) was specially designed for IEEE 802.11i; it presents the base of the CCMP. CCM combines the CTR mode with the ISO-standardized CBC MAC in order to guarantee the integrity and the arrival of the received packets in the correct order. First, a 64-bit MIC is computed and then attached to the MPDU. The counter contains a 2-byte CTR field that is initialized with 1, then incremented for each MPDU so that 2^{16} blocks of 128 bits can be encrypted using a different counter. After that, the MPDU is encrypted through the counter and the session key. The result is then divided into 128-bit blocks and XORed with the counter. Finally, the MAC header and the CCMP header are concatenated to the cipher text and the whole packet is transmitted. CCMP uses AES with 128-bit keys as the utilization of 192 or 256 bit keys is not yet justified (Soltwisch, 2004). The CCMP encryption process is illustrated by Figure 7.8. In fact, a packet number (PN) is incremented for each MPDU. Additional Authentication Data or AAD are then deduced from fields in the MAC header while a CCM nonce block (IV) is derived from the PN and the priority. After that, the CCMP header is created from the PN and the key. Finally, the cipher-text and the MIC are generated through processing the CTR mode by using the TK, AAD, nonce value, and MPDU data.

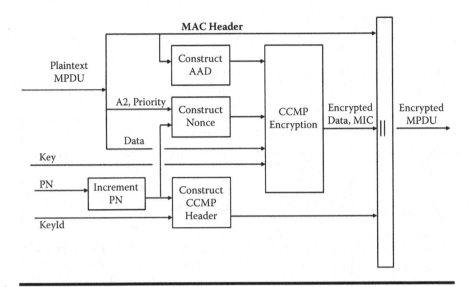

Figure 7.8   CCMP encryption.

The transmitted MPDU is made of the original MAC header, the CCMP header, the cipher-text, and the MIC.

The main difference between TKIP and CCMP is that AES does not require a per-packet encryption-key. Consequently, the per-packet key derivation function is no longer needed. Besides, CCMP uses the same AES key for guaranteeing confidentiality and integrity services so that the implementation complexity is reduced. However, using the same key in this context is safe as CCM guarantees that the space for the counter mode never overlaps with that used by the CBC-MAC initialization vector (Cam-Winget, 2003). CCMP also uses 8-byte MIC, which is far stronger than the Michael one. Table 7.3 summarizes the previously described security architectures that were adopted by IEEE 802.11 by comparing their respective encryption key sizes, lifetime, and the adopted mechanisms for guaranteeing integrity, replay detection, and key management (Cam-Winget, 2003).

7.6.1.3 Port-Based Authentication Protocol (802.1x) and Key Management

IEEE 802.11i implements a 802.1x architecture that uses EAP and mostly RADIUS servers to correctly meet its security goals. As described in the previous section, 802.1x adopts a port-based network access mechanism in order to authenticate clients and prevents the non-legitimate users from accessing the network resources. Besides, 802.1x offers the possibility of dynamically changing the encryption keys as the key distribution between the authentication server, the AP, and the mobile station is part of the 802.1x framework.

Table 7.3 WEP, TKIP, and CCMP Comparison

	WEP	TKIP	CCMP
Cipher Key size(s)	RC4 40- or 140-bit encryption	RC4 128-bit encryption, 64-bit authentication	AES 128-bit
Key lifetime Per-packet key	24-bit wrapping IV Concatenate IV to base key	48-bit IV TKIP mixing function	48-bit IV Not needed
Integrity Packet header	None	Source and destination addresses protected by Michael	CCM
Packet data Replay detection	CRC-32 None	Michael Enforce IV sequencing	CCM enforce IV sequencing
Key management	None	IEEE 802.1x	IEEE 802.1x

A mobile user wishing to access the mobile network should send an EAP-start message for its AP. Every packet that is not EAP encapsulated (e.g., a HTTP packet and a POP3 packet) will be automatically rejected by the AP when its transmitter is still unauthorized. Upon receiving the EAP start, the AP answers by a EAP request identity message. The client should now send an EAP response that presents its own identity. The AP relays the received identity to the authentication server, which verifies it and then returns a message indicating whether the client is authorized. Depending on the authentication server's response, the AP will send an EAP-success packet and enable the client's port for all packet types or it will send a reject packet and deny all traffic initiated by that user (Vollbrecht, 2002).

The key management of IEEE 802.11i implemented by IEEE 802.1x defines a Pair-wise Master Key (PMK) that is used by the AP and the mobile client to generate a Pair-wise Transient Key (PTK). It is the PTK that will secure the future communications. When the dynamic key management is activated and configured, a session key is transmitted by the authentication server to the AP along with the accept message. The AP uses this temporal key to build and secure an EAP key message before sending it to the client. Finally, the user generates the encryption-key from the received information. It is valuable to note that the session keys may be periodically changed while the authentication process may be frequently reprocessed (Vollbrecht, 2002). The key exchange procedure is based on a 4-way handshake and a group of key handshake, based on EAP messages. More precisely, the different keys used by the 802.1x architecture may be categorized as pair-wise keys or group keys as follows:

- *Pairwise Keys*
  - Master Key (MK): used when an access is needed
  - Pairwise Master Key (PMK): used for authorizing the access to the 802.11 medium
  - Pairwise Transient Key (PTK): includes three keys:
    - Key Confirmation Key (KCK): used for binding the PMK to the AP and to the station and also for verifying the possession of PMK
    - Key Encryption Key (KEK): used for Group Transient Key (GTK) distribution
    - Temporal Key (TK): used for securing data traffic
- *Group Keys*
  - Group Transient Key (GTK): the equivalent of a TK, used for securing multicast/broadcast traffic

The 4-way handshake guarantees that the AP and the mobile user will use fresh session keys while indicating that there is no man in the middle between the AP and the station with the same PTK when there has been no man-in-the-middle when using the PMK. The four-way handshake process is illustrated by Figure 7.9 (Soltwisch, 2004).

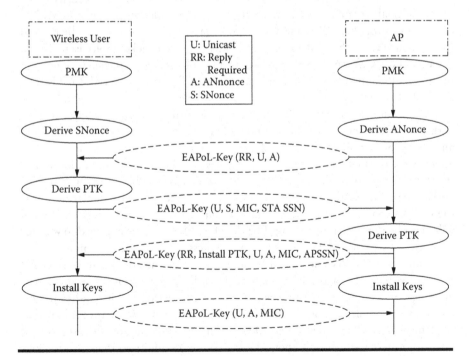

Figure 7.9   Four-way handshake.

In fact, the AP should first send a message with a nonce value (ANonce or Authenticator nonce) to the client. The client generates the PTKs and consequently the TK by using the PMK, its own nonce value (SNonce or Supplicant Nonce), the transmitter address, and his MAC address. The client then sends an EAPOL message with his SNonce and MIC for guaranteeing integrity. The AP uses the SNonce to deliver the PTKs and the EAPOL MIC key; it then sends a MIC-protected message containing an activation of the PTKs notice and an initial SN value. The client confirms with a last message that it has successfully generated the required key while activating the encryption of the current connection. Upon receiving this confirmation message, the AP checks the integrity and enables the PTKs.

7.6.2 WPA2 and 802.11i Vulnerabilities

IEEE 802.11i was initially designed to overcome WEP flaws and to offer a secure framework for wireless communications in LAN environments. The intermediate WPA solution that was built using the ready modules of IEEE 802.11i also suffered from vulnerabilities especially because TKIP was based on RC4 encryption. The complete specifications of IEEE 802.11i include the mandatory CCMP protocol in combination with AES in order to guarantee effective data confidentiality and integrity. However, it has been shown that the CCMP presents some security flaws. In fact, special attention must be paid to the counter choice when applying MPDU encryption through CTR mode and AES. In fact, that counter should be derived from a nonce value that differs for each sender and each packet (Soltwisch, 2004). Besides, the CCMP may be the target of pre-computation attacks as it uses an incremental PN to derive nonces and initializes the PN to 1 for every fresh TK (He, 2005). In fact, a malicious user may compute a table offline by choosing one nonce and 2^{64} possible keys. After that, he may observe the online messages encrypted with that specific nonce and an unknown key. On average, the attacker may find an overlap of keys after observing 2^{64} messages with that chosen nonce and different keys, thus deducing the TK of that session. Such pre-computation attack reduces the key space size from 2^{128} to 2^{64}, which is possible to be broken (He, 2005). Nevertheless, the impact of this attack is limited to one particular station as the CCMP constructs the nonce using the source MAC address. Besides, the PN will never be the same for the same TK; therefore, the attacker may need to wait for refreshed TKs of different sessions to observe messages with the same nonce.

Combining the PN and the MAC addresses leads the attacker to observe 2^{64} different sessions for a particular station in order to break one. The previously described pre-computation attack is not practical to perform as it requires important resources while having a limited impact. Consequently, we may conclude that implementing CCMP prevents attackers from breaking data confidentiality and integrity unless they already know the key. Note, in addition, that the attackers cannot obtain valuable information about the used key even if they analyze the cipher text while already knowing the corresponding plaintext (He, 2005).

IEEE 802.11i may also suffer from Security Level Rollback attacks when implementing Pre-RSNA and RSNA in the same WLAN. In fact, an attacker may impersonate the authenticator by generating false Beacon or Probe Response frame and then pretend that only the Pre-RSNA solution is supported. The attacker may also impersonate the supplicant by generating a false Association Request and then pretend that only the Pre-RSNA solution is supported. As a result, the Pre-RSNA will be adopted to guarantee the security of the communications even though the more robust RSNA could be supported. Consequently, the supplicant and the authenticator will not be able to confirm the cipher suite while the attacker will be able to discover the default keys by exploiting the WEP flows and then compromise the communication's security. The Security Level Rollback attack could be countered by configuring the authenticator and the supplicant to only support RSNA especially when a high security level is required. It is also possible to allow the wireless entities supporting both RSNA and Pre-RSNA; meanwhile, appropriate policies about when using each solution and how to use them should be defined. For example, the wireless users may choose to deny the Pre-RSNA before initiating a connection while the authenticator may limit Pre-RSNA connections to only insensitive data (He, 2005).

Reflection attacks may also target the IEEE 802.11i standard when the ad hoc mode is adopted. In fact, the four-way handshake mechanism relies on a pre-shared PMK in order to calculate correct MIC and compose valid messages. When the same device plays the role of the supplicant and the authenticator under the same pre-shared PMK, attackers may perform a common reflection attack targeting that device. More precisely, when that device initializes a four-way handshake as an authenticator, the attacker will initialize a second four-way handshake using the same parameters in order to let the device act as a supplicant. The victim device will compute a message as a supplicant while the attacker will use that message in order to respond to the first initialized four-way handshake, thus violating the mutual authentication. To address this issue, the wireless devices should play a unique role or use different pre-shared PMK when playing separate roles.

IEEE 802.11i was not designed to protect service availability. For instance, vulnerabilities inherited from the support of the IEEE 802.1x authentication mechanisms have led to serious DoS attacks. Moreover two DoS attacks called RSN Information Element (RSN IE) Poisoning and four-way handshake blocking were developed (He, 2005). As it was defined by IEEE 802.11i, the RSN IE field is used by the authenticator and the supplicant during the four-way handshake authentication in order to prevent an attacker from tricking them into using a weaker security scheme by forging the RSN IE negotiations. Nevertheless, the supplicant verifies the RSN IE values before the MIC value and aborts the handshake if the RSN IE is unmatched. An attacker may exploit this design vulnerability to abort the authentication process and prevent legitimate users from accessing the network resources. Similarly, the four-way handshake blocking attack takes profit of a vulnerability observed in the design of the authentication mechanism.

As it was described earlier, the first and the third messages M1 and M3 of the four-way handshake carry the nonce generated by the authenticator; the second message M2 carries the nonce generated by the supplicant, while the fourth message M4 acknowledge the success of the handshake. Meanwhile, 802.11i adopts a Temporary PTK (TPTK) to store the newly generated PTK until the third message is verified. A supplicant implementing the vulnerability handshake should accept all first messages it receives in order to guarantee that the handshake will succeed despite a possible packet loss. Nevertheless, an attacker could simply send a fake first message M1' with a different nonce value between the legitimate first message M1 and the third one M3, thus causing a PTK inconsistency between the supplicant and the authenticator and blocking the authentication process. To address this issue, either M1 should be authenticated or the same nonce should be used for all received M1 messages until the vulnerability handshake succeeds (He, 2005).

To conclude this section, let us notice that the IEEE 802.11i specifications tried to address the security flaws of the WEP and the WPA solutions. Although the IEEE 802.11i is considered as the most robust framework, it suffers from design vulnerabilities that may lead to security level rollback, reflection, and DoS attacks. Moreover, the implementation of IEEE 802.11i required hardware and software upgrades and was considered as "costly and complex." Subsequently, many wireless users have preferred WPA and rejected (temporarily) the IEEE 802.11i upgrade although it offers better security guarantees.

References

B. Aboba, L. Blunk, J. Vollbrecht, J. Carlson, and H. Levkowetz, Extensible Authentication Protocol (EAP), Internet proposed standard RFC 3748, June 2004. Available at http://tools.ietf.org/html/rfc3748.

B. Aboba and D. Simon, PPP EAP TLS Authentication Protocol, RFC2716, available at http://www.faqs.org/rfcs/rfc2716.html.

AirCrack. Available at http://www.grape-info.com/doc/linux/config/aircrack-2.3.html.

S. M. Bellovin, Problem Areas for the IP Security Protocols, in *6th USENIX Security Symposium*, San Jose, CA, July 1996. USENIX.

D. Blunk and A. Girardet. WLAN Open Source Linux Tool for Breaking 802.11 WEP Keys, WepAttack. Vers. 0.1.3., 2002. Available: https://sourceforge.net/projects/wepattack/.

N. Borisov, I. Goldberg, and D. Wagner, Intercepting Mobile Communications: The Insecurity of 802.11, *ACM SIGMOBILE*, July 2001, Rome, Italy, pp. 180–188.

N. Cam-Winget, R. Housley, D. Wagner, and J. Walker, Security Flaws in 802.11 Data Link Protocols, in *Communications of the ACM*, May 2003, Vol. 46, No. 5.

coWPAtty. Available at http://www.wirelessdefence.org/Contents/coWPAttyMain.htm.

N. Ferguson, Michael: An Improved MIC for 802.11 WEP. IEEE 802.11, doc 02-020r0, 2002. Available at <http://grouper.ieee.org/groups/802/11/Documents/DocumentHolder/2-020.zip>.

C. He and J. C. Mitchell, Security Analysis and Improvements for IEEE 802.11i, *The 12th Annual Network and Distributed System Security Symposium (NDSS'05)*, pp. 90–110, 2005.

IEEE, Draft Amendment to STANDARD FOR Telecommunications and Information Exchange between Systems—LAN/MAN Specific Requirements—Part 11: Wireless Medium Access Control (MAC) and physical layer (PHY) specifications: Medium Access Control (MAC) Security Enhancements, IEEE Draft 802.11i/D7.0, 2003.

IEEE, Standards for Local and Metropolitan Area Networks: Standard for Port Based Network Access Control. IEEE Std 802.1x-2001, available at http://standards.ieee.org/ getieee802/download/802.1X-2001.pdf, 2001.

K. Masica, Securing WLANs Using 802.11i Draft, Lawrence Livermore National Laboratory, February 2007, available at <http://csrp.inl.gov/Documents/Wireless802.11i RecPractice.pdf>.

R. Moskowitz, Weakness in Passphrase Choice in WPA Interface, ICSA Labs, 2003.

T. Newsham, Cracking WEP Keys, available at http://lava.net/~newsham /wlan/WEP_password_cracker.pdf, 2001.

W. Simpson (Ed.), The Point-to-Point Protocol (ppp), Request for Comments 1661, 1994, available at http://www.ietf.org/rfc/rfc1661.txt.

R. Soltwisch and D. Hogrefe, A Survey on Network Security, Institute fur Informatik, 2004, vailable at citeseer.ist.psu.edu/soltwisch04survey.html.

T. Takahashi, WPA Passive Dictionary Attack Overview, available at http://www.3c.com.vn/Uploaded/huyenctt/New%20Folder/WPA_Passive_Dictionary_Attack_Overview.pdf.

J. Vollbrecht, Wireless LAN Access Control and Authentication 802.11b Wireless Networking and Why It Needs Authentication. Available at http://www.interlinknetworks.com/images/resource/WLAN Access Control.pdf, Dec. 2002.

J. Walker, 802.11 Security Series Part II: The Temporal Key Integrity Protocol (TKIP). Technical Report, Intel Corporation, 2002, Available at http://cedar.intel.com/media/pdf/security/80211 part2.pdf.

Wi-Fi Protected Access: Strong, Standards-Based, Interoperable Security for Today's Wi-Fi Networks, Wi-Fi Alliance, 2003, available at <http://www.wi-fi.org/files/wp_8_WPASecurity_4-29-03.pdf>.

Chapter 8

Security of Ad Hoc Networks

8.1 Introduction

In mobile ad hoc networks, there is no fixed infrastructure involving base stations, access points, or switching centers. Such wireless networks consist of mobile devices, referred to as nodes, which can be created on the fly as the term ad hoc evokes it. The mobile ad hoc networks (or MANETs) have attracted a large interest in research, military, and industrial communities because of their autonomy, changing topology, self-configuration, ease of deployment, and self-maintenance capabilities. In a MANET, the mobile nodes that are within each other's radio range communicate directly, while those that are far from each other need to rely on the other nodes to forward the messages they want to exchange.

Due to their particular nature, mobile ad hoc networks put forward a set of new challenges to the security design of networks presenting open network architectures and variable topologies. While a lot of efforts to deploy mobile ad hoc networks have focused on solving problems such as multihop routing, wireless local access, and mobile localization, security has obtained too little development and has become a major concern in the protection of the communication between nodes, particularly when the existing security solutions for wired networks do not efficiently apply to mobile ad hoc networks. The fundamental objective of the security solutions to be provided for the mobile ad hoc networks is to make available security services such as authentication, confidentiality, integrity, and availability.

In addition, they might be able to provide several other security features to respond to specific requirements such as anonymity and privacy protection.

Mobile ad hoc networks generate challenges and provide opportunities to achieve the aforementioned objectives, since the use of wireless links allows a large spectrum of attacks to target these networks. These attacks range from passive eavesdropping to active attacks, including impersonation, message replay, message modification, and signal jamming attacks. Eavesdropping allows an adversary to get to sensitive information and violate security features such as data privacy whereas active attacks can allow a malicious adversary deleting messages flowing through it, inserting erroneous messages within a route it is a participant in, taking control of a node, and impersonating a node. Such attacks might violate availability, integrity, authentication, and non-repudiation. On the other hand, nodes roaming in a hostile environment with limited physical protection can be easily compromised and their resources can be destroyed.

An ad hoc network has a dynamic nature because of the frequent changes in its topology and the variability of the number of nodes connected to it. This means that nodes are mobile and can autonomously enter and leave the network. During topology variation, trust relationships between nodes may also change due to the detection of compromised nodes. Therefore, any security solution with a static configuration would not be sufficient to provide the same level of security, and should be made scalable to handle large sized networks, since an ad hoc network may consist of hundreds (or even thousands) of nodes. Security mechanisms should also not only consider malicious attacks from outside the ad hoc network but also take into account the attacks launched by compromised nodes within the network. They need to adapt on-the-fly behavior to network changes. In addition, security solutions should have a distributed architecture with no central entities to achieve high survivability. Indeed, introducing any central entity into the security solution could lead to a major vulnerability; that is, if this centralized entity is compromised, then the entire activity of the network is weakened.

8.2 Ad Hoc Networking

The major concept behind ad hoc networking is multi-hop relaying, which means that messages sent by a node to a node B are relayed by the other nodes if the target node is not directly reachable. Figure 8.1 illustrates an example of an ad hoc network containing six terminal devices: (a) two laptops $\{n_2, n_5\}$; (b) two GSM phones $\{n_3, n_6\}$; and (c) two personal digital assistants $\{n_1, n_4\}$. The dashed lines indicate wireless links built between the wireless devices, while the three arrows indicate a connection between n_1 and n_4. Since node n_1 and n_4 are in each other's radio coverage, n_1 cannot reach directly node n_4, nodes n_2 and n_3 are involved in the transmission of the data from n_1 to n_4.

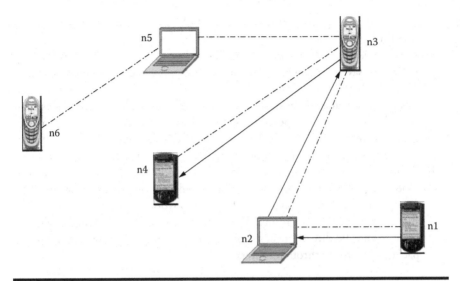

Figure 8.1 Example of ad hoc nodes establishing a connection.

Nodes in an ad hoc network can be of different types, including smart sensors, cell phones, PDAs, and laptops. Differences between ad hoc networks and infrastructure-based networks have been studied in the literature. These differences include the following and are limited to them:

- *Constrained processing*: Most mobile devices involved in an ad hoc network present a limited capacity and relatively involve slow processors, due to cost and size constraints and limited energy sources. In fact, most of the currently available PDAs have a processor's activity based on several-hundred-MHz clocks. Naturally, ad hoc networks need to be designed in such a way that the power consumption they have should be optimized.
- *Constrained bandwidth*: Nodes in ad hoc networks have significantly improved. They also increased the bandwidth and signal range they can handle; but, they still lack the performance of their counterparts' wired components. Nowadays, the bandwidth used by a mobile node to send information does reach the bandwidth provided by wired networks.
- *Variable topology*: As nodes in an ad hoc network are free to move arbitrarily, they may enter and leave the network and the network topology will change in an unpredictable manner, leading to unpredictable changes in the route tables and the multicast tables. Therefore, the topology in a mobile ad hoc network is highly dynamic and may induce the increase of the complexity of the network management.
- *Lack of central management*: Mobile ad hoc networks can be set up everywhere and at any time they are needed. Nodes in a MANET collaborate to perform their tasks, while maintaining a certain level of autonomy. Generally, there is

no central management available in the MANET, and one can also assume that limited information is shared among the nodes.

■ *Limited energy*: Mobile nodes in a mobile ad hoc network generally use battery power. To save energy while performing their tasks, mobile nodes may schedule properly their active and idle modes. During an idle period, the nodes are possibly not reachable or do not process traffic. In addition, the nodes may apply high-consuming operations such as coding, decoding signals, transmitting, and receiving data. However, the constrained energy makes it difficult to implement some complex operations such as the public key systems.

■ *Temporary connectivity and availability*: As nodes can be in active or idle modes, can stay connected or leave independently the ad hoc network, they may not be reachable at any time. For this to work, the nodes in a MANET should play the role of a router. The nodes located out of the range of a given node cannot be directly reached by that node. They can only be reached by packet forwarding through routers.

The following subsections will describe the major features of the infrastructure, routing, and multicast routing provided in ad hoc networks.

8.2.1 Ad Hoc Network Uses

Along with the development of the next generation of wireless communication systems, there will be a need for the rapid deployment of independent mobile users. Typical examples of scenarios where ad hoc networks might be useful include the following:

■ *Emergency services*: Emergency disaster relief personnel coordinating efforts after a natural incident (such as a typhoon or seismic activity) has occurred may need the rapid deployment of a communication network. Such networks cannot rely on centralized and organized connectivity, and can be built using the mobile ad hoc network paradigm. To this end, an autonomous collection of mobile users that communicate over relatively bandwidth–constrained wireless links can cooperate to set up a network, whose topology may change rapidly and unpredictably because of the mobility of the nodes operated by the emergency team. Using a MANET to set up a network infrastructure turns out to be easy and fast.

■ *Home networking*: Knowing that the utilization of wireless computers and applications is increasing in the home environment, the need for helping the coordination and management of home computers and applications is also growing. Using the paradigm of ad hoc networks and allowing the nodes to be self-configurable is truly helpful.

■ *Wireless sensor networks*: A wireless ad hoc sensor network consists of a number of sensors distributed across a geographical area. Each sensor is a small device that is able to gather information, that has a wireless communication capability, and that performs specific functions (e.g., supervision, tracking, or sensing), and some level of intelligence for signal processing and networking of the data. Typical applications involving sensor nodes include target tracking and measurement of the natural parameters such as air pollution, vibrations, and temperature. Putting these sensors together would deploy an ad hoc network that reports back to a central data collecting node.

■ *Video conferences*: In many situations the need for connecting and exchanging information between participants of a video conference is obvious. Examples of such situations include university students who want to participate in an interactive lecture using their portable computers, and business employees needing to share information during a meeting organized on an enterprise scale. A MANET, in that case, built on portable computers would achieve the objectives.

■ *Personal area networks*: Many objects that are tightly coupled to a single person can take advantage of being connected to each other forming an ad hoc network (called personal area network). The network itself is most generally mobile since people tend to move from one place to another in a limited region. This makes the interconnection of personal area networks useful as it allows people to share resources.

■ *Embedded systems*: An embedded system employs a combination of hardware and software entities to perform specific functions. It constitutes a part of a larger system that is more complex. It should be able to work in a reactive and time-constrained environment. Power, cost, and reliability are often regarded among the important features that influence the design of embedded systems. Examples of embedded systems include industrial process controllers, engine controllers, and control systems for consumer products. Real-time requirements include constraints on data collection, error detection, reliability, and communication. It can be seen that new services and applications will surely benefit from having ad hoc network support to the embedded systems.

Finally, let us notice that there are two ways to classify the communication in ad hoc networks, based on (a) whether the nodes are individually addressable and (b) whether the data in the network is aggregated. MANETs can be further classified into two large categories: homogeneous and heterogeneous networks. In homogeneous ad hoc networks all nodes are identical in terms of battery energy and hardware complexity. While in a heterogeneous ad hoc network, two or more different types of nodes with different features, resources, and functionalities are used.

8.2.2 Routing in Mobile Ad Hoc Networks

Due to the high variability of the topology in ad hoc networks and the lack of central management of the basic communication functions, the protocols used in a traditional network to find a path from a source node to a destination node cannot be directly used in mobile ad hoc networks. A large set of routing protocols for ad hoc networks has been developed in the recent past to provide alternative solutions. A routing protocol applying to a mobile ad hoc network should, however, have the following constraints: (a) it must be fully distributed; (b) it must be adaptive to topology changes; (c) it must involve a minimum number of nodes during route computation and maintenance; (d) it should minimize the number of packet collisions and provide a good level of quality of service; and (e) it must optimize the use of the limited resources of ad hoc networks.

Routing protocols for ad hoc networks can be classified according to different criteria, including the routing information update, the use of temporal information, topology information organization, and resource utilization. The classification dimensions are described by the following features.

8.2.2.1 Routing Information Update Mechanism

Routing protocols in this category can be driven either by a routing table or on demand. Using a routing table, the communicating node stores network information related to routing activity and updates it periodically. In order to build a route to a given destination, the node uses an appropriate path-finding algorithm to find the "closest" path. Typical routing protocols of this case are destination-sequenced distance-vector routing (Perkins, 1994), wireless routing protocol, Source-Tree Routing (Aceves, 1999), and Optimized Link State Routing (Clausen, 2001). With routing on demand, nodes do not need to maintain the network topology. They establish the route when they need it, by using a connection establishment process. Moreover, the node does not need to exchange routing information. Among the best-known protocols of this type are the Dynamic Source Routing Protocol (Johnson, 1996), the On-Demand Distance Vector Routing Protocol (Perkins, 1994), the Associativity-Based Routing Protocol (Toh, 1997), and the Preferred Link-Based Routing Protocol (Sisodia, 2002). Some protocols, such as the Core Extraction Distributed Ad Hoc Routing Protocol (Sinha, 1999), combine the two features.

8.2.2.2 Use of Temporal Information for Routing

Since the topology in ad hoc networks is highly dynamic, the use of temporal information to routes may help with the efficient building of the routes. Routing protocols can be classified based on the type of temporal information they use. Two subclasses can be distinguished: the subclass of protocols using past information and the subclass of protocols using information applicable in the near future.

■ *Past information*: Routing protocols using post temporal information about the past status of the wireless communication links or the status of links at the time of making routing decisions. Such protocols include the

Destination-Sequenced Distance-Vector Routing (Perkins, 1994), the Ad Hoc On-demand Distance Vector Routing (Perkins, 1994), and the Global State Routing (Chen, 1998).

■ *Predictive information*: Routing protocols using predictive temporal information, which report on the expected future status of the wireless communication links to make routing decisions. Examples of protocols predicting the future status are the Route-Lifetime Assessment-Based Routing (Agarwal, 2000) and the Link Life-Based Routing (Manoj, 2001).

8.2.2.3 Topology Information Organization

To route messages from one node to another, a mobile ad hoc network can be organized into a flat topology (basically, when the network has a small size) or a hierarchical topology. In the first case, the availability of a globally unique addressing mechanism for nodes in ad hoc wireless networks is assumed. Protocols deployed for the second topology make use of a logical hierarchy in the communication network and implement an addressing scheme that complies with the hierarchy. Examples of protocols using a flat topology include, but are not limited to, the Ad Hoc On-Demand Distance Vector Routing (Perkins, 1999), the Associativity-based Routing (Toh, 1997), and the Preferred Link-Based Routing Protocol. Examples using a hierarchical organization include the so-called CGSR protocol developed in (Chiang, 1997).

8.2.2.4 Utilization of Specific Resources

The protocols of this category can be further classified based on the resources it employs to achieve routing decisions. Specific resources include power energy and geographic information. Two types of routing protocols are of large interest. They are the following:

■ *Power-aware routing protocols*: Protocols belonging to this category attempt to minimize the battery power. A typical protocol is the power-aware routing protocol (Singh, 2000).
■ *Geographical information assisted routing*: Protocols belonging to this category aim at improving the performance of routing and at reducing the control overhead by utilizing the geographic information. A typical protocol using geographic information is the location-added routing protocol (Ko, 1998).

Another important classification of routing protocols considers a fifth dimension, say Unicat/Multicast communication. Most developments in the area of ad hoc network routing have focused on unicast communication. However, several protocols have been designed recently to address the multicast routing in ad hoc networks. Approaches range from simple to implement ideas such as the selective flooding (Ho, 1999) to more complex protocols that maintain knowledge of the network connectivity or perform real-time gathering of route information. Many multicast protocols manage the routing function in ad hoc networks by building and maintaining multicast trees (or meshes) to establish connectivity among group members (Royer, 1999). However, some of these protocols did not support faultless packet delivery and suffer from packet loss problems during tree reconfiguration. Some other multicast protocols have been proposed to overcome these limitations. They are typically based on anonymous conversation, that is, a technique that would proceed in two phases. In the first phase, packets are multicast to the group using any unreliable multicast protocol. In the second phase, periodic anonymous process takes place in the background and ensures that most of the reachable members receive the packets. This method can be set up on top of any tree-based (or mesh-based) protocol inducing small overhead (Chandra, 2001).

Due to the excessive variability of network topology, the bandwidth limitation, and the scarcity of other resources in ad hoc networks, adapting existing multicast routing protocols available in wired networks or developing new multicast protocols for ad hoc networks emerged as a challenging task. In fact, a good multicast routing protocol for ad hoc networks should satisfy at least the following requirements:

- *Control overhead*: The multicast protocol should keep the number of the control packets minimal due to the limited bandwidth in ad hoc networks and to the need to maintain the QoS unchanged. The main parameters for QoS are throughput, delay, delay jitter, and reliability. In addition, a multicast routing protocol should use minimum power and memory.
- *Scheme efficiency*: The multicast scheme should be efficient, in the sense that the ratio of the number of data packets received by the receivers to the total number of packets transmitted in the networks should be very high.
- *Independence with respect to the unicast routing*: A multicast routing protocol should be independent of all specific unicast routing protocols.
- *Scheme robustness*: The multicast scheme should be robust enough to maintain the mobility of the nodes and attain a high packet delivery ratio.

8.2.3 Cluster-Based Ad Hoc Networks

When the number of nodes in an ad hoc network is kept at a controllable size, the routing protocols discussed in the previous subsection may work well. However, if the network grows and the number of nodes becomes too large, then unnecessary overhead scaling may be generated and some parameters used in the routing decision may become too large to handle efficiently. To overcome this limit, the concept of clustering may be advantageous in many ways. It can be utilized to form groups of nodes into clusters, provide protocols for the election of cluster heads, and organize communication between nodes via their cluster heads. Consequently, congestions can be lowered and the overhead reduced. Clustering algorithms are often based on the use of the so-called "dominating sets" in the graph underlying the ad hoc network structure. A dominating set is a subset S of nodes satisfying the following property: Every node in the network is either an element of S or a neighbor of a node in S. A dominating set is said to be independent if it does not contain adjacent sets. An example of a dominating set in an ad hoc network that can be given is the set of clusters.

Clustering algorithms focusing on set domination can typically be of three types: (a) clustering with independent dominating sets, which produce a relatively small number of clusters in the network while guaranteeing that the cluster heads form an independent dominating set; (b) clustering with dominating sets, while relaxing the independence condition and keeping away from unnecessary modifications such those induced when two cluster heads move and become neighbors; (c) clustering with connected dominating sets, which ensures that the cluster heads form a virtual backbone that can reduce broadcast redundancy; and (d) clustering with weakly connected sets, which builds dominating sets using the concept of joining sub-networks having a star topology (also called scatternets) and where the central nodes are the elements in the dominating set (Chen, 2004). One can say, therefore, that a number of approaches to the clustering and routing activity can be made available in mobile ad hoc networks. However, various challenges have to be looked toward including the addressing problem, keeping routing tables updated, and guaranteeing loop freedom.

8.3 Major Routing Protocols in Ad Hoc Networks

In this section, we describe the major solutions designed for routing in ad hoc networks in order to overcome some of the routing problems. Our aim is to provide a short introduction that focuses on the different needs for an efficient routing. The section is also intended to give the foremost cases where vulnerabilities can be observed and attacks can be performed. To this end, we have selected four major routing protocols: the Optimized Link State Routing Protocol, the Destination-Sequenced Distance Vector (DSDV) Protocol, the Dynamic Source Routing (DSR), and the Ad Hoc On-Demand Distance Vector Routing.

8.3.1 The Optimized Link State Routing Protocol (OLSR)

OLSR operates as a table driven routing protocol. It is a proactive protocol, in the sense that it regularly exchanges topology information with other nodes of the ad hoc network. A node selects a set among its one-hop neighbor nodes and calls it the set of multipoint relays (MPR). A node selected as MPR is responsible for forwarding the control traffic intended for broadcast to all the nodes of the network. The MPRs form a useful mechanism for controlling the flooding traffic and reducing the number of transmissions required. They also have to declare link-state information for their MPR selectors to provide shortest path routes to all destinations. Additional available link-state information may be utilized for different purposes, such as for redundancy.

A node selected as multipoint relays by a neighbor node should announce periodically, in its control messages, that it has reachability to its selector. During route computation, the MPRs are utilized to form the route from a given source to any destination. The OLR protocol uses the MPRs to provide an efficient flooding of control messages in the network. Thus, establishing the route through MPRs will certainly avoid the problems associated with data packet transfer over unidirectional links (e.g., prevention of getting link-layer acknowledgments for data packets at each hop). In addition, OLSR is built and should work independently from other protocols (IETF-OLSR, 2001).

8.3.2 Dynamic Source Routing (DSR)

DSR is a demand-based routing protocol. It does not require any sort of periodic updates or announcement messages to build routes. Instead, the protocol DSR collects the needed routing information on demand. DSR is composed of two main mechanisms, *Route Discovery* and *Route Maintenance*, which collaborate together to allow nodes to discover and maintain source routes to arbitrary destinations in the ad hoc network (Johnson, 2002).

The discovery phase is initiated when a node needs to send data to a destination that is not available in its current path cache. The node broadcasts a special discovery packet containing the destination address and a unique identification number. The packet is then broadcasted to all nodes within the wireless transmission range. A node receiving a discovery packet adds its node address to the path in the packet header and retransmits the discovery packet, if it is not the destination. It deletes the packet if the node itself is mentioned in the path header. Finally, if the node is the destination, it returns a route reply. The packet containing the reply is sent to the source. Three ways can be used: (a) the use of routing cache, if a route to source is present in it; (b) the initiation of a new route discovery by the source; and (c) the reversing of source paths collected by the route discovery packet.

After a path has been discovered, the maintenance phase of the DSR protocol takes place. During this phase, all communications are done using previously found paths. Each node occurring on the path established between a source and destination is responsible for resending packets that are not acknowledged by the next hop node. However, an intermediate node is allowed to send back to the source node an error route message, after a maximum limit number of retries has been attempted.

Different enhancements have been provided to optimize DSR protocol including a mechanism allowing each intermediate node occurring in a path that is not actually originating a route discovery to cache some part of the messages that they see go by. This can accelerate route discovery and place, however, some overhead on the intermediate nodes and enhance the performance of the node.

8.3.3 Ad Hoc On-Demand Distance Vector Routing (AODV)

The AODV routing protocol is an on-demand protocol. It offers rapid adaptation to dynamic link changes, low processing, memory overhead, and low network utilization. Like OLSR, it uses two procedures, the route discovery and the route maintenance. It uses a destination sequence numbers to ensure loop freedom avoiding problems counting to infinity associated with classical distance vector protocols.

AODV establishes routes using a route request/route reply query cycle. When a source node wants to set up a route to a destination for which it does not know a route, it broadcasts a route request packet across the ad hoc network. Nodes receiving this packet update their information for the source node (as possible destination) and build pointers backwards to the source node in their route tables. In addition to the source node's IP address, current sequence number, and broadcast ID, the route request also contains the most recent sequence number for the destination of which the source node is aware. A node receiving the route request may send a route reply if it is the destination of the route request or if it knows a route to the destination with corresponding sequence number greater than or equal to that contained in the route request. If that is the case, it sends a route replay back to the source. Otherwise, it rebroadcasts the route request. Nodes keep track of the route request's source IP address and broadcast ID. If they receive a route request that they have already processed, they discard the route request and do not forward it. As the route reply message propagates back to the source, nodes receiving the reply set up pointers forward to the destination. Once the source node receives the route reply message, it may begin to forward data packets to the destination. If the source later receives a route reply containing a greater sequence number or the same number with a smaller hop count, it may update its routing information for that destination and begin to use the better route.

A route is considered active as long as there are data packets flowing from the source to the destination along that route. When the source stops sending data packets, the segments in route will time-out and be deleted from the routing tables of the intermediate nodes. As long as the route remains active, the maintaining phase keeps on maintaining it. If a link break occurs (due to a node move, for instance) while the route is active, the node upstream of the break propagates a route error message to the source node to inform it of the now unreachable destination(s). To eliminate the risk of flooding, a large network adds a lifetime to the route request, and therefore can limit the number of hops a route request can perform. The number of hops is gradually incremented until the destination is reached.

The AODV protocol can only handle symmetrical links. It can also be used in multicast. Multicast routes are set up in a similar manner. A node wishing to join a multicast group broadcasts a route request with the destination IP address set to that of the multicast group and indicates that it would like to join the group. Any node receiving this route request that is a member of the multicast group that has a fresh enough sequence number for the multicast group may send a route reply. As the route replies propagate back to the source, the nodes forwarding the message set up pointers in their multicast route tables. When the source node receives the route replies, it keeps track of the route with the newest sequence number and the smallest hop count to the next multicast group member. After the specified discovery period, the source node sends a multicast activation message to its selected next hop. This message serves to activate the route.

8.3.4 Destination-Sequenced Distance Vector Protocol (DSDV)

DSDV protocol is an adaptation to ad hoc networks of the universally known distance vector algorithm used in wired networks. To overcome the problems observed in the classic version of the algorithm, such as slow convergence and the appearance of loops in highly changing topologies, DSDV adds to the routing table entries unique sequence numbers, which are updated by the destination nodes when new links arriving to them are detected. In addition, a node detecting a broken link should send this information out along with an updated sequence number. The receiving nodes check for higher or equal sequence numbers. If a route update packet is received with a lower sequence number, it is discarded. In the case where a sequence number is already held, then the metric is checked to know whether it is better.

To save bandwidth, DSDV uses two types of route update messages. The first message is referred to as a *fulldump* that a node sends to all its neighbors. This message contains the complete routing table of that node. The second message is an incremental update, which only updates the routes that have changed since the last update. Acting this way would reduce the size of incremental information as much as possible and keep the messages small, conserving bandwidth and transmission time for active routes. The interval of time between the transmissions of two successive fulldumps can be reduced when the incremental updates are getting too big. The interval can be made longer if the update messages are small. In addition, to help with the system stability and reduce the probability of an oscillating system, update messages are only sent out after a delay. The delay is computed using a running, weighted average over the most recent updates. Moreover, synchronizing the different nodes is not necessary, because the update events are handled asynchronous. On the other hand, the information related to link failure is propagated by DSDV protocol in a faster way and uses the sequence number rules for updating distance vector values to guarantee loop-free paths to each destination at all times.

8.4 Attacks against Ad Hoc Networks

While a mobile ad hoc network is more flexible than a wired network, it is also more vulnerable to attacks. This is due mainly to the varying nature of the radio transmissions as explained in the previous chapters. On the other hand, while a mobile ad hoc network is more versatile than traditional wireless networks, the absence of communication infrastructure makes it more targeted by threats. An adversary is able to eavesdrop on all messages within a radio cell, by operating in a promiscuous mode and using a packet sniffer (and possibly a directional antenna). There is a wide range of tools available to detect, monitor, and penetrate an ad hoc network. Therefore, simply by being within radio range, the intruder can have access to the network and can easily intercept transmitted data without the sender even being aware of it. As the intruders are potentially invisible, they can store, alter, and then retransmit packets as they are sent by a legitimate sender. In addition, due to the limitations of the medium, communications can easily be disturbed. An attacker can perform such perturbation attacks by keeping the medium busy sending its own messages, or just by jamming communications with noise.

8.4.1 Attacks against the Network Layer

Two major classes of attacks will be identified: the attacks against the routing function and the attacks against packet forwarding. The routing attacks refer to any action of advertising routing updates that does not follow the specifications of the routing protocol. The specific attack behaviors are related to the routing protocol used by the MANET. For example, in the context of DSR, the attacker may modify the source route listed in the route request (RREQ) or route reply (RREP) packets by deleting a node from the list, switching the order of nodes in the list, or appending a new node into it. In the context of AODV, the attacker may announce a route with a smaller distance metric than its actual distance to the destination. It can also advertise routing updates with a large sequence number in order to invalidate all the routing updates from other nodes. Finally, in the context of on-demand ad hoc routing protocols, the attackers may target the route maintenance process and announce that an operational link is broken.

Attacks against routing protocols in ad hoc networks may attempt to modify the routing process so that traffic flows through a specific node controlled by the attacker, hampering or slowing down the formation of the network, making legitimate nodes store incorrect routes or perform wrong decisions, and modifying network topology. Attacks targeting the routing scheme can be classified into two major categories: incorrect traffic generation and incorrect traffic relaying.

8.4.1.1 Incorrect Traffic Generation

Attacks belonging to this class rely on sending false control messages, such as those sent on behalf of another node (e.g., identity spoofing), or control messages containing incorrect routing information, or outdated routing information. The consequences of incorrect traffic generation attack include degradation of network communication protocols, unreachability of nodes, and creation of routing loops. The major attacks are described as follows:

- *Cache poisoning*: This is a case of incorrect traffic generation, targeting the distance vector routing protocol, for example. An attacker node can advertise a zero metric for all destinations, which will cause all the nodes around the attacker to route packets toward it. Then, the attacker will drop all the packets arriving to it. By dropping these packets, the attacker causes a large part of the communications exchanged in the network to be lost. In a link state protocol, the attacker can falsely declare that it has links with distant nodes. This causes incorrect routes to be stored in the routing table of legitimate nodes.
- *Other DoS attacks*: The attacker can also try to perform a denial of service attack on the network layer by saturating the medium with broadcast messages, reducing the nodes' performance and possibly preventing the nodes from communicating. The attacker can even send invalid messages just to keep the nodes busy, wasting their CPU, and consuming their battery energy. DoS attacks can target all the communication layers. Therefore, they are difficult to counter. For instance, they can be easily launched on the physical layer (e.g., jamming or radio interference); in this case, they can be dealt with by using physical techniques.

It has been shown that on the transport layer, low-rate DoS attacks performed by sending short bursts repeated with a slow timescale frequency can be effective to put an overhead on the network (Kuzmanovic, 2003). When targeting TCP, these attacks choose frequency based on the Retransmission Time Out (RTO). In the case of network congestion occurring with TCP, the throughput (composed of legitimate traffic as well as DoS traffic) triggers the TCP congestion control protocol, so the TCP flow enters a timeout and awaits a RTO slot before trying to send another packet. If the DoS attack period is chosen close to the RTO of the TCP flow, the flow repeatedly tries to exit timeout state and may fail, producing no traffic. This attack is effective because the sending rate of DoS traffic is too low to be detected by anti-DoS countermeasures.

Another DoS attack performed on the transport layer is referred to as the *jellyfish* attack (Aad, 2004). This DoS attack can be carried out by utilizing three mechanisms: (a) using scrambled delivery of packets. In that case, the attack is performed by a node delivering all received packets, but in an incorrect order, compared to the initial order. Duplicate transmissions can be replayed from this malicious behavior, even if all sent packets are received; (b) dropping all packets for a very short duration at every RTO slot. In that case, the flow enters timeout at the first packet loss caused by the jellyfish attack, then re-enters the timeout state at every elapsed RTO slot; and (c) holding a received packet for a random time before processing it. This action may cause transport traffic (such as TCP) to be sent in bursts, which increases the anomalous collisions and losses. It also induces an excessive increase in the RTO value and may cause an incorrect estimation of the available bandwidth in congestion control protocols, which is based on packet delays.

By attacking the routing protocols, the attackers can attract traffic toward certain destinations in the nodes under their control and cause the packets to be forwarded along a route that is not optimal or even non-existent. The attackers can create routing loops in the network and introduce serious network congestion and channel contention. Multiple attackers may even cooperate to prevent a source node from finding any route to the destination, and divide the network in the worst case. The attackers may further subvert existing nodes in the network, or fabricate new identities and impersonate legitimate nodes.

8.4.1.2 Incorrect Traffic Relaying

Attacks can also be conducted against packet forwarding operations. Such attacks do not disrupt the routing protocol but they cause the data packets to be delivered in a way that is intentionally inconsistent with the routing states. For example, the attacker, along an established route, may drop the packets, modify the content of the packets, or duplicate the packets it has already forwarded. Another type of packet forwarding attack is the denial of service (DoS) attack via network-layer packet blasting, in which the attacker injects a large amount of useless packets into the network. These packets waste a significant portion of the network resources and introduce severe wireless channel contention and network congestion in the MANET. Network communications coming from legitimate and protocol-compliant nodes may be contaminated by misbehaving nodes. Attacks of utmost importance include the following:

■ *Blackhole attack*: To reduce the quantity of routing information available to the other nodes, an attacker can drop received routing messages, instead of relaying them as the routing protocol may require. This attack is called black-hole attack (Hu, 2003). The attack can be done selectively (e.g., drop routing packets for a specified destination, stop one packet every **n** packets, or a packet every **t** seconds) or by dropping all packets during a certain period of time. It may make the destination node unreachable or reduce communications in the network.
■ *Message tampering*: If no mechanism ensuring message integrity (i.e., a digital signature made on the payload) is provided by the protocol, an attacker can modify the messages originating from other nodes before relaying them.
■ *Replay attack*: An attacker can perform a replay attack by storing valid control messages and re-sending them to make other nodes update their routing tables with false routes. As topology changes, older control messages, that were valid in the past, describe a topology configuration that no longer exists. Replay attacks can be successful even when the control messages are protected by a digest or a digital signature (that does not include a timestamp).

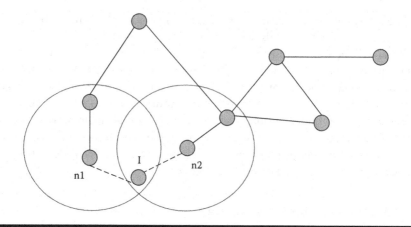

Figure 8.2   A wormhole created by node I.

■ *Wormhole attack*: The wormhole attack consists in copying traffic from one region of the network and replaying it in another (Hu, 2003). The attack is launched by an intruder node located within the radio range of legitimate nodes, say n_1 and n_2, that are not within each other's transmission range. The intruder node tunnels control the traffic between n_1 and n_2, without the modification presumed by the routing protocol, so that the intruder remains virtually invisible. This creates a link controlled by intruder I (as depicted by Figure 8.2). Intruder I can then delete the packets sent through the tunnel (or even break the link), when desired. The intruder node can also collaborate with another intruder node, to which it may be connected using various means, and can create a longer and more harmful wormhole.

   The wormhole attack is very effective and difficult to detect, even in a network where confidentiality, integrity, authentication, and non-repudiation is provided. In addition, the wormhole attack is likely to be used against routing protocols, such as the distance vector routing protocol, since it is easy to implement.

■ *Rushing attack*: This is an attack that can be performed against on-demand routing protocols. It acts as an effective denial of service and exploits the fact that, in an on-demand routing protocol, a node needing a route to a destination should flood the network with route requests. In addition, to limit the overhead of the flood requesting, each node forwards only one route request coming from any route discovery (Hu, 2003a). If the route requests for route discovery sent by the attacker are the first to reach the neighbors of the destination, then any new route discovered would probably include a hop through the attacker. To perform the attack, the attacker has only to send a large set of requests so that non-attacking requests will be discarded upon their arrival at the target's neighbors.

8.4.2 A Case Study: Attacks against OLSR

For the sake of clarity, we will discuss in the sequel the attacks targeting a special case of routing, the OLSR. Our objective is not to focus on the flaws of OLSR, but only to show the vulnerabilities of a routing protocol that did not include security measures in its design, like several other routing protocols. While some vulnerabilities are specific to OLSR, other attacks can be seen as instances of vulnerabilities found in common routing protocols. More attacks may be performed against almost any operating function of the protocol.

8.4.2.1 Incorrect Traffic Generation

A specific set of attacks generating incorrect traffic is based on the generation of control messages in a way that does not conform to the protocol rules. Attacks in this category include incorrect hello, incorrect TC message generation, and MID message generation. In the following we detail such attacks. Notice that TC (topology control) and MID (multiple interface declaration) are control messages sent by a node to advertise neighbors and inform that the node has multiple interfaces.

- *Incorrect HELLO message generation*: An intruder node, say I, may send HELLO messages with a spoofed originator address (set to the address of node C, for example). Subsequently, nodes n1 and n2 may announce reachability to node C through HELLO and TC messages. In addition, node I selects multipoint relays (MPRs) among its neighbors and it can signal this selection while pretending to have the identity of node C. Thus, the selected MPRs will advertise in their TC messages that they provide a last hop to C. Conflicting routes to node C, with possible connectivity loss, may then be induced by the intruder node.

    In addition, it is interesting to notice that a node, say I, can force its election as an MPR by setting the Willingness field to the WILL_ALWAYS constant in its HELLOs. According to OLSR, the node's neighbors will always select it as an MPR. Using this modification, a compromised node can easily gain, as an MPR, a privileged position inside the network. It can then perform a DoS attack. Node I can also misbehave by signaling an incomplete set of neighbors. Depending on their links with other nodes, the ignored neighbors might experience breakdown in connectivity with the rest of the network.

    On the other hand, another type of attack is also possible using identity spoofing. An intruder node can set the Willingness field to WILL_NEVER on its HELLO messages sent on behalf of node A. According to OLSR, nodes receiving the HELLO messages will never choose A as an MPR, which may result in connectivity loss. A misbehaving node X may perform link spoofing (or the signalization of an incorrect set of neighbors in a control message) in its HELLO messages advertising a link with non-neighbor node, say B. This will induce the storage of an incorrect 2-hop neighborhood (at the neighbors of X) and selecting a wrong MPR set.

■ *Incorrect TC message generation*: TC messages with a spoofed originator address may cause false neighbor relationships to be advertised in the network. For instance, an intruder node I can send a TC message on behalf of node C, advertising that a node, say A, is a neighbor. Another node, say D, upon reception of the TC message, will falsely conclude that C and A are neighbors. For this attack to succeed, the TC message must support an ANSN (Advertised Neighbor Sequence Number) value greater than the highest ANSN value referenced to C; otherwise D will discard the TC message.

TC messages with spoofed links can severely disturb the network topology when stored by legitimate nodes. An example can be given by an intruder node I that can generate HELLOs, be selected as an MPR by its neighbors, but that can refuse to generate TC messages (or generates TCs signaling an incomplete set of nodes). This leads to the fact that some nodes may not have their link state information disseminated throughout the network and may be disconnected.

■ *Incorrect MID message generation*: A misbehaving node can generate wrong MID (multiple interface declaration)/HNA (host and network association) messages, announcing interfaces that are not their own (link spoofing), or falsifying the source address of the message so that it declares interfaces that are not their own. In this case, nodes will have problems while attempting to reach these interfaces.

■ *ANSN attack*: The misbehaving node may listen to a TC message from node A and record the ANSN of the message; then it sends a TC with a spoofed source address (of node A) and an ANSN greater than the value recorded. According to the protocol specifications, nodes will ignore further TC messages from A, because these messages bear a smaller ANSN as that recorded in the Topology Set. Therefore, such messages are considered as arrived out of order. We call this an ANSN attack. If no further action is taken by the attacker, the ANSN attack is effective until the ANSN of node A reaches the value of the ANSN in the spoofed TC.

This attack can be marked as the spoofed TC accepts an ANSN that is much higher than that of the latest genuine TC message received from A (this means that the higher the difference between the two ANSNs is, the longer TCs from node A are ignored). However, the misbehaving node may perform this attack repeatedly, by forging each time spoofed TC messages with a slightly greater ANSN.

8.4.2.2 Incorrect Traffic Relaying

If the control messages are not properly relayed, the network may malfunction and have low performance. Incorrect relaying includes blackhole, replay, and wormhole attacks.

- *Blackhole attack*: If a node fails to relay TC messages, the network may experience connectivity problems. In networks where no redundancy exists, for example, a connectivity loss will certainly occur. In addition, if MID and HNA messages are not properly resent, additional information regarding multiple nodes interfaces and connections with external networks may be lost.
- *Replay attack*: As it has been mentioned in the foregoing discussion, replaying old control messages in the ad hoc network causes nodes to record out-of-date topology information. A control message can neither be replayed nor accepted by nodes that have already received it, because of the value of the message sequence number field (MSN). Therefore the attacker needs to increase the MSN of the message, causing a possible message loss. For a TC, the attacker must increase the ANSN too, indirectly causing an ANSN attack. Replayed HELLOs may have a lesser impact, because the link state advertised in HELLOs must be given in a well-defined order.
- *Wormhole attack*: An extraneous (A, B) link can be artificially created by an intruder node X by wormholing control messages between A and B. A longer wormhole can also be created by two colluding intruders X and Y.

  To successfully exploit the wormhole, the attacker must wait until A and B have exchanged sufficient HELLO messages (through the wormhole) to establish a symmetric link. Until that moment, other tunneled control messages are rejected because the OLSR protocol specifies that TC/MID/HNA messages should not be processed if the relaying node is not a symmetric neighbor. However, once created, the (A, B) link is at the mercy of the attacker.
- *MPR attack*: The "first transmit rule," described in the OLSR specifications, states that a node receiving a message in MPR flooding checks whether the sender is its MPR selector. If that is the case, the node retransmits the message. If the sender is not an MPR selector of the node, the latter will never retransmit the message. While this rule is established for performance reasons (i.e., to avoid messages travelling on large loops in highly dense networks), it could be utilized to slow down the correct relaying of control messages.

  The related misbehavior is referred to as *MPR attack*. Consider now the following scenario: node A sends a message to its neighbors B and X, where B is an MPR of A, X is not an MPR; and let C be an MPR of B. The misbehaving node X does not select its MPR set properly, and retransmits the message that is received by C. Node B retransmits the message to C. The crucial point is that C, even being an MPR, will not relay the message because it has already received it from X.

8.5 Securing Ad Hoc Networks

In this section, we review the state-of-the-art of security proposals for MANETs. Efficient security solutions have naturally considered both the network and link, because the MANET connectivity is provided through distributed protocols at the two link layers. Typically, two basic approaches can be used to secure a MANET: proactive and reactive. A proactive approach attempts to prevent security threats in the first place, generally through various cryptographic techniques. On the other hand, the reactive approaches try to detect threats *a posteriori* and react accordingly. Each approach has its own merits and is suitable for addressing different issues. For example, nearly all secure routing protocols adopt the proactive approach in order to secure routing messages exchanged between mobile nodes, while the reactive approach is widely used to protect packet forwarding operations.

Nonetheless, a complete security solution for MANETs should integrate both proactive and reactive approaches, and cover prevention, detection, and reaction. This is due to the absence of a clear line of defense and a clear delimitation of the frontier of the domain to protect. The prevention component deters the attacker by increasing considerably the difficulty of penetrating the system. However, it has been clearly shown that a completely intrusion-free system is infeasible, no matter how carefully the prevention mechanisms are designed. This is especially true in MANETs, consisting of mobile devices that are prone to compromise or physical capture. Consequently, the detection modules that discover the infrequent intrusions and the reaction components that implement countermeasures to avoid persistent undesirable effects, are essential for the security solutions to operate in the presence of limited intrusions.

In the MANET framework, the prevention process is mainly achieved by secure ad hoc routing protocols that prevent the attacker from setting up incorrect routing states at other nodes. These protocols are typically based on earlier ad hoc routing protocols such as DSR, AODV, and DSDV, and make use of different cryptographic primitives (e.g., HMAC, digital signatures, and hash chains) to authenticate the routing messages. The detection process discovers ongoing attacks through the identification of abnormal behavior shown by malicious nodes. Such misbehavior is detected either at the end node, or by the neighboring nodes through overhearing the channel and reaching collaborative consensus. Once an attacker node is detected, the reaction process makes adjustments in routing and forwarding operations, ranging from avoiding the node in route selection to collectively excluding the node from the network.

8.5.1 Security Challenges

Four distinctive vulnerabilities may cause security problems in MANETs (Lou, 2003). They are the following:

1. *Wireless medium*: Communication in an ad hoc network is performed on the basic of a broadcast that causes it to be vulnerable to both active and passive attacks. The signals propagate from the source over the open air to all directions. Therefore, it is very difficult to protect against passive eavesdropping

and traffic analysis. Even if security mechanisms, like encryption, can ensure a high degree of confidentiality and integrity of the messages sent over the network, we can never protect against traffic analysis. This can help malicious users to identify nodes that play central roles in the network and target them with denial of service attacks. Active signal interference and jamming are examples of denial of service attacks that ad hoc networks are vulnerable to.

2. *Routing based on cooperative algorithms*: As nodes cooperate with each other to route messages, they always depend on the goodwill of other nodes to make the network functional. Routing can be a serious weakness, if some of the nodes are operating maliciously. Consequently, the cooperative nature of routing protocols makes it easier for data tampering, impersonation, and denial of service attacks. Among such known attacks, we can cite the wormhole attacks and rushing attacks. In the wormhole attack, two malicious nodes n_1 and n_2 quickly forward a message from node S to T making T discard the original message that flows through external link (A, D). This way, node S is not updated on the actual route of A through D but thinks it has a direct connection to T.

3. *Lack of fixed infrastructure*: Since there is no fixed infrastructure in a mobile ad hoc network, there are no central points through which information is directed. This feature makes the traffic in ad hoc networks hard to monitor. In addition, because the nodes in ad hoc networks move freely, it is not uncommon that the network is divided into two or more parts or sub-networks during certain intervals of time. This obviously adds to the difficulty of monitoring the traffic. In addition, many services that can be used in traditional networks to ensure performance and monitoring cannot be implemented in ad hoc networks because of the lack of a fixed infrastructure. Public key cryptography schemes are hard to employ because management of the public keys usually involves a centralized trusted control point, the certification authorization (CA).

4. *Low capacity of the devices in MANETs*: The devices that compose the MANETs are generally very restricted in terms of limited memory, low battery power, limited bandwidth, and limited processing power, which hinders the deployment of computationally intensive security schemes. The limited bandwidth restricts the amounts of information that can be sent during a certain period of time. Hence, security schemes and communication protocols, which are very dependent on timing, are difficult to implement. Protocols also have to take into consideration the asynchronous nature of communication in ad hoc environments. To cope with this issue, we have to use schemes that assume that the connection will go down at any given moment.

8.5.2 Network Layer Security

The network-layer security objectives for MANETs seek to make sure that the routing messages exchanged between nodes are consistent with the protocol specification, and the packet forwarding behavior of each node is coherent with its routing states. Consequently, the existing proposals can be classified into two categories: secure ad hoc routing protocols and secure packet forwarding protocols. Before we describe these classes, we first introduce several cryptographic primitives for message authentication, the essential component in any security design, and analyze the trade-offs behind them.

8.5.2.1 Message Authentication Primitives

There are three cryptographic primitives widely used to authenticate the content of the messages exchanged among nodes:

1. *HMAC* (message authentication codes). If two nodes share a secret key K, they can efficiently generate and verify a message authenticator $h_K(\cdot)$ using a cryptographic hash function. The computation is very efficient. It is inexpensive for low-power end devices such as sensor nodes. However, a HMAC can be verified only by the intended receiver; this makes the solution unattractive for broadcast message authentication. Besides, establishing the secret key between any two nodes is a non-trivial problem. If the pairwise shared key is used, a total number of $n(n-1)/2$ keys should be managed in a network with n nodes.

2. *Digital signature.* Because digital signature is based on asymmetric key cryptography, it involves more computation overhead in signing/decrypting and verifying/encrypting operations. This method is less flexible against DoS attacks since an attacker may provide a victim node with a large set of false signatures to weaken the victim's computation resources in verification. Each node also needs to maintain a certificate revocation list of revoked public keys. Nonetheless, a digital signature can be verified by any node given that it knows the public key of the signing node. This makes the digital signature scalable to large numbers of receivers.

3. *One-way HMAC key chain.* Many cryptographic one-way functions exist. They guarantee that retrieving x, knowing the image $f(x)$ of x, it is computationally infeasible to determine x. By applying function f many times on x, one can obtain a chain of outputs $\{x_i\}_{i \geq 0}$ such that

$$x_0 = x; \; x_{i+1} = f(x_i) = f^i(x), i \geq 1.$$

Outputs $\{x_i\}_{i \geq 0}$ can be used in the reverse order of generation to authenticate messages: a message with a HMAC using x_{i+1} as the key can be proven to be authentic when the sender submits x_i. The one-way HMAC key chain is commonly used to authenticate messages. TESLA (Perrig, 2002) is an example of hash-chain-based protocol commonly used to authenticate broadcast messages.

The computation involved in one-way key chain-based authentication is weak, and one authenticator can be verified by a large number of receivers. However, this benefit comes at a certain cost: (1) the hash-chain-based authentication necessitates clock synchronization, which may need special hardware support; (2) the receivers need to buffer the messages to verify them when the key is revealed; (3) the delay in the verification of routing messages may significantly decrease the responsiveness of the routing protocol; (4) the release of the key involves a second round of communication and the timer has to be carefully estimated according to the context; and (5) the storage of the hash chain is non-trivial for long chains.

8.5.3 Securing Ad Hoc Routing

Secure ad hoc routing protocols improve the existing ad hoc routing protocols, such as DSR and AODV, with security extensions. They allow each mobile node to proactively sign its routing messages using specific cryptographic authentication primitives. This way, the receiving nodes can efficiently authenticate the legitimate traffic and differentiate the unauthenticated packets issued by attackers. However, an authenticated node may be compromised and controlled by an attacker.

In the following, we describe how the common routing protocols are secured.

8.5.3.1 Source Routing

The major challenge for source routing protocols, such as DSR, is to ensure that a node cannot add extra nodes (or remove existing nodes from) to a route. The basic security technique is to attach a per-hop authenticator for the source routing forwarder list so that any modification of the list can be detected either immediately or after the key is disclosed for HMAC key-chain-based authentication.

A secure extension of DSR is given by Ariadne (Hu, 2002). It uses a one-way HMAC key chain for the purpose of message authentication. Through key management and distribution, a receiver is assumed to own the last released key of the sender's key chain. The source node n_0 using DSR to connect to destination n_d through route $(n_0, n_1, ..., n_d)$ initiates the construction of a hash chain,

$$H\left(n_{d-1}, H\left(...H\left(n_1, \mathrm{HMAC}_{k_{0,d}}(m)\right)\right)\right)$$

where $k_{0,d}$ is a shared secret key between n_0 and n_d, $H(-)$ is the hash function, and $\mathrm{HMAC}_{k_{0,d}}(m)$ denotes the HMAC code generated by n_0 on message m using the shared key. The hash function $H(-)$ authenticates the contents in the chain, and $\mathrm{HMAC}_{k_{0,d}}(m)$ is responsible for the authentication of the source-destination relation.

8.5.3.2 Distance Vector Routing

The major objective for a distance vector routing protocol, such as DSDV and AODV, is to ensure that each intermediate node has to advertise the routing metric correctly. Supposing that the maximum hop count of a valid route is n, a node generates a hash chain of length n every time it initiates an RREP message, say (h_0, h_1, ..., h_n), defined by $h_{i+1} = H(h_i)$, for a common hash function $H(\cdot)$. The node then adds $h = h_0$ and h_n into the routing message, with Hop_Count set to 0. When a node receives an RREQ or RREP packet, it first checks whether

$$h_n = H^{n-\text{Hop\_Count}}(h)$$

where $H^{n-\text{Hop\_Count}}(h)$ is the iteration of H, n-Hop_Count times. Then the node sets $h := H(h)$, increments the Hop_Count by 1, updates the authenticator, and forwards the route discovery packet. This approach provides authentication for the lower bound of the hop count, but does not prevent a forwarder from advertising the same hop count as the one from another forwarder.

8.5.3.3 Link State Routing

Each node implementing the Secure Link State Routing (SLSP) tries to learn and update its neighborhood using Neighbor Lookup Protocol (NLP) and periodically floods Link State Update (LSU) packets to broadcast link state information. NLP is responsible for (a) maintaining mappings between the MAC and IP addresses of a node's neighbors; (b) identifying potential inconsistencies, such as the use of multiple IP addresses on a single link; and (c) measuring the control packet rates from each neighbor. Neighbors use one-hop hello messages to discover each other, and connectivity is assumed to be lost if a hello message is not received within a timeout. A node collects LSUs from all the nodes of the network in order to construct the global topology and calculate the route to any destination.

SLSP uses a digital signature approach in authentication. NLP's hello messages and LSU packets are signed with the sender's private key. Any verifier can use the public key contained in the sender's valid certificate to verify a message's authenticity. A certificate can be delivered to verifiers either as an attachment to an LSU packet or using dedicated public key distribution packets.

8.5.4 Securing Packet Forwarding

The protection of routing message exchange is only part of the network-layer security solution for MANET, since it is feasible for a malicious node to correctly participate in the route discovery phase but fail to correctly forward data packets. The security solution should ensure that each node should forward packets according to its routing table. This is naturally achieved by the reactive approach because attacks on packet forwarding cannot be avoided: an attacker may simply drop all packets passing through it, even though the packets are carefully signed. At the heart of the reactive solutions are a detection technique and a reaction scheme, which are described in the following.

- *Detection*: Because the wireless channel is open, each node can perform a localized detection by overhearing ongoing transmissions and evaluating the behavior of its neighbors. However, its accuracy is affected by a number of factors including channel error, interference, and mobility. A malicious node may also abuse the security solution and intentionally accuse legitimate nodes. In order to address such issues, the detection results at individual nodes can be integrated and refined in a distributed manner to achieve consensus among a group of nodes. An alternative detection approach relies on an explicit acknowledgment from the destination and/or intermediate nodes to the source so that the source can detect where the packet was dropped.
- *Localized detection*: A watchdog to monitor packet forwarding on top of source routing protocols can be set up. It assumes symmetric bidirectional connectivity: if A can listen to B, B can also listen to A. Since the whole path is well known, when node A forwards a packet to the next hop B, it knows B's next hop C. It then overhears the channel for B's transmission to C. If it does not listen to the transmission after a timeout, a failure counter associated with B is increased. If the counter exceeds a threshold bandwidth, A sends a report packet to the source notifying B's misbehavior.
- *ACK-based detection*: A fault detection mechanism can be developed based on explicit acknowledgments. The destination sends back acknowledgment messages (ACK) to the source for each successfully received packet. The source can begin a fault detection process on a suspicious path that has recently dropped more packets than an acceptable threshold. It sends out data packets piggybacked with a list of intermediate nodes, also called probes, which should send back acknowledgments. The source shares a key with each probe, and the probe list is "onion" encrypted. Upon receiving the packet, each probe sends back an ACK, which is encrypted with the key shared with the source. The source in turn verifies the encrypted ACKs and attributes the fault to the node closest to the destination that sends back an ACK.
- *Reaction*: When a malicious node is detected, several actions can be executed to protect the network against future attacks initiated by this node. The reactive component managing and implementing such actions is typically referred to as prevention component. Based on their scope, the reaction schemes can be categorized as global reaction schemes and end-host reaction schemes. In the former category, all nodes in the network react to a malicious node as a whole, meaning that the malicious node must be excluded from the network. In an end-host reaction scheme, each node may make its own decision on how to react to a malicious node (e.g., putting this node in its own blacklist). A

global reaction can be achieved among the nodes having reached the consensus that one of their neighbors is malicious. These nodes collectively perform the needed actions to isolate the malicious node so that it cannot participate in the routing or packet forwarding operations in the future.

8.6 Authentication in Ad Hoc Networks

Two different approaches are presented to provide authentication in mobile ad hoc networks. The first is based on a distributed model of trust and references. It performs much as a person acts toward another person, in different situations. The second approach is used to perform the more efficient verification, while not using public key cryptosystems as the main verification system. The second scheme fits better in broadcast environments and ad hoc networks, where some packets might not reach the destination due to sporadic network connectivity changes.

8.6.1 Light-Weight Authentication Model

Contrary to many of the other authentication protocols discussed so far, in this book, the light-weight authentication model does not base its foundation on strict mathematics. Instead, the model tries to simulate the human behavior. In this model, devices and their users are authenticated by the use of references. When a node wishes to communicate with another node, the destination node is asked to supply common knowledge such as a shared key. If this is not available, the source node checks its list of trustworthy devices and asks them if they can authenticate the identity of the target node. These nodes can do the same operation, recursively, to finally reach a "yes" or "no" decision. The source node can decide whether the result is good enough by using a threshold on the number of "yes"-decisions with respect to the total number of answers.

The initiating node can also ask the destination node for a list of references. The referencing nodes can be asked if they know the destination node. Also, using the trust relations as previously, each reference can be authenticated to make sure they are not set up to work on behalf of the destination node. The initiator can check each recommendation chain for suspicious nodes and ignore these chains during further evaluation. It is also possible, for the initiator, to put different weights on different trust chains. For example, the chains of the references given by the destination can be weighted lower than the trust chains acquired through the list of currently trusted nodes.

Once the authentication is finished, a secure channel can be set up. One way to realize it is to send a random value over a trustworthy channel, that is, via nodes that the initiating node trusts, explicitly or implicitly. The destination node sends another random value using a random path. Using these two values, both nodes are able to compute a shared secret key. Of course, this approach is vulnerable to random value overhearing by malicious nodes. If the nodes are able to use public key cryptography, a key agreement protocol based on this will be available for the two nodes. However, the aforementioned setup is vulnerable to active man-in-the-middle attacks. Furthermore, a feedback message that can be sent out if a node that has been authenticated has shown to be untrustworthy. The initiating node can send out this feedback to the trust chain that has informed that the node was trusted.

8.6.2 Timed Efficient Stream Loss-Tolerant Authentication

A general approach to authenticate broadcast messages is to use public-key systems allowing the sender to sign the message and all the receivers to verify the signature. This leads to high computation overheads, because signing and verifying take some time to perform. It also consumes bandwidth resources, since every verifying node needs to contact an on-line certificate authority to get the public certificate needed for verification. This might not be needed every time the verification is performed, but the security level may need a check periodically for certificate revocation. The Timed Efficient Stream Loss-tolerant Authentication (TESLA) protocol for broadcast authentication might solve these problems (Perrig, 2002). The basic operation of TESLA protocol is based on one-way chains. A one-way chain is generated by repeatedly applying a one-way hash function. Each value can be computed using the previous value in the chain. The chain of values is used in reverse order, thereby not revealing any secret information until later on.

As shown in Figure 8.3, the generation of a one-way chain is controlled by the sender. In a first step, a random value s_n is selected. From this value a one-way chain, denoted by $(s_0, s_1, \ldots, s_{n-1}, s_n)$, is computed using a one-way function F. The value s_0 is said to be a commitment to the entire chain meaning that any element s_j can be verified by applying the one-way hash function on s_j repeatedly until s_0 is reached. More formally, to verify that s_j is the element of index j in the chain, we only check whether

$$s_0 = F^j(s_j), \; n \geq j \geq 1 .$$

The TESLA protocol is based on the assumption of *loosely time synchronized nodes*. That is, the receivers need to be able to approximate an upper bound on the

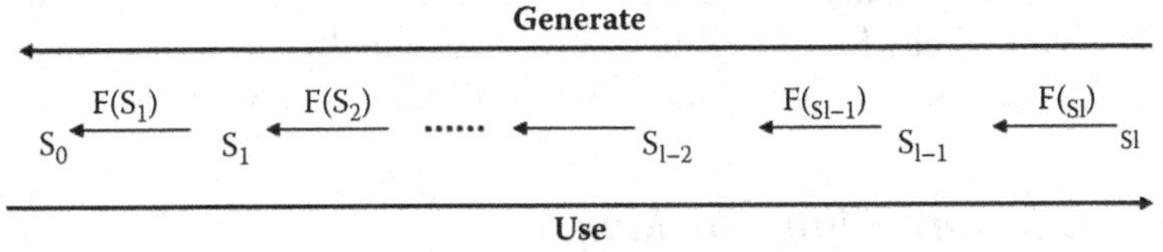

Figure 8.3 A one-way chain.

sender's clock. This can be accomplished using a time synchronization protocol. Each receiver performs the following steps to make their approximations of the maximum time synchronization error: The receiver sends a time synchronization message to the broadcast source storing the time at which it sends out the message, say t_R. The message includes a randomly generated number identifying the specific message. When the server receives the message, it notes its time; then it creates a reply message containing the sender time, t_S, and the received number. Before transmission back to the node that asked for it, the server signs the message using its private key. This allows the receiver to verify that the time message was not altered on its way back. Using the time of reception, the receiver node can compute the time synchronization data it needs.

The TESLA protocol divides the upcoming time into intervals of a specified length. Each of these intervals is assigned one key from the one-way chain of key values. The sender also decides on a disclosure time that states how many intervals it must go by before an interval key is disclosed. During broadcasting, the sender attaches a MAC to each message it transmits out. With each message, the key from the previously disclosed interval is attached. When a receiver sees such a message, it stores it, waiting for the key to be disclosed. If the message is delayed in some way and the receiver knows that the key has already been disclosed, it should no longer trust the packet. However, if the receiver can know that the secret key has not yet been disclosed, as it knows the key disclosure schedule, the packet is buffered. When a few intervals have elapsed and the interval key for a previous interval has been revealed, the receiver can verify that the message was really authentic.

The server does not use the actual interval key to sign the messages. Instead, a signing key is generated from the interval key by applying some known one-way function. Processing this way, the key is only used for one task and not for both signing and generating the one-way chain.

8.7 Key Management

The major problem of mobile ad hoc networks is the lack of an on-line key management service. Depending on the nature of the ad hoc network and its usage, different solutions might be available. In this section, four solutions are discussed for key management. The first solution considered addresses small nodes without any need for human input or feedback. The second and third solutions are based on threshold cryptography for some different settings. Finally, the fourth solution is a distributed scheme involving each node as part of the key service.

8.7.1 The Resurrecting Duckling

A duckling is a device that can have different states. It is in an unprintable state waiting for the mother duckling to imprint it. The process of imprinting is done using a secure channel such as direct electrical contact in which a shared secret is sent in plaintext. After the duckling has been imprinted, it cannot be imprinted by another device unless it "dies" or the mother duckling tells it to die. The duckling can be programmed to die on certain events or after periods of time. While holding the secret key, it can be used to communicate or authenticate the duckling with the mother duck using any cryptographic methods that are available in the two devices, and using the key. Since the secret key is a shared secret, it is possible to use symmetrical cryptography if the partners want to keep the data transmitted private (Stajano, 1999).

To include the ability to allow communications between ducklings, the mother duckling downloads an access control list that states what can be done on the device and by whom. Instead of using access control lists, the ducklings can use certificates signed by the mother duckling to trust each other. This, however, puts some additional demands on the devices such as secure clocks, to have validation periods on the certificates. This is an undesired constraint, since keeping a secure clock is a demanding process. However, there exist methods where the valid period is short enough so that the need for secure clocks in each device may not be important. This often implies that new certificates must be downloaded frequently. This adds the necessity of an on-line CA. The resurrecting duckling security policy model adds the ability to authenticate a node to its master.

8.7.2 Secure Key Distribution

We consider here a special case of key distribution applicable to ad hoc networks based on an ID-based threshold reinforced cryptographic scheme overcoming the common needs for a priori distribution of secret keys or public certificates (Khalili, 2003). More systems have been proposed in a previous chapter. The model is based on the concept of identity. Then, threshold cryptographic principals are used to build a private key generation system that each node can ask to get its private key. The use of each node's identity saves bandwidth, since there is no need to distribute public keys and signatures from certification authorities. Also, the identity is often shorter than public keys. Furthermore, the computational overhead in creating the public/private key pair is reduced. There is no need to compute any public key. The private key must be created by some distributed nodes. This should only be needed to be done once for each new node in the network.

The benefit of using a threshold system for the private key generation is of course the need to be independent of any special CA. In addition, the mobile nature of ad hoc networks allows for some robustness using the threshold system, since only a number of key servers need to be connected to the network. If not enough servers are available, the mobile node can simply move to another location after collecting those that are available at the previous one. This way, the mobile node will finally receive a complete private key.

The whole scheme consists of four algorithms. They are the following:

1. *Setup*: When the network is initially formed, the present nodes must agree on the security parameters, the threshold scheme, and other useful things. The setup phase takes as input these parameters and generates the master public/private keys for the system.
2. *Extract*: To get its private key, each node needs to contact the distributed private key generation service. This service extracts the node's private key using the master secret key and the node's identity as input. This phase is vulnerable to some attacks, since the identity might be easily spoofed.
3. *Encrypt*: The encryption is done by each node using the public master key, the recipient identity, and the message plaintext. The output from this algorithm is the cipher-text to transmit.
4. *Decrypt*: At the receiving end, the node uses the master public key, the incoming cipher-text, and a personal secret key to decrypt the cipher-text into the plaintext.

To address the problem of identities being spoofed, Khalili (2003) suggests using unpredictable identities chosen when the node enters the network. Additionally, they suggest that the private key generation service should refuse to process any private key to each identity more than once. This effectively solves the problem of spoofing, since no adversary will be able to guess the identity of any node in advance.

Each node in the network needs to know about the master public key to conduct the encryption and decryption processes. The master public key is easily distributed by broadcast or any other way. However, this may lead to vulnerabilities used by man-in-the-middle attacks targeting nodes joining the network. A malicious node can send out any key to a new node, pretending the key to be the master public key. This public key would serve to implement a private key generation service that any adversary can have control over. This implements a man-in-the-middle attack.

8.7.3 Distributed Key Management

We consider in this section a second solution that can be implemented in ad hoc networks to provide key management. Many other solutions can be discussed; some of them have been addressed in a previous chapter. The solution, threshold cryptography, has been developed by Zhou (1999). The service at large has a public key known to all nodes in the system. The nodes trust certificates signed by the service's private key. The service can be queried by the nodes in the network to get the other nodes' public keys. The nodes can also submit update requests to update their own public keys.

The key management service consists of n special nodes in a threshold $(t + 1, n)$ configuration, meaning that (a) the number n is equal to $3t + 1$; (b) the n nodes are the servers of the service. They may themselves also have public keys to the other servers, and thus be able to construct secure channels between themselves; (c) the threshold scheme allows for an adversary to compromise up to t servers in any period of time. However, this assumption requires that the adversary lacks computational power to break the cryptographic schemes that are used.

The servers share the ability to sign certificates in this key management service. When the service is requested, each available server partially signs the certificate. The partial signatures are combined by any combiner node to retrieve the complete signature. In the case of compromised servers the combiner can check the validity using the public key of the service, and if it is not valid the combiner node tries another set of t + 1 signatures. The combiner itself need not to know any secret and can thus be any node in the network. To be more robust the scheme employs share refreshing and allows for automatic adaptation of the threshold configuration. Thus, when non-compromised servers identify a compromised server, they can update the threshold scheme to exclude it. The share refreshment permits the servers to compute new shares from the old shares they already possess. It is therefore obvious that the new shares are independent of the old shares and that a compromised server cannot use the old shares in any combination with the new ones to sign a certificate.

8.7.4 Self-Organized Public Key Infrastructure

One way to overcome the centralized issues related to the PKI is to allow the mobile nodes themselves to provide certificates. However, the problem of distributing the certificates using public key-servers that have to be handled as centralized servers needs special treatment. To be really decentralized, the certificate storage must be accomplished without the need for any special work of ad hoc nodes. A solution to address this issue is called self-organized PKI, which stores the public certificates at each mobile node. However, the nodes are not able to know about every certificate and every other user. Instead, they hold a small number of certificates; when needed, they can merge these certificates with some other nodes to get enough certificates to facilitate the verification of a public key using a trust chain in the merged set of certificates.

A graph-based algorithm, called The Shortcut Hunter algorithm (Hubaux, 2001), in which each mobile node can compute a sub-graph from the complete trust graph, uses only the locally available information to it. The algorithm gives a scalable and decentralized solution to the problem of PKI certificate distribution. It has been shown that the probability that two nodes' merged certificate lists can hold a certificate chain between them was high, provided that some conditions hold on the number of certificates at each node and the size and connectivity of the complete trust graph. Unfortunately, it is possible to extract information from such trust graphs about "who knows who" or "who trusts who." This may be used in malicious ways by adversaries.

8.8 Intrusion Detection in Ad Hoc Networks

Intrusion detection is a process of monitoring activities in a computer network and the computing systems connected to it. The mechanism by which detection is performed is called an intrusion detection system (IDS). An IDS collects activity information and then analyzes it to determine whether there are activities that violate the security rules governing the networks. When an unusual activity or an attack occurs, the IDS generates an alarm to alert the security administrator. In addition, the IDS can also initiate a proper response to the malicious activity. Although several intrusion detection techniques have been developed for the wired networks, one can state that they are not suitable for wireless networks due to the differences in their characteristics. Therefore, those techniques must be modified or enriched by new techniques to make intrusion detection work effectively in MANETs. This section aims at describing the basics of intrusion detection in MANETs and how it differs from intrusion detection in wired networks. The IDS architectures that have been provided for MANETs are presented and some of the available IDSs for MANETs are discussed.

Intrusion detection makes two major assumptions in order to provide efficiency of IDSs. The first presumes that user and node activities are observable. The second assumption considers that normal and intrusive activities can be distinguished, as intrusion detection must capture and analyze the system activity to determine whether an attack has occurred. An IDS can be classified based as network-based (N-IDS) or host based (H-IDS). While an N-IDS collects and analyzes packets from the network traffic, the H-IDS uses the operating system or application logs in its analysis. The detection techniques belong to three major categories: (a) the *detection of anomalies*, which is based on the characterization of normal profiles of users. This technique detects any activity that deviates from the normal profiles as a possible intrusion; (b) the *detection of misuse*, which keeps patterns (or signatures) of known attacks and uses them to analyze captured data. Any matched pattern is treated as an intrusion; and (c) the *specification-based detection*, which defines a set of constraints that describe the correct operation of a program or protocol. Then, it monitors the execution of the program with respect to the defined constraints.

Typically, the medium in MANETs is wide open in a way that allows both legitimate and malicious users accessing it. Furthermore, there is no clear separation between normal and unusual activities in a mobile environment. Since nodes can move arbitrarily, false routing information could be sent from a compromised node or a node that has outdated information. Therefore, the IDS techniques developed for wired networks cannot apply accurately to MANETs. The optimal IDS architecture for a MANET may depend on the network infrastructure itself (Brutch, 2003). In a flat network infrastructure, all nodes perform the same actions while some nodes are considered different in the multi-layered infrastructure. Nodes may be partitioned into clusters with one cluster head for each cluster. Four architectures can be distinguished for MANET IDSs:

1. *Stand-alone intrusion detection systems*: In this architecture, an intrusion detection system is run on each node independently to determine intrusions. Every decision made is based only on information collected at that node, since there is no cooperation among nodes in the network. Therefore, no data is exchanged. Besides, mobile nodes in the network do not know anything about the situation of other nodes in the network as no alert information is passed. Although this architecture is not effective due to its limitations, it may be suitable in a network where not all nodes are capable of running intrusion detection functionalities. Such architecture is also more suitable for flat network infrastructure than for multi-layered network infrastructure. Since information on each individual node might not be enough to detect intrusions, this architecture has not been chosen in most of the IDS for MANETs.

2. *Distributed and cooperative IDSs*: Since the nature of MANETs is distributed and requires cooperation of other nodes, the intrusion detection system in MANETs can also be distributed and cooperative (as depicted in Figure 8.4).

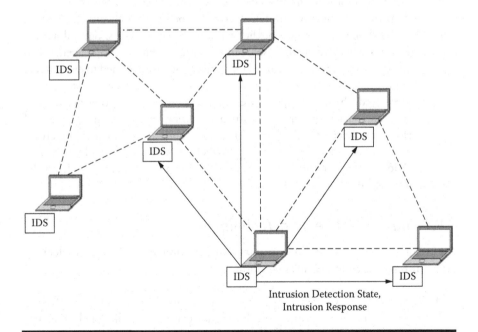

Figure 8.4   Distributed and Cooperative IDS in MANETs.

In such architecture, every node is expected to participate in the intrusion detection activity by implementing an IDS agent running on that node. An IDS agent is responsible for detecting and collecting local events and data to identify potential intrusions. However, neighboring IDS agents cooperatively participate in global intrusion detection actions when the evidence is inconclusive. Similarly to stand-alone IDS architecture, this architecture is more suitable for flat network infrastructure.

3. *Hierarchical intrusion detection systems*: Hierarchical IDS architectures extend the distributed and cooperative IDS architectures. They are suitable for multi-layered network infrastructures. The cluster heads act as control points implementing communication functions comparable to switches, routers, or gateways in wired networks. The multi-layering concept can be applied to intrusion detection systems to define hierarchical IDS architecture is proposed. Each node in a cluster is expected to implement an IDS agent, which is responsible for local monitoring (i.e., monitoring and deciding on locally detected intrusions). A cluster head is responsible locally for its node as well as globally for its cluster, meaning that it monitors the cluster activity, cooperates with the local nodes, and initiates a global response when an intrusion is detected.

4. *Mobile agent for intrusion detection systems*: Due to its ability to move from one node to another, a mobile agent is assigned to perform only one specific task, so that one or more mobile agents can be distributed into each node. This allows the distribution of the intrusion detection tasks. Several advantages can be noticed in the use of mobile agents. First, mobile agents help reduce processing overhead, since not all functions are assigned to every node. Second, they provide fault tolerance so that if the network is or some agents are destroyed, the mobile agents are still able to work. Third, they are scalable in large and varied system environments, as they tend to be independent of platform architectures. On the other hand, mobile agent-based architectures require a secure module, at the nodes, where mobile agents can be stationed to. Additionally, mobile agents must be able to protect themselves from the secure modules on remote hosts.

8.8.1 Intrusion Detection Techniques

The techniques used in the IDSs in intrusion detection are mainly dependent on the architecture used to provide detection:

■ *Distributed and Cooperative IDS*: The model used in this class of IDSs is organized into six modules: the local data connection, the local detection engine, the local response, the secure communication, the cooperative detection engine, and the global response. The local data collection module collects real-time audit data, which includes system and user activities within its radio range. The collected data will be analyzed by the local detection engine module for evidence of anomalies. If an anomaly is detected with dependable evidence, the local response module can determine that the system is under attack and initiates an alert of the local user. The global response module decides on the action to apply when an intrusion occurs. The response depends on the type of intrusion, the network protocol and applications, and the certainty of the evidence. If an anomaly is detected with weak evidence, the IDS can request the cooperation of the neighboring IDS using the cooperative detection engine module, which communicates to other agents through a secure communication module.

■ *Local Intrusion Detection System (LIDS)*: A LIDS is implemented on each node for local detection. It cooperates with other LIDSs to provide global detection (Albers, 2002). Two sorts of data are exchanged between two LIDSs: the security data, which are used to obtain complementary information, and intrusion alerts, which are used to inform other nodes of locally detected intrusion. The format of the exchanged data might be different from one local area to another, which makes it hard for LIDS to analyze the data it receives accurately. The major agent categories that may be used in a LIDS are the following:

– *Local agent*: This agent is responsible for local intrusion detection and local response. It also reacts to intrusion alerts sent from other nodes to protect itself against intrusions.

– *Mobile agent*: The mobile agents are scattered by a LIDS to collect and process data on other nodes. The results obtained after evaluation are either sent back to their LIDS or sent to another node for further investigation.

– *Local MIB agent*: To provide a means of collecting MIB variables for mobile agents or local agents, a local MIB agent acts as an interface with SNMP agent, if SNMP runs on the node, or with an agent developed specifically to allow updates.

■ *Dynamic Hierarchical Intrusion Detection Architecture*: Since nodes move arbitrarily across the network, a static hierarchy is not suitable for such dynamic network topology. A dynamic intrusion detection hierarchy can be structured in more than two levels. Nodes labeled 1 are the first level cluster heads while nodes labeled 2 are the second level cluster heads, and so on until an nth level. Members of the first level of the cluster are called leaf nodes.

A node labeled i is responsible for monitoring the nodes labeled $(i + 1)$ under its control (by accumulating counts and statistics), logging, analyzing (such as attack signature matching or checking on packet headers and payloads), responding to intrusions detected if there is enough evidence, and alerting or reporting to the cluster head labeled $(i - 1)$ linked to it. In addition, the cluster heads perform the following tasks:

- *Data fusion and reduction*: Cluster heads aggregate and correlate reports from the cluster's members. They may perform data reduction to avoid conflicting data and overlapping reports. They may also request additional information from the nodes under their control to correlate reports correctly.
- *Intrusion detection computations*: Since different attacks require different sets of detected data, data on a single node might not be able to detect an attack occurring under its control. Thus, cluster heads may need different computations to analyze and consolidate data before reaching a decision.
- *Security management*: The uppermost levels of the hierarchy have the responsibility for managing the detection and response capabilities of the clusters. They may coordinate signature updates, modify policies, and change the configurations for intrusion detection and response.

8.8.2 Node Cooperation for Detection

An IDS is not only able to detect an intrusion, but it sometimes has to identify the attack and trace the attacker, when feasible. Various types of features are evaluated from a sampling period by capturing the basic view of the network topology, routing operations, and traffic patterns, in a normal traffic. Features can be classified into two categories: (a) non-traffic-related, where statistics are determined based on the mobility and the trace logfiles (e.g., route add count, route removal count, and total route changes) and (b) traffic-related, where statistics are involved in routing and packet forwarding and can be computed by counting packets going in and out of a node (e.g., the number of packets received, the number of packets forwarded, the number of route reply messages).

8.8.2.1 Watchdog and Pathrater

These are two techniques proposed to work on top of the standard routing protocols such as SDR in ad hoc networks (Marti, 2000). The watchdog identifies the misbehaving nodes by eavesdropping on the transmission of the next hop. The pathrater helps to find the routes that do not contain those nodes. The routing information, in DSR, is passed together with the message through intermediate nodes until it reaches its destination. Each intermediate node in the path should know who the next hop node is going to be and should listen to the next hop's transmission.

The watchdog performs as follows in SDR: Assume that a mobile node, say S, wants to send a packet to node D, through nodes A, B, and C. Consider now that A has already received a packet from S destined to D containing a message and routing information. When A forwards this packet to B, it also keeps a copy of the packet in its buffer. Then, it promiscuously listens to the transmission of B to make sure that B forwards the packet to C. If the packet overheard from B matches the packet stored in the buffer, A deduces that B has forwarded to the next hop. Then, it removes the packet from the buffer. If no matching is observed during a certain interval of time, the watchdog increments the failures counter for node B. If the failure counter exceeds a given threshold, A concludes that B is misbehaving and reports to the source node S.

The pathrater performs the management of a metrics for each path, called *path metrics*. By keeping the rating of every node in the network that it monitors, the pathrater can compute path metrics by combining the node rating together with link reliability. Based on the path metrics for all available paths, the pathrater chooses the path with the highest metrics value. The system can therefore be effective for choosing paths to avoid misbehaving nodes. However, the pathrater does not provide any action against those misbehaving nodes, because they keep using resources of the network and receiving packets relayed by other nodes, while they forward packets for no one.

8.8.2.2 CONFIDANT

This technique is similar to Watchdog and Pathrater (Buchegger, 2002). Each node observes the behaviors of the neighbor nodes within its radio range and listens to them. When a node experiences a misbehaving node, it will send a warning message to other nodes in the network that it labels as friends based on trusted relationship. The process performed by CONFIDANT can be divided into two sub-processes: the process of handling its own observations and the process of handling reports from trusted nodes.

To perform the first sub-process, the monitor uses a so-called *neighborhood watch* to detect any malicious behaviors within its radio range such as no forwarding and unusual frequent route updates. This activity is close to the task performed by a watchdog. If a suspicious event is detected, the monitor then reports to the *reputation system*. Upon receiving the report, the reputation system performs several checks and updates the rating of the reported node in the *reputation table*. If the rating result is unacceptable, it passes the information to the path manager, which then removes all paths containing the misbehavior node. An ALARM message is also sent by the trust manager to notify other nodes that it considers as friends.

To perform the second sub-process, the monitor, upon receiving an ALARM message from another monitor, transmits it to the trust manager. The trust manager will first evaluate the received message for the trustworthiness of the source node. If the message is trustworthy, the ALARM message will be stored in the alarm table, together with the corresponding level of trust. All ALARM messages of the reported node will then be combined to check whether there is enough evidence to consider that it is malicious. If that is the case, the information will be sent to the reputation system, which is in charge to perform the functions described in the previous paragraph. Since this protocol allows nodes in the network to send alarm messages to each other, it could give more occasions for attackers to send false alarm messages that a node is misbehaving.

8.8.2.3 CORE

This third technique detects a particular type of misbehaving nodes, namely the selfish nodes. It forces these nodes to cooperate (Michiardi, 2002). In addition, it integrates a monitoring system and a reputation system.

The reputation of a node is rated based on past activity history and is initialized to zero. As for CONFIDANT, each node can receive a report from other nodes. CORE allows only positive reports to be passed. This can prevent false accusation and some denial of service attack. The negative rating is given to a node only from the direct observation when the node does not cooperate, which results in the decreased reputation for that node. The positive rating, in contrast, is given from both direct observation and positive reports from other nodes, which results in the increased reputation.

References

I. Aad, J.-P. Hubaux, and E. W. Knightly, Denial of Service Resilience in Ad Hoc Networks, in *Proc. of the 10th Annual Int. Conf. on Mobile Computing and Networking (MobiCom '04)*, Philadelphia, PA, USA, Sept. 26–Oct. 1 2004.

S. Agarwal, A. Ahuja, J. P. Singh, and R. Shorey, Route-Lifetime Assessment-based Routing (RABR) Protocol for Mobile Ad Hoc Networks, *Proc. of IEEE ICC 2000*, 3:1697–1701, June 2000.

P. Albers, O. Camp, J. Percher, B. Jouga, L. M., and R. Puttini, Security in Ad Hoc Networks: a General Intrusion Detection Architecture Enhancing Trust Based Approaches, *Proc. of the 1st Int. Workshop on Wireless Information Systems (WIS-2002)*, pp. 1–12, April 2002.

P. Brutch and C. Ko, Challenges in Intrusion Detection for Wireless Ad-hoc Networks, *Proc. of 2003 Symp. on Applications and the Internet Workshop*, pp. 368–373, Jan. 2003.

S. Buchegger and J. Le Boudec, Performance Analysis of the CONFIDANT Protocol (Cooperation of Nodes-Fairness In Dynamic Ad-hoc NeTworks), *Proc. of the 3rd ACM Int. Symp. on Mobile Ad Hoc Networking and Computing (MobiHoc'02)*, pp. 226–336, June 2002.

R. Chandra, V. Ramasubramanian, and K. P. Birman, Anonymmous Gossip: Improving Multicast Reliability in Mobile Ad-Hoc Networks, *Int. Conf. on Distributed Computing Systems*, 2001, pp. 275–283.

T. W. Chen and M. Gerla, Global State Routing: A New Routing Scheme for Ad Hoc Wireless Networks, *Proc. of IEEE ICC 1998*, pp. 171–175, June 1998.

Y. Chen and A. Liestman, and J. Liu, *Clustering Algorithms for Ad Hoc Wireless Networks in Ad Hoc and Sensor Networks*, Y. P. Y. Xiao (Eds), Nova science publisher, 2004.

C. C. Chiang, H. K. Wu, W. Liu, and M. Gerla, Routing in Clusted Multi-Hop Mobile Networks with Fading Channel, *Proc. of IEEE SICON 1997*, pp. 197–211, April 1997.

T. H. Clausen, G. Hansen, L. Christensen, and G. Behrmann,The Optimized Link State Routing Protocol, Evaluation Through Experiments and Simulation, *Proc. of IEEE Symposium on Wireless Personal Mobile Communications*, Sept. 2001.

T. Clausen, P. Jacquet, A. Laouiti, P. Minet, P. Muhlethaler, A. Qayyum, and L. Viennot, Optimized Link State Routing Protocol IETF Manet working group, (available at http://www3.ietf.org/proceedings/02mar/I-D/draft-ietf-manet-olsr-06.txt).

J. J. Garcia-Luna-Aceves and M. Spohn, Source-Tree Routing in Wireless Networks, *Proc. of IEEE ICNP 1999*, pp. 273–282, Oct. 1999.

C. Ho, K. Obraczka, G. Tsudik, and K. Viswanath, Flooding for Reliable Multicast in Multi-Hop Ad-Hoc Networks. MobiCom Workshop on Discrete Algorithms and Methods for Mobility, Aug. 1999.

Y. C. Hu, A. Perrig, and D. B. Johnson, Packet Leashes: A Defense Against Wormhole Attacks in Wireless Ad Hoc Networks, in *Proc. of the 22nd Annual Joint Conf. of the IEEE Computer and Communications Societies (INFOCOM 2003)*, San Francisco, CA, USA, April 2003.

Y. C. Hu, A. Perrig, and D. B. Johnson, Rushing Attacks and Defense in Wireless Ad Hoc Network Routing Protocols, in *Proc. of the 2003 ACM Workshop on Wireless Security*, pp. 30–40, San Diego, CA, USA, 2003.

Y. Hu, A. Perrig, and D. Johnson, Ariadne: A Secure On-demand Routing Protocol for Ad Hoc Networks, *ACM MOBICOM*, 2002.

J. P. Hubaux, L. Buttyan, S. Caplsun, The Quest for Security in Mole Ad Hoc Networks, in *Proc. of AEM Symp. on Mobile Ad Hoc Networking and Complexity* (MobiHoc 2001), pp. 146–155.

D. B. Johnson and D. A. Maltz, Dynamic Source Routing in Ad Hoc Wireless Networks, *Mobile Computing*, 353:153–181, 1996.

D. B. Johnson, D. A. Maltz, and Y. Hu, The Dynamic Source Routing Protocol for Mobile Ad Hoc Networks (DSR), IETF MANET Working Group (available at http://www3.ietf.org/proceedings/02mar/I-D/draft-ietf-manet-dsr-07.txt).

A. Khalili, J. Katz, and W. A. Arbaugh, Toward Secure Key Distribution in Truly Ad-Hoc Networks, *Tech. Rep.*, University of Maryland (College Park), 2003.

Y. Ko and N. H. Vaidya, Location-Aided Routing (LAR) in Mobile Ad Hoc Networks, *Proc. of ACM MOBICOM 1998*, pp. 66–75, Oct. 1998.

A. Kuzmanovic and E. W. Knightly, Low-rate TCP-targeted Denial of Service Attacks, in *Proceedings of the 2003 Conference of the Special Interest Group on Data Communication (SIGCOMM '03)*, pp. 75–86, ACM Press, Karlsruhe, Germany, 2003.

W. Lou and Y. Fang, A Survey of Wireless Security in Mobile Ad Hoc Networks: Challenges and Available Solutions, a chapter in book *Ad Hoc Wireless Networking*, Xiuzhen Cheng, Xiao Huang, Ding-Zhu Du, Kluwer (Eds.), pp. 319–364, May 2003.

B. S. Manoj, R. Ananthapadmanabbha, and C. S. R. Murthy, Link Life-Based Routing Protocol for Ad Hoc Networks. *Proc. of IEEE ICCCN 2001*, pp. 573–576, Oct. 2001.

S. Marti, T. J. Giuli, K. Lai, and M. Baker, Mitigating Routing Misbehavior in Mobile Ad Hoc Networks, *Proceedings of the 6th Annual Int. Conf. on Mobile Computing and Networking (MobiCom'00)*, pp. 255–265, Aug. 2000.

P. Michiardi and R. Molva, Core: A Collaborative Reputation Mechanism to Enforce Node Cooperation in Mobile Ad Hoc Networks, *Communication and Multimedia Security Conference (CMS'02)*, Sept. 2002.

C. E. Perkins and P. Bhagwat, Highly Dynamic Destination-Sequenced Distance-Vector Routing for Mobile Computers, *Proc. of ACM SIGCOMM 1994*, pp. 234–244, Aug. 1994.

A. Perrig et al., The TESLA Broadcast Authentication Protocol, *RSA CryptoBytes*, 5(2), (2002): 2–13.

E. M. Royer and C. E. Perkins, Multicast Operation of the Ad-hoc On-Demand Distance Vector Routing Protocol, *Proc. of the Fifth Annual ACM/IEEE Int. Conf. on Mobile Computing and Networking (MOBICOM'99)*. Aug. 1999, pp. 207–218.

S. Singh, M. Woo, and C. S. Raghavendra, Power-Aware Routing in Mobile Ad Hoc Networks, *Proc. of the IEEE WCNC 2000*, pp. 1220–1225, September 2000.

P. Sinha, R. Sivakumar, and V. Bharghavan, CEDAR: A Core Extraction Distributed Ad Hoc Routing Algorithm, *IEEE Journal on Selected Areas in Communications*, 17(8):1454–1466, Aug. 1999.

R. S. Sisodia, B. S. Manoj, and C. S. R. Murthy, A Preferred Link-Based Routing Protocol for Ad Hoc Wireless Networks, *Journal of Communications and Networks*, 4(1):14–21, Mar. 2002.

F. Stajano and R. Anderson, The Resurrecting Duckling: Security Issues for Ad-hoc Wireless Networks, in *Proceedings 7th International Workshop Security Protocols, (1999)*, B. Christianson, B. Crisp, and M. Roe (Eds).

C. K. Toh, Associativity-based Routing for Ad Hoc Mobile Networks, *Wireless Personal Communications*, 4(2): 1–36, Mar. 1997.

L. Zhou, and Z. J. Haas, Securing Ad Hoc Networks, *IEEE Network*, 13(6) (1999): 24–30.

SECURITY OF NETWORK-BASED SERVICES IN MOBILE COMMUNICATIONS

III

Chapter 9

Inter-System Roaming and Internetworking Security

9.1 Introduction

The growing number of wireless communication technologies supporting user mobility has called for solutions that enable interoperation between access providers and technologies. It also generated several security requirements and concerns, which have occurred mainly because of the level of the provider's responsibility, the wireless access nature, and the lack of physical protection mechanisms comparable to those provided in traditional static-user, wired-link, and fixed-topology networks. Roaming and handover procedures represent two forms of inter-operation that aim at providing for user mobility across interconnected networks. Roaming procedures allow wireless access in areas covered by network providers, with which the roaming mobile user does not have any prior administrative registration, and require as little additional effort as possible from the roaming terminals. On the other hand, the handover procedures enable the maintenance of ongoing connections between mobile users, while one among the users is moving across different wireless access points or networks.

Roaming is based on a major operation, called *pre-registration procedure*, during which a user establishes a special (including commercial) relationship with a unique wireless access network provider, called the *home network* (HN) of the mobile user, as explained in the previous chapters. During pre-registration, pre-shared credentials are established at the home network and the user's mobile device or station

(MS) to allow the HN and the MS user to authenticate each other (or, for some networks, to authenticate the user to the HN only) when the pre-registered MS requests an access to a communication service offered by the HN. Typically, the mutual authentication between MS and HN requires an interaction with the user based on the provision of a username/password combination. Additionally, the credentials are managed at a Security Center (denoted by SCHN, or more simply SC) in the home network. The most largely used credential types in wireless access networks belong to three classes:

1. A secret key, which is a long-term secret key shared between the MS and the SC. As stated in Chapter 5, the GSM network utilizes this type of credentials;
2. A public-key certificate. With this type of credentials, both the MS and the HN are assumed to own a pair of public and secret keys. They both must own a public key certificate binding them to their public keys that they should store for verification needs; and
3. A public-key certificate mixed with a username/password. With this type of credentials, the HN has a public/secret-key pair and a certificate signed by a trusted certificate authority CA. The HN generates a username and password for the pre-registered user and keeps secret this pair in the SC. The MS stores the public-key certificate corresponding to CA that has signed the HN's certificate.

To support large mobility to its subscribers, a mobile network operator sets up the so-called *roaming agreements* with other network operators. A roaming agreement between two operators allows the user of each network to access the other network without pre-registration to that network. This access includes being able to receive incoming connections as well as being capable of placing phone calls between the two interconnected networks (the HN and the FN). Therefore, roaming includes updating the location information for the roaming user, re-routing incoming user's traffic to the user's new access network, and providing the level of security supported by the user's home network. In the early years, mobile devices were equipped with only one technology reducing inter-provider roaming. Nowadays, mobile devices attempt to integrate multiple technologies, including GSM, UMTS, WIMAX, and WIFI.

On the other hand, a handover procedure re-routes, to the user's new access point, the incoming and outgoing traffic, without causing any communication interruption. This requirement puts severe constraints on the efficiency of the handover process, which may also lead to the setup of routing paths different from those occurring in the roaming phase. In addition to the increasing motivation to protect against unauthorized accesses and eavesdropping in mobile networks, the technical issues of supporting roaming and handover take part in the challenges that aim at providing secure solutions for roaming and handover in mobile networks. The security challenge on inter-provider roaming is to enable a mobile device and a foreign network to authenticate each other, negotiate security mechanisms, and set up cryptographic keys to secure the network access without any prior direct trust relationship.

The aim of this chapter is four-fold: (a) to discuss the security challenges imposed on the infrastructure-based wireless access networks by inter-provider and inter-system roaming and handover procedures; (b) to analyze the current solutions providing inter-system roaming and handover; (c) to highlight the features of the new security solutions for roaming; and (d) to analyze in detail the roaming and handover procedures within and between known technologies such as GSM and UMTS. We present also a systematic threat analysis of the security solutions that are based on security-information transfers during inter-provider handover, and we discuss the security requirements on internetwork connection based on this analysis. As opposed to the traditional handover procedures, which do not meet the aforementioned requirements, we present some other procedures that are history-enriched and policy-based to enhance the security-context transfers on inter-provider and inter-system handover. But, before we start, let us briefly describe the latter concepts.

9.2 Roaming

The main concern in roaming between providers and technologies is to allow a foreign network and a mobile station to authenticate each other and agree on the cryptographic keys for the encryption and integrity protection of the communication they establish without the need to share any pre-established credentials. Roaming protocols can be classified according to three dimensions: (a) the type of control they give to the home provider to adapt changes in roaming agreements and roaming profiles; (b) the interaction required between the mobile station (MS), the foreign network (FN), and the home network (HN); and (c) the knowledge of confidential information gained by each party. Moreover, the following design goals for security solutions should be met: (a) the roaming authentication and key-agreement protocol should minimize the authentication traffic required between a foreign network and the home provider, and between a foreign network and the MS; (b) the roaming must permit an easy handling of changes in roaming profiles and agreements; and (c) the roaming must allow the derivation of cryptographic keys at the FN, where they are used.

9.2.1 Establishing Temporary Residence Abroad

The basic statement of user mobility assumes that a user has a unique home (the HN). The user's home is the administrative domain where he is registered on a long-term basis (the network where the pre-registration was initially performed).

Usually, it is the place where accounting and billing information is collected. As a mobile user migrates from one network to another, he often enters in a new and foreign domain with the goal to access some services via the FN. A user may be simply transiting in the foreign domain or planning to delay leaving for some time. In either case, he must establish a temporary residence with the foreign network and get service locally. This process is similar to the mobility scenario in the real-world, where a person physically traveling from one country to another must engage in some form of an administrative procedure whose aim is to establish a temporary physical residence in the new country.

The procedure of establishing a temporary residence can vary from one network to another. In some cases, it may be sufficient for a mobile user to possess a universal credential that the FN can verify and accept. However, while the credential's authenticity is verifiable, its current status may not be verifiable in general. In other words, it is naturally advantageous for a foreign network to establish that the newly-arrived and visiting mobile user is presently in good status (in terms of commercial engagement to the HN). This cannot be accomplished without some cooperation of the user's home network, since only the home network is able to report on the user's current status. Therefore, a basic solution aiming at providing mobile users with a global user certification, such as the one certificate authority, is not applicable to the general mobile-user environments. In addition, the authorization to access a service for a mobile user cannot be addressed with a digital certificate only. Nonetheless, certificates can be used for identifying users properly.

Even if the user's ability to pay for services can be provided without contacting the home network (by using a mobile payment, for example), two issues remain unsolved: the location tracking security and the access control. Since a mobile user is normally registered with a HN on a long-term basis, any user desiring to establish a communication with the MS will have to consult the MS's HN first. This implies that a HN should track the locations of its registered users. Moreover, if the HN implements a mandatory access control, a user may be required to contact his home network, before establishing a temporary residence within a FN. The second issue can be important from the point of view of charging for the use of services. If a user accesses some services while roaming under a foreign network, later the FN bills for it, and the mobile user's HN may refuse to pay if the expenses are not authorized. These two issues call for the involvement of the HN in the process of establishing temporary residence in a foreign domain.

However, there is a difference between a user appearing in a foreign network and a user moving between two adjacent foreign networks. While the user moves from one foreign network to another, his trustworthiness must be confirmed with every crossing of network boundaries. In general, we cannot assume that the path taken by a mobile user is continuously connected to only one FN. A particular behavior takes place in a wireless environment where, instead of staying under the coverage of the FN, in which the mobile engaged in a conversation has roamed, he may decide to switch connection to another foreign network and then may get back to the first FN. This type of real-time inter-network transition demands a real-time knowledge of the user's state including information on the current session activity data and authentication/authorization information.

By and large, the TMSI (i.e., the Temporary Mobile Station Identity) is used by the network operators to identify a mobile station operating within the network. It has attracted a lot of interest because of two reasons. First, a large signaling capacity can be saved because the TMSI utilizes few bits to identify the mobile station. Second, a TMSI really enhances mobile station identification confidentiality. The use of a TMSI makes it more difficult for a malicious mobile user to obtain the mobile identification number by monitoring exchanged signals. The fraudulent user may obtain the mobile station's TMSI, but the TMSI will most likely remain valid for only a short period of time.

The management of TMSIs in a mobile network may present some deficiencies. Indeed, if a MS is assigned a TMSI in a first mobile switching center (MSC) and then moves into the service area of a second MSC operating with the same System Identity (SID), the mobile station attempts to register in the second MSC utilizing the TMSI assigned by the first MSC. Unfortunately, the second MSC cannot recognize the first TMSI. Therefore, the second MSC cannot identify the MS and does not know which HLR to access to retrieve the subscriber information. The unsuccessful registration attempt results in a registration reject from the second MSC pushing, in so doing, the mobile station to reattempt the registration utilizing its permanent mobile identification number.

Several solutions have been proposed to overcome the aforementioned deficiency. One approach recommends to pass the mobile station's TMSI from the serving MSC to the neighboring MSCs. This defines the TMSI in such a way that it can be reused by a new serving MSC, causing additional intersystem signaling to convey TMSI information from one MSC to another, and to the home location register (HLR). In addition, it appears unsuitable for privacy protection, since the network keeps track of each mobile station's TMSI as the mobile station moves from one MSC to another. A second approach aims at disclosing a method of structuring a TMSI based on a tree-hierarchy of the MSCs and conveying the TMSI information from children to parent only.

9.2.2 Roaming Modeling

In typical roaming procedure, a network provider has a roaming agreement with other network providers that we refer to as its roaming partners. In such a roaming agreement, two roaming partners agree on which services the roaming users should be able to use. They optionally agree on the procedures to apply for authentication and key agreement along with the cipher suites to be used and the way they are derived. The roaming agreement additionally sets up the charging and billing activities and some legal obligations between the providers. When entering a roaming agreement, two roaming partners exchange credentials and establish a trust relationship. On the mobile user's side, the user and his home provider establish the user's roaming profile during registration. The roaming profile includes the set of foreign networks as well as the services a user will be able to access when it roams. During the registration step, the user and its home provider may perform the following tasks: (a) additionally exchange information on their roaming security policies; (b) store the exchanged information on the user's mobile station; and (c) fix the roaming charges during registration.

The authentication performed between the FN and the MS differs from mutual authentication between a MS and its HN in the sense that the credentials exchanged by the HN and the MS during registration allow them to uniquely identify each other and then verify the binding between their identities and the credentials while on roaming; the MS and the FN are not guaranteed of each other's identity, but rather of the HN's authorization of the roaming occurrence. For this, the FN has to prove to the MS that it is authorized by the HN to offer the requested service. The MS also has to prove to the FN that it is authorized by the HN to get FN's services. In the case of commercial network providers, the HN's authorization of the roaming operation confirms to the FN that the HN is willing to pay it back for service provisioning to the MS. Apart from the authentication procedure itself, the FN and MS may have to agree on a master session key for the roaming needs, such that they can derive encryption and integrity protection keys in order to protect data and control traffic established between them. During connection establishment, the MS and the FN have to negotiate the roaming authentication protocol, the roaming key-agreement protocol, and the cipher suite to use. This security-suite negotiation may require HN's involvement.

On roaming to a FN, current solutions only take the roaming policies of the FN and the MS into consideration. However, this may not be sufficient and the involvement of the HN may be needed. Indeed, as the HN's authorization of the roaming occurrence is given to the MS and to the FN, it may be held responsible for attacks launched against the security of the MS or the FN. In particular, the HN may have to reimburse the FN for service provisioning even if the roaming authentication between MS and FN was broken and an unauthorized party got access to the FN on behalf of the MS. Nevertheless, whether or not the HN is held responsible for attacks performed involving the roaming user, the HN may give its pre-registered users assurance on the level of protection provided by the FN where they may roam. One can therefore assume that not only the MS and the FN should have roaming policies with respect to the security suites to be used on roaming, but also the HN should use its own. Typically, the MS, the HN, and the FN express their policies by pre-defining subsets of rules.

The fact that the current roaming procedures do not enforce the HN's policy with respect to the security mechanisms allowed to be used during the roaming activity constitutes a threat that the HN can be held responsible for. On the other hand, if the HN's policies are to be selected during roaming, either the HN can be engaged in the negotiation protocol itself, or certain policies may be fixed in the roaming agreement as part of the roaming scheme for the MS. In the following, we briefly describe three security negotiation methods that consider the use of the HN's policy. The first method does not take into consideration any preference among the MS, FN, and HN. The second method considers some established preference between them. The third method allows the HN to be engaged into an online negotiation. The latter method is more flexible with respect to the frequency of changing policy. However, all the three methods assume that the HN has to trust at least the FN or the MS to comply with its policy and that the FN (respectively, the MS) alone cannot modify the negotiation result in a way that does not comply with its policy without the MS (respectively, the FN) detecting the modification (Meyer, 2005).

- *First method*: If the HN is not engaged online in the negotiation, it can reveal its policy to the MS in the pre-registration phase and to the FN in the roaming agreement. Both the MS and the FN then use the intersection of their own policy rules with the HN's policy shown, rather than exclusively use their own rules. The MS and the FN can then negotiate a security suite. Let us notice that, in this case, the HN has to trust the MS and the FN to comply with its policy. Nevertheless, revealing its policy to both parties ensures that the FN and the MS would have to ignore HN's policies in order to choose a mechanism whose use is not allowed by the HN.
- *Second method*: If the MS and the FN are engaged in the negotiation online and the HN is not to be taken into account, then the MS and the HN (respectively, the HN and the FN) can reconcile their policies during registration (respectively, on entering a roaming agreement). The policy reconciliation mechanism helps the roaming partners deriving a total preference order on the intersection of their policies. The MS can then store the reconciled policy together with the combined preference order. The same operation holds for the FN. Upon roaming, the MS and the FN start using the stored reconciled policy to negotiate the security suite.
- *Third method*: Engaging the HN online in the negotiation is particularly easy to achieve if the HN is required to be part of the roaming authentication and key-agreement procedures. The way the security-suite negotiation between the HN, the MS, and the FN is integrated into the connection establishment depends largely on the manner the HN is engaged in the authentication. To give just one case, the MS could send its roaming policy expression to the FN in a first step. The FN could then compute the intersection of the MS's policy and its own policy and forward it to the HN together with a request for authentication information. The HN could then select one among the suites in the received intersection and command the FN and the MS to use it. To prevent the FN from changing the selected security suite, the HN may need to add an authentication token to the selected suite that can be verified by the MS.

9.3 Roaming Authentication and Key Agreement

The authentication protocol, the key agreement, and the security-mechanism negotiation on roaming can be implemented in many ways. The classification of the existing roaming solutions can be made according to the following dimensions.

■ *Public-key-based versus non-public-key-based*: Roaming authentication and key-agreement protocols can be based on public-key certificates or they can be non-public-key based. In general, public-key-based authentication protocols have the advantage of allowing the two partners to mutually authenticate each other without requiring any prior trust relationship to be set up between them. A public-key-based method seems to be among the methods of choice for authentication upon roaming. However, such a method raises several difficulties. First, many current wireless access technologies do not support public-key certificates. Second, when the MS and the FN authenticate each other based on their certificates, both have to validate these certificates during the authentication process. The validation includes verifying certification authority's signature on each other certificates and checking the revocation status of the certificates. The validation is particularly difficult to perform by the MS, as it has to check the FN's certificate and needs for a network connection to access remote resources for status checking. Third, the fact that the MS may not trust the CA that has signed the FN's certificate might make the validation more difficult to perform. In that case, the MS has to obtain a chain of certificates with a trusted certificate as a root and validate each certificate in this chain in order to validate FN's certificate.

One approach to overcome the aforementioned difficulties assumes that on roaming, the MS should delegate the validation of the FN's certificate to a trusted third party (Bayarou, 2004). In this case, the MS has to be only sure of the revocation status of the certificate of the trusted third party. However, this method causes additional authentication traffic between the MS and the trusted third party (to which the FN is forwarding the information). Thus, it increases the overall load on the network and delays the authentication completion. Additionally, this approach does not address the following problem: If the relevant information regarding the roaming profile of the MS is included in MS's certificate (in the form of attributes, for example), then a new certificate has to be issued for the MS every time the roaming profile of the MS changes. Similarly, the FN's certificate should change if the relevant information on the roaming agreement between the FN and the HN is encapsulated in the FN's certificate. Therefore, using public-key certificates without online engagement of the HN in the authentication process cannot easily accommodate changes in the roaming schemes and agreements.

To avoid the above shortcomings, public-key-based authentication protocols that engage HN online have been proposed. In the most extreme case of engagement, the HN and the MS mutually authenticate each other based on individual certificates and agree upon a master key, while the FN only serves as a forwarder of the authentication traffic between them. The HN then uses the credentials pre-established with the FN to securely transfer this master key to it. Then, the MS and the FN mutually assure each other of their authorization by proving indirectly possession of the master key. In this case, the MS has to validate only the HN's certificate. In other cases of roaming authentication, the HN engages in the authentication in a different way and rather assists the FN and the MS in authenticating each other, than authenticating both parties itself (Meyer, 2005).

■ *Online/offline engagement of the HN*: Typical non-public-key-based authentication and key-agreement protocols require the HN's engagement in the roaming authentication (Salgarelli, 2003). The HN's interaction can either be online prior to (or during) the authentication. In the offline case, the HN typically provides the FN with some security-related information with the help of which the MS and the FN can achieve mutual authentication. The offline engagement of the HN generally trades an efficient authentication protocol that assures the freshness of the HN's authorization of the roaming process. Some other roaming protocols try to minimize the round-trip message exchanges with the HN and guarantee the freshness of HN's authorization. They can, for example, require that the first in a series of authentications between the FN and the MS engages the HN online, while the subsequent ones should involve only the FN and the MS.

■ *Key derivation by the HN or the FN*: To perform any public-key-based or non-public-key-based authentication and key-agreement protocols, the MS and the FN have to establish a master key K. Using key K, they subsequently derive the integrity and the encryption key to protect their connection. If the HN is involved in the roaming authentication, the master key K can be derived by the HN or by the FN. If the HN has to derive it, it may do it offline before the actual roaming authentication or online during the authentication. If the HN has to derive K, it should establish a secure channel to the FN for the purpose of transferring K to it. In the case where the FN is in charge of deriving K, the HN may not be able to get knowledge of K. However, the HN and the FN still require establishing a secure channel between them, since the HN may need to transfer the secret information contributing to the FN during or before authentication.

9.3.1 Roaming Procedures

Three types of roaming procedures can be distinguished in the current mobile communication networks:

1. The type 1 roaming procedures require that the MS and the FN mutually authenticate each other using their public-key certificates and validate each other's certificates during authentication. The information needed for certificate validation should be made available by the HN or a trusted third party prior, during, or subsequently to the authentication. The MS and the FN negotiate the security suite to use without interacting with the HN. However, the HN can attempt to ensure its policies offline by revealing them to the MS during pre-registration and to the FN upon entering the roaming agreement.

The interest of type 1 procedures is three-fold. First, the FN is in charge of the master key generation and the HN does not get knowledge of it. This means that no secure channel for key transfer is needed between the HN and the FN. Second, this method does not require any traffic on the network other than the traffic related to the revocation status of certificates. Third, the method cannot easily handle changes in roaming plans or agreements, which makes it inflexible. In addition, one can notice that it requires particularly adapted solutions to enable the MS to validate the certificate presented by the FN. Figure 9.1 depicts the security-mechanism negotiation as well as authentication and key agreement upon roaming to FN, in this case. For simplicity, the possibly required certificate revocation's status checks before, during, or after authentication are not illustrated.

The protocols depicted in Figure 9.1 are the roaming authentication *ra* and the key-agreement protocol *rka* between the FN and the MS. As the FN and the MS authenticate each other based on individual public-key certificates, a master key K is generated in the authentication server of the foreign network (AS$_{FN}$) and the MS. It is transferred to the FN's Encryption and Integrity-Protection Endpoint (EIPEFN) using a key-transfer mechanism, *kt*. Since

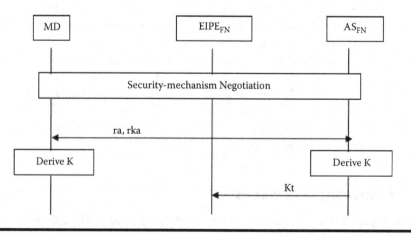

Figure 9.1 Type 1 roaming procedures.

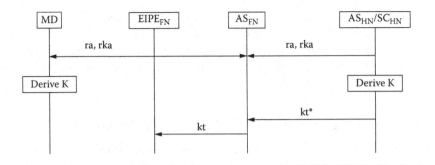

Figure 9.2   Type 2 roaming procedure.

the AS_{HN} does not get knowledge of K, no secure channel between AS_{HN} and AS_{FN} needs to be shown in Figure 9.1.

2. A type 2 roaming procedure can be public-key-based or non-public-key-based. Figure 9.2 illustrates this type of roaming procedure. The procedure assumes that the FN forwards the authentication and key-agreement protocol messages between the MS and the HN. As a result, the HN's engagement in the roaming procedure is clearly made online. The MS and the HN authenticate each other based on their pre-established credentials. Then they agree on a master key K and the AS_{HN} transfers it to the AS_{FN} after successful authentication, using key transfer protocol kt^* and a secure channel previously established as part of their roaming agreement.

The necessity of a secure channel represents one of the disadvantages of the type 1 roaming procedures. Another disadvantage is related to the fact that the AS_{FN} has to forward all authentication traffic between the MS and the HN, which may result in an acceptable load. Nevertheless, an important advantage can be noticed: the HN can easily control each roaming instance and has real-time interaction to the changes observed in the MS's roaming profile or in the roaming agreement with FN. It can also control the use of encryption and integrity-protection mechanisms after authentication completion.

3. Type 3 roaming procedures require HN's online or offline interaction to provide more than just information on the revocation status of certificates submitted; rather, they divide the activity load of the authentication between the HN and the FN. Type 3 procedures can be public-key-based or non-public-key-based. In some of the type 3 procedures, the HN generates the master key and transfers it to the FN, thus requiring a secure channel between the HN and the FN. In other type 3 procedures, the FN derives the master key itself while requiring some secret input from the HN to achieve it. In a third subset of procedures, the FN may be able to derive the master key without HN's interaction. For the sake of simplicity and the similarity with the other types, no figure is provided for type 3. It is worthy to note that the type 3

roaming procedures make no general statements on when or how negotiating security mechanisms can be operated and that the MS is required to communicate with both AS_{HN} and AS_{FN} at the same time in the procedures suggested for the authentication across different technologies in wireless overlay networks (Chen, 2003).

9.3.2 Roaming across Different Access Technologies

The first generations of the wireless devices were equipped with only one technology. This has limited the inter-operation of communication systems to providers supporting the same access technology. However, this does not cope with the major goal of the inter-system roaming support that aims at making it feasible for the mobile users to access different networks operating with different technologies with a unique registration process (mainly, with their home provider) and only one bill. Another concern is related to the type of used credential. In some wireless technologies, the roaming authentication and key-agreement protocols use only one particular type of credentials. To allow the MS to roam across different technologies, the home provider has to supply the MS with a set of credentials of the right type for the available technologies. A single set of credentials would be satisfactory, provided that each candidate technology supports an authentication and a key-agreement protocol based on one credential within the set. To prevent impersonation attacks in that case, it is necessary for the HN to instantaneously report all of its MSs about the revocation of a roaming credential as soon as it notices that it has been compromised. Then it has to provide a new credential and distribute it.

A home provider may want to issue a set of credentials to its pre-registered mobile users and utilize it to authenticate the users wherever they roam. This is the case if the HN provider has already invested in an expensive authentication infrastructure and needs to reuse it while contracting new roaming agreements with providers operating different technologies. Moreover, the network provider may simply want to avoid disturbing his users to obtain an additional set of credentials to use new technologies; rather, it allows them to reuse the one he already has. This may call for new roaming authentication procedures to allow roaming across technologies based on the set of pre-established credentials with the HN.

To be able to roam across different technologies, a mobile user may register for a certain set of technologies with his HN. The home provider assigns one or more credentials to the user such that the user obtains a suitable set of credentials in a way that one credential is available for at least one (roaming) authentication and key-agreement protocol for each technology for which he registers. The same type of credentials may be usable by different technologies such that the number of technologies may exceed the number of credential types assigned for the user. This set may be indirectly defined, for example, using a geographical roaming region. The MS and its HN then determine the set of services the MS may use upon roaming for each technology, and optionally also exchange information on their roaming policies. In case of commercial providers, the MS and the HN may additionally agree on the roaming charges.

Roaming authentication and key-agreement protocols for mobile users roaming between different technologies can be defined in exactly the same way as in the inter-provider case. However, the FN may implement a technology the home provider does not support. If engaged in the authentication and key agreement at all, the HN here supports the home provider back-end of the roaming protocols for FN's technology. The same approach holds for the roaming key-agreement protocol. Finally, let us notice that a MS roaming across different technologies has to specify a roaming policy for every technology it supports. The MS specifies a subset security policy that includes all roaming security suites it allows to be used on roaming to this technology. Similarly, each FN expresses its policy with respect to the roaming security suites it allows to be used. Then, the home provider specifies its subset for each technology it has a roaming agreement with. Additionally, it may have to set policies for technologies it does not support itself.

9.4 Inter-Provider Roaming within the GSM and UMTS

The main 2G mobile communication standards competing with the GSM are the American IS-95, which is used in the US and in some other countries, and the PDC system that is used in Japan and Korea. The incompatibilities of these 2G systems were among the reasons that led to the vision of the third generation mobile communications standard that aimed at allowing for an easy worldwide roaming (see Chapter 6 for more details). The standards family IMT-2000 defined by the ITU is the result of this vision and consists of several compatible 3G standards that allow multi-mode MSs to roam between all standards in IMT-2000, as explained in a previous chapter. The Universal Mobile Telecommunications Standard (UMTS) standardized by 3GPP allows for roaming between UMTS and GSM.

9.4.1 GSM Intra-Provider Roaming

Inter-provider roaming from a GSM network to another causes a new authentication between the new network and the MS. The MS is also authenticated any time the provider of the visited network requests a new authentication. When moving out of the range of the serving BTS and getting into the range of a new one controlled by the same MSC, the encryption key Kc negotiated during the last authentication is transferred from the MSC to the new BTS. On the other hand, when the MS roams beyond the control of its serving MSC, the old MSC first transfers Kc to the new MSC. The new MSC subsequently transfers Kc to the BTS. Therefore, the MS is indirectly re-authenticated by the new BTS, as the unique (and legitimate) mobile user to know the current key Kc. As a BTS cannot differentiate between correctly and incorrectly encrypted data traffic, and as no integrity protection is used in GSM, the new BTS cannot detect impersonated MSs based only on this re-authentication method. Nevertheless, trying to impersonate the MS to the new BTS is useless for an attacker without knowing the current key Kc, as long as the new BTS enables encryption. In this context, it is interesting to note that, if encryption was disabled between the MS and the old BTS, it would not be enabled after roaming to the new BTS. As a result, the fast re-authentication in idle mode brings in impersonation attacks.

During registration, the home provider allocates an International Mobile Subscriber Identity (IMSI) for the MS and generates a long-term secret key Ki of 128 bits. The credential pair (IMSI, Ki) is stored in HN's AuC as well as on the subscriber identity module (SIM). Aside from the credential pair, the SIM also contains the provider-specific algorithms A3 and A8. When the registration is performed, the mobile user registers for a certain geographical roaming region. Upon roaming, the FN is assured of MS's authorization to roam to the FN but the MS does not authenticate the FN. This, unfortunately, enables network impersonation attacks to be launched against the MS. The roaming authentication and key-agreement protocols require the HN's interaction on the first authentication of the MS to the FN. On the first authentication, the FN requests the authentication data from the HN. The HN then provides it with one or more authentication vector(s), say ($RANDG_i$, $RESG_i$, Kc_i), $i < p$ for a given p, each containing a random challenge $RANDG_i$, an authentication response $RESG_i$, and an encryption key Kc_i.

As the aforementioned vectors are generated using the HN-specific algorithms A3 and A8, one can say that, theoretically, the GSM support as many authentication and key-agreement protocols as there are GSM operators. In fact, only a few A3 and A8 implementations are in use. In addition, a secure channel between the HN and the FN is needed to secure the transfer of authentication vectors. Therefore, the GSM inter-provider roaming authentication and the key-agreement protocols are of type 3; they require a secure channel between the HN and the FN; and they are secret-key-based. On the other hand, a SIM card supports exactly one pair of A3 and A8 algorithms, so then no authentication and key-agreement protocols have to be negotiated during connection establishment. In addition, the GSM does not support integrity protection and does not use any key-establishment process between the MS and the BTS; each cipher suite in the GSM consists of only one encryption mechanism. The MS and the FN negotiate the encryption mechanism on roaming without HN's interaction.

On connection setup with the FN, the MS sends to the FN its security capabilities, a list of A5 algorithms that the MS is able to support. As stated by the standards, the MS is authorized to support A5/0, A5/1, and A5/2 and the FN is required to reject the connection if it receives security capabilities from the MS that do not include A5/0. The FN chooses one of the algorithms among those supported by the MS and acknowledges its choice to the MS using a GSM security mode command message. It is interesting to note that the HN has no authority on the FN's choice of the encryption algorithm. In particular, it cannot interdict the use of A5/0 or the weak encryption mechanism A5/2. Similarly, the MS cannot enforce the use of one among the stronger algorithms A5/1 and A5/3.

9.4.2 UMTS Inter-Provider Roaming

As opposed to the GSM, the UMTS offers encryption and integrity protection. It also provides for mutual authentication between the MS and the visited network. A UMTS user registers with a dedicated HN operated by its home provider. On registration, the HN allocates an IMSI and a long-term secret key *Ku* for the user, similarly to the GSM networks (see Chapter 6). The pair (*IMSI, Ku*) is stored on a Universal Subscriber Identity Module (USIM) and in HN's AuKuC. The USIM also contains five key-generation functions $f_1, ..., f_5$. During registration, a user furthermore registers for a certain geographical roaming region. Moreover, the UMTS standard makes use of the same mechanism of temporary identities to protect a UMTS user's IMSI (Niemi, 2003).

After authentication completion, the FN is assured that the HN authorizes the MS's roaming to it and the mobile user is guaranteed of the HN's authorization of FN's service provisioning to the MS. The roaming authentication and key-agreement protocols are implemented together and require HN's interaction on the first authentication. However, an unspecified number of subsequent authentications can then take place without HN's interaction. On the first authentication, the FN requests the appropriate authentication data related to the MS from the HN. Then the HN provides the FN with a number of authentication vectors, say ($RANDU_j$, $RESU_j$, CK_i, IK_i, $AUTN_i$), $j < p$ for a given p. Each authentication vector contains a random challenge $RANDU_j$, an authentication response $RESU_j$, an encryption key CK_j, an integrity-protection key IK_j, and an authentication token $AUTN_j$. To build these vectors, the functions $f_1, ..., f_5$ is used as shown in Chapter 6. The parameter *AUTN* is a sequence number that is integrity-protected with a message authentication code generated using the long-term secret key *KU*.

The choice of the implementation of the key generation functions $f_1, ..., f_5$ is left to the UMTS providers. Nevertheless, the UMTS specifies a sample set of functions called MILENAGE (3GPP, 2004). Therefore, the UMTS theoretically supports as many authentication and key-agreement protocols as there are UMTS providers. It is, however, expected that almost all the providers will use MILENAGE. As a consequence, the FN will not be able to generate the data-protection keys *IK* and *CK*. In the UMTS, the HN and the FN require a secure channel between each other to protect the transfer of authentication vectors. The encryption and integrity-protection mechanism are implemented in the MS and not on the USIM. The UMTS can accommodate up to 16 different encryption mechanisms. Presently, only two UMTS encryption algorithms are specified, namely UEA_0 (which does not provide encryption) and UEA_1, which is a stream cipher based on the block cipher KASUMI. Similarly, only one of 16 possible UMTS integrity algorithms, namely UIA_1, is specified in the standard. It is based on the block cipher KASUMI (see Chapter 6 for details).

One can notice that no mechanism to restrict the lifetime of an encryption key has been standardized in the GSM. On the opposite, the UMTS standard overcomes this weakness, in that the lifetime of the keys IK and CK is limited using a threshold set up by the HN based on how much data may be protected using the same key pair. The threshold is stored on the USIM. It is checked every time a radio resource connection (RRC) is released. If a new RRC is established and the threshold was reached during the last RRC, a new authentication and key agreement have to be initiated (Niemi, 2003).

Similar to GSM, the visited network and the MS, in UMTS, negotiate only the encryption and the integrity-protection mechanisms to use. The HN is not engaged in the negotiation. On connection establishment, the MS sends to the FN the list of all the encryption and integrity-protection mechanisms it supports, that is, the security capabilities. After the successful authentication of the MS, the FN's MSC decides which mechanisms the MS and the RNC are allowed to utilize. Then, it sends the list of allowed mechanism pairs to the RNC. The latter selects one of among those pairs and acknowledges its choice to the MS using a security mode command message. This message is integrity protected and repeats the security capability received from the MS on connection setup. On the receipt of this message, the MS verifies the integrity protection and compares the security capability with the one it has previously sent. Following this way, the MS can detect every operation on its security capabilities. Furthermore, the integrity protection of the security command message guarantees to the MS that the FN is authorized by the HN to provide service to the MS. This stands due to the fact that only the HN can generate a valid integrity-protection key for a given $RANDU$ value and present the authentication vectors only to authorized FNs. Finally, it is important to note that the MS is currently allowed to support the no encryption mechanism UEA_0. Consequently, neither the MS nor the HN can make compulsory encryption to be enabled. As an alternative, the visited network provider is made responsible of the choice for whether the encryption is used.

Similar to intra-provider roaming in GSM networks, the encryption and integrity-protection keys IK and CK agreed upon during the last authentication and key agreement are transferred from one RNC to the next if an idle mode MS roams within a UMTS network. If the encryption was disabled by the last serving RNC, it stays disabled after roaming to the new RNC. However, the UMTS security standard does not specify whether the same UEA is used after roaming to the new RNC or another UEA is negotiated. It does not also specify how a new UEA can be selected when the new RNC does not support the UEA that was used between the source RNC and the MS.

9.4.3 Roaming between GSM and UMTS

The UMTS standard allows for SIM-equipped users that have pre-registered to GSM services to roam to the UMTS network. It allows radio access networks

UTRAN and BSS to be simultaneously operated with a single backbone network and a single hierarchy of MSCs. To this end, the UMTS MSCs are made capable of controlling UTRANs and the GSM BSSs. On the contrary, only the BSSs can be connected to the original GSM MSCs. To highlight this difference, we will refer in the sequel: (a) to networks that allow BSSs to be connected to 3G MSCs as mixed-mode networks; (b) to networks that only operate UTRAN as UMTS networks; and (c) to networks that only operate GSM BSSs and 2G MSCs as GSM networks.

The advantage of allowing a SIM equipped user to roam is linked to the fact that a user can subscribe to UMTS and still keep his old SIM card. This makes easy the process of subscribing to UMTS and saves operators from handing out new smart cards. Vice-versa, a USIM-equipped user can roam to GSM networks and mixed-mode networks. In the transition phase, this type of roaming is crucial for user acceptance and satisfaction, as users at least obtain GSM services in areas that are not covered by UMTS.

Taking into account the different types of smart cards and the nature of serving radio access network, three intersystem roaming authentication scenarios can be distinguished. These scenarios are the SIM-equipped MS roaming to UMTS scenario, the USIM-equipped MS roaming to GSM scenario, and the SIM/USIM-Equipped MS Roaming to a Mixed-Mode Network scenario.

9.4.3.1 Scenario 1 (A SIM-Equipped MS Roaming to UMTS)

When a SIM-equipped MS connects to the UMTS network the node B involved in the connection forwards all GSM traffic transparently. The MSC of the FN requests a GSM-authentication vector $(RAND_G, Kc, RES_G)$ from the MS's HN. On receiving the vector, it sends $RAND_G$ to the MS (via Node B). Then the MS generates the authentication response *RESG* and the encryption key *Kc* from $RAND_G$ and the long-term secret key K_i. The MS sends *RESG* back to the visited MSC, which compares RES_G to RESG. The authentication is declared successful if the two values match. After a successful authentication, the MS and the MSC transform the established GSM key *Kc* into UMTS keys *CK* and *IK* using functions c_4 and c_5 as follows:

$$CK = c_4(Kc) = Kc \| Kc$$

$$IK = c_5(Kc) = (Kc_1 \oplus Kc_2) \| Kc \| (Kc_1 \oplus Kc_2),$$

where $\|$ is the concatenation operation and Kc_1 and Kc_2 are 32-bit numbers such that $Kc = Kc_1 \| Kc_2$. The visited 3G MSC transfers keys *IK* and *CK* to the RNC. The UMTS keys *CK* and *IK* are subsequently used to encrypt and protect the integrity of the communication between the RNC and the MS.

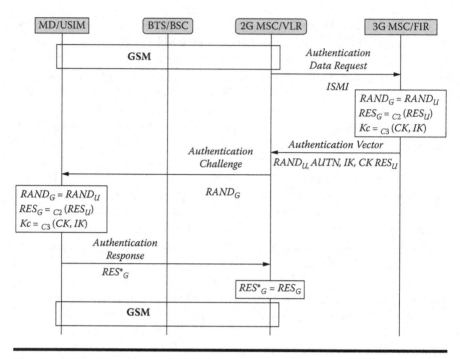

Figure 9.3 UMTS-GSM roaming process.

9.4.3.2 Scenario 2 (A USIM-Equipped MS Roaming to GSM)

A MS equipped with a USIM card connects to a GSM BTS that we assume to be connected to a 2G MSC. Since the 2G MSC does not support UMTS authentication, the MS can be authenticated by the 2G MSC only if the USIM provides a transformation of UMTS-authentication vectors into GSM-authentication vectors. The MS presents its identity to the visited network as depicted in Figure 9.3. The visited network requests a GSM-authentication vector from the HN. For this, the HN first generates a UMTS-authentication vector under the form (*RANDU, RESU, CK, IK, AUTN*) and then translates it to a GSM-authentication vector, say (*RANDG, RESG, Kc*), using the following identities and function c_2 and c_3:

$$RANDG = RANDU$$

$$RESG = c_2(RESU) = RESU_1 \oplus RESU_2 \oplus RESU_3 \oplus RESU_4$$

$$Kc = c_3(CK, IK) = CK_1 \oplus CK_2 \oplus IK_1 \oplus IK_2$$

where $CK = CK_1 \| CK_2$ and $IK = IK_1 \| IK_2$ are split into four segments CK_1, CK_2, IK_1, and IK_2 of 64 bits length and RES_G is generated from the 128-bit UMTS-authentication response *RESU* by splitting the UMTS response into four 32-bit values, such that

$$RESU = RES_{U_1} \| RES_{U_2} \| RES_{U_3} \| RES_{U_4}$$

The home network forwards the GSM-authentication vector to the FN. Then, the visited 2G MSC sends the authentication challenge to MS, which itself generates the GSM-authentication vector from the UMTS-authentication vector. GSM-encryption key *Kc* is then transferred to BTS and BTS sends the GSM cipher mode command to MS. Subsequently, MS and BTS use *Kc* and the negotiated encryption mechanisms to protect data traffic between them.

9.4.3.3 Scenario 3 (SIM/USIM-Equipped MS Roaming to a Mixed-Mode Network)

In this scenario, a USIM-equipped MS connects to a GSM base station (BTS), which is controlled by a 3G MSC. Since the 3G MSC and the USIM support the UMTS connection establishment, the GSM BTS forwards the related traffic transparently. After completing this, the MS and the visited 3G MSC convert the generated UMTS keys *IK* and *CK* into a GSM key *Kc*:

$$Kc = c_3(CK, IK)$$

The MS and the GSM BSS go on a GSM authentication. In particular, the BTS acknowledges its choice of the encryption mechanism to the MS in the GSM cipher mode command. The key *Kc* and the negotiated GSM-encryption mechanism are subsequently used to encrypt data traffic between the MS and the BTS.

Finally, in case where a SIM-equipped MS roams to a GSM BTS that is controlled by a 3G MSC, the 3G MSC acts in exactly the same way as on a regular GSM roaming.

9.5 Man-in-the-Middle Attack on UMTS

In this subsection, we present a man-in-the-middle attack that can be launched on UMTS using GSM. This attack can be launched because of a weakness observed in the inter-system roaming procedure between GSM and UMTS (Meyer, 2004). The attack allows an intruder to impersonate a valid GSM base station to a UMTS subscriber, despite the fact that UMTS authentication and key agreement are used and that mutual authentication between the MS and the network is enabled. The potential targets of this attack are the MSs supporting the UTRAN and the GSM air interface simultaneously.

To mount the man-in-the-middle attack, the attacker may have to impersonate a valid network to the mobile user. To do so, the attacker can mount the impersonation attack since the current GSM base stations do not support integrity protection. On the receipt of the authentication token *AUTN*, the MS first extracts the sequence number SQN and checks whether it is in the right range. If that is the case, the MS is guaranteed that *AUTN* was issued recently by the MS's HN. Otherwise, the MS knows that either *AUTN* is a copy of an old value or the synchronization of the sequence number has failed. The MS then checks the message authentication code (MAC) included in the token *AUTN*. A correct MAC indicates that the authentication token was originally generated by the HN. It is important to note that the correctness of the MAC and SQN do not provide assurance to the MS that the token was in fact received directly from the authorized network and not relayed by an attacker. It is only the combination of that property with an additional integrity protection of the control messages that is able to prevent network impersonation.

The security mode command message is integrity-protected and it includes the security mode capabilities that the MS originally announced to the network on radio connection establishment. By checking the correctness of the integrity protection, the MS is guaranteed that this message was generated by a network entity having the right integrity key. Furthermore, incorporating the security capabilities of the MS in the integrity-protected security mode command message prevents both the mobile device as well as the network from being taken into using no encryption or weak encryption by an attacker. In order to succeed in his attack, the attacker needs to forge the integrity protection on the security mode command message, by replacing the original (not integrity-protected) security capabilities with his own and making the valid network integrity-protect the security mode command with the new capabilities. Therefore, the attacker can claim on behalf of the victim MS to support only the mandatory encryption algorithms (instead of its original security capabilities). Then, the attacker can inform the MS of the choice of no (or weak) encryption by the network (using the security mode command).

If a UMTS subscriber roams to a GSM BTS controlled by a 3G MSC, as described in the previous subsection, the cipher mode command message is neither integrity-protected nor does it repeat the security capabilities previously announced by the MS on radio connection establishment. This limitation is due to the fact that GSM does not currently support integrity protection. Consequently, the message can be easily forged by the attacker.

In the following, we will detail a man-in-the-middle attack exploiting this shortcoming. We assume that the attacker knows the IMSI of the victim. This is a reasonable assumption, since the attacker can easily get hold of the IMSI, for example, by initiating an authentication procedure prior to the attack and disconnecting from the MS after receiving the IMSI. Note that by doing so, the attacker also learns the security capabilities of MS. The attacker can force a victim MS to connect to a BTS operated by himself instead of a UMTS Node B operated by a legitimate network provider. The attack works in two phases: During the first phase, the attacker acts on behalf of the victim MS to obtain a valid authentication token AUTN from any real network by executing the following four-step process:

1. During the connection setup, the attacker sends the security capabilities of the victim MS to the FN. The attacker sends the TMSI of the victim MS to the visited network.
   - If the current TMSI is unknown to the attacker, he sends a fake TMSI (which eventually cannot be resolved by the network).
   - Since only UEA_0 and UEA_1 and one integrity-protection mechanism UIA_1 are defined, and all three are mandatory, currently.
   - The security capabilities are currently always the same and could not be sent down even if no integrity protection was used.
2. If the network cannot resolve the TMSI, it sends an identity request to the attacker who replies with the IMSI of the victim.
3. The HN requests the authentication information for the victim MS from its HN. Then the HN provides the authentication information to the visited network. Finally, the network sends RANDU and AUTN to the attacker.
4. The attacker disconnects from the visited network.

Let us notice that none of the messages sent during the first phase is protected; the network cannot recognize the presence of the attacker. Consequently, the attacker obtains an authentication token that he can use during the second phase of the attack to impersonate a network to the victim MS.

In second phase, the attacker impersonates a valid GSM base station to the victim MS by executing the following steps:

1. The victim MS and the attacker establish a connection. Then the MS sends its security capabilities to the attacker.
2. The victim MS sends its TMSI (or IMSI) to the attacker. The attacker sends to the MS the authentication challenge *RANDU* and the authentication token *AUTN* he has received from the real network (in the first phase).
3. The victim MS successfully verifies the authentication token and replies with the authentication response.
4. The attacker decides to use no encryption (or weak encryption). Then he sends to the MS the GSM cipher mode command including the chosen encryption algorithm.

In order to allow for a regular use of the connection by the victim unit, the attacker has to establish a regular connection to a real network to forward the traffic it receives from the MS. In addition, the attacker has to overcome two difficulties to impersonate a valid network to the UMTS subscriber: First, he has to forward a valid authentication token to the victim MS. Then he has to ensure that no encryption is used after the authentication. The attacker solves the first problem by impersonating the victim MS to a real network to obtain a valid authentication token. This task is possible since none of the respective messages is encrypted or integrity-protected. Requesting no or weak encryption is more difficult, as the radio access network decides which encryption algorithm is used in the GSM and the UTRAN. The decision strongly depends on the security capabilities of the MS, which are sent to the network during connection setup. However, both radio access networks allow no encryption.

As a consequence of the non-support of integrity protection in GSM, the corresponding cipher mode command message is not integrity-protected, thus allowing an attacker to easily forge this message and misleading the victim MS into using either no encryption or a weak encryption algorithm. Thus, the attacker is able to eavesdrop on all mobile-initiated communication. The attack only works as long as the time gap between Phase 1 and Phase 2 is too small for any other authentication between the victim MS and another FN to take place. Otherwise, the sequence number within the authentication token might be out of range.

In UTRAN, the security mode command message that informs the MS about which algorithm to use is integrity-protected. The integrity protection alone, however, does not protect against network impersonation. As UEA_1 is mandatory and currently only UEA_0 and UEA_1 are defined, a go-down is currently not possible even if the integrity protection was forged. Furthermore, in the integrity-protected security mode command message, the network sends the security capabilities it received back to MS. Unless the attacker can forge the integrity checking, the MS would thus detect the attack. Thus, the attack does not work against mobile equipment that is capable of the UTRAN interface only. Yet, in the transition phase from GSM to UMTS, most users are expected to use equipment that is capable of both the UTRAN radio interface and GSM.

9.6 Inter-Provider Handover

A handover procedure generally allows a user to move from one radio cell to another without loss of the services he is currently using. During a handover procedure, the mobile device switches from a connection with one AP (the old) to another one (the new AP). From the network perspective, a handover procedure makes necessary the re-routing of incoming and outgoing data traffic through the new AP. In its simplest form, a handover takes place between two access points within the same network (as managed by a network provider). This type of handover is referred to as *intra-provider handover*. The handover procedures between different providers (referred to as *inter-provider handover*) aim at offering faultless services to mobile users in a larger coverage area than the one offered by the home provider. On an inter-provider handover, the source and the destination APs belong to different wireless access networks (Rohr, 2003; Wang, 2004).

The inter-provider handovers are expected to be regulated via handover agreements established between the network providers that handle the mobility management, the security-related issues, and the terms for accounting and billing, if needed. The handover agreements are required to play a similar role in the wireless network access provider environment as roaming agreements do nowadays in mobile communications networks. The mobile users then register for handover to a set of foreign providers as a component of the registration process with their home providers.

On inter-provider handover, a mobile device starts to utilize a service while it is connected to a certain network AP, referred to as the *anchor AP* of the handover procedure. The network to which the anchor AP belongs is called the *anchor network* (AN), while the provider that operates the anchor network is referred to as the *anchor provider*. A handover from an anchor AP to a destination AP is called the *first-order handover*. A first-order handover can take place within the network of one provider or between two different providers. A handover from the destination network, denoted by DST_1, of a first-order handover to another network, say DST_2, is referred to as the second-order handover. Iteratively, to handover from DST_{n-1} to DST_n is defined to be the nth-order handover. However, the subsequent handover differs from the first-order handover in the mobility management made from the network perspective. The re-routing of data traffic to the destination AP can be managed by the source network of the handover, the HN of the MS, or by the AN.

In the following subsection, we describe the major types of inter-provider handover procedures (first-order as well as subsequent). The description considers the handover procedures used in UMTS, GSM, and WLAN. The description will be made in a technology-independent way. More prominently, the description makes a differentiation between the first-order handover and higher-order handovers. By introducing the AN and the subsequently serving source and destination networks, the description model explicitly addresses subsequent handover and handover after roaming. Furthermore, the security challenge imposed by first-order and subsequent inter-provider handover is described. We also discuss state-of-the-art approaches to address this challenge. The first approach is based on an authentication and key agreement between the MS and DST_k during handover. The second approach generalizes the so-called pre-authentication method for the intra-provider handover adopted for the new security architecture 802.11i (Pack, 2002b). The third approach, called the security-context transfer (SCT) with key derivation, generalizes the solutions currently used to support intra-provider handover, e.g., in GSM, UMTS, and CDMA2000. In this approach, the master session key used after handover is derived from previously used master session keys (Soltwisch, 2004; Wang, 2004).

9.6.1 Basic Handover Phases

A handover procedure consists of three basic phases. In the first phase, a reason for handover is detected. A large range of reasons to initiate a handover procedure can be distinguished. The obvious reason occurs when the MS moves out of the transmission range of the currently serving network access point (AP). A second handover reason indicates that a handover is desirable but not required for a seamless use of service, because of a stronger signal while the currently serving AP is still available. A third reason to initiate handover occurs when there is a need for load balancing: this is the case when the MS is in the transmission range of more than one AP and the currently serving AP is overloaded. A fourth example for a handover reason is the proximity of the AP: this reason is somehow opposite to load balancing. A mobile device connects to the closest network access point to save battery power by reducing the necessary control power.

In the second phase, the handover algorithm takes the collected measurement data as input. It then outputs a decision on whether the handover should take place. Typically, the measurement data includes, but is not limited to, the current load on the serving AP, the current received signal strength, the signal to interference ratio, the bit-error rate, and the carrier interference ratio. On the detection of handover reason, a new NAP (called the destination AP) is selected. Usually, the choice of the destination AP depends on the signal characteristics of the candidate APs mentioned above and the availability of resources to serve the MS after handover in these APs.

Finally in the third phase (called the execution phase), the MS disassociates from its serving AP, which is the source AP of the handover, and connects to a new AP, the destination AP. The execution phase also allows the management of mobility (on the network side) to provide for the re-routing of incoming and outgoing data traffic over the new AP.

The handover procedures can be classified into three categories, the mobile-initiated, the network-initiated, and the mobile-assisted handover procedures. In a mobile-initiated handover procedure, the MS detects a handover reason, while in the network-initiated case, the currently serving network detects the reasons. In a mobile-assisted handover procedure, the MS provides the network with the measurement data related to the reception level of signals from the surrounding APs. The network processes the measurement data and determines the handover reasons. Another classification can be considered depending on who selects the destination network and initiates the execution of the handover. Thus, the handover procedures can be classified into mobile-controlled and network-controlled handovers (Zdarsky, 2004).

A third classification of the handover procedures considers that they can be hard and soft procedures. In a hard handover procedure, the MS can only be connected to one AP at a time. In the execution phase, the MS first disassociates from the source AP before it associates with the destination AP. Consequently, a discontinuity of the incoming and outgoing data traffic occurs. To provide faultless use of service, the hard handover procedures have to be fast, such that the interruption does not lead to a disruption of the services used by the MS. On the contrary, a soft handover procedure allows the MS to be associated with several APs at the same time. In the execution phase of a soft handover, the MS is first connected to the source AP only, then connected to both the source and the destination AP for some time before it disconnects from the source AP. The soft handover procedures have the advantage that data traffic arriving to the MS or departing from it can be sent to and received from both the source and the destination AP, as long as the MS is connected to the two APs. Consequently, the soft handover procedures can easily support uninterrupted service use.

Hard handover procedures are used in mobile networks such as the GSM and IEEE 802.11. On the other hand, the soft handover procedures are used in UMTS and CDMA2000.

9.6.2 Hard and Soft Networked-Initiated Handover

In this subsection, we discuss the major types of hard and soft handover, mobile-assisted handover, and network-initiated handover. We also present examples of hard and soft mobile-initiated, mobile-controlled handover procedures. For the sake of simplicity, we refer to mobile-assisted, network-controlled handover procedures simply as network-initiated handover procedures and to mobile-initiated, mobile-controlled procedures simply as mobile-initiated handover procedures.

9.6.2.1 First-Order Handover with the HN as Anchor

Figure 9.4 describes a network-initiated first-order handover procedure of the MS from its HN to some destination, say DST. In a network-initiated handover, the network collects measurement data related to the quality of the link layer connection between the MS and the currently serving AP. While the network may measure parts of the data itself, the MS can also help in this activity (in the case of mobile-assisted handover). In this situation, the MS measures reception parameters of the surrounding network access points, including the currently serving one. It sends measurement reports to the HN. Then, the HN processes these reports. The measurement data may also include some estimate on the current location of the mobile device.

Based on the collected measurement data, the HN detects a handover reason. A handover algorithm takes the collected measurement data as an input and outputs whether or not a handover should take place. If the network component collecting the measurement data has knowledge of the network topologies of its own and the surrounding networks, then knowing the location of the mobile device means knowing whether or not the mobile device is in the transmission range of other network access points of the home network or foreign networks.

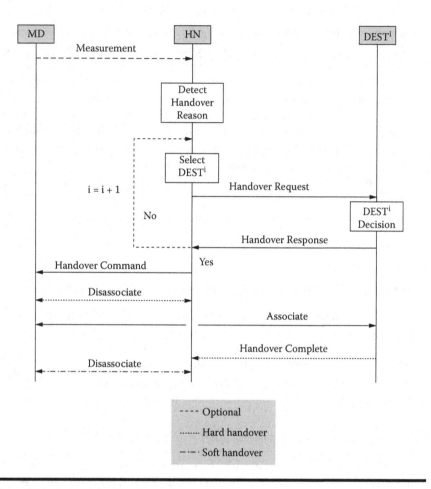

Figure 9.4   Inter-provider handover scenario.

The HN subsequently uses the collected measurement data to generate an ordered list L,

$$L = \{DST_1, ..., DST_n\},$$

of candidate destination networks for handover. This list also specifies the network access points in these candidate destination networks. We do not here make any assumptions about the algorithm used by the HN to determine the list of candidate networks here, but assume its availability.

As long as the MS receives the signal of an AP belonging to the HN with sufficient quality and with free capacity, the HN itself will be DST_1 and the HN initiates an intra-provider handover. However, in this subsection we focus on the inter-provider handover. For this, we assume that all destination networks in L are foreign networks. The HN chooses the candidate destination network DST_1 in the list L and sends a handover request to it. This request includes the identities of the MS, its HN and DST_1. It may also include a list of allowed subsequent handover destinations and other security-related information. On receiving a handover request, DST_1 decides whether or not to accept it and answers accordingly with a positive or negative handover response. If DST_1's handover response is positive, the HN sends a handover command message to the mobile device commanding a handover to DST_1. If the handover response of DST_1 is negative, the HN keeps trying the candidate destination networks until it receives a positive answer from the candidate network, say DST_i, in the list L. Then, the HN selects DST_i as the next destination network and sends a handover command to the MS including DST_i's identity. If all the destination networks in L send a negative response, a handover is not possible and the HN has to drop the connection.

In the case of a hard handover procedure, the MS disassociates from the HN as soon as it receives a handover command to a destination network, DST, and the HN typically keeps the resources allocated for the MS until it receives a handover-complete message from the destination network, indicating a successful handover. As soon as the MS successfully associates with DST, the latter sends a handover-complete message to the HN. If the association fails, the MS tries to re-associate with the HN. In the case of a soft handover procedure, the MS associates with DST upon reception of a handover command. The MS disassociates from the HN if and only if it has successfully associated with DST.

9.6.2.2 k^{th}-Order Handover with the HN as Anchor

Assume that the mobile device has established a connection with its HN and started using a service of a service provider and that it has subsequently been handed over from HN (= SRC_1) to a destination network, DST_1, by applying a first-order handover procedure, as described above. After k subsequent handover operations from SRC_j ($2 < j < k$) to DST_j, the MS is connected to the destination network DST_k of the kth-order handover. DST_k is the source network SRC_{k+1} of the $(k+1)^{th}$-order handover. Notice that new authentication between the HN and the MS resets the handover chain to HN (= SRC_1).

Two control types for subsequent handover procedures with HN as the anchor can be distinguished: the HN-controlled handover and the SRC-controlled handover. These controls reflect different types of handover agreements between the wireless access networks. In an HN-controlled k^{th}-order handover, the source network SRC_k determines that a handover reason has occurred and informs the HN, which selects the candidate destination network and initiates the actual handover. The MS accepts the handover commands originated from its HN. A handover to DST_k can take place if the HN and DST_k have a handover agreement. It is the role of the HN to guarantee to DST_k that the MS is authorized to be handed over to DST_k.

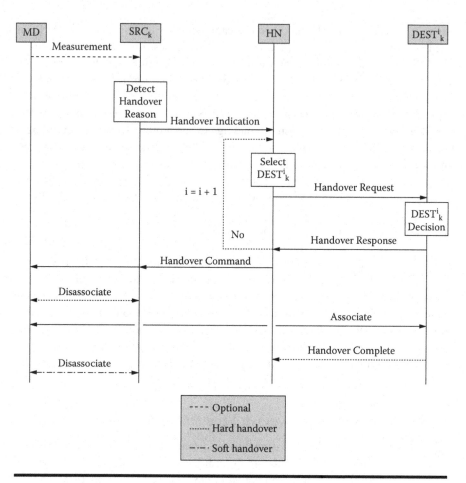

Figure 9.5   HN-controlled subsequent handover.

Figure 9.5 depicts the HN-controlled k^{th}-order handover procedure. The source network SRC_k of a kth-order handover sends a handover indication to the HN as soon as it detects a handover reason. SRC_k includes the necessary measurement data in the indication to give the HN the power to process it. The HN proceeds as in the first-order handover case with the generation of a list of candidate destination networks and sends handover requests to them. On receiving a positive handover response from one of them, the HN sends a handover command to SRC_k including the identity of the selected DST_k, SRC_k forwards the handover command to the MS.

In the SRC-controlled k^{th}-order handover, the source network SRC_k detects the handover reason, selects the candidate destination network, and initiates and authorizes the actual handover. A handover of the MS from SRC_k to DST_k takes place if SRC_k and DST_k have previously accepted a handover agreement. The MS's

HN is involved in the kth-order handover procedure only if it is the source or the destination network of the procedure (i.e., HN = SRC_k or HN = DST_k). As a result, the HN should delegate the control of the second-order handover to DST_1 (= SRC_2), and for all j, $2 < j < k - 1$, SRC_j delegates control of a $(j+1)^{th}$-order handover to DST_j (= SRC_{j+1}). The MS is assured of SRC_k's authorization of the handover as soon as it receives a handover command message from SRC_k. Similarly, SRC_k's authorization of the handover is given to DST_k in the form of a handover request. For the MS, a SRC-controlled handover implies a transitive trust in the network providers: the MS trusts HN's authorization by means of the initial authentication between the MS and the HN. Consequently, an SRC-controlled subsequent handover procedure can be described by replacing the HN with SRC_k in the procedure depicted by Figure 9.5. In the SRC-controlled case, SRC_k determines the list of candidate destination networks itself. SRC_k sends a handover request for MS to the candidate destination networks and commands MS to associate with DST_k upon receiving a positive handover response from a destination network.

9.6.2.3 Handover with the FN as Anchor

On inter-provider handover after roaming, the MS roams to a FN and starts using a service connected to the FN. If a handover reason is detected by the FN, a first-order inter-provider handover with the FN as anchor is initiated. The selection of the destination network and the initiation of the first-order handover itself can be conducted by the FN or by the HN. Five situations can be distinguished:

- *HN-Controlled Handover.* In this case, the HN selects the destination network, and initiates and authorizes the actual handover. An HN-controlled first-order handover procedure with the FN as anchor can be depicted by Figure 9.5, provided that SRC_k is replaced with the FN.
- *FN-Controlled Handover.* In this case, the FN selects the destination network, and initiates and authorizes the actual first-order handover. An FN-controlled first-order handover procedure with FN as anchor can be depicted by Figure 9.4, where HN is replaced by the FN.
- *FN-Controlled Subsequent Handover.* Subsequent handover with FN as anchor generates in a chain of subsequently serving networks:

$$FN = AN = SRC_1,$$

and

$$DST_j = SRC_{j+1}, 1 < j < k.$$

A full (roaming) authentication between the MS and the FN or between the MS and another FN or the HN resets the handover chain. Consequent handover with FN as anchor can be SRC-controlled, HN-controlled, or FN-controlled. Subsequent FN-controlled procedures with FN as anchor can be described by replacing HN with FN in Figure 9.5. In this case, the FN selects each subsequent destination network, and initiates and authorizes the subsequent handover.

■ *HN-Controlled Subsequent Handover.* Subsequent HN-controlled handovers with FN as anchor are the same as subsequent handover with HN as anchor and can be described by the same procedure. In this case, the source network SRC_1 of a first-order handover is the FN. The HN selects the destination network of a subsequent handover, and initiates and authorizes the actual handover.

■ *SRC-Controlled Subsequent Handover.* Subsequent SRC-controlled handover with FN as anchor are executed in exactly the same way as subsequent SRC-controlled handover.

9.6.3 Hard and Soft Mobile-Initiated Handover

When it comes to mobile-initiated handover, two scenarios can be distinguished based on who is the anchor, the HN or the FN.

9.6.3.1 HN Anchor-Based Handover

In the first-order mobile-initiated handover, the MS detects a handover reason and selects the destination network. Two different approaches can be used to notify the HN of the upcoming handover. In the first, the MS sends a notification message to the HN before it associates with the destination network or DST_1; while in second, it sends a notification message to the HN after it associates with the destination. In both cases, the handover procedure is initiated when MS detects a handover reason. The MS measures the signal and other quality of signal indicating parameters related to its serving NAP and the surrounding available APs. Consequently, the MS generates an ordered list of candidate destination networks $L = \{DST_1, ..., DST_n\}$. The HN can support the MS in generating the list L. For example, it may send a list of candidate destination networks to the MS at any time prior to the handover reason detection.

In a case where the MS is in charge of notifying the HN of the upcoming handover, it sends a handover indication message to the HN including the selected destination's identity, DST, immediately after selecting it from the list of L. The HN decides whether or not to allow the handover and indicates its authorization to DST in the form of a handover-indication message that includes its identity. When a hard handover procedure is executed, the MS then disassociates from HN and tries to associate with DST. After successful association, the MS sends a handover request to DST, which answers back the MS with a handover-response message indicating its decision. On the other hand, the MS disassociates from the HN only

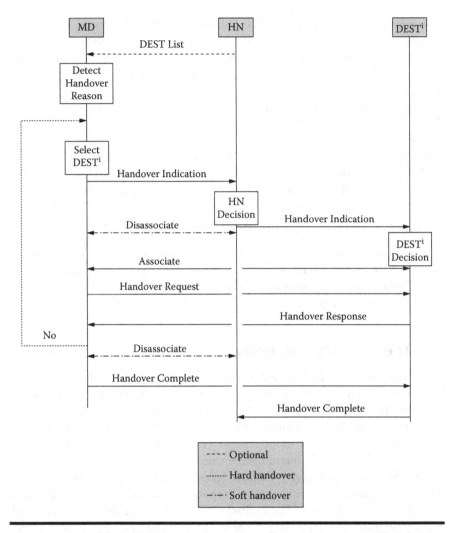

Figure 9.6 Mobile-initiated, HN as anchor, first-order, HN notified by MS handover.

after receiving a positive handover response from DST, in case of a soft handover procedure. Figure 9.6 depicts a procedure for mobile-initiated first-order handover, where the HN, notified by the MS, serves as an anchor.

In the case where the destination network is in charge of notifying the HN of the upcoming handover, the MS tries to associate with the first destination network of its list L of candidates. Then, it sends a handover-indication message to DST. If DST does not want to accept the handover, it immediately transmits back a negative handover response to MS; otherwise, it forwards MS's handover indication to the HN. In that situation, the HN replies positively or negatively in its handover-request message to DST. Subsequently, the latter sends its handover response to the MS. In a hard handover procedure, the MS disassociates from the HN before associating with DST, whereas in a soft handover procedure, the MS disassociates from the HN only after receiving a positive handover response from DST.

On the other hand, the subsequent handover with the HN as anchor network can be controlled either by the HN, in which case the HN is engaged in every handover procedure, or it can be SRC-controlled. In the latter case, a subsequent handover procedure can be illustrated by replacing the role of the HN with the one of SRC_k. In a HN-controlled handover, DST_k sends an additional handover indication to the HN. Then, the HN replies with a positive or negative handover request.

9.6.3.2 FN Anchor-Based Handover

As in the network-initiated case, a subsequent handover with FN as anchor can either be HN-controlled, FN-controlled, or SRC-controlled. The source network SRC_k forwards MS's handover notification to HN or DST_k notifies the HN after the MS has associated to DST_k.

9.7 The Security Solutions

In the following, we discuss three different security solutions for authentication and key agreement. The first one is based on a new run of an authentication and key agreement between the MS and DST_k during handover. It is typically discussed to motivate the need for new solutions. The second solution generalizes the so-called pre-authentication method introduced in (Pack, 2002b) for the intra-provider handover in WLAN to the inter-provider case.

The third solution, Security-Context Transfer (SCT) with key derivation, generalizes the solutions currently used to support intra-provider handover in GSM, UMTS, CDMA2000, and WLAN (Pack, 2002a; Mishra, 2004; Zeadally, 2004) to the inter-provider case. Although the use of SCT with key derivation has previously been suggested in the inter-provider context in (Soltwisch, 2004; Wang, 2004), they do not explicitly address the subsequent handover or distinguish between handover with HN and FN as anchor. Consequently, none of them identifies and discusses the differences between HN-controlled, AN-controlled, and SRC-controlled subsequent handover.

9.7.1 Full Authentication between MS and DST_k

With a full authentication, the MS and DST_k negotiate a security suite, authenticate each other, and agree on a master session key in a way similar to the one achieved in roaming before the mobility management redirects data traffic to the MS using network access point of DST_k (that we denote by AP_{Dk}). Figure 9.7 depicts the

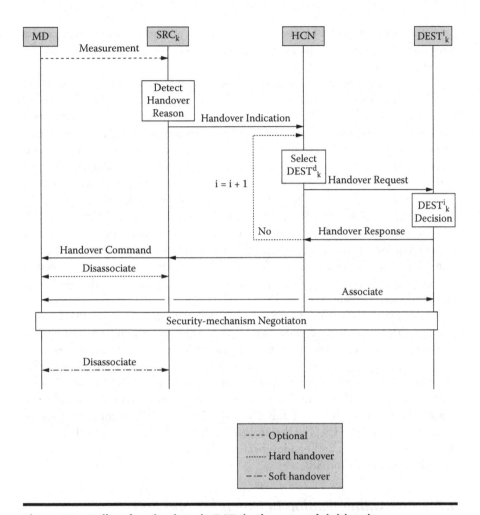

Figure 9.7   Full authentication via DST$_k$ in the network-initiated.

actions this solution applies when integrated into a network-initiated handover procedure. The major actions in this figure are as follows. On association during handover, the MS and DST$_k$ negotiate the security suite to use, and then mutually authenticate each other using the negotiated authentication protocol *NAp*, agree on a master session key *MSk* by means of the key agreement *KAg*, establish fresh data-protection keys using protocol *DPk*, and consequently use the encryption and integrity-protection mechanisms *EPm* and *IPm* on which they agreed to protect the data and control the traffic exchanged between them. Note that the HN, in this solution, is responsible for authorizing the handover as part of the initial authentication. However, the selection of the next destination network DST$_k$ of a kth-order handover and the initiation of the actual handover can be controlled either by the HN, the AN, or SRC$_k$.

On a mobile-initiated handover, the setup of a new secure connection can be integrated into the association between MS and DST_k the same way as detailed for the network-initiated case.

In the case where the hard handover procedures are supported, the setup of the new secure connection via AP_{Dk} has to be achieved after the MS has disassociated from SRC_k. Therefore, the time required to set up a new secure connection generates a service disruption. Consequently, the aforementioned solution is commonly known to be too inefficient to support seamless use of real-time services under hard handover procedures. On the other hand, when soft handover procedures are supported, the MS can be associated with the AP_{DSk} AP of SRC_k (that we denote by AP_{Sk}) and the AP_{Dk} at the same time. Consequently, on soft handover, the establishment of a new secure connection can take place via the AP_{Dk} before the MS disassociates from the AP_{Sk} such that it does not add to the disruption time. Even in the soft handover case, a secure connection has to be established as long as the MS is in the intersection of the cells of AP_{Dk} and AP_{Sk}. Consequently, the intersection between the cells of AP_{Dk} and AP_{Sk} has to be sufficiently large and the MS has to move sufficiently slowly.

Each time the setup of a new secure connection in the above described way is possible, the connection between the MS and DST_k after handover completion is protected in the same way as on roaming to DST_k. The security of the new connection does not depend on the security of the connection before handover. The HN has as much control over each handover instance as the initial authentication and key agreement allow for. However, new security threats occur induced by the handover procedures that use a full authentication over AP_{Dk} compared to roaming and accessing the HN. The threats include denial of service attacks that exploit the handover messages and handover-specific behavior of the MS, DST, handover controlling network, and SRC.

9.7.2 Pre-Authentication between MS and DST_k

An approach that uses a full authentication and key agreement aims at executing these protocols between the MS and DST_k via AP_{Sk}. In this case, SRC_k forwards the traffic related to the security suite negotiation, authentication, and key agreement between MS and DST_k, prior to the handover execution. A pre-authentication like this one can be operated just before handover controlling network (HCN) sends a handover request to DST_k. However, it can also take place pro-actively before a handover reason is detected.

A pro-active pre-authentication has the advantage that it does not add delays to the overall latency of the handover. To allow for pro-active pre-authentication, upcoming handover events have to be predicted and candidate destination networks have to be determined in advance. To allow for pro-active pre-authentication, the upcoming handover events have to be predicted and candidate destination networks have to be determined in advance. The way the pre-establishment of a new security context can be integrated in a network-initiated handover procedure. Note that in this security solution, the HN authorizes the handover by means of the initial authentication between the MS and DST_k. However, the selection of DST_k and the initiation of the actual handover may be controlled by the HN or DST_k.

In the mobile-initiated case, pre-authentication-based establishment of a new security context can be integrated into the procedure in the same way as on network-initiated handover, and then does not need to be detailed. The pre-authentication-based solution is limited by the following factors:

- SRC_k has to forward all traffic related to security-suite negotiation and authentication and key agreement between DST_k and the MS. The traffic has to be encapsulated in the messages exchanged between SRC_k and the MS. This can be difficult if the authentication and key agreement protocols are implemented as part of the MAC layer. The authentication and key-agreement protocols may then have to be adapted to support pre-authentication via SRC_k.
- The execution of the authentication and key-agreement protocols has to be made possible without causing the currently used data-protection keys to be replaced by new keys immediately. This will guarantee that the keys used before and after handover are different. Additionally, in case the pre-authentication is to be achieved just in time, the cell intersection has to be large enough to allow for a pre-authentication before the MS leaves the radio cell. The required size of the intersection again depends on the rapidity of the MS to react and the path from SRC_1 to the final DST_k.
- In case the pre-authentication is to be achieved just in time, the cell intersection has to be large enough to allow for a pre-authentication before the MS leaves the cell.

Furthermore, carrying out the pre-authentication, before a handover reason is detected, may result in many unnecessary authentications that put an unnecessary load on both networks, as well as on MS. This problem has recently been addressed by integrating mechanisms that more precisely predict the AP for the next handover using movement patterns of users.

9.8 Attacks against Inter-Provider Handover

By launching an attack against a handover, the attacker can aim at (a) violating the confidentiality of the air interface between the MS and a wireless AP, (b) violating the integrity protection between the MS and a wireless AP, (c) mounting denial of service attacks against MSs or networks, and (d) conducting impersonation attacks against mobile networks on behalf of a victim MS. We particularly discuss different interesting attacks that a malicious user can attempt to achieve the aforementioned objectives. Specifically, we will discuss how an attacker can take advantage of some weaknesses in the message exchange needed by the handover execution to launch impersonation and DoS attacks against mobile users or/and mobile networks. Particularly, the weaknesses observed in the detection of handover reasons, the selection of candidate destination networks, the key relations, and the initial security suite will be considered.

To analyze the different attacks, two concepts can be utilized to specify the attacks. The first concept is the so-called attack tree, which provides a formal method of specification of the system security based on varying attacks starting from root attack scenario down to the initial steps an attacker has to achieve in order to launch an attack in the root scenario (Schneier, 1999). The second concept identifies recurring modules the attacker can combine to mount sophisticated attacks. Once these modules are identified, the protection mechanisms can be easily evaluated. It is by far easier to execute a defense strategy in three steps: First, construct attack trees. Second, identify recurring attack modules. Finally, design protection mechanisms that protect against the identified attack. This allows using the advantages of each specification method.

The following attacks against a system are represented in a tree structure as follows: The root of the attack tree describes the goal the attacker wants to achieve. The children of a node represent different ways to achieve the parent node. Each node in the tree thus becomes a subgoal for an attacker that wants to achieve the goal at the root of the tree. The tree has two types of nodes, the AND and OR nodes. While the OR nodes represent different alternatives to achieve the parent node (attack), the AND nodes represent steps that have to be taken to achieve the parent node (Schneier, 1999).

For the sake of simplicity and the lack of space we only develop in the following a few sets of root attack scenarios we consider for an inter-provider first-order handover with HN as anchor independent of the initiation type of the handover procedure, despite the difference of the attack trees representing the root scenario for the network-initiated and the mobile-initiated handover procedures. Then we consider some developments of the first attack to show how the subgoals can be addressed.

In the following four scenarios of attacks, the objective of the attacker is to violate the confidentiality of data or control traffic:

1. *ATC1*: The attacker recovers the plaintext of some encrypted data or control traffic he has intercepted on the air interface between the MS and the HN after a handover of the MS from the HN to an authorized destination has taken place.

   In the attack ATC1, the attacker intercepts and records encrypted data or control traffic on the air interface between MS and HN before a handover takes place. Some time later, the MS is handed over from the HN to an authorized destination network, say DST. The attacker tries to exploit this handover procedure to gain access to the plaintext of previously recorded traffic.

2. *ATC2*: The attacker recovers the plaintext of data or control traffic exchanged between the MS and an authorized destination, DST, after a handover of the MS from the HN to destination takes place. The scenario ATC2 is restricted to hard handover procedures. The MS is about to be handed over from the HN to some destination network DST. The HN sends the handover command to the MS, which disassociates from the HN and tries to associate to DST. An attacker tries to prevent the handover of the MS by simulating a handover failure, impersonates the HN to the MS when the latter tries to re-associate with the HN, and then tries to recover the plaintext of data or control traffic sent by the MS.

3. *ATC3*: The attacker obtains access to the plaintext of data or control traffic sent by the MS by impersonating the network access point of the DST network (NAP_D) on an actual handover of the MS made from the HN to DST. In ATC3, the MS is handed over from the HN to an authorized destination network DST. An attacker intercepts data or control traffic on the air interface between the MS and DST after handover. He tries to exploit the handover procedure to recover the plaintext of the intercepted encrypted traffic after handover.

4. *ATC4*: The attacker gains access to the plaintext of data or control traffic sent by the MS by simulating a handover from the HN to destination and impersonating AP_{D1} to the MS.

In the fourth scenario, the attacker simulates a handover to a fake AP_{D1}. The next two scenarios allow the attacker to perform by an attack that aims at violating the integrity of data or control traffic. The root attack scenario considered here is the following:

Root: *When the MS is handed over from HN to an authorized DST, an attacker tries to modify the data or control traffic exchanged between the MS and DST.*

1. *ATI-1*: The attacker modifies data or control traffic between the MS and an authorized DST after or during handover.

   Note that the attacker may be able to decrypt data or signaling traffic exchanged between the MS and the HN before handover due to the knowledge he gained during or after the handover. However, the attacker cannot use any knowledge gained during or after handover that allows him to modify the traffic exchanged between the HN and the MS before handover. In addition, the attacker can try to obtain access to the HN or DST and to use services on behalf of a victim MS.

2. *ATI-2*: An attacker tries to gain access to HN's network on behalf of a victim MS exploiting an actual handover procedure. He can also gain access to DST's network on behalf of a victim the MS on an actual handover of MS from HN to DST.

On the other hand, an attacker can easily jam MSs or network APs. It can send out false requests and other traffic to keep the APs and other network components busy. Three DoS scenarios can be described targeting the MS, the HN, and DST while they participate in a first-order handover procedure with HN as anchor. The attacker can (a) prevent the MS from continuously using HN's or DST's service by interfering with the handover procedure; (b) use the handover procedure to block resources (like a channel allocated for MS) in the HN; and (c) use the handover procedure to overload an authorized DST.

Now let us consider how we build an attack tree for the first of the identified root attack scenarios, ATC-1, discussed for a network-initiated first-order handover procedure. The purpose of this example tree is to show how the attack modules can be extracted and described. It is important to note that the leaves in our attack trees themselves are attack modules against the network components or protection mechanisms used in between the MS and the HN before handover or between the MS and DST after handover. Whether or not an attacker can launch, the attack modules described as the leaves of the attack trees depicted by Figure 9.8 have to be analyzed for each wireless access technology separately. That is why we omit to go further in the specification of the leaves.

Figure 9.8 reads from the top to the bottom as follows: The data or control traffic is encrypted with the encryption mechanism Em_0 and the encryption key EK0. In addition, we can state the following:

- To recover the plaintext of encrypted traffic, the attacker has either to find some means to recover the encryption key EK0 (node $n_{1,1}$) or to be able to recover the plaintext without the encryption key. The second alternative is referred to as partially break em0 (node $n_{1,2}$).
- The attacker can recover EK0 with or without the knowledge of K0 (nodes $n_{2,1}$ and $n_{2,2}$).
- To recover EK0 with any knowledge of K0, the attacker must be able to recover K0 and to reconstruct EK0 from K0 by means of reconstructing the key-establishment process ke0. An attacker can reconstruct ke0 if he can recover EK and IK from K (node $n_{3,1}$); otherwise, he has to recover K0 without the knowledge of EK0 and IK0 (node $n_{3,2}$).
- The attacker can recover EK0 without any knowledge of K0 by totally breaking em0 (node $n_{3,3}$) or by compromising EIPE-HN (node $n_{3,4}$).
- The attacker can recover K0 without knowledge of EK0 and IK0 in two ways: either with K1 or without knowledge of K1 (nodes $n_{4,2}$ and $n_{4,1}$).
- How the attacker can reconstruct ke0 depends on whether the key-establishment process is static or dynamic (nodes $n_{4,3}$ and $n_{4,4}$). In the former case, reconstructing ke0 is consequently equivalent to knowing the long-term key K0. In the dynamic case, the attacker has to recover the ke0 traffic between the MS and HN's AP.

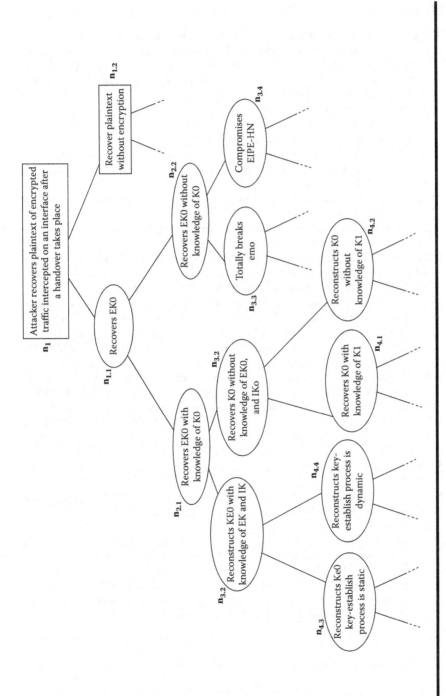

Figure 9.8 Partial description of attack scenario ATC-1.

References

3GPP Technical Specification. 3GPP TS 35.206, V6.0.0, Release 6, Third Generation Partnership Project; Technical specifications group services and system aspects; 3G Security; Specification of the MILENAGE algorithm set: An example algorithm set of the authentication and key generation functions $f_1, f_1^*, f_2, f_3, f_4, f_5, f_5^*$, document 2: Algorithm specification, Dec. 2004.

K. Bayarou, M. Enzmann, E. Giessler, M. Haisch, B. Hunter, M. Ilyas, S. Rohr, and M. Schneider, Towards Certificate-Based Authentication for Future Mobile Communications, *Wireless Personal Communications*, 29, 2004.

H. Chen, M. Zivkovic, and D.-J. Plas, Transparent End-User Authentication Across Heterogeneous Wireless Networks, in *Proceedings of the IEEE Vehicular Technology Conference (VTC'03-Spring)*, 2003.

U. Meyer, Secure Roaming and Handover Procedures in Wireless Access Networks, Technical University at Darmstadt, 2005.

U. Meyer and S. Wetzel, A Man-in-the-Middle Attack on UMTS, in *Proc. of the ACM Workshop on Wireless Security (WiSe04)*, October 2004.

A. Mishra, M. H. Shin, N. L. Petroni, Jr., T. C. Clancy, and W. A. Arbaugh, Proactive Key Distribution Using Neighbor Graphs, *IEEE Wireless Communications*, February 2004.

V. Niemi and K. Nyberg, *UMTS Security*. John Wiley & Sons, 2003.

S. Pack and Y. Choi, Pre-Authenticated Fast Handoff in a Public Wireless LAN Based on IEEE 802.1X Model, in *IFIP TC6 Personal Wireless Communications*, October 2002a.

S. Pack and Y. Choi, Fast Inter-AP Handover Using Predictive-Authentication Scheme in a Public Wireless LAN, in *Proc. of Networks 2002 (Joint ICN'02 and ICWLHN'02)*, 2002b.

S. Rohr, K. M. Bayarou, C. Eckert, A. R. Prasad, P. Schoo, and H. Wang, Feasible and Meaningful Combinations of Access and Network Technologies for Future Mobile Communications, in *Proc. of the Wireless World Research Forum (WWRF'03)*, 2003.

L. Salgarelli, M. Buddhikot, J. Garay, S. Patel, and S. Miller, Efficient Authentication and Key Distribution in Wireless IP Networks, *IEEE Wireless Communications Magazine*, 10(6), 2003.

B. Schneier, Attack Trees: Modeling Security Threats, *Dr. Dobb's Journal* 24(12): 21–29, 1999.

R. Soltwisch, X. Fu, and D. Hogrefe, A Method for Authentication and Key Exchange for Seamless Interdomain Handover, in *Proc. of the IEEE Int. Conf. on Networks (ICON 2004)*, 2004.

H. Wang and A. R. Prasad, Fast Authentication for Inter-Domain Handover, in *Proc. of the Int. Conference on Telecommunications (ICT'04)*, LNCS Vol. 3124, 2004.

F. A. Zdarsky and J. B. Schmitt, Handover in Mobile Communication Networks: Who Is in Control Anyway? In *Proc. of the IEEE EUROMICRO Conference*, 2004.

S. Zeadally and S. Naduri, Fast Secure Handover in Public Wireless LANs, in *Proc. of the IEEE Vehicular Technology Conference (VTC'04-Spring)*, 2004.

Chapter 10

Securing Mobile Services

10.1 Introduction

Web services are nowadays emerging as a major technology for achieving automated interactions between distributed entities using heterogeneous applications. Various technologies are behind this achievement including the Web Service Definition Language (WSDL, 2000), the Universal Description, Discovery, and Integration (UDDI, 2002), and the Simple Object Access Protocol (SOAP, 2000). Enterprises that want to decide to develop Web services (or e-service) need to address first the issue of their legacy information systems, which are defined as applications of critical value (to their business) that have been deployed for reasonable periods of time. Two approaches can be used to this end: (a) the integration and (b) the gradual migration of the legacy application. The outcome of the integration is a composite system where the old applications are not replaced, while the migration produces a new system that takes the place of the old one, possibly by using intermediate and partial integrations.

An e-service is specified by a Service Description (SD), which typically describes the service in terms of the device type, the service type, and some attributes (such as the location of user). An e-service is an application component that an organization provides in order to be assembled and re-used in a distributed manner, on an Internet-based network. The term *component* used here refers to a set of object oriented classes assembled together to be deployed as a single software unit, with explicit and well-specified interfaces for the services it provides and for the services it expects from other components. An e-service can be used in a portal, in an e-commerce application, to offer services in a public context (e.g., a service allowing

367

citizens to access and to manage information about their retirement plans) or more generally in an e-application. An e-application is a distributed application that integrates, in a cooperative way, the e-services offered by some entities.

The typical development of an e-service, for an enterprise, goes through four steps. The first step develops a simple Web site, in which the only e-service offered is advertising using Web pages. The second step adds to the Web site the opportunity of exploiting simple e-service components, such as the retrieval of information from back-end databases of the enterprise and remote data entry of some information (about customers, for example). The third step integrates the legacy applications of the organization with the Web front-end, allowing customers to effectively use the e-services. Finally, the fourth step updates the enterprise's network and allows the heterogeneous information systems of the enterprise to cooperate in order to offer services to customers.

It is worthy to notice that, in the first two steps, the enterprise is completely independent of the other organizations and available Web services; the only cooperation with other organizations consists possibly in the use of links allowing access to the enterprise sites. The technologies used for integration are mainly the same in the last steps; however, a major difference can be noticed: In the third stage the integration is mainly intra-organization, while it is completely inter-organizations in the fourth stage. The technology used for the integration and inter-operation of the back-end systems can be based on a set of middleware technologies including the integration of mediators, access to legacy applications, application servers, and transaction monitors. Therefore, one can conclude that an e-service is developed independently from computing paradigms and specific platforms. It should be easily composable with other e-services, and its integration with other e-services should not require the development of complex actions.

It happens that a user needs to access an e-service using a mobile device or that he desires to delay the actions related to an ongoing access to an e-service because of the lack of resources available on his mobile terminal. Therefore, it becomes essential that the mobility of the user, the management of applications and resources on the user's terminal, and the overall guarantee of security of the operations performed to provide e-service need to be taken care of to extend the use and offer of e-services to the mobile entities. These issues have made the emergence of a new concept, the mobile service (or m-service).

Two definitions can be provided for an m-service (Maamar, 2002). The first definition allows a mobile user to trigger remotely, transact with, and complete a session with an e-service from a mobile device. In that case, the e-service is considered as an m-service, since it allows remote access from mobile terminals. The second definition requires that an appropriate system is designed to transfer an e-service through a wireless channel from its hosting site to a mobile device, where it is completely (or partly). In that case, the e-service is considered as an m-service if it satisfies the following requirements:

1. It allows the mobile user to search additional facilities, when needed, and to get transparently these additional services.
2. It is transportable through wireless networks, executable on mobile terminals, and composable with other m-services.
3. It is adaptable according to the computing and display capabilities of the user's mobile terminals.

In addition, we assume that the m-service is capable of providing security services such as user authentication, user privacy, brokering protection, e-service authorization, and negotiation security. Today's m-services threats are often the same as those suffered by standard Web applications, such as the denial of service attacks, for example. However, with m-services, a more sophisticated platform is made available to a wider audience. In addition, the m-service interfaces should be human readable and easier to use, to a great extent, than the previous integration and messaging technologies. To just mention some, security breaches can include the access, via the wireless network or via e-service authorization procedures, to confidential information and the misuse of funds involved with an m-service provided by an organization. In addition, the detection of their attacks can be highly difficult, because the enterprise employees are the most familiar with internal systems. On the other hand, unintentional compromises are also possible. If an interface is not secure, an employee may accidentally access user's information that they are not intended to view. With additional interfaces and access to data, more sophisticated compromises can occur.

M-services infrastructures create complex challenges for managing threats because they allow service composition typically comprised of heterogeneous systems with decentralized administration. Getting consistent picture information is a difficult challenge, since each composing system may have its own logging data structures and its own timestamping procedures. Putting together information to determine unauthorized activity with an m-service is another difficult task.

10.2 Basics on E-Services

E-services are modular applications that can be described, published, located, and invoked through a communication network. E-services enable application developers to respond to a specific need described by a user or an organization, by using appropriate services published on the network, rather than building it from scratch, while offering potential benefits as compared to the traditional applications. Furthermore, the e-service paradigm simplifies the business applications and interoperation of the applications deployed on a public network. Additionally, it significantly serves end-user needs by enabling them, using browser-based interfaces, to choose, configure, and compose the e-services they want.

An e-service classification can be provided using the type of business activity that the e-service supports and the type of participants involved in the provision and the use of the e-service. Typically, four classes of e-services can be distinguished:

1. Business-to-Customer (B2C) e-services: This class considers a company providing e-service to an individual directly.
2. Business-to-Business (B2B) e-services: This category includes business models that can be used by an organization to provide a service to another peer organization.
3. Government-to-Business (G2B) e-services: This category includes services provided by governmental agencies to business organizations.
4. Government-to-Citizen (G2C) e-services: This represents services provided by the governmental agencies to citizens.

The end-product that the e-service supports can be used to present another classification. Three classes can be considered for this dimension (Tiwana, 2001):

■ The physical product processing e-service, which is concerned with the design, assembly, delivery, and tracking of physical goods. Selling books on the Internet represents a traditional example.
■ Digital product delivering e-services, which processes digital goods that exist typically in electronic form (e.g., software package and online music).
■ Pure service delivery applications, which provide provision of true service that has not a tangible form and do not deliver a tangible product to the service consumer (e.g., the e-voting service and the online tax filing and payment).

10.2.1 Actors, Models, and Tools of E-Services

Three major actors can be identified in the provision of e-services and the security services they may require while involved in e-service usage. They are the service provider, the service requester, and the service broker. The typical role of each actor is described as follows:

■ *E-service provider (ESP)*: An ESP is the entity (i.e., an individual, organization, or application acting on behalf of a user) that sets up applications for specific needs and services. The service providers publish, remove (or unpublish), and update e-services. The ESP is supposed to be the owner of the e-service, from the business point of view. However, from the architectural point of view, it is the platform that provides the implementation of the e-service.
■ *E-service requester (ESR)*: An ESR is the entity that has a specific need that can be fulfilled by a e-service available on the communication network. From the business point of view, a service requester is the user that requires

certain functions to be executed. From the architectural perspective, a service requester is the networked application that is searching for and invoking an e-service.

■ *E-service broker (ESB)*: An ESB is the entity that acts as a mediator between requestors and providers to provide a searchable repository of e-service descriptions where the ESPs publish (or inform about) their e-services and the ESRs find the e-services they are searching for, and obtain the binding information to access these services. The ESB should satisfy some requirements such as having the capability to enable interaction, negotiation, and selection of e-services.

Since the ESP, the ESR, and the ESB interact with each other, they need to use standardized technologies for e-service descriptions and communication. In this perspective, we are currently witnessing the rapid development of interrelated standards that are defining the e-services infrastructure, along with various development tools. Figure 10.1 depicts the typical model that describes the interaction between the three actors. Steps 1 and 2 depicted in Figure 10.1 show the registration of the service (or its update) with a broker. The registration can be made by the ESP through the exchange of two messages with the service broker. Steps 3 and 4 allow searching the registries, available with the broker, for e-services required by the requester (the ESR). The registry can be used to locate the appropriate component at the ESP side (steps 5 and 6). The service can be accessed using steps 7 and 8.

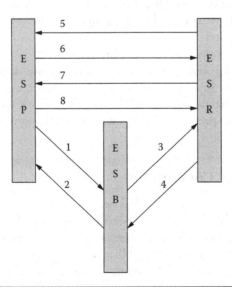

Figure 10.1 The e-service model.

Various e-service tools and platforms have been made available for developers of e-services. They include the following typical tools and techniques:

- *E-service description languages*: Description languages of e-services provide for the specification of service properties and interfaces by allowing the depiction of specific features such as the e-service it can provide, the way it is invoked, and the place where it is located. More specific features can be provided in the description including the service content, the commercial conditions, and the cost to access the e-service. Advanced semantic aspects can be incorporated into the description, as well.
- *E-service development tools*: Development tools allow the creation of platforms to design, implement, test, and secure e-services. The tools include software development tools and e-service composition tools, which provide the opportunity for defining value-added composite e-services using basic e-services and other composite e-services that can be searched for on the ESP's sites.
- *E-service publishing/unpublishing mechanisms*: Publishing procedures are responsible for making an e-service known (in other words, they make them available on the communication network) through ad hoc ESPs for potential users. Specific registries can be distributed across Web sites and be used to contain business information, service information, and binding information. The unpublishing procedures are in charge of the removal of an e-service in the case where the e-service is no longer offered.
- *E-service discovery tools*: Discovery tools allow e-service potential users to specify their needs at a high abstract level (e.g., by expressing a set of characteristics and defining the values of parameters that they are willing to accept from the e-service) and make sure that their needs conform to what an e-service available on the network can provide.
- *E-service binding and invocation tools*: The binding task refers to the time and manner the interoperation between e-services is established. Invocation allows a potential user to request the execution of an e-service by simply triggering it. It may provide a QoS framework that supports reliable delivery and security constraints.

10.2.2 E-Service Composition

In addition to the aforementioned classification, specification model, and useful tools to describe, design, and operate e-services, some operations are of uttermost importance. Among these operations, one can mention the composition, the negotiation, and the discovery, and the publication of e-services. In the following, we discuss the major features of the composition operations. The negotiation will be addressed as follows.

An e-service can be atomic, meaning that it is an operation with input and output parameters, and possibly preconditions and effects. It can be also interactive, meaning that it shows a "long-running" behavior. The behavior can be message-based or activity-based. In the first case, the e-service behavior is any possible message that the e-service may send or receive. In the second case, the behavior is any possible sequences of atomic actions that the e-service may perform.

The problem of e-service composition can be typically stated as follows: Let Λ be a set of available e-services and Σ be the specification of the desired behavior of the requested e-service, then a new e-service, called composite e-service, is built by joining pieces of Λ that comply with Σ using special operations such as the tradition logic operations. The result of composition can be of two types: (a) a one-use e-service that carries out a specific request of a particular end-user, assuming that the result of composition cannot be used for other requests of other end-users; and (b) a reusable-service that responds a standard request of a generic end-user.

Various forms of composition requiring different reasoning support and presenting diverse degrees of complexity have been developed. Among these forms, the one-use composition involving only atomic e-services represent the simplest form of the composition problem, whereas the most difficult form is the determination of a reusable composition implying interactive e-services. Several use cases can be identified, as they require different composition methods depending on how the description of the requested behavior of the composite e-service is made. In particular, it is the specification of the desired behavior of the composite e-service that characterizes the degree of reusability of the composite e-service. It is noteworthy, however, that two entities are engaged in the composition process—the end user and the service designer. Specifically, if this is the case of one-shot composition, the behavior of the composite e-service is defined by the end-user; if the case of reusable composition, it is specified by the e-service designer.

To explain how the composition can be performed, let us consider the following use case: An end-user wants to organize a trip to attend an event. The user can register to the event and set up the accommodation and the transportation using the available e-services. Let us assume that three e-services are available: (a) the registration e-service, which provides for the registration of end users that wants to attend specific events (e.g., a business meeting, or a tourist event); (b) the travel arrangement e-service, which negotiates with airlines for flight tickets; and (c) the accommodation e-service, which negotiates with hotels to book rooms. The end-user may be involved in this use case according to different roles. Each role characterizes different needs of the end-user, in terms of the order in which the various e-services are executed. The end user can attend a business event and needs first to arrange the travel so that his stay can be optimized. He can also want to be present at the event and desires first to arrange for the accommodation for himself and his spouse, for example.

■ *Case 1 (Available e-Services: atomic; Result of composition: one-shot)*: In this case, the end-user submits a request to the composition agent, who synthesizes a composite service by simply defining an order of execution of the three atomic services; then, he executes a composite e-service by coordinating the results of the available e-services. Let us consider for example the following request:

> **register_event** *(Event_name, date_start, date_end);*
> **book_plane** *(departure_place, date_leave, destination, date_back);*
> **book_hotel** *(hotel_name, city, check_in, check_out),*

assuming that some additional constraints on the unspecified parameters can be added. The composition agent can then state that the composite e-servce is given by

> **composite e-service** = *< registration e-service; travel arrangement e-service; accommodation e-service >,*

and executes the new service on the provided input and checks the result with the specification presented by the end user. If the composition agent cannot satisfy the end-user request, it may ask the client to relax some of the constraints.

■ *Case 2 (Available e-Services: interactive; Result of composition: reusable)*: We assume now that the use case contains, instead of the three atomic e-services, the following three e-services:

1. **Register_event**: This e-service is similar to the previous atomic and is an uninterruptible registration. Its input is a 3-tuple (event_name, start_registr_date, end_registr_date), and its output informs about the success or failure of the registration operation.

2. **Book_airtrip**: This e-service is assumed to have an interactive behavior allowing selecting companies, changing airports, and reserving cars.

3. **Book_accommodation**: This e-service has an interactive behavior allowing selecting a hotel (and the hotel city) based on specific end user requirements.

The effects from the e-services are to have an event registered, a flight booked, and a room reserved, respectively. To provide the composite e-service, we assume that the designer specifies the desired composite e-service as a finite state machine (FSM). In fact, any finite state based formalism can be used and the interactive e-services provide a discovery phase that incorporates several aspects regarding the precondition and effects of the e-service, the behavior of the e-services, and handling incomplete specification on the available e-services. Figure 10.2 depicts a FSM that provides a general composite service that resolves the use case.

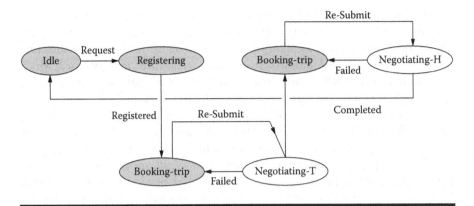

Figure 10.2 Composite e-service FSM.

The composition operation applies to m-services. However, different issues need to be solved depending on the mobility nature of each actor involved in the discovery, brokering, and use of the m-services. In particular, the location of the composing services should be managed properly since the composition structure may be recomposed when a composing service becomes unreachable, for example, or when the mobile is getting into an area where alternative services can offer better quality.

10.3 M-Services Discovery

Discovery helps a mobile user (or application) to discover m-services, data, and computation sources, which may be found in the neighborhood of the mobile device or somewhere on the network. Due to resource constraints and mobility, the mobile terminals may not have complete knowledge about all m-service locations and computing sources currently available. The discovery operation assumes that the network can help in establishing an end-to-end connection with other entities in the network and provide the involved network layers with the knowledge and context of available m-services and computing sources. Three entities are involved in a service discovery system: the manager (or ESP) owns the service descriptions, SDs; the User (or requester) has a set of requirements for the e-services it needs; and a registry caches available services so that the users, or applications acting on behalf, can discover the services through queries to the registry. A mobile node can play three roles: the user, the manager, or the registry.

10.3.1 Discovery Architectures

Service discovery architectures are frequently structured based on how the users discover the m-services. A discovery architecture can be of two types: It can be a lookup-registry-based discovery or a peer-to-peer discovery. Lookup-registry-based discovery works by registering information about the e-services to a registry. The users query this registry to obtain knowledge about the m-service sources (including information such as its location and how to invoke it). This type of architecture can be further subdivided into two categories: the centralized registry-based, which contains one centralized registry, and the distributed registry-based architecture, which consists of multiple registries distributed across the mobile network.

Peer-to-peer discovery approaches query each node in the network to discover the available e-services on that node. Broadcasting of requests and advertisements to peers is an easy task, but inefficient. An example of distributed peer-to-peer service discovery protocol (SDP), using caching, that significantly reduces the need to broadcast requests and advertisements has been developed in Chakraborty (2002). Another example of peer-to-peer service discovery protocol was given by Bluetooth. It uses 128-bit unique identifiers to represent services. In particular, the protocol does not provide any information on how to invoke the service. It only provides information on the availability of the service on a given device.

In the registry-based architecture, registries can be deployed by a system administrator, or automatically elected by the nodes in the system. Once the registry is available, the other nodes in the system will have to discover it before m-services can be registered and queried. In the non-registry-based architecture, the users and the managers can perform multicast queries and service advertisements to other nodes. Therefore, unlike the registry-based architecture, the system is not vulnerable to single point of failure issues; but, since extensive multicast is used, an overhead of network traffic can be experienced. This may cause scalability problems.

Service discovery entities often communicate with each other through a logical topology, which is used to optimize the way the system propagates and processes messages, and thus optimize the following items: communication cost, system scalability, energy efficiency, and query effectiveness. Four basic logical topologies can be distinguished based on the nature of query propagation: (1) the meshed topology, where messages are sent to all listening entities; (2) the clustered topology, where messages are sent to clusters of entities formed by registries and/or mobile users, based on the type of service they provide; (3) the tree topology, where messages are propagated along a hierarchy of registries; and (4) the unconnected registries topology, where messages are sent to any discovered registry. The non-registry architecture is typically based on the meshed and the clustered topologies. The members of a cluster communicate only with each other and limit their service advertisements and queries to the cluster. The meshed topology improves the chances of discovering a service and the availability of the service, because it is not vulnerable to single point of failure issues. Nevertheless, it increases communication cost and presents some scalability because of the extensive use of multicast operations. Whereas the clustered topology makes the system more scalable, however, it tends to increase the complexity of the system, because managers and users will have to establish clusters and dynamically add and remove members.

The registry architecture uses the four types of logical topologies: In the unconnected registry topology, the registries do not communicate with each other, but the users may associate themselves with multiple registries; in the meshed registry topology, registries communicate with each other and forward queries and copies of their caches to their peers; in the tree-based registry topology, registries form a parent-child relationship. A child registry forwards queries to its parent when it does not find matching services within its own cache; finally, in the clustered registry topology, the query processing is done only by a few registries.

The unconnected registry topology does not require registries to synchronize registration data among each other. Therefore, adding a new registry is not a complicated task. Redundant registries also provide increased robustness against communication and registry failures. However, when the users and the managers redundantly communicate with each discovered registry, communication cost is increased, and scalability is reduced. The other three topologies allow the manager and the user to communicate with a single registry only, keeping the service discovery task simple on the user side. The meshed registry topology is not practical for large systems that are deployed on wide areas. The tree-based registry topology allows each registry to store only a part of the available services, therefore allowing load balancing for systems with high density of nodes. However, the system becomes more vulnerable to single points of failure, since the registries will need to check continuously on the availability of the parent and child registries.

10.3.2 Discovery Functions

The service discovery aims at accomplishing two objectives: to discover the services that match the request of a user and to detect any change in the availability of a service or its attributes. Four major functions are needed to achieve these objectives: the configuration discovery, the configuration update, the service registration, and the service description discovery. The term "configuration" refers to the entities in the system. The Configuration Discovery and Service Registration (for registry-based architectures) functions are required for entities in the system to collect information on the availability of nodes and services.

10.3.2.1 Configuration Discovery

This function allows the registries to be set up and the identities of registries and cluster members to be discovered. There are two sub-functions of configuration discovery:

1. *Registry auto-configuration*: This sub-function allows the system to configure registries automatically through registry election or registry reproduction, where a parent registry generates a child registry based on some criteria (such

as load threshold or service location). Registry auto-configuration is done dynamically, without supervision.

2. *Entity discovery*: This sub-function allows entities in the system to discover a registry through active discovery, where the nodes initiate the discovery by sending announcements; or passive discovery, where the nodes discover the required entities by listening for announcements.

10.3.2.2 Service Registration

This function allows the managers to register their services at a registry. It includes the unsolicited registration, where the nodes request the registry to register their services, and the solicited registration, where the registries request new nodes to register. The registry keeps a cache of available service descriptions (SDs) and updates them according to requests from the managers.

10.3.2.3 SD Discovery

This function allows the users to obtain SDs that satisfy their requirements. The users may cache the discovered SDs to reduce access time to the service and reduce bandwidth utilization. There are two sub-functions in SD Discovery:

1. *Query*: This is a pull-based model where the users initiate query to a registry. The query specifies the requirements of the user. The registry that holds the matching SD replies to the query.
2. *Service notification*: This is a push-based model, where the users receive notification of new services by the registry or multicast service advertisements by the managers.

10.3.2.4 Configuration Update

This function monitors the node and service availability, and changes to the service attributes. It contains two sub-functions:

1. *Configuration Purge*: This allows the detection of disconnected entities using leasing and advertisement time-to-live (TTL). In leasing, the manager requests and maintains a lease with the registry, and refreshes the lease periodically. The registry assumes that the manager who fails to refresh its lease has left the system and, hence, purges the related information. With the advertisement TTL, the user monitors the TTL on the advertisement of a discovered manager. Again, if the manager fails to advertise before its TTL expires, then the user assumes that the manager has left the system and deletes the related information.

2. *Consistency Maintenance*: This allows users and registries to detect updates on cached SDs. Updates can be propagated using a push-based update notification or a pull-based polling for updates by the user to the registry or the manager for a fresher SD.

10.3.3 Matching and Location Management

The typical service discovery approach uses simple interfaces, attributes, or unique identifiers based matching techniques to locate the appropriate sources. The drawbacks of these techniques include the lack of rich representation of e-services, the inability to specify constraints on service descriptions, and lack of inexact matching of service attributes. Semantic matching is an alternative technique that alleviates these limits. In particular, the Bluetooth Semantic Service Discovery Protocol (SeSDP) uses a semantically rich language, denoted by DAML (DARPA Agent Markup Language), to describe and match services and data. Semantic descriptions of services and data allow greater flexibility in obtaining a match between the query and the available information. Matching can be inexact, meaning that parameters such as functional characteristics and device characteristics of the e-service provider need to be used in addition to determine whether a match can occur.

Location management aims at providing location information to mobile devices about the m-services that can be reached by the discovery process. Location information dynamically changes with mobility of the device receiving and the mobility of nodes involved in the m-service provision. Most location techniques, relevant to m-services, consider the position determination, with respect to some global (latitude/longitude) or local (distances with respect to specific objects) information.

Service management represents an important component in the development of m-services. It mainly consists of service discovery monitoring, service invocation, service execution management, and service fault management. The service management performs various functions depending on the m-service architecture. In the client-driven architecture, most of the management (e.g., service composition and execution) is done by the server acting on behalf of the user. Mobile users mostly manage the m-service invocation, notifications, alerts, and monitoring of the local resources needed to execute a request. In the broker-based architecture, most of the management is done at the broker side. Disconnections are managed by tracking the state of execution of an m-service and retransmitting data once connection is established. Another important issue of the service management is to manage composite m-services that might require interaction of the composing e-services to provide a reply.

Most of the existing e-service management platforms for composite queries are centralized and oriented toward services in the fixed wired infrastructure. Distributed broker-based architectures for service discovery, management, and composition in wireless ad hoc environments are under development (Chakraborty, 2002a).

10.4 Basic Examples of M-Services and Challenges

Mobile services differ from traditional e-services in their ability to provide service offerings regardless of temporal and spatial constraints. They are also different from other types of e-services, such as wireless online services, where the service delivery is associated to a fixed local area network or specific location. Nowadays, a large spectrum of m-services has been made available for use by individuals, organizations, and governmental agencies through communication networks. To illustrate the large opportunities to build m-services and the technologies involved in their development and security, we consider three important (composite) m-services. Some of these m-services will be used in the sequel to analyze the main concepts used in m-service provisioning, scenario of attacks, and protection schemes. They are: the *real-time stock m-service*, the *virtual bank account m-service*, and *travel agency s-service*. More advanced m-services can be given by the government m-services and the m-government. These two fields will be addressed in the following sections.

10.4.1 Common E-Service

A large set of common services has been developed and provided and offered on the mobile networks. For the sake of clarity of the m-service concept and operation, we consider the following basic services:

10.4.1.1 Real-Time Stock M-Service

This is an e-service that can be used in an application offering real-time stock market information for mobile users integrating mobile brokers. A similar e-service has been provided in Tsalgatidou (2002) for processing by static actors. The m-service can be accessed through an ad hoc portal and mobile devices offering and integrating a series of different services including, but not limited to, the following:

- *A real-time view of the stock market*: These services provide information to mobile users on the real-time status of a given set of stocks handled at different places.
- *A stock quote service*: This service enables investors to retrieve a quote in a specific currency, given the ticket-symbol of any publicly trade stock.
- *A portfolio management service*: This service allows an investor to track the performance of his/her shares and perform appropriate transactions.
- *A tool for account balance check*: This service checks the investor's account in order to make sure that he has the necessary account for the completion of a mobile transaction.
- *A tool for currency conversion*: This tool can be accessed by a mobile investor needing to get the actual conversion between currencies of interest for him.
- *A news service*: This service provides the headlines of the latest financial news related to stock market and investment opportunities.

Implementing the real-time stock m-service can be achieved using three approaches: (a) developing the system from a complete specification; (b) customizing an existing e-service available on the Internet, provide the appropriate interfaces and displays for mobile devices; and (c) composing e-services and m-services available with other vendors or business partners. A specific method can choose to develop the news service from scratch, while it assumes that the money conversion service can be acquired, and the portfolio management service be composed. The security issues to be addressed would mainly consider that sending messages should be made confidential, when they contain sensitive information, and be signed when the information is used as a basis for transactional decisions.

10.4.1.2 Virtual Bank Account M-Service

Real accounts and virtual accounts (that are used by an anonymous payment) are identified by their account numbers and a pin number. Actions in this system are the deposit, transfer, payment, and balance check. Our model for a virtual bank account m-service is assumed to satisfy the following rules:

■ For each virtual account, a digital certificate identifies the account and establishes a unique link between the owner and the virtual account.
■ Each account holder can issue signed payment orders, which authorize the payment of, or the transfer of money to, a third party's account (such an m-service provider's account). A payment order authorizes to perform only one payment and cannot be transferred.
■ A mobile holder of an account can submit a transaction to deposit the amount of money related to a valid payment order signed by another account owner.
■ A mobile user presenting a valid payment order can obtain the opening of a virtual account with an initial deposit amount equal to the amount of the signed order. Therefore, the requester should be able to operate on the new virtual account (the order may, for example, include a digital certificate for this purpose).

The design of the virtual bank account m-service can be based on the integration of the following five m-service components:

1. *Account balance m-service*: This service allows checking the virtual client's account in order to make sure that he has the necessary amount for the completion of a payment order.
2. *Digital certificate generation m-service*: This service is invoked by a potential client to start the process of creating a virtual account and generating the appropriate naming certificate.
3. *Payment order generation m-service*: This service is invoked by an account holder when he needs to generate a payment order for the benefit of a third

party. The actions it involves include at least filling the order, signing it, and (possibly) the expected balance after payment.

4. *Certificates revocation e-service*: This service is used to invalidate a digital certificate issued and close the related account. This service may need to handle appropriately the remaining amount of money on closing the account.

5. *LDAP-based verification e-service*: This service is invoked by a mobile user or any other entity when there is a need to check the validity of a certificate or an order submitted with a transaction to be executed.

The security needs of the virtual bank account m-service require mainly the LDAP used to check the validity of certificates to be available and secured, and that the infrastructure used for certificate and order generation be operated following a well-defined security policy. An m-service can be added to provide an anonymous payment service.

10.4.2 Challenges in Security of M-Services

Challenges facing m-services occur first during the analysis of requirements and the description of user needs and security requirements. Moreover, challenges can occur in almost all phases of the m-service life cycle, where security is a major issue. Typically, an m-service life-cycle is a six-phase process, which are represented by the description, publishing, invocation, integration, and management of m-service. The major technical challenges that have attracted the attention of m-service developers are the description, discovery, brokering, composition, publishing, reliability, management, accountability, testing, and traceability of services. Figure 10.3 depicts

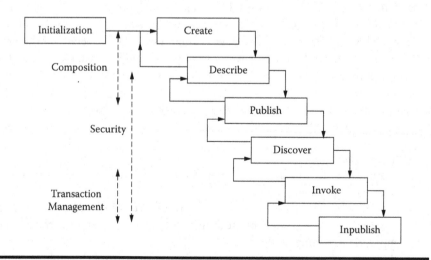

Figure 10.3   The m-service life-cycle.

the life-cycle activity of an m-service and main technical challenges. Starting with the creation challenges, testing is an important activity to perform (specially, when security should be provided). Description challenges address the specification of service based on user needs. They address also the syntax and semantic definition of system functionality, reliability, capabilities, and security guarantees.

There is a set of basic security requirements that are important to have trustable m-services. These requirements include authentication, data confidentiality, data integrity, non-repudiation, confidence in the reliability, identity validity, user anonymity, user location traceability, transaction traceability, privacy, and security dependability. Advanced m-services (e.g., m-health and m-government) may have more restrictive requirements in addition to the aforementioned requirements. They may also impose additional restrictions (as it will be shown in Sections 10.5 and 10.6). Therefore, the security is a major challenge for publishing, discovery, and invocation. Important issues in security related to discovery address protecting exchanged messages, protecting registries, and securing the activity of brokers. Finally, one can notice that transaction management involves monitoring service states, security levels, and transaction protections. Additionally, all activities, when applied to a simple or a composite m-service, are expected to expose their functionality at two different management levels: at a syntactic level, where implementation aspects are addressed, and at the semantic level, where the conceptual aspects of services are facilitated.

Major security challenges, however, are encountered when addressing brokering, reliability, monitoring, and transaction protection. We briefly discuss these challenges and highlight the role of security and the complexity of the related technical challenges. We discuss in the following sections how security requirements are addressed when composing m-services, registering services, and exchanging messaging.

10.4.2.1 Brokering

As explained, brokering is the activity of mediating between requestors and providers in order to match customer's desires and providers' offerings. Brokering is a more complete activity than discovery since the broker must be able to facilitate service-to-service interactions, negotiations, and selection of services based on the optimality criteria. This activity is mainly dependent on the way the services are modeled, described, and protected. If the service specification contains the QoS required (including security requirements), then the broker can construct a query that specifies which data are asked for and which QoS requirements are acceptable. The broker can in turn access the available registries to find out which m-services are able to provide the required QoS. The broker then orders the replies and selects the optimal offer. Different contributions in the area of service quality and automatic service selection via service brokering are available in the literature such as in Scannapieco (2002).

Brokering discusses the appropriateness of m-services and finds the optimal solution among the m-services presented by the suppliers having the same or similar objectives. Security problems related to brokering address the protection of message and the specification. In order to secure a message, a broker should consider two types of threats: (a) the message can be modified or read by a malicious adversary and (b) an adversary may send messages to a service that, while well-formed, lacks appropriate security claims to secure transaction processing. Message integrity should be provided by using signature to ensure that messages are transmitted without modifications. The integrity mechanisms may be designed to support multiple signatures and should be extensible to support various signature formats.

10.4.2.2 Reliability

Reliability of an m-service is the capability of the service or its components to perform its functions and interoperate with brokers and requesters under stated conditions for a specified period of time. Stated conditions may include privacy protection, intrusion tolerance, and service availability. Based on the protection mechanisms they implement, the m-service providers can offer more reliable m-services than other providers. Therefore, the availability mechanisms allowing measuring reliability and controlling the security system state should be used. Transmitting reliability information helps necessarily improve the reliability of composite m-services. It is also needed to provide reactive measures to improve reliability and recover an m-service locally when the provider system goes off-line temporarily, caused by backup or maintenance actions or even by damage due to security breaches.

Furthermore, it should be required to specify what happens when a service goes off-line during the execution of a transaction involving the m-service. Moreover, the m-service providers may have to implement appropriate processes for disaster recovery and the migration of all their business partners to new m-service platforms. Finally, since a provided (or requested) QoS typically could be expressed using a set of parameters including throughput, response time, and cost, an interesting contribution to m-service reliability can include the development of models, techniques, and mechanisms to allow each service to have a set of QoS metrics that gather information on the expected QoS of the m-service execution. These requirements, however, may affect the level of security protections since they have a nature opposed to performance-related parameters.

10.4.2.3 Monitoring

Monitoring takes place during m-service management (including invocation and operation). Once a service requester and an e-service provider have been linked together in the provision or the access to an m-service, the m-service execution needs to be continuously monitored. This is a difficult task to realize because a large set of sensors and profiles may be needed. In fact, modifications are required to be incorporated in a real-time manner without affecting the operations executing at the requester's site. This becomes a challenging issue for at least three reasons: (a) the m-service may be running on a system that is not under control of the requester, (b) the m-service requester cannot interoperate with the running operating system, and (c) the security countermeasures are hard to determine.

M-service monitoring becomes more complex to perform in the case of composite m-services. Typically, the properties of a composite m-service are dependent on those of its component services and the managers of these components. Composing e-services may need to be coordinated in a constraining way. One solution to this is to outsource the management of some m-services to other providers. Monitoring may have to address the quality of service provision and security requirements to be observed while invocation task is processed. In this case, the monitoring process should integrate the appropriate tools for handling events and related metrics.

10.4.2.4 Compositions

Composite m-services need to be more adaptive and more dynamic to cope with the dynamic features of m-service environment and the continuous need to add new services. They consider security as a key issue since services do not reside within a protected single site and the sensitivity of resources is different from one site to another. In order to meet the requirements imposed by the need for efficiency and reliability of composite e-services, several models and techniques have been developed for the description, analysis, and optimization of composite e-services. Most of these techniques apply to m-services and were workflow-based. Composition of services is addressed by eFlow (Caseti, 2000) and by the service model CMI (Schuster, 2000). eFlow supports the dynamic composition of simple e-services. It allows configurable composite services, where different requesters invoke the desired subset of features that this composite service can offer. The service model of CMI provides another solution that also enables dynamic composition of e-services at a semantic level by separating the service interfaces from service implementation and defining the service interfaces in terms of service state machines and its input/output parameters.

10.4.2.5 Discovery Security

A secure service discovery system must support confidentiality, message integrity, and availability. Methods to address these concerns include authentication of communicating entities, access control, protection of sensitive service attributes by hiding the value, data integrity, and detection and blacklisting of malicious nodes. The challenge for a secure service discovery system is to maintain self-configuration of the system, because the owner of the devices will most probably be required to provide authentication and access control. Security also consumes resources due to encryption algorithms. Most service discovery systems assume participating nodes are secure by delegating security to the application layer. However, the full deployment of a service discovery system will eventually require some secure measures integrated into the service discovery functions, particularly when it is done on ad hoc mobile networks.

10.5 M-Government

The m-government is a subset of e-government. Its objectives aim at getting the public IT systems and services available to interoperate with citizens' mobile devices. The m-government actors must have the capability to use each other's data so that citizens, companies, and officers do not have to provide special interfaces. Its setup involves the utilization of all types of wireless and mobile technologies, services, applications, and devices for improving benefits to the parties involved in e-government including citizens and businesses. The m-government, however, is in its early stage of development. It is a subset of e-government. Its challenges and security issues are closely related to those of e-government.

Several issues have been identified to be addressed for a successful and large deployment of m-government services (Kushchu, 2003; Lallana, 2004). These issues include the following:

- *Easy use*: The mobile government, like the e-government, requires the ease of use and composition of mobile services. The Governments, for example, need to offer easy registration and access to m-government information in various forms. For instance, the use of video and voice communications may be an efficient tool to encourage citizen participation and provide citizen-oriented services, when they are possible. Scalability should be built into the m-government services, and the maintainability of both the functionality and efficiency of the service it provides should be made possible if changes are realized. Modularity and scalability must relate to the nature and scope of the offered services, and to the number of citizens or transaction volume.
- *Mobile payments infrastructures*: These infrastructures are essential to the large deployment of m-government. They should be trusted by the citizens and the businesses since the very first barrier for the citizens to pay online is a feeling of mistrust in sending their credit card information over the mobile phone. Several solutions for m-payment offering a greater security have been constructed for e-government purposes.
- *Infrastructure development*: The information technology infrastructure must be at a satisfactory level of quality and coverage. The infrastructure refers to the technology, equipment, and network required for the setup of m-government. Organizational policies and software that make m-government transactions possible are also essential. The openness of the infrastructure is considered

at several levels: the standards, the interfaces, and the source codes. In principle, public authorities should use open, formal standards, but where this is not possible or attractive, the advantages and disadvantages of using open de facto standards should be considered.

■ *Privacy and security*: Because of possibility of data interception in all traffics, malicious outsiders can attack on wireless network providing m-government services to access sensitive information and temper with documents and files. The typical issue is to convince the citizens that the government can overcome the mistrust and assure mobile users that their privacy is protected, their information will not be delivered to third parties, and their mobile phone numbers will not be traced when they submit their data and inquiries to the e-government services. Security functions provided to this purpose can be organized in such a way that the citizens/business needs for security can be met to an extent that is acceptable in the given application scenario. The solution also has to be adjustable for new requirements.

■ *Compatibility and interoperability*: Compatibility of systems is a technical difficulty that might arise from compatibility of the mobile systems with the existing e-government systems. This may get even more serious in the cases of government offices having legacy systems that are not easy to integrate. The solution lies in implementing open systems using open standards. On the other hand, interoperability is based on bilateral agreements where coherent policies for communication and accesses are defined for each new system that is integrated. The main issue of interoperability is the specification of common (or compatible) data models and protocols for exchanging data. The data exchange format is based on the XML standards. In many cases, however, the need for security can be seen as a conflicting requirement with interoperability and openness.

■ *Legal issues*: Governments have an important responsibility in the deployment, acceptance of e-government, and the promotion and operation of m-government services. They may play an arbitrary and authoritative role. In addition, they should offer a trusted environment allowing the different actors (such as the citizens and the service providers) to adopt this new type of service. All of the e-government activities can be organized by specific laws. As a part of e-government, the m-government should be addressed by these laws.

10.5.1 Monitoring Security States in M-Government

Typically, the e-government (and more particularly the m-government) security monitoring process architecture integrates four interoperable processes: (a) the security policy definition and security mechanism selection and implementation to secure fixed and mobile access and transfer; (b) the metrics definition to measure the effectiveness of security mechanisms and detect attempts of attacks; (c) the

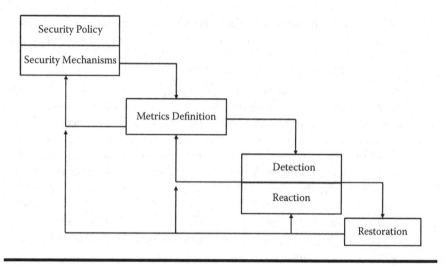

Figure 10.4 Monitoring architecture.

attacks detection, attack identification, and reaction to attacks; and (d) the system restoration from failure induced by attacks. Figure 10.4 depicts the monitoring process and the interoperation between the four processes.

The security policy aims to define a complete set of rules capable of keeping the system state at an acceptable level of security if traffic features, accesses, and service offering comply with these rules. The rules are collected through questionnaires submitted to service and security systems administrators. Security mechanism selection and implementation provides the efficient enforcement of the above security policy. Such mechanisms can be characterized and acquired off-the-shelf or designed and implemented.

The metrics can be an effective tool for security managers to measure the effectiveness of various components of their security procedures and the staff ability to address security issues of which they are responsible. Indeed, the metrics can help managers to identify the level of risks in their organization. They allow the managers to make the best and customized decisions concerning their security plan according to the security pre-defined strategies. The choice of metrics for an e-government system depends strongly on the offered services, the types of potential users, and the type of assets and their values. Examples of metrics for m-governments include the following:

- The number of attacks observed within a period of time on a radio link. Using such metrics, one can differentiate the origin of the attacks (insiders or outsiders), the moment of the attacks, and the resource targeted.
- The number of accesses on specific assets and resources. Such metrics can be used to detect denial of service attacks and to localize congestion areas.

■ The estimation of the number of residual attacks (occurred on e-government systems but not observed). These metrics can be used to trigger special measures to protect assets, to redefine new policies.

■ For each type of attack (and each service), the mean time to get the system repaired. This metric can be used to adapt the decision support to provide response to attacks.

These metrics can be determined by two estimating techniques (Benabdallah, 2002): The *simulation methods* provide the estimation of real time variables using a pre-defined evolution of the metrics. The *rule-based approach* offers a qualitative estimation of the performance parameters based on rules made available by the expert. The definition process of metrics addresses the construction of a complete set of mechanisms in a way such that any evolution of the e-government system security level toward an unacceptable situation is notified. Reducing the computational complexity of the management of the set of metrics may need to be addressed in this step. To help reasoning about and classification system states, multi-objective function can be developed to measure the likelihood occurrences of attacks.

The attack detection is an intelligent process integrating the different values computed for the multi-objective functions involving critical metrics, attacks features, and damage assessment. At the end of the detection process, one can assume that a specific attack (or a reduced set of attacks) is identified. The reaction process will help choosing the appropriate reaction to the attack(s) based on the reaction cost, the service availability, and the efficiency of the reaction.

Restoration is a major task that may induce security policy remodeling, system architecture redefinition, security mechanisms redesign, damage recovery, and metrics modification. In the following subsections we present different tools that are useful to the monitoring process.

10.5.2 Monitoring Tools

To achieve its role in the protection of m-government services, the monitoring process requires the implementation of a number of useful tools. The following list is not exhaustive; it presents the major tools:

■ A set of collecting procedures/equipments that are implemented on the appropriate components in order to detect all occurrences of specific events useful for the computation of the implemented metrics. Equipments may include mechanisms allowing the capture of signal and the determination of signal fingerprints.

■ A group of security administrators whose role is to build the security policy, manage the e-government system, and perform the manual part of the recovery actions. This group can be accompanied by several incident response teams that are responsible for conducting attack identification, reaction against identified attacks, audit the e-government systems, and propose security solutions and policies.

■ A set of routines that perform automated recovery actions along with a set of databases, where all information required to detect attacks or to select measures are stored. Such information includes intrusion signatures, intrusion scenarios, abnormal behaviors, and rules for the selection of decisions.

■ A set of event journals reporting on all events detected by the security monitoring process. These journals are efficiently located at their appropriate places.

■ A set of alarms that can be triggered by the occurrence of particular events or system behaviors, including those considered as prioritized notifications to systems or service administrators.

■ A decision support system that will help achieving the right decision when an attack is identified or suspected.

■ A set of security mechanisms including preventing tools such as the distributed firewalls, access controllers, authorization systems, and cryptographic systems; and reactive mechanisms including intrusion detection systems and intelligent systems that help for the selection of the appropriate measures.

In addition to the management of metrics, the monitoring process manages a set of multi-objective functions, which combines metrics, risks, security rules, events histories, asset semantics, and asset dependencies, if any. These functions are used to check the security of the e-government system states or for decision process reaching. The main purposes of these functions are

■ To implement features that help addressing anomalies and unexpected behaviors,

■ To determine the attack types that have the highest probability to be those attacks for which metrics move to critical states,

■ To help select the appropriate actions to react against an attack.

An example of a multi-objective function is given by

$$f(m_1, \ldots, m_n)(E) = \Sigma_{i<1<n} \, w_i m_i(E),$$

where m_i, $i < n$, are n metrics considered by the detection process of an attack, w_i, $i < n$, are predefined weights that put relative importance between the metrics, and E is a set of event collected by the monitoring process on a given set of assets related to the m-service. We note that the second term of this equation concerns only metrics on which a security failure has been detected. The others are not included.

10.6 M-Commerce

E-commerce has been defined as any form of business transaction in which the parties interact electronically rather than by physical exchanges or direct physical contact. The e-commerce is considered as the process of performing business electronically involving automation of various business-to-business and business-to-consumer transactions. Similar to e-commerce, m-commerce is a process involving a chain of operations (or interoperating services). Transactions in m-commerce typically involve several actors: the customers, merchants, the banks, mobile network operators, and possibly other entities (Figure 10.5 depicts the major actors and adds specific actors). M-commerce involves services, mobile devices, middleware, and wireless networks. It is defined as a means of conducting commercial transactions via a "mobile" telecommunications network using a communication, information, and payment device such as a mobile phone or a personal device.

M-commerce technology is different from e-commerce where processing is accomplished through the Internet via a browser and being connected through wired Internet connection. To explain the difference let us first describe the m-commerce primary actors, discuss its functions, its vast range of services, and illustrate its market segments. Functions such as the transport, basic enabling service, transaction support, presentation service, personalization support, user application, and content aggregators are the seven main functions in the mobile business value chain (as illustrated in Table 10.1). Based on the value chain, we can identify seven actors in mobile

Table 10.1   Operation Chain of M-Commerce

Function	Responsibility
Basic enabling service	To provide basic services such as data backup, server hosting, and service integration
User applications	To execute m-commerce transactions for mobile consumers
Presentation service	To translate the content of Internet-based applications to wireless standards suitable for the display on mobile terminals
Transport	To operate and maintain the infrastructure to ensure data communication between mobile users and service providers
Transaction support	To provide assistance for transactions execution, for transaction security, and for billing users
Personalization support	To collect users' personal information, which enables personalizing services for mobile users
Content aggregators	To provide information to help users find their needs on the network

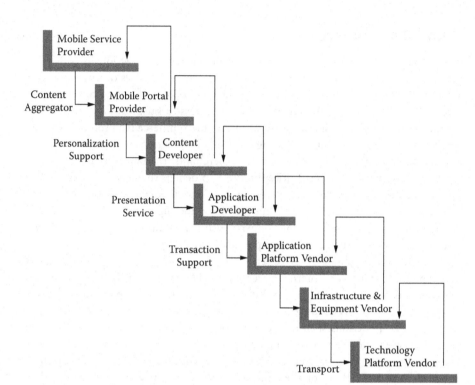

Figure 10.5   Actors in the m-commerce value chain.

commerce as shown in Figure 10.5. Among these actors, the following four entities are the main actor in m-commerce involved in the management of a transaction:

1. The *customer*. The customer is able to initiate a transaction in one location, move to another location, and receive the service. Then he completes the transaction in a third location.
2. The *content provider*. It makes available customers specific content, which can be transmitted through a WAP Gateway or through a portal.
3. The *mobile portal*. It offers the customers services with a high level of personalization and localization.
4. The *mobile network provider*. It plays different roles in m-commerce varying from a simple mobile network provider to an intermediary, portal, or trusted third party, depending on where it stands in the m-commerce value chain.

M-commerce can be divided into three basic categories (Liebmann, 2000), namely the internal business operations, the business-to-business applications using an extranet, and the Web-based consumer services. The first category is probably the most common form currently in use.

10.6.1 Technical Challenges in M-Commerce

Many challenges make the practical use of m-commerce complicated. Challenges include the technical restrictions of mobile devices and wireless communication, the business concerns, and the security and legal constraints. The challenges mainly address the application and network levels.

At the application level, the challenges include the following:

■ *Mobile devices limitations*: Mobile terminals exhibit a large degree of mobility and flexibility; they present, however, reduced features when compared to personal computers. Because of the need to remain physically small and light, the input and output mechanisms of these terminals slow down the development of user-friendly interfaces and graphical applications for common mobile terminals. Mobile handsets are also limited in computational power, memory capacity, disk volume, and battery life. These shortcomings do not support complex applications and transactions, and consequently may limit the deployment of m-commerce needing complicated management environments.

On the other hand, to conduct business via mobile terminals, enterprises must be able to manage and support a large base of mobile customers or employees. They also must deal with the logistics and asset management issues surrounding large numbers of mobile devices and software. These two facts induce a challenge to the traditional helpdesk and customer care function and make customer care far more complex and harder to manage.

■ *Customer distrust*: Each party involved in a transaction needs to be able to authenticate its partners, to make sure that received messages are not tampered with, to keep the exchanged data confidential, and to ensure that the received messages are issued by legitimate senders. Due to the inherent vulnerability of the mobile environment, users in m-commerce are more concerned about security issues involved with mobile transactions. In particular, m-commerce users need to be assured that their financial information is secure and that wireless transactions they are involved in are protected. The mass adoption of m-commerce in supply chain management will not be realized until users begin to trust m-commerce systems (Siau, 2002).

■ *Strategy changes*: To stay competitive and realize real productivity profits from m-commerce, many businesses actually need to be reengineered. Indeed, they will have to make essential changes in their organizational behavior to eliminate the inefficiencies of the current organizational structures, and to develop new business models. The process of redesigning is a demanding activity and should integrate efficient protection methods. Actually, reengineering the way an enterprise does its business requires rethinking how business will be organized from the perspective of its customers and reengineering how to perform its functions according to the needs of its customers. Taking into account the way m-government is organized and protected will certainly help in getting the right changes.

At the network level, one can distinguish three major challenges. They are the following:

- *Heterogeneity of networks*: As previously discussed, one can notice and a large number of complex and competing protocols exist in today's mobile communication networks. These different protocols have resulted in the global incompatibility of cellular terminals. The network incompatibility poses problems for organizations to communicate and cooperate with their suppliers, distributors, and customers. Roaming is a technique that aims at reducing the difficulties to keep access to m-services continuous.
- *Bandwidth access*: The scarcity of radio frequencies has conducted the radio communication authorities to reserve limited frequency bands for use by cellular network operators across the world. In order to encourage competition, some radio authorities (and regulators) have even prohibited cellular operators from owning a large part of the radio spectrum in a given geographic region (e.g., the Federal Communications Commission has prohibited it in the United States). This regulation imposes barrier for cellular network operators who can be attempted to implement the new high-bandwidth m-services.
- *Security concerns*: As previously shown in this book, most wireless data networks now provide reasonable levels of encryption and security; however, the technologies under use do not ensure transmission security in the network infrastructure. Therefore, data packets can be lost due to different reasons, including mobile terminal malfunctions and ongoing transactions alteration. In addition, the mobility provided by composite m-commerce services create many more challenging security tasks. Serious consideration must be given to the issue of security of the transport and routing of m-commerce traffic, transaction, and computation.

Separately from the application and network challenges, the development and deployment of m-commerce services face other concerns, namely, the legal framework. The application of traditional law to the mobile Internet is not always an appropriate and a straightforward process. Legal issues should take care about how to maintain privacy, how to protect intellectual property, and how to treat Internet taxation (Deitel, 2001).

The interaction between the partners in this m-commerce incorporates negotiation at various steps. To cope with entity's mobility and the dynamic trading of m-services, m-commerce architectures must consider issues such as scalability, micro transactions, failures, network performance degradation, and fraud prevention and detection. The negotiation in m-commerce environments is achieved based on the constraints and requirements imposed by the domain of commerce. For example, transactions issued from mobile devices would normally be involved in short negotiations, while those executed on fixed hosts can take part in continuous and computationally expensive negotiations. Moreover, the features of mobile telecommunications determine the suitable negotiation mechanism and are often inter-related; the quality of the network may itself be partially defined in terms of the quality of service measurements. The quality of the network can vary with changes in bandwidth, frequency of disconnections, costs of connection, and data integrity and security.

10.6.2 Security Issues in M-Commerce

The emerging area of *m-commerce* creates new security and privacy concerns because of the challenges raised by the technologies under use, the protection requirement, and the need to handle properly activities such as service discovery, service composition, and negotiation. In this section, we discuss some important issues related to the security and trust issues raised by two special cases of m-service provision: (a) the disconnected case, where both the client and the merchant are disconnected from the payment infrastructure and communicate with each other directly using a local link, for example; and (b) the connected case, where the client and the merchant are connected on-line to the payment server. Besides the characteristics of the individual entities involved in these cases, the security exposures and protocols are mainly imposed by the type of connectivity between the entities.

10.6.2.1 The Disconnected Cases

In these cases, neither the merchant M nor the client device C can have access to the latest state information of the m-server. Additionally, neither the user nor the merchant can update the information in the server. This case raises a number of security threats including the following:

- *Double spending.* Since the latest state is not accessible, the merchant cannot check whether the e-money (or the coupon, in certain examples) presented by C has already been spent with someone else before it is presented to M. This may generate double spending. However, it is worth it to notice that, in some examples, the protocol used can detect this attack a posteriori. In some cases it also identifies the attacker.
- *Lack of updates.* The merchant cannot inform online the server that a particular amount of e-money has been spent by a given client. The state of client account is not immediately updated and lacks freshness since disconnection causes several difficulties as it can be characterized as a lack of freshness.
- *Credential freshness.* The provision of m-commerce services requires the authentication of the client and the merchant during a transaction or prior to the access to an m-service. When the authentication is performed by presenting a certificate that binds a public key with the identity of the authenticated

entity, neither the client nor the merchant can verify whether the certificate is still valid and has not been revoked.

■ *State integrity.* A particular exposure of the m-commerce services is the possibility of *rollback*. In fact, due to the limited physical security of the mobile devices and the potentially exposed nature of merchants, the internal state of M and C can be manipulated by rolling back the effects of a transaction, if appropriate countermeasures are missing. On the other hand, because of the lack of connection, a delay may exist between synchronizing this internal state with remote storage (at the payment server, for example), and because of the lack of physical security, this uncommitted state may be subject to manipulation. Cryptographic techniques can prevent some types of manipulation. This creates a set of security and privacy challenges, typified by the following attack, referred to as the *Active Double Spending by Rollback*: Before spending an amount of money (e.g., e-cash) at the vending machine, the client can record the complete internal state of his mobile device, then restore this stored state afterwards and spend this e-money again.

■ *Privacy of state.* A particular consequence of the lack of physical security of mobile devices is characterized by the fact that the state at the device can be read and cloned. This exposure occurs in authentication. Typically, if one entity cannot be trusted to keep its authentication information secret, then the other entity can compensate by checking with some global state. Unfortunately, this feature cannot be provided in the disconnected case, since online checks are not possible.

10.6.2.2 The Connected Cases

These cases provide the ability to overcome some of the major shortcomings observed with the disconnected case. We assume that both the user and the merchant have some portion of safe trusted storage in the infrastructure, and then the entity that has a link to this infrastructure can also serve as a proxy extending this link to the other partner. Three particular cases can be distinguished among the connected case. They are the server-centric case, where an external server can route the communications issued by C to M, and conversely. The client-centric case can carry out process of forwarding the messages to simulate full connectivity. The merchant centric cases carry out a process of forwarding messages. The main security issues which arise in this context are the following:

■ *Privacy and authenticity of communications.* The intermediate entities located on the routes between C and M and the server can try to launch attacks against the communications established. In fact, the message paths often bring in more points of failure and more opportunity of attacks. For example, if a portion of the path involves a WLAN, then other devices on this network might carry out many types of attacks. Typical attacks involve altering the contents or the order of messages and replaying messages sent earlier, traffic analysis, and denial of service. On launching a DoS, an intermediate entity can suppress messages meant for the other parties and stop sending commit messages. When performing traffic analysis, the intermediate entity does not actively modify the messages passing through, and it will be aware of the frequency and length of the messages being exchanged. This can potentially lead to a substantial breach of privacy.

- *Reduced availability.* The occurrence of more points of attack (on a path to the server, for example) can increase the likelihood of failure of communications during a transaction. This can be a serious problem, since standard fault-tolerance techniques are not always suitable security protocols; in fact, their application may actually subvert protocols.

- *Active translation.* Typically, the server connected to the client and the merchant needs to translate between the user's protocols and those of the server since the client devices do not speak the same protocols implemented by the server and the merchant.

- *Client-kiosk identification.* Typical authentication techniques aim at checking whether one party is in fact communicating with another party with a given identity. However, these techniques are effective only if one knows the identity in the first place. In many m-commerce interactions, the pairing (C, M) of a client and a merchant occurs because, sometimes, of co-location and user needs. For example, the user wants the system to know that he needs to acquire from the vending machine close to him and does want to pre-load the serial number of a particular vending machine. Direct C–M interaction techniques provide an easy way to address this problem, for example. However, the server-centric case may make this very difficult to provide. This situation can create some interesting risks. For example, a server trying to match a user to the closest vending machine may not necessarily be aware of the locked door separating the user from the physically closest machine.

- *Privacy provision drawbacks.* In server-centric case, the provision of an m-service based on cellular networks may create privacy risks, since the server is able to know the physical location of the client device. With current technology, interactions between the client and the server appear, without doubt, to draw in this technology.

10.7 M-Service Message Protection Mechanisms

The use of the SSL protocol can provides a basic security for m-services. However, various services may need a high level of granularity that is not provided with that service, particularly when the service is composite. Therefore, it may be useful how the messages and transactions exchanged during the use of an m-service can be protected by technique suitable for m-service. Available standards such as XML encryption and XML signature seem to be able to be integrated into m-service systems. Particularly, ebXML can handle digital signature and digital encryption (ebTA, 2001).

10.7.1 Security Needs

Because m-services may contain sensitive information with respect to providers and customers, the security of m-services is an essential issue for the service provider and the service requester. The request, payment, and delivery of an m-service, for example, usually involve the exchange of messages between the actors involved: the customer and the provider. Thus, the protection of these messages is essential. The security assessment for m-service systems is also essential to enhance the trust the clients, citizens, and businesses can feel for the use of m-services. The assessment can be realized to satisfy the following goals:

- to look at the security related causes of m-service failure; and
- to investigate how far the reliability of m-service monitoring and access control could compromise the m-service system if they allow attackers to take advantage of a security breach.

The best way to operate a risk analysis is to identify the assets and determine their values, list the threats targeting to these assets, and contain their vulnerabilities. Then the analysis has to provide a protection that presents the highest ration of quality/cost. Several methodologies have been proposed to this end for e-services and can be adapted to m-services. Failure modes and analysis of effect attempt to look at individual components and functions of the system and investigate their possible modes of failure. They consider the possible causes for each failure mode, while analyzing their potential consequences. The effects of failure can be determined for each m-service component, and then evaluated for the complete m-service. Countermeasures can be proposed after these tasks.

To protect messages, two types of messages, denoted by *request* and *response*, should be distinguished. They are used to allow service requesters to ask for a remote service and authorize e-service providers to respond to requests. Each message consists of two parts: a header and a payload. The header may carry auxiliary information for authentication and transactions. The payload of the message contains, in the case of request for remote services, the method name, arguments, and m-service target address. The response method is structured just like the request, except that payload should contain the encoded content result.

To provide a maximum of security, the m-service must provide secure message exchange and support a large spectrum of security models including X.509 and SPKI, when the mobile device of the user is well equipped to handle their complexity or when a mobile agent is acting on behalf of the mobile user to access the m-service and report to him securely; or including mobile PKI, when the mobile device is involved directly in the execution of the m-service and has limited resources.

Message exchange security should provide support for multiple formats, multiple trust domains, multiple signature formats, and multiple encryption technologies. It is also helpful that the secure exchange takes into consideration how keys are derived, how trust is established, how security policy is agreed on (when actors involved operate with different platforms, for example), how authentication is performed, and how non-repudiation is provided. In the following subsections, two major contributions are discussed: the SOAP message security and the ebXML message security.

10.7.2 SOAP Message Security

The SOAP message security takes care of the protection of messages against threats aiming at changing the flowing messages and sending well formed messages to an e-service that lacks appropriate security of the claims they contain. SOAP message security provides specific solutions for the following issues: the message newness to protect against replay attacks and limit delays; the man-in-the-middle attacks; the integrity of security elements used to provide confidentiality, integrity, and authentication; and the correct use of digital signature.

An abstract message security model is built in terms of security tokens combined with digital signatures to protect and authenticate SOAP messages. A security token represents a list of protected claims (i.e., information included in the token by an entity including the name, the identity, the group, the security keys, and privileges). A signed token is a security token that is cryptographically signed by a trusted authority. The typical examples of signed security tokens contain the Kerberos tickets and the X.509 digital certificates. Signed security tokens can be used to guarantee bindings between authentication elements and the principal identity (i.e., key owner's identity). In the absence of approval by a trusted third party, a security token provides no guarantees of the claims it contains. In this case, it is the responsibility of the token recipient to accept or reject the claims made in the token.

Message headers can be used as a mechanism for conveying security information. They allow for security tokens to be directly inserted into the header. Security tokens have different types and are attached to message headers. Three types of security tokens can be distinguished: user *name tokens*, binary *security tokens*, and *XML tokens*. User name tokens are introduced as a way to provide a username and optional password information. The additional security property induced by the use of password is made by appending two optional elements, a nonce or a timestamp. A password-digest is computed by the following formula:

$$Password\_Digest = SHA\ (nonce + timestamp + password).$$

where the hashed elements are assumed available. However, if only one of these elements is available, it must be included alone in the digest. This helps hiding the password and offers a basis for preventing replay attacks.

The digital signatures are naturally used to verify message origin and message integrity. They are also used by the message producers to demonstrate knowledge of the key used to prove the claims made in the token and bind their identities to the message they produce. The following example illustrates the form of a secure SOAP message where the body of a message is only signed. The example shows where to place the used security token:

```
<SOAP envelope
<secured message header
  <Timestamp creation time>
  <security token>
  <signature
      <signature-information>
      <signature-value>
      <key-information>>
<message body>>
```

In this example, we assume that the message producer uses a security token and a signature. The token contains a data transporting a symmetric key, which is assumed to be properly authenticated by the recipient. The message producer uses the symmetric key with a hash function to sign the message body, while the recipient uses the same key and hash function to validate the signature. Moreover, the field *signature-information,* in this example, describes what is being signed in the message and the type of standards being used to normalize the data to be signed. It includes indications on the signature method, the hash method, the elements that are signed, and how to hash them. The field *signature-value* specifies the signature value of the normalized form of the data as defined in WSSE (2004). Finally, the *key-information* field provides information as to where to find the security token associated with this signature. This may indicate, in addition, that the token can be accessed and retrieved from a specified URL.

To achieve its requirements, SOAP message security necessitates that a message recipient should be able to reject messages with missing necessary claims or whose claims have improper values. In addition, since it is important for the addressee of a message to be able to determine the freshness of a message and protect it against replay attacks, time stamping can be used to provide such services. The specification offered in WSSE (2004) defines and illustrates the use of time references in terms of what is defined in the XML schema. The *timestamp* module provides a mechanism for expressing the creation and expiration times of the security elements in a message. It assumes that all time values should be written in *UTC format* as specified by the XML schema type. The expiration time, however, is relative to the requestor's clock. In order to evaluate the expiration time, a recipient needs to recognize that requestor's clocks may not be synchronized to its clock. The recipient, therefore, should make an estimation of the level of trust to be placed in the requestor's clock and estimate the degree of clock alteration. The aforementioned example places in the *timestamp* field the creation data of the secured message.

Finally, it is worthy to notice that there are many situations where an error can occur while processing the security information associated with a sent message. This may occur in the case of unsupported type of security token, invalid security token, invalid signature, decryption failure, or unreachability of the referenced token archive. To this end, a SOAP fault mechanism can be made available so that the errors can be reported to it. The SOAP fault mechanism can help in detecting the cause associated with the reported error, provided that a specific file is available for this task.

10.8 Securing Registry for M-Services

As previously discussed, an m-service registry offers a set of services that facilitate sharing of information about m-service offerings between interested parties. M-service registries use mechanisms to guarantee that any alteration of the content submitted by an entity can be detected. These mechanisms must support explicit identification of the responsible entity for any registry content. To this end, registry clients may have to sign any content before submission, otherwise the m-service registry should be able to reject the content. The signature ensures that any alteration (e.g., changes or tampering) of the content's authenticity can be discovered using its association with the requesting entity. A registry client, named C_A, wishing to check whether a given client C_B has really published a specific content, that is available at the registry service, can rely on the following components of the response he gets from the registry service:

- the payload of the response containing the content that has been published by the client C_B;
- the public key for validating C_B's payload signature, while using appropriate information so that the C_A can retrieve that key; and
- the signature element containing the header signature of the response (as made by the registration authority).

The aforementioned type of protection has been implemented by various methods. Next, we briefly describe how the ebXML integrates this approach and provides security mechanisms for the ebXML registry service (ebRS, 2002). For the lack of space in this chapter and the similarity of methods, we do not consider the other aforementioned schemes.

10.8.1 ebXML Registry Security

The ebXML registry service specifies the ebXML scheme definition and provides the security solutions for the access control of ebXML registries. It provides an authorization mechanism and proposes another mechanism for the confidentiality of registry content. In addition, it also develops a message header signature, presents a payload signature, and discusses a key distribution scheme. On one hand, the access control in ebXML registries is provided by creating a default access control policy that grants the default permissions to registry customers. Three policies are defined and the related permissions are set up. They are (a) the content holders are authorized to access all the content they own, (b) the registry administrators are allowed access to all contents, and (c) the unauthorized users (or guests) of registry can only access all read-only content. Moreover, the access control assumes that any user can publish content, provided that he/she is a registered user; that any user can access the content without entering an authentication process; and that, on content submission, the submitting entities are assigned the default content owner role as authenticated by the credentials in the submission message.

The signature process involved with the ebXML registry takes care of two issues: the payload signature and the header signature. The payload signature is put together with payload forming a multipart message encoded in MIME, where the first body part contains an XML signature. Message headers are signed to provide data integrity while the message is flowing through the network. This is achieved by a hash digest computation of payloads. Header signature requires that algorithm be identified using an algorithm attribute. Signature in the header is not included in the signature calculation.

To check the validity of digital signature, the receiver of the signature requires the public key associated with the signer's private key. For this purpose, the participants may use a key-information field or distribute the public keys using a trusted third party (or certification authority). Moreover, the ebXML registry service has to assume that the registration authority and the registry client have their digital certificates, that the registry client should register its certificate with the registration authority, and that a registry client obtains the registration authority's certificate and stores it in its own protected archive.

In summary, one can say that the ebXML registry is able to authenticate the identity of the entity involved with a client request by checking the validity of the message header signature with the entity's certificate, which may be included in the message itself or provided by the registration authority (using unspecified ways). The authentication of a payload is achieved by the signature associated with the payload. It is made on a per message basis, since all the messages can be considered independent and there is no need to establish a connection to send a request. The verification includes the identification of the privileges that the entity associated with the message has with respect to specific contents in the registry.

10.8.2 Service-Side Protection of Registries

It is noticeable that protecting an m-service registry server from attacks implies considering issues including security policy, user management, security auditing, system configuration, and log management. An overview of these issues is discussed in the following.

The utilization of a security policy helps significantly to prevent against configuration inconsistencies and react efficiently to security incidents. A security policy is a dynamic document that is intended to protect registry related systems, delimit acceptable uses, elaborate personnel training plans for security policy enforcement, and enforce security measures. The dynamic character of the security is dependent on various features including the m-service nature, the type of users, and the evolution of the m-service. It should be customized to the enterprise need to cope with its business activity, and resource sensitivity. Registry security policies need to be updated on a regular basis, and also on the occurrence of situations where new vulnerabilities can be a critical concern. These situations may require changing security rules, adding new rules, or developing new strategies.

To avoid damages and cut down the resulting risks of their uses, requesters should be managed effectively. Default configurations of m-services and registries often include default accounts with known passwords, as well as active guest accounts. These default objects may be used by the attackers to launch critical attacks. Assigning full trust to a registry server user, who has been authenticated, may also be harmful since the unauthorized user can, for example, take control of an active session while its legitimate user is not in control. To overcome this threat, the registry operating system can be configured to lock any user session after an idle period. Such a solution seems to be completely insufficient in highly sensitive contexts.

After a server risk mitigation plan is enforced and a security policy, including procedures and configuration settings, is implemented on a registry, a security audit needs to be continuously conducted on the registry since it remains necessary to test whether the security policy is being enforced correctly. The auditing aims at checking that no breaches in the registry configurations have been initiated during system operation and that the server is correctly protected. To check whether security measures set up to protect a registry server have effectively protected the server, the security audit should attempt to identify whether the server has been targeted by intruders. To this end, the set of available assets to be examined, statistical analysis performed, anomalies detected, and alerts correlated in order to identify attack scenarios and decide the appropriate countermeasures to mitigate them.

Log files represent the typical sources for detecting suspect behaviors and intrusion attempts. The efficiency of these files can be considerably reduced by any failure or attack on the related data collection mechanisms and any weakness in protecting their output. Logs protection is very important in the sense that intruders may access these files or the tools that manage them, in order to remove, alter signs of malicious activity, or even add erroneous information to these files. Log file protection can be altered due to bad configuration of access permissions, insecure transfer to remote hosts, and storage in public areas, in addition to any successful intrusion on the system that hosts the log files. Logging data locally is easy to configure; it allows instantaneous access, but it is less secure as the log content may be lost whenever the registry is compromised. On the other hand, the remote log storage is well protected, but requires strengthening the communication security medium using, for example, a dedicated channel or an encryption mechanism.

Another important concern that registry security administrators have to face is the guarantee of registries' availability. Registry overload should be efficiently managed to avoid being a victim of QoS deterioration, where a client response time increases over an acceptable level and the target reaches a denial of service. To protect servers from overload, a simple bounding of the flow rate is inefficient. Various other actions should be attempted, including traffic shaping, load controlling, and policy management. Traffic should shape the flow that meets the server performance. This is done by delaying traffic excess using buffering, queuing mechanisms, and request rejection. Classifying traffic, however, is not enough to reduce or avoid overload on the monitored registry. To this end, it should be complemented by a load controlling mechanism based on reliable traffic metrics. Once a message is linked to its category, the set of load metrics associated to that category are assessed to decide which action should be taken in response to that message. Policy management is the administratively configurable part of the overload protection component. It defines the metrics values, specifies the registry reactions, and specifies whether specific measures should be taken any time the load controller metrics exceed their thresholds.

The denial of service attacks try to keep the systems and applications from providing their services. As a result, a server victim of such attack will appear unreachable to its users. Recently, DoS attacks have changed to more intense and damaging attacks including Distributed DoS attacks (DDoS) that use many compromised servers to launch coordinated DoS attacks against single or multiple targets. To protect servers against denial of service attacks, many proposals have been made without completely solving the issue. Protection mechanisms include packet filtering, automated attack detection, and security vulnerability fixing. With packet filtering, end-routers apply ingress packet filtering to allow only servers supported protocols, deny security-critical services and suspicious identified source IP domains and services, and forged IP addresses. Routers should be monitored to update their filtering rules, as intruders' techniques and behaviors evolve.

References

T. Bellwood et al., *UDDI Verssion 3.0*, Published specification, July 2002 (available at http://uddi.org/pubs/uddi-v3.00-published-20020719.html.

S. Ben Abdallah, N. Boudriga, and S. Guemara-ElFatmi, Security Issues in E-Government Models: What Government Should Do, *IEEE International Conf on Systems, Man, and Cybernetics (ICSMC2002)*, Oct. 6–9, 2002, Hammamet, Tunisia.

D. Chakraborty, A. Joshi, T. Finin, and Y. Yesha, GSD: A Novel Group-Based Service Discovery Protocol for MANETS. In *4th IEEE Conf. on Mobile and Wireless Comm. Net. (MWCN'02)*, pp. 301–306, Stockholm, Sweden, Sept. 2002.

D. Chakraborty, F. Perich, A. Joshi, T. Finin, and Y. Yesha, A Reactive Service Composition Architecture for Pervasive Computing Environments, in *7th Personal Wireless Communications Conference (PWC)*, pp. 53–62, Singapore, October 2002b.

D. Chalmers and M. Sloman, A Survey of Quality of Service in Mobile Computing Environments, *IEEE Communications Surveys*, 2(2), 1999.

F. Caseti, S. Ilinicki, L. Jin, V. Krishnamoorthy, and M. Chan, Adaptive and Dynamic Service Composition in eFlow, in *Proc. of Int. Conf. on Advanced Information Systems Engineering (CaiSE'00)*, Stockholm, Sweden, 2000.

H. M. Deitel, P. J. Deitel, and K. Steinbuhler, *e-Business and e-Commerce for Managers*, Prentice Hall, New Jersey (2001).

ebXML Registry Services Specification v2.0, April 2002, available http://www.ebxml.org/committees/regrep/documents/2.0/spects/ebRS.pdf.

ebXML Technical Architecture Specification v1.0.4, February 2001 (available at http://www.ebxml.org).

I. Kushchu, and H. Kushchu, (2003), From E-Government to M-Government: Facing the Inevitable, in *Proceedings of European Conference on E-Government (ECEG 2003)*, Trinity College, Dublin.

E. Lallana (2004), e-Government for Development, M-Government Definitions and Models, Web: http://www.egov4dev.org/mgovdefn.htm.

L. Liebmann, Preparing for m-commerce. *Communication News,* 37(9), (2000). Available at: http://www.libfind.unl.edu:2020/journals/iris/busis.html.

Z. Maamar, W. Mansoor, and Q. H. Mahmoud, Software Agents to Support Mobile Services, in *Proceedings of the First International Joint Conference on Autonomous Agents & Multi-Agent Systems (AAMAS'2002)*, Bologna, Italy, 2002.

D. Mennie and B. Pagurek, An Architecture to Support Dynamic Composition of Service Components, in *5th International Worshop on Component-Oriented Programming*, June 2000.

Nadalin et al., Web Services Security: SOAP Message Security, OASIS standard 200401, March 2004 (available at http://docs.oasis-open.org/wss/2004/1/oasis-soap-message-security-1.0). M. Scannapieco, V. Mirabella, M. Mecella, and C. Batini, Data Quality in e-Business Applications, *LNCS* 2512, pp. 121–138, 2002.

H. Schuster, D. Georga Koupoulos, A. Cichocki, and D. Baxter, Modeling and Composing Service-Based and Reference-Based Multi-Enterprise Processes, CAISE 2000, pp. 247–263.

K. Siau and Z. Shen, Building Consumer Trust in Mobile Commerce, *Communications of the ACM*, 46(4)(2002): 91–94.

SOAP, http://www.w3/TR/SOAP.

A. Tiwana and B. Ramesh, e-Services: Problems, Opportunities, and Digital Platforms, *Proc. of the IEEE 34th Hawaii Intern. Conf. on System Sciences*, 2001.

A. Tsalgatidou and T. Pilioura, An Overview of Standards and Related Technology in Web Services, *Distributed and Parallel Databases*, 12, 135–162, 2002.

WSDL, http://msdn.microsoft.com/xml/general/wsdl.asp.

Chapter 11

Security of Mobile Sensor Networks

11.1 Introduction

Often, the term *sensor network* is used to refer to a heterogeneous system combining tiny sensors with general purpose computing elements. Sensor networks may integrate hundreds (and even thousands) of low-power, low-cost sensor nodes. The sensors are possibly mobile but more likely at fixed locations. They are largely deployed to monitor a specific environment. Sensor networks often have one or more points of centralized control called *base stations*. A base station is typically a gateway to another network, a powerful data processing and storage center, or an access point for human interface. They can be used to distribute control information into the network or collect useful information from it.

Wireless sensor networks (WSN) share multiple similarities with the ad hoc wireless networks. The dominant communication method in both networks is multi-hop networking, but several important differences can be observed between the two networks. For example, the ad hoc networks support routing between any pair of nodes, whereas the sensor networks have a more dedicated communication pattern. In fact, three (overlapping) categories of traffic can be distinguished in the sensor networks:

- *Many-to-one traffic*: Multiple sensor nodes send information collected by the sensors to a base station or to an aggregation point in the network.
- *One-to-many traffic*: A single node, in the WSN, floods a query or control information to a group of sensor nodes.

■ *Local traffic*: The nodes in a limited area send localized messages to discover, for example, the neighboring nodes and coordinate with each other. A node may broadcast messages intended to be received by all neighboring nodes or it may send messages intended for a single neighbor.

The nodes in ad hoc networks have usually been seen as having limited resources, but sensor nodes are more constrained, particularly when energy is considered. Moreover, the nodes in a sensor network often implement trust relationships beyond the relationships that can be found in an ad hoc network. Neighboring nodes in WSNs often perceive the same environmental events. If each node sends a packet to the base station in response, valuable energy and bandwidth can be lost. To reduce some of the redundant messages, decrease the generated traffic, and save energy, a WSN may require in-network processing, function aggregation, and duplicate removal. This often necessitates the management of trust relationships between nodes that are not typically utilized by the ad hoc networks.

The applications of Wireless Sensor Networks, which cover both the civil and military domains, are continuously expanding. The WSNs can be used to gather and analyze information about vehicular movement, the control of a large spectrum of parameters such as the humidity, the temperature, and the pollution character, as well as target traces. However, the large potential of WSNs can be addressed only if the corresponding infrastructures are adequately protected. Security issues in ad hoc networks are similar to those in sensor networks and have been well enumerated in the literature, but the protection mechanisms developed for the ad hoc networks are not directly applicable to the sensor networks due to a large set of reasons mainly related to the differences observed between sensor and ad hoc networks.

Violating one or more security properties would lead to erroneous decisions in wireless sensor networks. Subsequently, this will induce wrong reactions for the applications implemented by the WSNs. Thus, the security should rank at the top of the issues that should be addressed during the design phase of a WSN. This highlights the fact that WSNs are, by nature, mission-critical, meaning that they are developed for sensitive tasks where error-tolerance is very small. The importance of security in the WSN context is worsened by a set of crucial factors that include the following:

■ Sensor nodes have limited storage, computation, and power resources. The implemented security mechanisms should cope with the limitation of the embedded resources and the WSN capabilities.
■ The WSN does not have a fixed infrastructure and does not have a static topology either. The WSN architecture variability makes the use of the existing robust cryptographic mechanisms more difficult than in ad hoc networks.

■ The sensing and communication tasks are often performed in a hostile environment where the gathered events are subjected to numerous threats that might affect the final decision.

■ The detected events are forwarded through the sensor nodes themselves, preventing the application of strong communication security mechanisms.

This chapter surveys the major security features, issues, and challenges of the WSNs. More precisely, the following aspects will be discussed:

■ *WSN issues*: Several WSN basic issues are addressed to highlight the security challenges. The components, architecture, topology, routing, mobile target tracking, and alert management will be discussed among other items.

■ *WSN security objectives*: Traditional security goals should be extended to fit the requirements of WSNs. Several particular concepts are introduced at this level. For instance, confidentiality, authenticity, and integrity, which have been customarily associated to data and node identity, should be extended to cover node location. However, this poses several new security challenges in the WSN context.

■ *Attacks against WSNs*: This chapter describes the most important attacks techniques concerning WSN routing and other protocols. Attacks are classified according to the basic security properties they violate. Particularly, the classification of the attacks addressed here is based on four major activities: (a) the attacks against transmitted information; (b) the attacks against the network architecture, structure, and protocols; (c) attacks against the localization framework, and (d) the attacks targeting the functions performed by the WSN.

■ *Counteracting against attacks*: The security solutions that allow counteracting the aforementioned attacks are discussed. The countermeasures are classified according to the level at which they act (e.g., link level, routing, application).

11.2 Wireless Sensor Networks

Due to the advances witnessed in wireless communications and electronics over the last few years, the development of networks of low-cost, low-power, multifunctional sensors has received increasing attention. These sensors are small in size and able to sense, process data, and communicate with each other, typically over an RF (radio frequency) channel. A sensor network is designed to detect events or phenomena, collect and process data, and transmit sensed information to interested entities in the network. The basic features of the sensor networks are the following:

- *Self-organizing capabilities*: The WSNs are able to cope with topology variability and infrastructure variations.
- *Short-range broadcast communication and multihop routing*: The sensor nodes have reduced radio ranges and should cooperate to achieve complete routing of information.
- *Dense deployment and cooperative effort of sensor nodes*: The shortage of the radio range and the need to have efficient sensing call for a dense deployment of sensors.
- *Limitations of energy, transmit power, memory, and computing power*: WSNs cope with limitation of resources and frequent changes of topology due to fading and node failures.

The aforementioned characteristics, particularly the last three, make sensor networks different from other wireless ad hoc or mesh networks. Clearly, the idea of mesh networking is not new; it has been suggested for some time for wireless Internet access or voice communication. Similarly, small computers and sensors are not innovative per se. However, combining small sensors, low-power computers, and radios makes for a new technological platform that has numerous important uses and applications.

11.2.1 WSN Features

The wireless sensor networks present an uttermost interest from an engineering perspective because they generate a number of serious challenges that cannot be adequately addressed by the existing technologies. A non-exhaustive list of challenges includes the following:

- *Extension of lifetime*: As mentioned above, WSN nodes will generally be severely energy constrained due to the limitations of batteries. A typical alkaline battery, for example, provides about 50 watt-hours of energy; this may translate to less than a month of continuous operation for each node in full active mode. Given the expense and the potential infeasibility of monitoring and replacement of batteries for a large WSN, significantly longer lifetimes would be desired. In practice, it will be necessary for many applications to provide guarantees that a network of unattended wireless sensors can remain operational without any replacements for very large periods of time.
- *Responsiveness*: A simple solution to extending network lifetime is to operate the nodes in a duty-cycled manner with periodic switching between sleep and wake-up modes. While synchronization of such sleep schedules is challenging in itself, a larger concern is that arbitrarily long sleep periods can reduce the responsiveness and the effectiveness of the sensors. In applications where it is critical that certain events in the environment be detected and

reported rapidly, the latency induced by sleep periods must be kept within strict bounds.

■ *Robustness*: The use of large numbers of inexpensive devices characterizes WSNs. Unfortunately, inexpensive devices can often be unreliable and prone to failures. Rates of device failure will also be high whenever the sensor devices are deployed in harsh or hostile environments. Protocol designs must therefore have built-in mechanisms to provide robustness. It is important to ensure that the global performance of the system is not sensitive to individual device failures. Further, it is often desirable that the performance of the system degrades as gracefully as possible with respect to component failure.

■ *Synergy*: Moore's law-type advances in technology have ensured that device capabilities in terms of processing power, memory, storage, radio transceiver performance, and even accuracy of sensing improve rapidly (given a fixed cost). However, if economic considerations dictate that the cost per node be reduced drastically, it is possible that the capabilities of the individual nodes will remain constrained to some extent. The challenge is therefore to design synergistic protocols, which ensure that the system as a whole is more capable than the sum of the capabilities of its individual components. The protocols must provide an efficient collaborative use of storage, computation, and communication resources.

■ *Scalability*: For many envisioned applications, the combination of fine granularity sensing and large coverage area implies that the WSN have the potential to be extremely large scale. Protocols will have to be inherently distributed, involving localized communication, and sensor networks must utilize hierarchical architectures to provide such scalability. However, visions of large numbers of nodes will remain unrealized in practice until some fundamental problems, such as failure handling and reprogramming, are addressed even in small settings (involving tens to hundreds of nodes). There are also some fundamental limits on the throughput and capacity that impact the network performance.

■ *Heterogeneity*: One would expect the heterogeneity of device capabilities (with respect to computation, communication, and sensing) in realistic settings. This heterogeneity can have a number of important design consequences. For instance, the presence of a small number of devices of higher computational capability along with a large number of low-capability devices can dictate a two-tier, cluster-based network architecture, and the presence of multiple sensing modalities requires pertinent sensor fusion techniques. A key challenge is often to determine the right combination of heterogeneous device capabilities for a given application.

■ *Self-configuration*: Because of their scale and the nature of their applications, wireless sensor networks are inherently *unattended* distributed systems. Autonomous operation of the network is therefore a key design challenge.

From the very start, nodes in a wireless sensor network have to be able to configure their own network topology; localize, synchronize, and calibrate themselves; coordinate inter-node communication; and determine other important operating parameters.

■ *Privacy and security*: The large scale, prevalence, and sensitivity of the information collected by wireless sensor networks (as well as their potential deployment in hostile locations) give rise to the final key challenge of ensuring both privacy and security.

11.2.2 Power Scarcity

Base stations are typically many orders of magnitude more powerful than the sensor nodes. They might have workstation or laptop class processors and memory, storage capacity, and high bandwidth links for communication between each other. On the other hand, the sensors are constrained to use lower-power, lower bandwidth, shorter-range radios. Therefore, it is envisioned that the sensor nodes would form a multi-hop wireless network to allow sensors to communicate to the nearest base station. However, a base station may request a continuous stream of data to comply with the needs of an application. To reduce the total number of messages sent and received by the sensors and thus save energy, sensor readings from multiple nodes may be processed at one among many possible *aggregation points*. An aggregation point collects sensor readings from neighboring nodes and forwards a single message representing an aggregate of the values.

The aggregation points are typical regular sensor nodes, and their selection is not necessarily static. They might be chosen dynamically for each query or event, for example. It is also possible that every node in the network functions as an aggregation point, delaying transmission of an outgoing message until a sufficient number of incoming messages have been received and aggregated. Figure 11.1 illustrates a representative architecture for sensor networks. A hierarchical WSN (HWSN) is shown in Figure 11.1(a); the hierarchy among the nodes is based on their capabilities: base stations, cluster heads and sensor nodes.

Power management in sensor networks is critical. Subsequently, if the designer of a sensor networks wants it to last for very long period of time, it is essential that they run at very low frequencies, moving between idle and active modes. Similarly, since the power consumption of the radio is approximately three orders of magnitude higher when transmitting or listening than when in idle mode, it is crucial to keep the radio in idle mode the overwhelming majority of the time. It is clear that when it comes to security, the wireless sensor networks differ from other distributed systems in important ways. Indeed, the resource-starved nature of sensor networks poses great challenges for security. These devices have very little computational power, and therefore, the public-key cryptography is so expensive as to be unusable. Even fast symmetric-key ciphers must be used scarcely. Also, communication bandwidth is extremely precious: each bit transmitted consumes

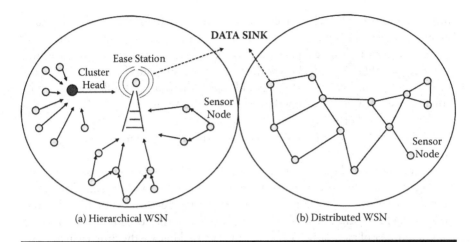

Figure 11.1   Hierarchical and distributed wireless sensor networks.

about as much power as what is needed to execute 800 to 1000 local instructions. Subsequently, any message expansion caused by security mechanisms comes at significant cost.

This leaves us with a very challenging problem characterized by the need of providing reliable security features under such tight constraints. With sensor networks being foreseen for use in critical applications such as building monitoring, intruder alarms, and emergency response, these networks are at high risk, if efficient solutions are provided to overcome the lack of physical security for hundreds of exposed sensors and vulnerabilities induced by the use of wireless links in communications.

11.2.3 Routing Protocols

One of the major concerns with respect to wireless sensor networks applications is the design and implementation of an energy-efficient and secure routing protocol that may operate hostile environments. The protocol should ensure that connectivity in a network is maintained for as long as possible, and the energy status of the entire network should be of the same order. This is in contrast to energy optimizing protocols that find optimal paths and then consume the energy of the nodes along those paths, inducing a wide disparity in the energy levels of the nodes of the WSN. Therefore, one can find that many protocols and algorithms, like the DSR and AODV (see Chapter 8), have been proposed for traditional wireless ad hoc networks. But they are not well suited for the unique features and application requirements of sensor networks because of many differences between sensor networks and ad hoc networks.

Nowadays, the multipath mechanism is widely used in WSN routing protocols in order to increase network lifetime as well as resistance to nodes failure. The literature on both single path and multipath routing solutions is important. The directed diffusion protocol (DDP) is a data-centric multipath routing protocol that is designed to meet the robustness, scaling, and energy-efficiency requirements. DDP uses data driven routing coupled with application specific in-network processing. It can help establish energy efficient data dissemination paths between sources and sinks. In addition, it allows the design of localized algorithms for flexible path construction and recovery, enabling these systems to be robust to dynamics. However, DP is vulnerable. For example, an adversary can influence the path taken by source and sink node by spoofing positive and negative reinforcements and bogus data events.

Various other multipath routing protocols can be found. They all use multipath routing to rapidly find alternate paths between the source and the sink. To recover from failure of the primary path, which is the best path measured by different metrics like low latency or minimal number of hops, they construct and maintain a small number of alternative paths that can be used in case the primary path fails. However, some major weaknesses can be observed with these multipath routing protocols in WSN. One can see that very few protocols have considered security. They only focus on how to efficiently use energy on each node. In addition, they bring additional maintenance overhead, such as the energy expended for setting up and maintaining multipaths.

To secure multipath routing in WSN (Perrig, 2004) has presented two optimized building block security protocols, called SNEP and μTESLA. SNEP provides confidentiality, authentication, and freshness between nodes and the sink, whereas μTESLA provides authenticated broadcast, which is an important mechanism for sensor networks. However, some attacks pose challenges to secure routing protocol design and there are not effective countermeasures against these attacks that can be applied after the design of a protocol has completed.

Another routing protocol, called SEEM (Secure and Energy-Efficient Multipath Routing) protocol, is proposed (Nasser, 2007). SEEM uses the Client/Server concept. It assumes that the base stations are responsible for the route discovery, route maintenance, and route selection. Instead of maintaining a single path, the base station periodically selects a new path from multipath based on current energy level of nodes along each path. SEEM considers energy-efficiency and security simultaneously and takes full advantage of the predominance of the base station, in the sense that the base station select a path from multipaths for the source and sink pairs of nodes. As a result, the network throughput, communication overhead, and network lifetime are improved. It also presents a good protection against some attacks.

The protocol SEEM is neither a proactive nor reactive routing protocol. Each node implementing SEEM does not have and store consistent up-to-date routing information to every destination in the WSN. The only information kept in the routing tables is the routes to the base station. This list contains the neighbors that can forward packets to the base station for the node. There is not an explicit route discovery process provided by SEEM. It is based on the knowledge that base stations have. It also builds on the fact that base stations are typically many orders of magnitude more powerful than common sensor nodes.

SEEM develops three phases: Topology Construction, Data Transmission, and Route Maintenance. The topology construction takes place right after the deployment of all sensor nodes and can be launched at any time by the base station during the lifetime of the network. To initiate the topology construction, the base station first broadcasts a neighbor discovery (ND) message to the whole network. Each node receiving this message does the following two steps:

- It records the address from which the current node receives the message and stores it in a list in ascending order of message received time.
- It checks if the broadcast message has been received. If the message has already been received once, then it drops this ND and does not rebroadcast it.

By receiving and re-broadcasting the ND message, each node knows its neighbors and stores them for using in the following phases. The base station waits for a short time to ensure that the ND broadcast can be flooded through the WSN. Then the base station broadcasts another message NC (Neighbors Collection) in order to collect the neighbors information of each node gathered during the previous broadcasting. In addition to notify nodes to send back neighbors information to the base station, broadcasting NC message can also help nodes recognize neighbors that are not collected when broadcasting the ND messages.

11.3 Security Issues of WSNs

Similar to the ad hoc networks, the wireless sensor networks build on radio links and should assume that they are insecure. At the very least, attackers targeting a WSN can eavesdrop on the radio transmissions, inject bits in the channel, and replay previously heard packets. It is obviously clear that if the network deploys many sensor nodes, then the attackers will likely also be able to deploy, within the WSN, a few malicious nodes with similar hardware capabilities as the legitimate nodes. The attackers may acquire these malicious nodes separately off-the-shelf, or by corrupting a few legitimate nodes by capturing them and physically overwriting their memory. We assume that the attackers might have control on an important set of sensors and the malicious nodes might cooperate to attack the WSN. In addition we can suppose that, in some cases, the cooperating malicious nodes might have high-quality communications links available for coordinating their attack. A final natural assumption supposes that if an adversary compromises a sensor, he can extract all key material, data, and code stored on that sensor.

11.3.1 Trust and Threats in WSN

Since the base stations interface the wireless sensor network to the outside world, the compromise of a significant number of the base stations can cause the entire network to be useless. For this reason, it is worthy to assume that base stations are *trustworthy*, meaning that they can be trusted, when needed. Most, but not all, routing protocols depend on sensor nodes to trust messages from the base stations.

The aggregation points may be seen as trusted components for the routing protocols. The nodes may rely on routing information from aggregation points and trust that messages sent to aggregation points will be accurately combined with other messages and forwarded to a base station. The aggregation points are often regular sensor nodes. It is worthy to notice that it is possible that the attackers may try to deploy malicious aggregation points or attempt to turn currently compromised nodes into aggregation points. For this reason aggregation points may not necessarily be trustworthy.

In addition, an important distinction should be made between the *mote-class attackers* and the *laptop-class attackers*. While the former attackers can have access to only a few sensor nodes with similar capabilities, the laptop-class attacker may have access to more powerful devices connected to the WSN, like laptops or their equivalent. Thus, in the latter case, malicious nodes have an advantage over legitimate nodes: they may have greater battery power, a more capable CPU, a high-power radio transmitter, or even a sensitive antenna. A laptop-class attacker can do more than an attacker with only ordinary sensor nodes. For example, an ordinary sensor node might only be able to jam the radio link in its immediate neighborhood, while a laptop-class attacker might be able to jam the entire sensor network using a stronger transmitter. A single laptop-class attacker might be able to eavesdrop on an entire network, while mote class attacker nodes would ordinarily have a limited range. Moreover, the laptop-class attackers might have a high bandwidth, low-latency communications channel not available to ordinary sensor nodes, allowing such attackers to coordinate their efforts.

A second distinction may be needed to address trust it considers that the attacks can be *outsider attacks* and *insider attacks*. We have so far been discussing outsider attacks, where the attacker has no special access to the sensor network. One may also consider insider attacks, where an authorized participant in the wireless sensor network has been compromised. Insider attacks may be mounted from either compromised sensor nodes running malicious code or adversaries who have stolen the secret key material, the useful code, and the data from legitimate nodes, and who then is able to use one or more laptop-class devices to attack the WSN.

11.3.2 WSN Security Challenges

WSNs are characterized by many constraints compared to the wireless ad hoc networks. Due to these particular constraints, the application of existing network security approaches do not fulfill the required security properties. Hence, appropriate security needs and techniques should be defined for the WSN environments while borrowing concepts from the currently used security mechanisms. In the following, we highlight the most relevant, from the security point of view, of the WSN features.

11.3.2.1 Resource Limitations

Security mechanisms and processes necessarily require a certain amount of processing, power, storage, and memory resources, even though the sensor nodes are often resource-poor. In the following, we detail the challenges imposed by the basic resource limitations characterizing WSNs.

- *Processing limitations*: A custom processor for the sensor nodes should essentially have a low-power sleep mode, allowing reducing energy consumption, and a low-overhead wakeup mechanism, preventing the occurrence of network congestion and reducing signalling messages. It can be shown that the processing speed offered by most of the available micro controllers ranges under 400 Mips (million instructions per second). Even though this performance allows implementing the communication functions, it turns out to be not sufficient to support advanced security mechanisms, in particular, when a heavy traffic is exchanged across the WSN. As a result, novel security algorithms should be considered to keep up with the sensor node processing limitations.
- *Limited memory and storage space*: A sensor is a tiny device with only a small amount of memory and storage space for the code. In order to build an effective security mechanism, it is necessary to limit the code size of the security algorithm. For example, one common sensor might have a limited CPU (e.g., 16-bit, 8 MHz RISC CPU) with reduced memory RAM, program memory, and flash storage. With such a limitation, the software built for the sensor must be quite small. For example, the total code space of the *TinyOS*, which represents the de-facto standard operating system for wireless sensors, is approximately 4Ko (Hill, 2000), while the core scheduler occupies only 178 bytes. Therefore, the code size for all the security related codes must be kept reduced.
- *Power limitation*: Often, one can assume that once the sensor nodes are deployed in a WSN, they cannot be easily replaced (high operating cost) or recharged (high cost of sensors). Therefore, the battery charge taken initially to the field must be conserved to extend the life of the individual sensor node and the entire sensor network. In particular, the energy impact must be considered when implementing a cryptographic function. When adding security code to a sensor node, the impact that security has on the lifespan of a sensor should be estimated along with the extra power consumed by the sensor node due to the processing required for security functions, the energy required

to transmit the security related data or overhead (e.g., initialization vectors needed for encryption/decryption), and the energy required to store security parameters in a secure manner (e.g., cryptographic key storage).

11.3.2.2 Data Loss

Obviously, the unreliability of communication is another threat to sensor security. The security of the network relies heavily on the defined protocol, which in turn depends on communication mode. The data loss should be reduced to its minimum by overcoming the unreliability feature of the wireless communication.

- *Unreliable transfer*: Normally the packet-based routing of the WSN is connectionless based. Thus, it is inherently unreliable. Indeed, the packets may get damaged due to channel errors or dropped at highly congested nodes. This results in the loss of retransmission of missing packets. Furthermore, the unreliability of the wireless communication channel also results in damaged messages. Higher channel error rate also forces the software developer to devote computation resources to error handling. More significantly, if the protocol lacks the appropriate error handling, critical security packets can be lost. The loss may include, for example, an exchanged cryptographic key.
- *Collisions*: The WSNs impose strict requirements on the medium access protocol used on the wireless links. This is basically due to the ad hoc architecture characterizing WSNs as well as the long network lifetime needs. Moreover, as data is broadcasted over the radio link, packets may collide resulting in decreasing the channel throughput. Depending on the medium access and the transport protocols, the information loss can reach a certain degree such that the analysis nodes in the WSN become no longer able to identify the events corresponding to the gathered data.
- *Latency*: Multi-hop routing, network congestion, and node processing can lead to an important latency in the WSN, thus making it difficult to achieve synchronization among sensor nodes. The synchronization issues can be critical to sensor security where the security mechanism relies on critical event reports and cryptographic key distribution. Therefore, real-time communications in wireless sensor networks should be efficiently addressed to limit any induced latency.

11.3.2.3 Uncontrollable Behavior

Depending on the function that a particular wireless sensor network implements, the sensor nodes may be left unattended for long periods of time. There are three main caveats to unattended sensor nodes:

- *Exposure to physical attacks*: The sensor node may be deployed in an environment open to adversaries, bad weather, and unsuitable environment characteristics. The likelihood that a sensor suffers a physical attack in such an environment is therefore considerably higher than the likelihood observed with typical PCs that are located in a secure place and mainly face attacks from a network only.
- *Managed remotely*: Remote management of a sensor network makes it virtually impossible to detect physical tampering and physical maintenance issues (e.g., battery replacement). By far, the most extreme and significant example, in this case, is a sensor node used for remote reconnaissance missions for a military purpose WSN that is left behind the enemy lines. In such a case, the node may not have any physical contact with friendly forces, once deployed.
- *No central management point*: A sensor network should be a distributed network without a central management point. This will increase the vitality of the sensor network. However, if designed incorrectly, it will make the network organization difficult, inefficient, and fragile.

11.3.2.4 Routing

In the ideal world, a secure routing protocol should guarantee the integrity, authenticity, and availability of messages in the presence of adversaries of arbitrary power. Every eligible receiver should receive all messages intended for it and be able to verify the integrity of every message as well as the identity of the sender. In our view, protection against eavesdropping is not an explicit security goal of a secure routing algorithm. Secrecy is usually most relevant to application data, and it is arguably not the responsibility of a routing protocol to provide it. However, we do consider it the responsibility of a routing protocol to prevent eavesdropping caused by misuse or abuse of the protocol itself. Eavesdropping achieved by the cloning or rerouting of a data flow should be prevented, for example. Similarly, we believe protection against the replay of data packets should not be a security goal of a secure routing protocol. This functionality is best provided at the application layer because only the application can fully and accurately detect the replay of data packets (as opposed to retransmissions, for example).

In the presence of only outsider adversaries, it is conceivable to achieve these idealized goals. However, in the presence of compromised or insider attackers, especially those with laptop-class capabilities, it is most likely that some if not all of these goals are not fully attainable. Rather, instead of complete compromise of the entire network, the best we can hope for in the presence of insider adversaries is *graceful degradation*. The effectiveness of a routing protocol in achieving the above goals should degrade no faster than a rate approximately proportional to the ratio of compromised nodes to total nodes in the network.

11.3.3 WSN Security Requirements

A sensor network is a special type of network. It shares some commonalities with an ad hoc network, but also poses unique requirements of its own. Therefore, we can think of the requirements of a wireless sensor network as encompassing both the typical network requirements and the unique requirements suited solely to wireless sensor networks.

11.3.3.1 Data Confidentiality

Data confidentiality is the most important issue in network security. Every network with any security focus will typically address this problem first. In sensor networks, the confidentiality relates to the following (Perrig, 2004):

- A sensor network should not leak sensor readings to its neighbors. In particular, the data stored in the sensor node may be highly sensitive in a military application.
- In many applications nodes communicate highly sensitive data, such as the key distribution; therefore, it is extremely important to build a secure channel between any pair of nodes in a wireless sensor network.
- Public sensor information, such as sensor identities and public keys, should be encrypted to some extent to protect against traffic analysis passive attacks.
- The standard approach for keeping sensitive data secret is to encrypt the data with a secret key that only intended receivers possess, thus achieving confidentiality on the communication channels.

11.3.3.2 Data Integrity

With the implementation of confidentiality, an adversary may be unable to steal information. However, this doesn't mean the data is safe. The adversary can change the data, so as to send the sensor network into disarray. For example, a malicious node may add some fragments or manipulate the data within a packet. This new packet can then be sent to the original receiver. Data loss or damage can even occur without the presence of a malicious node due to the harsh communication environment. Thus, data integrity ensures that any received data has not been altered in transit.

11.3.3.3 Data Freshness

Even if confidentiality and data integrity are assured, we also need to ensure the freshness of each message. Informally, data freshness suggests that the data is recent, and it ensures that no old messages have been replayed. This requirement is especially important when there are shared-key strategies employed in the design.

Typically shared keys need to be changed over time. However, it takes time for new shared keys to be propagated to the entire network. In this case, it is easy for the adversary to use a replay attack. Also, it is easy to disrupt the normal work of the sensor, if the sensor is unaware of the new key change time. To solve this problem a nonce, or another time-related counter, can be added into the packet to ensure data freshness.

11.3.3.4 Availability

Adjusting the traditional encryption algorithms to fit within the wireless sensor network is not free and will introduce some extra costs. Some approaches choose to modify the code to reuse as much code as possible. Some approaches try to make use of additional communication to achieve the same goal. What's more, some approaches force strict limitations on the data access, or propose an unsuitable scheme (such as a central point scheme) in order to simplify the algorithm. But all these approaches weaken the availability of a sensor and sensor network for the following reasons:

- Additional computation consumes additional energy. If no more energy exists, the data will no longer be available.
- Additional communication also consumes more energy. What's more, as communication increases so too does the chance of incurring a communication conflict.
- A single point of failure will be introduced if using the central point scheme. This greatly threatens the availability of the network.

The requirements of security not only affect the operation of the network, but also are highly important in maintaining the availability of the whole network.

11.3.3.5 Self-Organization

A wireless sensor network is a typically ad hoc network, which requires every sensor node be independent and flexible enough to be self-organizing and self-healing according to different situations. There is no fixed infrastructure available for the purpose of network management in a sensor network. This inherent feature brings a great challenge to wireless sensor network security as well. For example, the dynamics of the whole network inhibits the idea of pre-installation of a shared key between the base station and all sensors (Eschenauer, 2002). Several random key pre-distribution schemes have been proposed in the context of symmetric encryption techniques (Eschenauer, 2002; Hwang, 2004). In the context of applying public-key cryptography techniques in sensor networks, an efficient mechanism for public-key distribution is necessary as well. In the same way that distributed sensor networks must self-organize to support multihop routing, they must also self-organize to conduct key management and building trust relation among sensors. If self-organization is lacking in a sensor network, the damage resulting from an attack or even the hazardous environment may be devastating.

11.3.3.6 Time Synchronization

Most sensor network applications rely on some form of time synchronization. In order to conserve power, an individual sensor's radio may be turned off for periods of time. Furthermore, sensors may wish to compute the end-to-end delay of a packet as it travels between two pair-wise sensors. A more collaborative sensor network may require group synchronization for tracking applications. In Ganeriwal (2005), the authors propose a set of secure synchronization protocols for sender-receiver synchronization (when the nodes are within a single hop distance), multi-hop sender-receiver synchronization (for use when the pair of nodes are not within single-hop range), and group synchronization.

11.3.3.7 Secure Localization

Often, the utility of a sensor network will rely on its ability to accurately and automatically locate each sensor in the network. A sensor network designed to locate faults will need accurate location information in order to pinpoint the location of a fault. Unfortunately, an attacker can easily manipulate non-secured location information by reporting false signal strengths, replaying signals, etc.

A technique called verifiable multi-lateration (VM) is described in Capkun (2006). In multi-lateration, a device's position is accurately computed from a series of known reference points. The authenticated ranging and distance bounding are used to ensure accurate location of a node. Because of distance bounding, an attacking node can only increase its claimed distance from a reference point. However, to ensure location consistency, an attacking node would also have to prove that its distance from another reference point is shorter. Since it cannot do this, a node manipulating the localization protocol can be found. For large sensor networks, the SPINE (Secure Positioning In sensor NEtworks) algorithm is used. It is a three-phase algorithm based upon verifiable multilateration.

In Lazos (2005), SeRLoc (Secure Range-Independent Localization) is described. Its novelty is its decentralized, range-independent nature. SeRLoc uses locators that transmit beacon information. It is assumed that the locators are trusted and cannot be compromised. Furthermore, each locator is assumed to know its own location. A sensor computes its location by listening for the beacon information sent by each locator. The beacons include the locator's location. Using all of the beacons that a sensor node detects, a node computes an approximate location based on the coordinates of the locators. Using a majority vote scheme, the sensor then computes an overlapping antenna region. The final computed location is the centroid of the overlapping antenna region. All beacons transmitted by the locators are encrypted with a shared global symmetric key that is pre-loaded to the sensor prior to deployment. Each sensor also shares a unique symmetric key with each locator. This key is also pre-loaded on each sensor.

11.3.3.8 Authentication

An attacker is not just limited to modifying the data packet. He can change the whole packet stream by injecting additional packets. So the receiver needs to ensure that the data used in any decision-making process originates from the correct source. On the other hand, when constructing the sensor network, authentication is necessary for many administrative tasks (e.g., network reprogramming or controlling sensor node duty cycle). From the above, we can see that message authentication is important for many applications in sensor networks. Informally, data authentication allows a receiver to verify that the data really is sent by the claimed sender. In the case of two-party communication, data authentication can be achieved through a purely symmetric mechanism: the sender and the receiver share a secret key to compute the message authentication code of all communicated data.

To provide authentication, a key-chain distribution system is defined based on μTESLA (Perrig, 2002). The basic idea of the μTESLA system is to achieve asymmetric cryptography by delaying the disclosure of the symmetric keys. In this case a sender will broadcast a message generated with a secret key. After a certain period of time, the sender will disclose the secret key. The receiver is responsible for buffering the packet until the secret key has been disclosed. After disclosure the receiver can authenticate the message, provided that the message was received before the key was disclosed. However, one limitation can be observed with the μTESLA-based authentication scheme: Some initial information must be sent to each sensor node before authentication of broadcast messages can begin.

11.3.3.9 Key Distribution

As the sensor nodes in a WSN should communicate securely, key distribution should be provided to allow a protected distribution of keys among the sensor nodes. The requirements to be provided by a key distribution mechanism include the following:

■ *Scalability*: This is the ability to support large scale WSNs. The key distribution mechanism must support the distribution large networks, and must be flexible against significant growth in the size of the network.
■ *Resource efficiency*: Storage, processing and communication limitations on sensor nodes must be considered during the key distribution. The amount of memory required to store security credentials should be limited.

- *Processing and communication complexity*: The amount of processor cycles required to establish a key should be reduced. The number of messages exchanged during a key generation process should also be kept small.
- *Key connectivity* (probability of key-share): Probability that two (or more) sensor nodes store the same key or keying material. Enough key connectivity must be provided for a WSN to perform its intended functionality.
- *Resilience*: Resistance against node capture. Compromise of security credentials, which are stored on a sensor node or exchanged over radio links, should not reveal information.

11.4 Attacks against WSNs

Sensor networks are particularly vulnerable to several key types of attacks. Attacks can be performed in a variety of ways, most notably as denial of service attacks, but also through traffic analysis, privacy violation, physical attacks, and so on. Denial of service attacks on wireless sensor networks can range from simply jamming the sensor's communication channel to more sophisticated attacks designed to violate the 802.11 MAC protocol or any other layer of the wireless sensor network (Perrig, 2004).

Due to the potential asymmetry in power and computational constraints, guarding against a well orchestrated denial of service attack on a wireless sensor network can be nearly impossible. A more powerful node can easily jam a sensor node and effectively prevent the sensor network from performing its intended duty. We note that attacks on wireless sensor networks are not limited to simply denial of service attacks, but rather encompass a variety of techniques including node takeovers, attacks on the routing protocols, and attacks on a node's physical security. In this section, we first address some common denial of service attacks and then describe additional attacking, including those on the routing protocols as well as an identity based attack known as the Sybil attack.

11.4.1 Denial of Service Attacks

A standard attack on wireless sensor networks is simply to jam a node or set of nodes. Jamming, in this case, is simply the transmission of a radio signal that interferes with the radio frequencies being used by the sensor network. The jamming of a network can come in two forms: constant jamming and intermittent jamming. Constant jamming involves the complete jamming of the entire network. No messages are able to be sent or received. If the jamming is only intermittent, then nodes are able to exchange messages periodically, but not consistently. This too can have a detrimental impact on the sensor network as the messages being exchanged between nodes may be time sensitive. Attacks can also be made on the link layer itself.

One possibility is that an attacker may simply intentionally violate the communication protocol and continually transmit messages in an attempt to generate collisions. Such collisions would require the retransmission of any packet affected by the collision. Using this technique it would be possible for an attacker to simply deplete a sensor node's power supply by forcing too many retransmissions. At the routing layer, a node may take advantage of a multihop network by simply refusing to route messages. This could be done intermittently or constantly with the net result being that any neighbor who routes through the malicious node will be unable to exchange messages with, at least, part of the network. The transport layer is also susceptible to attack, as in the case of flooding. Flooding can be as simple as sending many connection requests to a susceptible node. In this case, resources must be allocated to handle the connection request. Eventually, a node's resources will be exhausted, thus rendering the node useless.

11.4.2 Traffic Analysis Attacks

Wireless sensor networks are typically composed of many low-power sensors communicating with a few relatively robust and powerful base stations. It is not unusual, therefore, for data to be gathered by the individual nodes where it is ultimately routed to the base station. Often, for an adversary to effectively render the network useless, the attacker can simply disable the base station. To make matters worse, two attacks have been conceived; they can identify the base station in a network (with high probability) without even understanding the contents of the packets (if the packets are themselves encrypted).

A rate monitoring attack simply makes use of the idea that nodes closest to the base station tend to forward more packets than those farther away from the base station. An attacker needs only to monitor which nodes are sending packets and follow those nodes that are sending the most packets. In a time correlation attack, an adversary simply generates events and monitors to whom a node sends its packets. To generate an event, the adversary could simply generate a physical event that would be monitored by the sensor(s) in the area (turning on a light, for instance).

11.4.3 Wormhole Attacks

In a wormhole attack, an attacker receives packets at one point in the network, "tunnels" them to another point in the network, and then replays them into the network from that point. For tunnelled distances longer than the normal wireless transmission range of a single hop, it is simple for the attacker to make the tunneled packet arrive with better metric than a normal multihop route, for example, through use of a single long-range directional wireless link or through a direct wired link to a colluding attacker. It is also possible for the attacker to forward each bit over the wormhole directly, without waiting for an entire packet to be received before beginning to tunnel the bits of the packet, in order to minimize

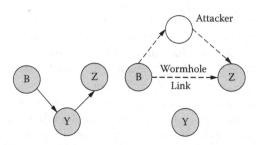

Figure 11.2 Wormhole attack.

delay introduced by the wormhole. Due to the nature of wireless transmission, the attacker can create a wormhole even for packets not addressed to it, since it can overhear them in wireless transmission and tunnel them to the colluding attacker at the opposite end of the wormhole.

If the attacker performs this tunneling honestly and reliably, no harm is done; the attacker actually provides a useful service in connecting the network more efficiently. However, the wormhole puts the attacker in a very powerful position relative to other nodes in the network, and the attacker could exploit this position in a variety of ways. The attack can also still be performed even if the network communication provides confidentiality and authenticity, and even if the attacker has no cryptographic keys. Furthermore, the attacker is invisible at higher layers; unlike a malicious node in a routing protocol, which can often easily be named, the presence of the wormhole and the two colluding attackers at either endpoint of the wormhole are not visible in the route. A wormhole attack is depicted in Figure 11.2. The wormhole attack is particularly dangerous against many ad hoc network routing protocols in which the nodes that hear a packet transmission directly from some node consider themselves to be in range of (and, thus a neighbor of) that node.

11.4.4 Sybil Attack

In many cases, the sensors in a wireless sensor network might need to work together to accomplish a task, hence they can use distribution of subtasks and redundancy of information. In such a situation, a node can pretend to be more than one node using the identities of other legitimate nodes

Figure 11.3 depicts a Sybil attack against a WSN. This type of attack where a node forges the identities of more than one node is the Sybil attack (Newsome, 2004). A Sybil attack tries to degrade the integrity of data, security, and resource utilization that the distributed algorithm attempts to achieve, and can be used to attack distributed storage, routing mechanisms, data aggregation, voting, fair resource allocation, and misbehavior detection.

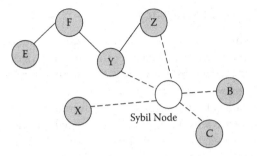

Figure 11.3   Sybil attack.

Basically, any peer-to-peer network (especially wireless ad hoc networks) is vulnerable to Sybil attack. However, as WSNs can have some sort of base stations or gateways, this attack could be prevented using efficient protocols. It has been shown that, without a logically centralized authority, Sybil attacks are always possible except under extreme and unrealistic assumptions of resource parity and coordination among entities. However, the detection of Sybil nodes in a network is not so easy.

11.4.5 Blackhole/Sinkhole Attack

In this attack, a malicious node acts as a blackhole to attract all the traffic in the sensor network (Culpepper, 2004). Especially in a flooding based protocol, the attacker listens to requests for routes and then replies to the target nodes that it contains the high quality or shortest path to the base station. Once the malicious device has been able to insert itself between the communicating nodes (for example, sink and sensor node), it is able to do anything with the packets passing between them. In fact, this attack can affect even the nodes that are considerably far from the base stations. Figure 11.4 shows the conceptual view of a blackhole/sinkhole attack.

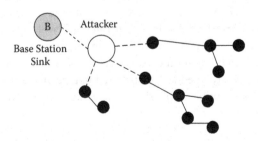

Figure 11.4   Conceptual view of blackhole attack.

11.4.6 Attacks against Privacy

Sensor network technology promises a vast increase in automatic data collection capabilities through efficient deployment of tiny sensor devices. While these technologies offer great benefits to users, they also exhibit significant potential for abuse. Particularly relevant concerns are privacy problems, since sensor networks provide increased data collection capabilities. Adversaries can use even seemingly innocuous data to derive sensitive information if they know how to correlate multiple sensor inputs. For example, in the famous "panda-hunter problem" (Ozturk, 2004), the hunter can imply the position of pandas by monitoring the traffic. The main privacy problem, however, is not that sensor networks enable the collection of information. In fact, much information from sensor networks could probably be collected through direct site surveillance. Rather, sensor networks aggravate the privacy problem because they make large volumes of information easily available through remote access. Hence, adversaries need not be physically present to maintain surveillance. They can gather information in a low-risk, anonymous manner. Remote access also allows a single adversary to monitor multiple sites simultaneously. Some of the more common attacks against sensor privacy are the following:

- *Monitor and eavesdropping*: This is the most obvious attack to privacy. By listening to the data, the adversary could easily discover the communication contents. When the traffic conveys the control information about the sensor network configuration, which contains potentially more detailed information than accessible through the location server, the eavesdropping can act effectively against the privacy protection.
- *Traffic analysis*: Traffic analysis typically combines with monitoring and eavesdropping. An increase in the number of transmitted packets between certain nodes could signal that a specific sensor has registered activity. Through the analysis on the traffic, some sensors with special roles or activities can be effectively identified.
- *Camouflage*: Adversaries can insert their node or compromise the nodes to hide in the sensor network. After that these nodes can masquerade as a normal node to attract the packets, then misroute the packets, e.g., forward the packets to the nodes conducting the privacy analysis.
- *Physical attacks*: Sensor networks typically operate in hostile outdoor environments. In such environments, the small form factor of the sensors, coupled with the unattended and distributed nature of their deployment, make them highly susceptible to physical attacks, i.e., threats due to physical node destructions.

Unlike many other attacks mentioned above, physical attacks destroy sensors permanently, so the losses are irreversible. For instance, attackers can extract cryptographic secrets, tamper with the associated circuitry, modify programming in the sensors, or replace them with malicious sensors under the control of the attacker. Recent work has shown that standard sensor nodes, such as the MICA2 motes, can be compromised in less than one minute. While these results are not surprising given that the MICA2 lacks tamper resistant hardware protection, they provide a cautionary note about the speed of a well-trained attacker. If an adversary compromises a sensor node, then the code inside the physical node may be modified.

11.5 Attacks against Specific Sensor Network Protocols

Obviously, all the proposed protocols for sensor network routing are highly susceptible to damaging attacks. Malicious attacker can attract or keep away traffic flows, increase latency, or disable the entire network with as little effort as sending a single packet. In this section, we discuss attacks targeting some of the proposed sensor network routing protocols.

11.5.1 TinyOS Beaconing Protocol

This protocol constructs via a "breadth first strategy," spanning a tree rooted at a base station. Periodically the base station broadcasts a route update. All nodes receiving the update mark the base station as its parent and rebroadcast the update message. The algorithm continues recursively with each node marking its parent as the first node from which it hears a routing update during the current time period. All packets received or generated by a node are forwarded to its parent (until they reach the base station).

The TinyOS beaconing protocol is highly susceptible to attacks. Since routing updates are not authenticated, it is possible for any node to claim to be a base station and become the destination of all traffic in the network. Authenticated routing updates will prevent an adversary from claiming to be a base station, but a powerful laptop class attacker can easily cause damage. An attacker interested in eavesdropping on, modifying, or deleting packets in a particular area can do so by mounting a combined wormhole/sinkhole attack. The attacker first creates a wormhole between two colluding laptop-class nodes, one near the base station and one near the targeted area. The first node forwards authenticated routing updates to the second through the wormhole. The latter participates normally in the protocol and rebroadcasts the routing update in the targeted area. Since the "wormholed" routing update will likely reach the targeted area considerably faster than it normally would have through multihop routing, the second node will create a large routing subtree in the targeted area with itself as the root. As depicted in Figure 11.5, the traffic in the targeted area will be channeled through the wormhole, enabling a powerful selective forwarding attack.

If a laptop-class attacker has a powerful transmitter, it can use a HELLO flood attack to broadcast a routing update strong enough to reach the entire network,

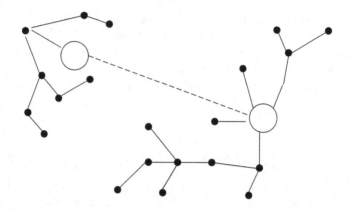

Figure 11.5   A laptop-class adversary using a wormhole to create a sinkhole in TinyOS beaconing.

causing every node to mark the attacker as its parent. Most nodes will be likely out of normal radio range of both a true base station and the attacker. Almost all the nodes are trapped and are likely to send their packets into nowhere. Due to the simplicity of this protocol, it is unlikely there a simple extension can be made to recover from this attack. A node that realizes that its parent is not actually in its range (by using link layer acknowledgments, for example) can delete the wrong information and inform its neighbors to do so, since the neighbors will likely have the adversary marked as its parent as well.

On the other hand, routing loops can be easily created by mote-class attackers spoofing routing updates. To explain this, assume that the adversary can determine that a sensor node S_1 and a sensor node S_2 are within the radio coverage of each other. The attacker can send a forged routing update to node S_2 with a spoofed source address indicating it came from node S_1. Node S_2 will then mark node S_1 as its parent and rebroadcast the routing update. Node S_1 will then hear the routing update from node S_2 and mark S_2 as its parent. Messages sent to either S_1 or S_2 will be forever forwarded in the loop.

11.5.2 Directed Diffusion Protocol

This is a data-centric routing algorithm for drawing information out of a sensor network (Intanagonwiwat, 2000). The protocol assumes that the base stations flood interests for named data, setting up needs within the network designed to data matching the interest. The nodes that are able to satisfy the interest disseminate information along the reverse path of interest propagation. The nodes receiving the same interest from multiple neighboring nodes may propagate events along the corresponding multiple links. Interests initially specify a low rate of data flow, but once a base station starts receiving events it will reinforce one (or more) neighbor in order to request higher data rate events. This process recursively proceeds until it reaches the nodes generating the events, causing them to generate events at a higher data rate. Alternatively, paths may be negatively reinforced as well.

Due to the robust nature of flooding, it may be difficult for an attacker to prevent interests from reaching targets able to satisfy them. However, once sources begin to generate data events, the attacker targeting a data flow might have one among the four following goals:

- *Suppression*: Flow suppression is an instance of denial of service. The easiest way to suppress a flow is to spoof negative reinforcements.
- *Cloning*: Cloning a flow enables eavesdropping. After an adversary receives an interest flooded from a legitimate base station, it can simply replay that interest with herself listed as a base station. All events satisfying the interest will now be sent to both the adversary and the legitimate base station.
- *Path influence*: An adversary can influence the path taken by a data flow by spoofing positive and negative reinforcements and bogus data events. For example, after receiving and rebroadcasting an interest, an adversary interested in directing the resulting flow of events through herself would strongly reinforce the nodes to which the interest was sent while spoofing high rate, low latency events to the nodes from which the interest was received. Three actions can result: (a) the data events generated upstream by legitimate sources will be drawn through the adversary because of her artificially strong positive reinforcements, (b) the alternate event flows will be negatively reinforced by downstream nodes because the adversary provides (or spoofs) events with the lowest latency or highest frequency, and (c) the adversary's node will be positively reinforced due the high quality spoofed and real data events she is able to provide. With this attack, an adversary is able to force any flow of events to propagate through himself on the way to the base station that originally advertised the associated interest.
- *Selective forwarding and data tampering*: By using the above attack to insert herself onto the path taken by a flow of events, an adversary can gain full control of the flow. She can modify and selectively forward packets of her choosing.

A laptop-class adversary can exert greater influence on the topology by creating a wormhole between sensor node S located next a base station and sensor node S′ located close to where events are likely to be generated. Interests advertised by the base station are sent through the wormhole and rebroadcast by node S′. The node S′ then attracts data flows by spoofing strong positive reinforcements to all neighboring nodes while node S broadcasts spoofed negative reinforcements to its surrounding nodes. The combination of the positive and negative reinforcements pushes data flows away from the base station and toward the resulting sinkhole centered at node S'.

The multipath version may appear more robust against these attacks, but it is just as vulnerable. A single adversary can use the Sybil attack against her neighbors. A neighbor will be convinced it is maximizing diversity by reinforcing its next most preferred neighbor not on the primary flow when in fact this neighbor is an alternate identity of the adversary.

11.5.3 Geographic Routing

Geographic and Energy Aware Routing (GEAR; Govindan, 2001) and Greedy Perimeter Stateless Routing (GPSR; Karp, 2000) leverage nodes' positions and explicit geographic packet destinations to efficiently disseminate queries and route replies. GPSR uses greedy forwarding at each hop, routing each packet to the neighbor closest to the destination. When holes are encountered where greedy forwarding is impossible, GPSR recovers by routing around the perimeter of the void. One drawback of GPSR is that packets along a single flow will always use the same nodes for the routing of each packet, leading to uneven energy consumption. GEAR attempts to remedy this problem by weighting the choice of the next hop by both remaining energy and distance from the target. In this way, the responsibility for routing a flow is more evenly distributed among a "beam" of nodes between the source and base station. Both protocols require location (and energy for GEAR) information to be exchanged between neighbors, although for some fixed, well-structured topologies (a grid for example) this may not be necessary.

Location information can be misrepresented. Regardless of an adversary's actual location, she may advertise her location in a way to place herself on the path of a known flow. GEAR tries to distribute the responsibility of routing based on remaining energy, so an appropriate attack would be to always advertise maximum energy as well. Without too much additional effort, an adversary can dramatically increase her chances of success by mounting a Sybil attack. As depicted in Figure 11.6 an adversary can advertise multiple bogus nodes surrounding each target in a circle (or sphere), each claiming to have maximum energy. By intercepting transmissions sent to each of the bogus nodes, the adversary maximizes her chances for placing herself on the path of any nearby data flow. Once on that path, the adversary can mount a selective forwarding attack.

In GPSR an adversary can forge location advertisements to create routing loops in data flows without having to actively participate in packet forwarding. Consider the hypothetical topology in Figure 11.6 and flow of packets from sensor node B to location (3,1). Assume the maximum radio range is one unit. If an adversary forges a location advertisement claiming B is at (2,1) and sends it to C, then after B forwards a packet destined for (3,1) to C, C will send it back to B because it can believe that B is close to the ultimate destination. Therefore, the loop can start.

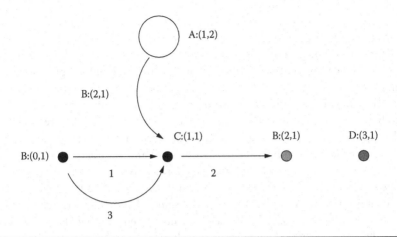

Figure 11.6   Creating routing loops in GPSR by forging a location advertisement.

11.6 Countermeasures against Attacks

In this section, we describe the measures that aim at protecting the wireless sensor network from attacks and complying with the security requirements established. We start with the key establishment mechanism provided for the wireless sensor networks. It lays the foundation for the security in a wireless sensor network, and then we describe the techniques used for defending against DoS attacks: secure broadcasting and multicasting; protecting against attacks on routing protocols; combating traffic analysis attacks; defending against attacks on sensor privacy, intrusion detection and secure data aggregation; and finally, defending against physical attacks and trust management.

11.6.1 Key Management Fundamentals

Key management issues in wireless networks are not unique to wireless sensor networks. Indeed, key establishment and management issues have been studied in depth outside of the wireless networking arena. Traditionally, key establishment is done using one of many public-key protocols. One of the more common is the Diffie-Hellman public key protocol, but there are many others (as stated in Chapter 2). Most of the traditional techniques, however, are unsuitable in low power devices such as wireless sensor networks. This is due largely to the fact that typical key exchange techniques use asymmetric cryptography. In this case, it is necessary to maintain two mathematically related keys, one of which is made public while the other is kept private. This allows data to be encrypted with the public key and decrypted only with the private key. The problem with asymmetric cryptography, in a wireless sensor network, is that it is typically too computationally intensive for the individual nodes in the sensor network. This is true in the general case; however, Malan (2004) show that it is feasible with the right selection of algorithms.

Symmetric cryptography is therefore the typical choice for applications that cannot afford the computational complexity of asymmetric cryptography. Symmetric schemes utilize a single shared key known only between the two communicating hosts. This shared key is used for both encrypting and decrypting data. The use of DES, however, is quite limited due to the fact that it can be broken relatively easily. In light of the shortcomings of DES, other symmetric cryptography systems have been proposed including 3DES (Triple DES) and elliptic cryptography AES. One major shortcoming of symmetric cryptography is the key exchange problem. Simply explained, the key exchange problem derives from the fact that two communicating hosts must somehow know the shared key before they can communicate securely. So the problem that arises is how to ensure that the shared key is indeed shared between the two hosts who wish to communicate and no other rogue hosts who may wish to eavesdrop the exchange. How to distribute a shared key securely to communicating sensors is a non-trivial problem since pre-distributing the keys is not always feasible.

11.6.1.1 Key Establishment

One security aspect that receives a great deal of consideration in wireless sensor networks is the area of key management. Wireless sensor networks are unique (among other embedded wireless networks) in this aspect due to their size, mobility, and computational/power constraints. Indeed, researchers imagine wireless sensor networks to be orders of magnitude larger than their traditional embedded counterparts. This fact, coupled with the operational constraints described previously, makes secure key management an utter necessity in the design of most wireless sensor networks. Because encryption and key management/establishment are so crucial to the defense of a wireless sensor network, with nearly all aspects of wireless sensor network protections relying on solid encryption, we first begin with an overview of the unique key and encryption issues surrounding wireless sensor networks before discussing more specific key distribution protocols.

Random key pre-distribution schemes have several variants. The key pre-distribution scheme proposed in Eschenauer (2002) relies on probabilistic key sharing among nodes within the sensor network. The system developed works by distributing a key ring to each participating node in the sensor network before deployment. Each key ring should contain a set of randomly chosen keys from a significantly larger pool of keys generated offline. Enhancements to this technique utilizing multiple keys have been described in Liu (2005) and Hwang (2004). Using the random key pre-distribution scheme, it is not necessary that each pair of nodes share a key. However, any two nodes that share a key may use it to establish a direct link to one another. In addition, such a technique can be extended to key revocation, re-keying, and the addition/deletion of nodes.

The LEAP protocol described in Zhu (2003) takes an approach that utilizes multiple keying mechanisms. The protocol proposes to use four different keys depending on whom the sensor node is communicating with. Sensors are preloaded with an initial key from which further keys can be established. As a security precaution, the initial key can be deleted after its use in order to ensure that a compromised sensor cannot add additional compromised nodes to the network.

In the PIKE system, a mechanism is described for establishing a key between two sensor nodes that is based on the common trust of a third node somewhere within the sensor network (Chan, 2005). The nodes and their shared keys are spread over the network such that for any two nodes A and B, there is a node C that shares a key with both A and B. Therefore, the key establishment protocol between A and B can be securely routed through C.

On the other hand, Huang et al. (Huang, 2003) propose a hybrid key establishment scheme that makes use of the difference in computational and energy constraints between a sensor node and the base station. Starting from the fact that an individual sensor node possesses far less computational power and energy than a base station, the authors propose placing the major cryptographic load on the base station where the resources tend to be greater. On the sensor side, symmetric-key operations are used in place of their asymmetric alternatives. The sensor and the base station authenticate based on elliptic curve cryptography. Elliptic curve cryptography is often used in sensors due to the fact that relatively small key lengths are required to achieve a given level of security. In addition, certificates are also used to establish the legitimacy of a public key. The certificates are based on an elliptic curve implicit certificate scheme (Huang, 2003). Such certificates are useful to ensure both that the key belongs to a device and that the device is a legitimate member of the sensor network. Each node obtains a certificate before joining the network using an out-of-band interface.

11.6.2 WSN and Public Key Cryptography

Two of the major techniques used to implement public-key cryptosystems are the RSA and the elliptic curve cryptography (ECC). Traditionally, these methods have been thought to be a too heavy weight for use in wireless sensor networks. Recently, however, several works have successfully implemented public-key cryptography in wireless sensor networks. Some works report that both RSA and elliptic curve cryptography are possible using 8-bit CPUs, with ECC demonstrating a performance advantage over RSA. Another advantage is that ECC's 160 bit keys result in shorter messages during transmission compared to the 1024 bit RSA keys. In particular, it has been demonstrated that the point multiplication operations in ECC are an order of magnitude faster than private-key operations performed within RSA, and are comparable to the RSA public-key operation.

The Diffie-Hellman key exchange algorithm used in Malan (2004) is detailed in the following: Let G be a point selected from an elliptic curve E, and assume that both *G* and *E* are public. A random integer *KA* is selected and will act as the private key. The sender's public key *TA* is then $TA = KA* G$. The receiver performs a similar set of operations to compute $TB = KB* G$. Both peers can now easily compute the shared-secret using their own private keys and the public keys that have been exchanged. In this case, the sender computes the following:

$$KA* TB = KA* KB* G,$$

while the receiver computes the equality $KB* TA = KB* KA* G$. Because $KA * TB = KB * TA$, the sender and the receiver now share a secret key.

11.6.3 DoS Countermeasures

Since denial of service attacks are so common, effective protections must be made available to combat them. One strategy in defending against the classic jamming attack is to identify the jammed part of the sensor network and effectively route around the unavailable portion. The approach made in Wood (2002) describes a two phase approach where the nodes along the perimeter of the jammed region report their status to their neighbors who then collaboratively define the jammed region and simply route around it. To handle jamming at the MAC layer, nodes might utilize a MAC admission control that is rate limiting. This would allow the network to ignore those requests designed to exhaust the power reserves of a node. This, however, is not fool-proof as the network must be able to handle any legitimately large traffic volumes.

Overcoming rogue sensors that intentionally misroute messages can be done at the cost of redundancy. In this case, a sending node can send the message along multiple paths in an effort to increase the likelihood that the message will ultimately arrive at its destination. This has the advantage of effectively dealing with nodes that may not be malicious, but rather may have simply failed as it does not rely on a single node to route its messages. To overcome the transport layer flooding denial of service attack one can suggest using the client puzzles posed by Aura (2001) in an effort to discern a node's commitment to make the connection by utilizing some of their own resources. The authors of Aura (2001) advocate that a server should force a client to commit its own resources first. Further, they suggest that a server should always force client to commit more resources up front than the server. This strategy would likely be effective as long as the client has computational resources comparable to those of the server.

11.6.4 Detecting Node Replication Attacks

Two algorithms are of utter interest: the randomized multicast and the line-selected multicast. Randomized multicast is an evolution of a node broadcasting strategy. In the simple node broadcasting strategy each sensor propagates an authenticated broadcast message throughout the entire sensor network. Any node that receives a conflicting or duplicated claim revokes the conflicting nodes. This strategy will work, but the communication cost is far too expensive. In order to reduce the communication cost, a deterministic multicast could be employed where nodes would share their locations with a set of witness nodes. In this case, witnesses are computed based on a node's ID. In the event that a node has been replicated on the network, two conflicting locations will be forwarded to the same witness who can then revoke the offending nodes. But since a witness is based on a node's ID, it can easily be computed by an attacker who can then compromise the witness nodes. Thus, securely utilizing a deterministic multicast strategy would require too many witnesses and the communication cost would be too high.

11.7 Mobile Target Tracking Using WSNs

With the advances in the fabrication technologies that integrate the sensing and the wireless communication systems, tiny sensor motes can be densely deployed in the desired field to form a large-scale WSN. In a given field of interest, there is a varying number of targets. They arise in the field at random locations and at random times. The movement of each target follows an arbitrary but continuous path, and it persists for a random amount of time before disappearing in the field. Detecting and tracking mobile targets in a specific geographic area introduces several difficulties such as (a) the need for secure collaborative communication and computation among multiple sensors (the information generated by a single node is usually incomplete or inaccurate), (b) the power capabilities available at the elementary components are so limited that traditional target tracking methods based on complex signal processing algorithms may not be applicable to the nodes, and (c) the amount of energy available at each node does not allow a permanent and ubiquitous activity of the network unless the sensor density is changed. As a result, theories and models are needed to cope with these challenges.

11.7.1 Coverage Approaches

The radio coverage is among the most important aspects that affect the performance of tracking schemes. In fact, the probability that a target, which is expected to be moving in the monitored area, is detected depends essentially on the average number of sensor nodes covering every point of the region of interest. In the literature, many coverage optimality concepts have been defined. The classification provided in Gage (1992) distinguishes three coverage classes:

■ *Blanket coverage*: Basically, a blanket coverage scheme deals with performing an arrangement of sensor nodes that maximizes the detection rate of targets appearing in the sensing field.

- *Barrier coverage*: This scheme is based on performing an arrangement of sensor nodes that minimizes the probability of undetected penetration through the barrier.
- *Sweep coverage*: In this scheme, sensor nodes are dynamically deployed in the monitored region.

It is worth mentioning that the problem of area coverage seeks to determine the minimum number of sensors that can be placed in an environment, such that every point in the environment is monitored under resource constraints, presence of obstacles, noise, and varying topography. To address this issue, the notion of degree of coverage of a WSN is introduced. It is the minimum number of sensors, in the WSN, that are able to sense any particular point in the sensing field. It has been observed and postulated that different applications would require different degrees of coverage in the sensing field. For instance, a military surveillance application would need a high degree of coverage, because it requires that even if some nodes fail to function, the security of the region will not be compromised, as other nodes will still continue to function.

One important issue for being able to deploy an efficient sensor network is to develop an optimal node placement strategy taking into consideration that some of the nodes can be malicious and the nodes can be mobile. In most of the practical contexts, since the sensors are randomly scattered, the major challenge is to determine the sensor density (i.e., number of sensors per unit of surface). The sensor density can be considered as a measure of Quality of Service because it allows assessing how well each point in the sensing field is covered by the sensing ranges (Gosh, 2006). However, we should be aware that coverage also affects some network performance parameters such as routing complexity or congestion rate. It also adds complexity to the security challenge. These statements are motivated by the fact that sensor nodes have two major built-in functionalities: gathering physical events and forwarding securely the information to the analysis center. Obviously, a dense configuration would convey optimal sensing capabilities, but it would necessarily increase the congestion probability. It also adds more effort in signing and detecting.

The main deployment approaches can be summarized as follows:

- *k-coverage techniques*: A region A is said to be k-covered if every point belonging to A is within the sensing range of k sensors. Thus, stating that A is covered requires enumerating all sub-regions resulting from the intersection of different sensor node-regions and verifying if each of these is k-covered.
- *k-connectivity techniques*: These techniques are applied when the network is k-connected, meaning that at least k nodes are within the transmission range of each sensor nodes. This coverage condition has been used in the literature to find, for a number of sensors (N), a sensing range assignment that ensures k-connectivity (Gupta, 1998).

The aforementioned coverage approaches do not consider the mobility and the dynamic activation of the sensor nodes. When positions are no longer static with respect to time, the minimum number of sensors that allow optimal target coverage should change. The same statement is valid when a sensor node automatically switches between the "SLEEP" and "ACTIVE" states. In the literature, several mobility models have been investigated. At this stage, our objective is just to illustrate the impact of mobility on coverage. The hypothesis that a sensor moves is isotropic (i.e., zero-knowledge about direction) is used to determine the greatest number of sensors that would be needed when mobility is introduced (Hamdi, 2006).

11.7.2 Architectural Issues

Many existing systems and protocols attempt to solve the problem of determining a nodes location within its environment. The approaches taken to solve this localization problem differ in the assumptions that they make about their respective network and device capabilities. These include assumptions about device hardware, signal propagation models, timing and energy requirements, network makeup (homogeneous vs. heterogeneous), nature of the environment (indoor vs. outdoor), time synchronization of devices, communication costs, error requirements, device mobility, and level of security.

The Time Difference of Arrival (TDoA) technique for ranging (estimating the distance between two communicating nodes) has been widely proposed as a necessary ingredient in localization solutions for wireless sensor networks. TDoA also relies on extensive hardware that is expensive and energy consuming, making it less suitable for low power sensor network devices. To complement TDoA and Time of Arrival (ToA) technologies, an Angle of Arrival (AoA) technique has been proposed that allows nodes to estimate and map relative angles between neighbors. However, AoA estimates require additional hardware to be used in large scale sensor networks.

A Received Signal Strength Indicator (RSSI)-based technique such as SpotOn has been proposed for hardware constrained systems (Hightower, 2000). In the RSSI techniques, theoretical or empirical models are used to translate the signal strength into distance estimates. The problems observed with this technique include multi-path fading, background interference, and irregular signal propagation characteristics. They make range estimates inaccurate. To mitigate such errors, some works have been realized aiming at making robust range estimation, refinement positioning, and parameter calibration. However, while solutions based on RSSI have demonstrated efficacy in controlled laboratory environments, the premise that distance can be determined based on signal strength, propagation patterns, and fading models remains questionable.

In sensor networks, errors can often be masked through fault tolerance, redundancy, aggregation, or by other means. Depending on the behavior and requirements of protocols using location information, varying granularities of error may be appropriate from system to system. In Bulusu (2000), a heterogeneous network containing powerful nodes with established location information is considered. In this work, anchors beacon their position to neighbors keeping an account of all received beacons. Using this proximity information, a simple centroid model is applied to estimate the listening nodes location. An alternate solution, called DV-HOP (Niculescu, 2003), assumes a heterogeneous network consisting of sensing nodes and anchors. Instead of single hop broadcasts, anchors flood their location throughout the network maintaining a running hop-count at each node along the way. Nodes calculate their position based on the received anchor locations, the hop-count from the corresponding anchor, and the average-distance per hop, a value obtained through anchor communication. Once a node can calculate the distance estimation to more than 3 anchors in the plane, it uses triangulation to estimate its location.

Like DV-Hop, an Amorphous Positioning algorithm proposed in Nagpal (2003) uses offline hop distance estimations, improving location estimates through neighbor information exchange. Once anchor estimates are collected, the hop distance estimation is obtained through local averaging. Each node collects neighboring nodes hop distance estimates and computes an average of all its neighbor values. Half of the radio range is then deducted from this average to compensate for errors caused by low resolution. The Amorphous Localization algorithm takes a different approach from the DV-Hop algorithm to estimate the average distance of a single hop.

11.7.3 Target Tracking Protocols

To track mobile targets, it is first essential to develop algorithms to locate the target and track their paths of mobility. Traditional tracking methods make use of a centralized database or computing facility. As the number of sensors increases in the network, the central facility becomes a bottleneck both as a resource and in terms of the network traffic directed toward it. Therefore, this approach lacks scalability and is not fault-tolerant. Another distinguishing feature of traditional tracking approach is that usually the sensing task is performed by any one node in the network at a time. These techniques are therefore computationally intensive on that one node.

11.7.3.1 Information-Driven Dynamic Sensor Collaboration for Tracking Applications

An information-driven dynamic sensor collaboration technique has been proposed for tracking applications (Hightower, 2000). This approach allows the detection, classification, and tracking of objects and events to require aggregation of data among the sensor nodes. However, not all sensors may have useful information;

hence, an informed selection of sensors that have the best data for collaboration will save both power and bandwidth cost. In addition, flooding can be avoided and tracking reports can be more accurate. The metrics used to determine the participant nodes (who should sense and whom the information must be passed to) are (a) the detection quality, which includes detection resolution, sensitivity and dynamic range, misses, false alarms, and response latency; (b) the track quality, which includes tracking errors, track length, and robustness against sensing gaps; (c) the scalability in terms of network size, number of events, and number of active queries, (d) the survivability, meaning fault tolerance; and (e) the resource usage in terms of power/bandwidth consumption.

There is one leader node that is active at any moment. It is in charge of selecting and routing tracking information to the next leader. The tracking protocol functions as follows. A user sends a query that enters the sensor network. The query is guided toward the region of potential events. The leader node generates an estimate of the object state and determines the next best sensor based on sensor characteristics such as sensor position, sensing modality, and its predicted contribution. It then hands off the state information to the newly selected leader. The new leader combines its estimate with the previous estimate to derive a new state, and selects the next leader. This process of tracking the object continues, and periodically the current leader nodes send back information to the querying node using a shortest-path routing algorithm.

11.7.3.2 Tracking Using Binary Sensors

Binary sensors are so-called because they typically detect one bit of information. This one bit could be used to indicate whether the target is (a) within the sensor range or (b) moving away from or toward the sensor. Two approaches to the problem of target tracking using binary sensors can be distinguished. The first technique uses a centralized method, while the second one is a distributed protocol. In the centralized method, each sensor node detects one bit of information stating whether an object is approaching or moving away from it. This bit is forwarded to the base station along with the node identity. The detection is performed as follows. Each sensor performs detection and compares its measurement with a precomputed threshold. If the probability of presence is greater than the probability of absence, the detection result is positive. The model assumes that sensors can identify whether a target is moving away from or toward it and that the sense bits are made available to a centralized processor. It also assumes that the base station knows the location of each sensor and that a secondary binary sensor can be used in conjunction with this sensor to discover the precise location of the target.

In the distributed approach to target tracking, sensors determine whether the object is within their detection range and then collaborate with neighbor node data to predict the trajectory of the object. This cooperative tracking scheme improves accuracy by combining information from several nodes rather than relying on one node only. Assuming that sensors are uniformly distributed in the environment, a sensor with range R will (1) always detect an object at a distance of less than or equal to $R - e$ from it, (2) sometimes detect objects that lie at a distance ranging between $R - e$ and $R + e$, and (3) never detect any object outside the range of $R + e$ where $e = R/10$ but could be user-defined. Objects can move with arbitrary speed and direction. Hence the trajectory of the object can be linearly approximated to a sequence of line segments along which the object moves with constant speed.

Each node records the duration for which the object is in its range. Neighboring nodes then exchange the timestamps and their locations. For each point in time, the object's estimated position is computed as a weighted average of the detecting node locations. The weights assigned are proportional to a function of the duration for which the target is within range of a sensor. The target will remain within range of sensors closer to the target path for a longer period. A line fitting algorithm (least-squares regression) is executed on the resulting set of points. The object path is predicted by extrapolating the target trajectory to enable asynchronous wakeup of nodes along that path. In this technique, it is assumed that nodes know their locations and that their clocks are synchronized. Note that the density of sensor nodes should be high enough for sensing ranges of several sensors to overlap for this algorithm to work, and also sensors should be capable of differentiating the target from the environment.

11.8 Key Exchange in WSN

In WSNs, sensor nodes use pre-distributed keys directly, or use keying materials to dynamically generate pair-wise and group-wise keys. The challenge of key distribution is to find an efficient way of distribution of the keys and keying materials to sensor nodes prior to their deployment. The solutions provided for this challenge can be classified into three classes: (1) probabilistic, (2) deterministic, or (3) hybrid. In probabilistic solutions, keys are randomly selected from a key-pool and distributed to sensor nodes. In deterministic solutions, deterministic functions are used to design the key-pool and the keys to provide better key connectivity. Finally, hybrid solutions use probabilistic approaches on deterministic solutions to improve scalability and robustness.

11.8.1 Pair-Wise Key Pre-Distribution Schemes

Pair-wise key pre-distribution schemes consist of three tasks in general: (a) key setup prior to deployment, (b) shared-key discovery after deployment, and (c) path-key establishment if two sensor nodes do not share a key. Several classes of pair-wise key pre-distribution schemes can be found in the literature. They can include the following:

■ *Pair-wise key pre-distribution solutions.* The trivial solution in terms of resource usage is to deploy a single master key to all sensors. Since an adversary may capture a node and compromise the key very easily, it has very low resilience. The other extreme is to use distinct pair-wise keys for all possible pairs in the WSN. Although such an exhaustive solution creates unnecessary storage burden on a sensor node, it has very good key resilience.

■ *Random pair-wise key scheme.* This solution addresses unnecessary storage (Chan, 2003). Each sensor node stores a random set of N_π pair-wise keys to achieve probability π that two nodes are connected. During the key setup phase, each node identity is matched with N_π other randomly selected node IDs with probability π. A pair-wise key is generated for each ID-pair, and is stored in both nodes along with the ID of other node. Each sensor uses $2N_\pi$ units of memory to store its keys. During the shared-key discovery phase, each node broadcasts its ID and receives one message from each node within its radio range. Neighboring nodes can tell if they share a common pair-wise key.

■ *Location-based pair-wise keys pre-distribution scheme* is an alternative to random pair-wise key scheme (Liu, 2003). It takes advantage of the location information to improve the key connectivity. Sensor nodes are deployed in a two dimensional area and each sensor has an expected location that can be predicted. The idea is to have each sensor share pair-wise keys with its c closest neighbors (for a predefined value of c). At key setup phase, a unique key K_i and c closest neighbors $n_{i,1}$, ..., $n_{i,c}$ are associated with each node n_i. In addition, for each pair of nodes (n_i, n_j) a pair-wise key $K_{i,j} = PR(Kj, IDi)$ is generated, where PR is a shared pseudo random generation function. Node n_i stores all pair-wise keys, whereas node n_j only stores the key K_j and the PR function.

The third solution allows an easy extension when deploying new nodes. A new node n_t can be loaded with the pair-wise keys for c sensor nodes in its expected location. This solution decreases the memory use and preserves a good key connectivity if deployment errors are low. The second solution is more scalable in the sense that it efficiently uses the memory spaces and helps support larger WSNs. However, it sacrifices key connectivity to decrease the storage usage.

11.8.2 Element-Based Key Pre-Distribution Solutions

While the aforementioned schemes discuss pre-distribution of pair-wise keys, other schemes pre-distribute some elements needed to construct the pair-wise keys during deployment or operation. The element can be a master key, a matrix, a polynomial, or any other object that can be securely stored. Two classes can be distinguished:

11.8.2.1 Master Key-Based Schemes

These schemes aim at having a single master key that is pre-deployed to sensor nodes. Various schemes can be distinguished. In the broadcast session key negotiation protocol (Lai, 2002), a pair of sensor nodes (Si, Sj) exchanges random values Rnd_i and Rnd_j. They use master key Km to establish session key $K_{i,j}$ using a pseudo random function PR and the following formula:

$$K_{i,j} = PR(K_m, Rnd_i, RN_j).$$

The broadcast session key negotiation protocol has very low resilience, since it is possible to derive all pair-wise keys once the master key is compromised. The lightweight key management system proposes a solution with slightly better resilience where more than one master key is employed (Dutertre, 2004). It assumes the WSN implementing the scheme is built progressively, by adding at each step a new set of N sensor nodes. Each sensor node is requested to store a group authentication key k_1 and a key generation key k_2. If two sensor nodes Si and Sj are deployed at the same step, they authenticate each other by using the authentication key k_1. They exchange random nonce values Rnd_i and Rnd_j and establish the session key

$$K_{i,j} = PR(k_2, Rnd_i, Rnd_j).$$

If the two nodes are deployed at different steps, the older sensor, say n_i, is requested to store a random nonce Rnd_i and a secret $S_{i,t}$ for each new generation t (sensors added at step t). The secret $S_{i,t}$ is used to authenticate sensor nodes from new generation t. The younger node S_j (assumed to be in generation t) can authenticate itself by generating the secret $S_{i,t} = PR(gk_t, Rnd_i)$, given RNA. Secret gk_t is only known to nodes added in step t. Once authenticated, both sensors use $S_{i,t}$ as the key generation key to generate the pair-wise key $K_{i,j}$. It can be noticed that a malicious node can log the messages flowing in the network to process later when the required credentials are compromised.

11.8.2.2 Random Key-Chain Based Key Pre-Distribution Solutions

The basic schemes rely on probabilistic key sharing among the nodes of a random graph. In key setup phase, a large key-pool of K keys and their identities are generated. For each sensor, k keys are randomly drawn from the K without replacement. These k keys and their identities form the key-chain for a sensor node. Thus, probability of key share among two sensor nodes can be shown to be equal to

$$((K-k)!)^2 / (K - 2k)! K!.$$

During the shared-key discovery phase, two neighbor nodes exchange and compare lists of identities of keys in their key-chains. Basically, each sensor node broadcasts one message, and receives one message from each node within its radio range where messages carry key ID list of size *k*. The cluster key grouping scheme provided in Hwang (2004) offers to divide the key-chains into a predefined number, say n, of clusters, assuming that each cluster has a start key ID. Remaining key IDs within the cluster are implicitly known from the start key ID. Thus, only start key IDs for clusters are broadcasted during the shared-key discovery phase; this means that the messages carry key ID list of size n instead of *k*.

11.8.3 Key Distribution in Hierarchical WSN

In a hierarchical WSN, the base stations are computationally robust and may act as a key distribution center. Initially, base stations may share a distinct pair-wise key with each sensor nodes. These keys can be used to secure the setup of other keys. Various schemes have been proposed. We discuss here some among the major solutions.

11.8.3.1 Pair-Wise Key Distribution Schemes (PKDS)

The PKDS schemes for distributed environments are straightforward: a base station can share a distinct pair-wise key with each sensor node in that environment. The base station can intermediate the setup of a pair-wise key between any pair of sensor nodes. The localized encryption and authentication protocol (LEAP) proposes that each sensor node can set up pair-wise keys with its immediate neighbor (Zhu, 2003). During the key setup phase of this scheme, the nodes receive a general key *K*. A sensor node Si can use the key *K* and one-way hash function *H* to generate its master key $K_i = H_K(IDi)$. During the shared key discovery phase, the node *Si* broadcasts a message m containing its identity and a random nonce value, say *m* = (ID_i, Rnd_i). A neighbor *Sj* will respond to this message by sending the message

$$(IDj, MAC_{Kj}(Rnd_i, IDj)).$$

The node *Si* can then generate the key $K_j = H_K(IDj)$ and both nodes *Si* and *Sj* can generate the session key $K_{i,j} = H_{Kj}(IDi)$.

To achieve a global coverage of the WSN, multi-hop pair-wise keys may be required to reach remote clusters to authorize a sensor S_i to reach a remote cluster *C*. In that case, the sensor node S_i generates a secret $K_{i,C}$ and finds m intermediate sensor nodes. It divides the secret into m shares:

$$K_{i,c} = s_i \oplus \ldots \oplus s_m$$

and sends each share through a separate intermediate node N_j ($1 < j < m$).

11.8.3.2 TESLA-Based Solutions

The Timed Efficient Stream Loss-tolerant Authentication (TESLA) is a multicast stream authentication protocol. TESLA uses a delayed key disclosure mechanism, where the key utilized to authenticate the ith message is disclosed along with the $(i + 1)$th message. SPINS (Perrig, 2004) is based on a version of TESLA to provide a scheme for key distribution. It uses a base station as key distribution center. Mainly it works as follows: The base station randomly selects the last key K_n of a chain and applies a public hash function H to generate the rest of the chain K_0, K_1, ..., K_{n-1} by applying the formula

$$K_i = H(K_{i+1}).$$

Given key K_i, every sensor node can generate the sequence K_0, K_1, ..., K_{i-1}. However, given K_i, no attacker can generate K_{i+1}. At the ith time slot, the base station sends authenticated message $(M, MAC_{K_i}(M))$. The sensor nodes store the message until the base station discloses the verification key in the $(i + 1)$th time slot. Then they verify disclosed verification key K_{i+1} by using the previous key K_i.

References

T. Aura, P. Nikander, and J. Leiwo, DoS-resistant Authentication with Client Puzzles. *8th International Workshop on Security Protocols*, Springer-Verlag, Vol. 2133, pp. 170–177, 2001.

N. Bulusu, J. Heidemann, and D. Estrin, GPS-less Low Cost Outdoor Localization for Very Small Devices, *IEEE Personal Communications Magazine*, 7(5): 28–34, October 2000.

S. Capkun and J. P. Hubaux, Secure Positioning in Wireless Networks. *IEEE Journal on Selected Areas in Communications*, 24(2) (2006): 221–232.

H. Chan and A. Perrig, PIKE: Peer Intermediaries for Key Establishment in Sensor Networks, in *Proc. of IEEE INFOCOM 2005*, vol. 1, pp. 524–535.

H. Chan, A. Perrig, and D. Song, Random Key Pre-Distribution Schemes for Sensor Networks, in *IEEE Symposium on Research in Security and Privacy*, Orlando, FL, USA, 2003.

B. J. Culpepper and H. C. Tseng, Sinkhole Intrusion Indicators in DSR MANETs, *Proc. First International Conference on Broad Band Networks*, pp. 681–688, 2004.

B. Dutertre, S. Cheung, and J. Levy, Lightweight Key Management in Wireless Sensor Networks by Leveraging Initial Trust. Tech. Rep. SRI-SDL-04-02, System Design Laboratory, Apr. 2004.

L. G. Eschenauer and V. D. Gligor, A Key-Management Scheme for Distributed Sensor Networks. *9th ACM Conf. on Computer and Communications Security*, Washington, D.C., USA, 2002.

D. W. Gage, Command Control for Many-Robot Systems, *Proc. of the 19th Annual Annual Technical Symposium and Exhibition of the Association for Unmanned Vehicle Systems*, Huntsville, AL, USA, June 1992.

S. Ganeriwal, S. Capkun, C.-C. Han, and M. B. Srivastava, Secure Time Synchronization Service for Sensor Networks, *Proc. 4th ACM Workshop on Wireless Security*, pp. 97–106, New York, USA, 2005.

A. Gosh and S. K. Das, Coverage and Connectivity Issues in Wireless Sensor Networks, in *Mobile, Wireless, and Sensor Networks: Technology, Applications, and Future Directions*, Rajeev Shorey et al. (Eds.), John Wiley & Sons, 2006.

P. Gupta and P. R. Kumar, Critical Power for Asymptotic Connectivity in Wireless Networks, in W. M. McEneany et al. (Eds.), *Stochastic Analysis, Control, Optimization and Applications*, Birkhauser, pp. 547–566, 1998.

M. Hamdi, N. Boudriga, and M. S. Obaidat, Designing a Wireless Sensor Network for Mobile Target Localization and Tracking, IEEE GlobCom Conf., SatComm Symp., San Francisco, CA, Nov. 2006.

J. Hightower, R. Want, and G. Boriello, SpotON: An Indoor 3D Location Sensing Technology Based on RF Signal Strength, TR Univ. of Washington CSE #2000-02-02, Feb. 2000.

J. Hill et al., Tiny OS: Operating System Design for Wireless Sensor Networks, *Proc. of ASPLOS 2000*, pp. 93–104, Cambridge, MA, USA, 2000.

Q. Huang, J. Cukier, H. Kobayashi, B. Liu, and J. Zhang, Fast Authenticated Key Establishment Protocols for Self-Organizing Sensor Networks, *Proc. of 2nd ACM Conf. on Wireless Sensor Networks and Applications*, pp. 141–150, San Diego, CA, USA, 2003.

J. Hwang and Y. Kim, Revisiting Random Key Pre-Distribution Schemes for Wireless Sensor Networks, *Proc. of the 2nd ACM Workshop on Security of Ad Hoc and Sensor Networks*, pp. 43–52, New York, USA, 2004.

C. Intanagonwiwat, R. Govindan, and D. Estrin, Directed Diffusion: A Scalable and Robust Communication Paradigm for Sensor Networks, in *Proceedings of the Sixth Annual International Conference on Mobile Computing and Networks (MobiCOM '00)*, August 2000.

B. Karp and H. T. Kung, GPSR: Greedy Perimeter Stateless Routing for Wireless Networks, in *Mobile Computing and Networking*, 2000, pp. 243–254.

B. Lai, S. Kim, and I. Verbauwhede, Scalable Session Key Construction Protocol for Wireless Sensor Networks, in *IEEE Workshop on Large Scale RealTime and Embedded Systems (ARTES)*, Austin, TX, USA, 2002.

L. Lazos and R. Poovendran, SERLOC: Robust Localization for Wireless Sensor Networks, *ACM Transactions on Sensor Networks 1(1)*, (2005): 73–100.

D. Liu and P. Ning, Establishing Pairwise Keys in Distributed Sensor Networks, *Proc. of the 10th ACM Con. on Comp. and Comm. Security*, Washington DC, pp. 52–61, 2003.

D. Liu, P. Ning, and R. Li, Establishing Pairwise Keys in Distributed Sensor Networks, *ACM Transactions on Information Systems Security 8(1)*, (2005): 41–47.

D. J. Malan, M. Welsh, and M. D. Smith, A Public-Key Infrastructure for Key Distribution in TinyOS Based on Elliptic-Curve Cryptography, *1st IEEE on Sensor and Ad Hoc Communications and Networks*, Santa Clara, CA, USA, 2004.

R. Nagpal, H. Shrobe, and J. Bachrach, Organizing a Global Coordinate System from Local Information on an Ad Hoc Sensor Network, LNCS 2634, *Proc. of the 2nd Int. Workshop on Information Processing in Sensor Networks (IPSN '03)*, Palo Alto, April, 2003.

N. Nasser, Y. Chen, SEEM: Secure and Energy-Efficient Multipath Routing Protocol for Wireless Sensor Networks, *Computer Communications J.* 30 (2007), 2401–2412.

J. Newsome, E. Shi, D. Song, and A. Perrig, The Sybil Attack in Sensor Networks: Analysis & Defenses, *Proc. of 3rd IEEE/ACM Information Processing in Sensor Networks (IPSN'04)*, pp. 259–268, 2004.

D. Niculescu and B. Nath, DV-Based Positioning in Ad hoc Networks, *Journal of Telecommunication Systems*, 22 (1–4), (2003): 267–280.

C. Ozturk, Y. Zhang, and W. Trappe, Source-Location Privacy in Energy-Constrained Sensor Network Routing, *Proc of the 2nd ACM Workshop on Security of Ad Hoc and Sensor Networks (SASN 2004)*, pp. 88–93, Washington, D.C., USA, Oct. 2004.

A. Perrig, J. Stankovic, and D. Wagner, Security in Wireless Sensor Networks, *Communications of the ACM*, 47(6), (2004): 53–57.

H. Shuster, D. Georga Kopoulos, A. Cichocki, and D. Baker, Modeling and Composing Service-Based and Reference Process-Based Multi-Enterprise Processes, *Proc. CAISE*, pp. 247–263, Stockholm, Sweden, 2000.

A. D. Wood and J. A. Stankovic. Denial of Service in Sensor Networks, *IEEE Computer*, 35(10), (2002): 54–62.

Y. Yu, R. Govindan, and D. Estrin, Geographical and Energy Aware Routing: A Recursive Data Dissemination Protocol for Wireless Sensor Networks, University of California at Los Angeles Computer Science Department, Tech. Rep. UCLA/CSD-TR-01-0023, May 2001.

S. Zhu, S. Setia, and S. Jajodia, LEAP: Efficient Security Mechanisms for Large-Scale Distributed Sensor Networks, *Proc. of the 10th ACM Conference on Computer and Communications Security*, pp. 62–72, Washington, D.C., USA, 2003.

Chapter 12

Security of Satellite Services

12.1 Introduction

While the use of satellite networks as a constituent of the Internet backbone started two decades ago, the utilization of satellites to provide high-speed network access is relatively recent. The success of the new satellite networks in delivering high-speed access has shown the ability of the communication protocols they set up to operate efficiently in the satellite-based network. However, such an environment is characterized by longer propagation delays, at least for some types of satellite communications systems. Another type of communication satellite networks presents a rapidly time-varying network topology. Nowadays, a large competition to deploy global satellite networks has taken place. Several providers have even proposed to use satellite-based systems to deliver mobile services over very large geographical areas. The cost of an omnipresent coverage of these areas will be considerably low compared with the cost to achieve the same coverage by terrestrial-based communication systems.

Satellite networks are expected to provide voice, data, and video services. This means that they have a flexible multiservice architecture, but do not necessitate all services being delivered over the same satellite platform. They are expected to be available everywhere and should be affordable. This implies a global coverage for handheld, car, ship, train, and airplane use and that satellite-based services are based on open standards that enable multiple manufacturers and end service providers to enter a competitive market with compatible products. The satellite networks are also reliable, meaning that most satellite systems are built with proven, stable technology using a standards-based approach in order to ensure equipment, network, and service compatibility.

Three categories of communication satellite systems can be distinguished. They are classified based on the nature of the orbits the satellites involved in these systems describe: the geostationary orbit (GEO) satellite systems, the medium Earth orbit (MEO) satellite systems, and the low Earth orbit (LEO) satellite systems. A geostationary satellite orbits the Earth directly over the equator, at a distance close to 22,000 miles. It performs one complete trip around the Earth in 24 hours by allowing the satellite to remain fixed over the same spot on the Earth's surface at all times. Therefore, a single geostationary satellite can see a very large part of the Earth's surface and stay fixed in the sky with respect to any point on the surface from which it can be seen. Communication GEO satellite systems are very common today, due their efficient uses in very important services such TV and radio broadcast, weather forecast, and transport of phone calls.

The GEO satellites present several advantages including the following:

- Tracking the satellite by its Earth stations is very easy. The traffic senders and receivers can simply use fixed antenna positions and do not need to adapt them (in terms of position and direction).
- The satellite has a very large coverage. In fact, at a distance of 22,000 miles from the Earth the satellite can communicate with at least ¼ of the surface of the earth.
- The mobile users communicating via a GEO satellite system do not need to perform handover operations provided that they stay within the very large footprint of the satellite.

On the other hand, the GEO satellites present several disadvantages. In particular, the signal received from the satellite gets weak after travelling over a long distance and the transmission quality of the signal is limited by the shading caused by high buildings and the natural elevations. In addition, Northern or Southern regions of the earth situated at above latitude of 60° have serious problems receiving these satellites due, for example, to the elevations found there. To overcome these limitations, large antennas are adopted. Thus, it appears that GEO satellites are not suitable for small mobile devices, due to requirements made on the transmission power of the antenna and the high latency observed by the transmitted phone calls (about 0.25 seconds on each way of transmission). An example of GEO network is depicted by Figure 12.1. It shows a

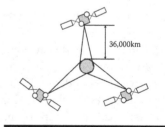

Figure 12.1 Example of GEO satellite network.

GEO satellite providing communication capabilities to mobile users, vehicles, and boats. Traditional networks are also connected through appropriate gateways.

A low Earth orbit (LEO) satellite system utilizes a large constellation of satellites, and each of these satellites is placed in a circular orbit at a constant altitude of a few hundred miles from the earth (Maral, 1991). Some satellite orbits take the satellites near the geographic poles. Each revolution takes a period of time varying from 90 minutes to a few hours. The constellation is organized in such a way that, at any time, any place on the Earth surface can see at least one satellite from the constellation. The entire system operates in a manner similar to the way a cellular telephone functions. Three main differences with the GEO networks can be observed: First, the coverage cells are very large with GEO networks. Second, the transponders or wireless receiver/transmitters are mobile in the LEO networks rather than fixed. Third, these transponders are placed on the satellites rather on the earth base stations. It appears also that a well-designed LEO network makes it possible for any mobile user to access universal services such as the Internet via wireless link from any point on the Earth.

The LEO satellites present several advantages that can help establish high quality mobile communications. In particular, a received LEO signal is stronger than a GEO signal for the same transmission power; the small footprints of LEOs allow for better frequency reuse, in a similar way to the cellular networks. A LEO satellite provides bandwidth for mobile terminals with uni-directional antennas using a low transmit power (in the range of 1W). LEO networks can provide reduced propagation delay, better global coverage, and higher elevation in polar regions. However, several disadvantages can be noticed in the use of LEO satellites communications. In particular, the transmission provided by LEO satellites is relatively reduced (even in the presence of compression schemes); a large number of LEO satellites are needed to provide a continuous wide coverage; additional mechanisms for handover between different satellites are required because of the short time of visibility with a satellite at high elevation. A number of studies have been reported in the literature, where the possibility of integrating LEO satellites with terrestrial Internet backbones was exploited (Hu, 2001). Two scenarios for the integration of satellite constellations have been addressed: (a) considering each satellite as an Internet node and adapting the existing terrestrial protocols to the satellite nature and (b) considering the satellites as a completely different network that has its specific architecture and protocols.

Figure 12.2 depicts an example of LEO network allowing GSM and UMTS users to communicate with each other. Most of the major components in the depicted network have been introduced in the previous chapters. They are described as follows: (a) the Radio Network Controller (RNC). It controls the radio resources. It is equivalent to the BSC-Base Station Controller of GSM and may be co-located with a gateway; (b) the Node B. This is a base station or a set of base stations. The 3GPP specification for the Node B may need to be adapted to cope with the movement of satellites in LEO constellations; and (c) the Iu and Uu interfaces. While

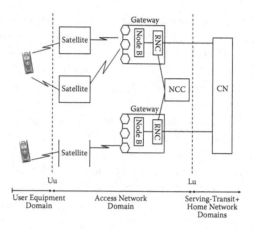

Figure 12.2   An example of LEO network integrating GSM and UMTS.

the first is the interface between the RNS and the core network (CN), the second is the air interface located between the user terminal and the satellite. Finally, a Network Control Center (NCC) has been introduced to co-ordinate the use of satellite resources among all gateways involved in the LEO network.

A medium Earth orbit (MEO) satellite is operating in an orbit placed within a distance ranging from a few hundred miles to a few thousand miles above the earth's surface. The orbital period of an MEO satellite ranges from 2 to 12 hours. A constellation of MEO satellites, with appropriately coordinated orbits, can provide global wireless communication coverage. Because MEO satellites are closer to the earth than geostationary satellites, Earth-based transmitters with relatively low power and ordinary-sized antennas can access the system. While some MEO satellites have near perfect circle orbits, and therefore have constant altitude and move at a constant speed, some other MEO satellites have more elongated orbits. In addition, the coverage area on the earth's surface is larger for the MEO satellites than LEO satellites because the MEO satellites operate at higher altitudes than LEO satellites. Compared to communication systems based on LEO or GEO satellites, the MEO-based systems present several advantages and disadvantages. Among the advantages, one can notice that a global-coverage constellation of MEO satellites may use fewer satellites than a global-coverage constellation based on LEO satellites. Thus, it requires fewer handovers, since a MEO satellite covers larger regions and moves slowly relatively to the Earth's rotation. Among the disadvantages, one can notice that the satellites require higher transmission power from the mobile devices and special antennas for smaller footprints.

Mobile satellite services (MSS) refer to networks of communications satellites intended for use with mobile and portable wireless telephones. Three major types of MSS systems can be distinguished: The aeronautical MSS (AMSS), the land MSS (LMSS), and the maritime MSS (MMSS). Provided that there are enough satellites in the system properly spaced around the globe, an MSS can link any pair of mobile devices at any time, no matter where in the world they are located. MSS systems are interconnected with land-based cellular networks. A telephone connection using MSS can be set up similarly to call connection in a cellular network, except that the repeaters are in orbit around the earth, rather than on the surface. MSS repeaters can be placed on geostationary, medium earth orbit, or low Earth orbit satellites.

Mobile satellite services are used by government agencies, administrations, and industries for many purposes, including public safety, natural resource conservation, transportation, and national defense functions, for the government; and electronic news gathering, aeronautical public telephony, and biomedical telemetry, for the private sector. On the other hand, satellites are expected to play an increasingly essential role in efficiently providing broadband Internet services over long distances. Future satellite networks will be hybrid in nature and will have terrestrial nodes interconnected by satellite links. Figure 12.3 depicts a hybrid system with one GEO satellite, one node of control, and three local areas networks connected to gateways. Security is an important concern in satellite networks, especially when they are hybrid, since the satellite segment is vulnerable to attacks including eavesdropping, session hijacking, and data corruption.

Four main *categories* of terminals have been distinguished in satellite systems: the handheld terminals, the transportable terminals, the vehicular terminals, and the broadcast only receivers. The characteristics of each of these categories have an impact on the security techniques provided to protect the access to the network and use of a service. Handheld terminals may provide dual mode, allowing the terminal to connect to terrestrial networks and supporting terminal mobility. Services accessible by these terminals comprise, but are not limited to, telephony, Web browsing, audio streaming, non-real-time video streaming, and location-based services. Unlike handheld and transportable terminals, the vehicular terminals are expected to be modular, meaning that components such as the antenna, front-end, and user interfaces may be distributed within the vehicle. Finally, broadcast only receivers are simple receivers of delivery services and do not transmit messages through the satellites.

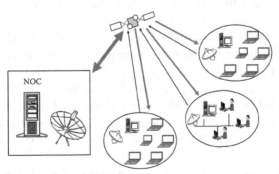

Figure 12.3   Example of hybrid network.

12.2 Examples of Satellite Networks

12.2.1 Communication Satellite Systems

The communication satellite systems provide point to point communications. Examples of such systems include the Globalstar, Iridium, and Inmarsat systems.

12.2.1.1 Globalstar

This is a satellite-based cellular telephone system using Code Division Multiple Access (CDMA) transmission. Globalstar is a LEO satellite network providing voice and data services. It allows mobile users to talk from anywhere in the world between 70° north and south latitudes with a quality higher or equal to what is provided by terrestrial cellular systems. It also transports digital and FAX data. It can also operate as a paging system and allow leaving short messages with one user or a group of users. It is composed by 48 low-orbiting satellites distributed in eight orbits (at a 414-km altitude), with six satellites in each orbital plane. Globalstar system authorizes interconnection between the user terminals and fixed networks through appropriate gateways. The interconnection between two Globalstar terminals is established through the satellite, processed by the gateway, forwarded to the PSTN, and then delivered back through the satellite.

12.2.1.2 The Iridium System

The Iridium system uses 66 satellites placed in 6 planes in a LEO (at a distance of 780 km) constellation to provide global coverage. The Iridium satellites perform on-board processing and use the inter-satellite links to route calls between the satellites, while minimizing the terrestrial cost of a connection to a PSTN subscriber. The call processing is based on the GSM architecture. A satellite has four neighbor satellites to which it is linked. Unlike Globalstar, Iridium employs an intra-satellite link architecture such that the satellites are able to communicate both with gateways and between themselves, reducing the need for regional gateways. The intra-satellite link architecture also supports end-to-end communications between any pair of users after initial call setup, without passing through ground stations. A routing table, set up in each satellite, shows how to reach a specific satellite that can deliver any requested call to a user. The gateways are used to connect the terrestrial network. Terminal to terminal calls can be routed directly using a path built over one or more satellites. A gateway includes a GSM MSC (mobile services switching center) in addition to the satellite system component in charge of handling the GSM BSS functions.

The Iridium system computes terminal positions. The location information is stored in the HLR and VLR (as in GSM). The services provided by Iridium include voice, data at 2.4 kbps, and some additional services such as call forwarding, call waiting, and conference calling. Each user has a single subscriber number to be used whether being registered in Iridium or roaming into a cellular network. Mobile calls, from the fixed network or another Iridium user, are routed via the home gateway that knows the location of the user. If the user roams to a gateway region, the home gateway is responsible for determining the user location (VLR) and sending a signal to the visited gateway to ring the user.

12.2.1.3 The Inmarsat System

This system offers a wide range of modern communications services to maritime, land-mobile, aeronautical, and other users. Inmarsat provides a variety of services tailored for specific range of applications. The main satellite constellation in Inmarsat consists of four satellites positioned at a geostationary orbit. The satellites, global beams provide overlapping coverage of the whole surface of the Earth (except for the polar regions) and are controlled from the satellite control center (SCC), which is responsible for keeping the satellites in position above the Equator and ensuring that the onboard systems are perfectly functional at all times. The flow of communications traffic through the Inmarsat network is monitored and managed by the Network Operations Center (NOC). On the other hand, a large number of land earth stations (LES) are made available. An LES acts as an Inmarsat gateway into the terrestrial telecom networks.

Typically, a call from any terminal is routed via the satellite system to a LES and then into the terrestrial phone and data networks. The Mobile Packet Data service provided by Inmarsat can route packets over any public network, such as the Internet and ISDN. It also allows setting Virtual Private Networks. This offers the enterprise's network to be extended to mobile users, while keeping access control and addressing management within the corporate infrastructure. Using a VPN authorizes the setup of secure tunnels, which involves encrypting data before sending it through the public network and decrypting it at the receiving end.

12.2.2 Mobile Satellite Broadcast

A large range of information data and entertainment services have started to be available for drivers and passengers of mobile vehicles. Such services include audio, video, and back-seat entertainment as well as traffic and travel related information services. The provision of broadcast services via a satellite system is attractive as wide coverage can be achieved using a single transmitter. A number of commercial mobile satellite broadcast networks have recently been deployed and some projects are running for the design of promising next-generation systems.

The objective of the broadcast system considered in this chapter is to deliver content data to mobile receivers. The delivery of the content is assumed to satisfy certain quality of service requirements such as service availability, maximal delay, bandwidth, and jitter variation. It should keep the service cost as low as possible.

Two categories of data are of interest to broadcast systems, namely the streams and files. A stream would provide a continuous service like traditional audio and video.

A file distribution service can provide traffic and other travel information. Files can be used to distribute audio and video content that can be played out depending on the preferences of the end user. Streaming services attempt to make use of the available data transportation capabilities efficiently. Therefore, variable rate source coders have been developed (Faller, 2002). The channel encoder can be seen as consisting of two parts. The first part is an encoder that applies an error correcting code to protect the content data from noise and channel fades. The second part is the multiplexer.

Various mobile hybrid satellite-terrestrial broadcast networks have been built, including the following systems:

12.2.2.1 Sirius Satellite Radio

The Sirius Radio System has been conceived to provide a Digital Audio Radio Service (DARS) directly from satellites to vehicles across the Continental United States region. It employs three satellites moving in elliptic orbits at an inclination of 63.4°, with two satellites in visibility at any moment. Terrestrial repeaters are present in major urban areas to permit continuous reception also in the presence of obstacles that might block the space-based signal. The access method used by satellites is TDM (Time Division Multiplex) while terrestrial repeaters use the OFDM multiplexing technique.

The Sirius system allows a large selection of music formats and program types using a specific technique to efficiently encode the audio signal. The satellite signals are received by different mobile platforms, including automobiles. The in-car receivers have three components. The first component is the digital radio box. The second is a compatible head-unit, fitted with either a CD or a cassette. The third is a compact digital aerial.

12.2.2.2 XM Satellite Radio

This Satellite Digital Audio Radio service provides (over the United States) high-quality compressed audio, as well as text and other digital data to mobile cars, home, and mobile personal receivers via two geostationary satellites and a network of terrestrial repeaters for re-broadcasting. Each satellite transmits the same content so that a receiver can construct the service signal from either satellite signal.

12.2.3 Global Navigation Satellite Systems

A Global Navigation Satellite System (GNSS) provides autonomous geo-spatial positioning with global coverage. It allows small electronic receivers to determine their location (longitude, latitude, and altitude) within a few meters using time signals transmitted along a line of sight by radio from satellites. Receivers on the ground with a fixed position can also be used to calculate the precise time as a reference for scientific experiments. Currently, the United States NAVSTAR Global Positioning System (GPS) is the only fully operational GNSS. The European Union's Galileo positioning system is a GNSS in initial deployment phase, scheduled to be operational in 2013. Some other systems have been developed or are under development in Russia, China, and India.

Disruption can affect the integrity of a GNSS. Two types of disruption can be distinguished: unintentional and intentional. The unintentional disruptions are typically caused by signal interference (in space or on the ground), transmission errors, and hardware failures. They can be alleviated by various systems, including a number of privately operated services providing improved accuracy and integrity. However, these systems do not effectively mitigate the intentional disruptions. In fact, intentional disruption can present a real threat against safety-critical applications such as hazardous-materials tracking. Intentional disturbance targeting GPS applications, for example, include various attacks such as jamming and spoofing of the GPS signal.

12.2.3.1 GPS

The GPS consists of up to 32 medium Earth orbit satellites placed in six different orbital planes. Operational since 1978 and globally available since 1994, GPS is currently the world's most utilized satellite navigation system. A GPS receiver computes its position by carefully timing the signals sent by the constellation of the GPS satellites. Each satellite continually transmits messages containing the time the message was sent, a precise orbit for the satellite sending the message (the ephemeris), and the general system health and rough orbits of all GPS satellites (the almanac). The receiver uses the arrival time of each message to measure the distance to each satellite, from which it determines the position of the receiver. However, the method of calculating position can be affected by several errors. One among the most interesting errors is related to GPS receiver clock. It affects the estimated distances from the GPS receiver to the satellites.

Techniques that can be performed to protect commercial GPS applications cannot use cryptographic authentications, since these mechanisms are not currently available for civilian GPS users. However, non-cryptographic validation techniques can be used to check the signal for irregularity detection. Next enhancements of GPS will provide a number of new services for civilian users with improved accuracy, integrity, and security. These new services will make possible the development of systems that offer guaranteed levels of security.

12.2.3.2 Galileo

This is a global navigation satellite system currently being built by the European Union. It is expected to use 30 MEO satellites placed in three orbital planes. It is intended to provide more precise measurements and better positioning services than available through GPS. It will provide four different navigation services: the Open Service that will be free for anyone to access. This service will allow receivers to achieve an accuracy lower than 4 m horizontally and 8 m vertically; the encrypted Commercial Service will be available commercially and will offer an accuracy of better than 1 m; the encrypted Public Regulated Service and Safety of Life Service will both provide an accuracy comparable to the Open Service. The last two services aim at providing robustness against jamming and reliable detection of problems within 10 seconds.

GNSS security services are used for many purposes, including providing strong integrity and quality of service guarantees to applications. Three categories of security services for GNSS can be distinguished (Wullems, 2005):

1. *Navigation data authentication and cryptographic integrity protection*: A Navigation Message Authentication (NMA) scheme adds authentication messages to the navigation message stream, authenticating the source and providing cryptographic integrity protection of the navigation data.
2. *Signal access control*: This mechanism prevents access to the signal by unauthorized users. Access to the signal can be restricted through Spreading Code Encryption (SCE), in which the spreading code is protected using cryptography. Only users with the appropriate cryptographic keys are able to obtain the secret spreading code, which then allows de-spreading of the signal. GPS and Galileo signals use Direct Sequence Spread Spectrum and Code-Division Multiple Access.
3. *Navigation data access control*: A navigation data access control scheme is able to facilitate restriction of access to a part or the entire navigation data stream modulated over a given signal using encryption.

Let us now notice that, nowadays, there is no indication that any security services will be provided to civilian users through the open service and that the Galileo high-level mission has indicated that NMA may be integrated into the OS.

12.3 Reliable Transport in Mobile Satellite Communications

The main characteristics of the end-to-end path that affect transport of a data packet in a satellite network (or a constellation of satellites) are latency, bandwidth, packet loss due to congestion, transmission errors, or security attacks. Compared to the wired networks, these characteristics can vary significantly. Three main components are important to latency. They are the propagation delay, the transmission delay, and the queueing delay. In a GEO satellite, the propagation delay is expected to be the most prevailing component, since it is typically on the order of 270 ms, for a one-way propagation delay to (or from) the satellite. It may be higher depending on the mechanisms used for error correction. On the other hand, this delay can be an order of magnitude less in the LEO networks. For example, satellites at an altitude of 1000 km will require only 20 ms for a one-way delay for a single hop. However, additional satellite hops will add delays to the latency.

Let us notice that some of the security attacks, such as some DoS attacks, have been built to extend the experienced latency or packet loss to an unacceptable level. These attacks operate mainly on the transport protocol such as the TCP or its variants.

12.3.1 TCP Flow Control in Satellite Communications

It is well known that the TCP protocol is a connection-oriented and end-to-end reliable transport protocol. It integrates three major mechanisms, the flow control, the congestion control, and the error control. It uses source and destination port numbers combined with IP source and destination addresses to uniquely identify each TCP connection. To achieve flow control, a TCP receiver sets the receive window field (RWND) so that the sender will never overflow the receiver's buffer and the TCP sender maintains a state variable CWND for congestion window. While RWND prevents the sender from overflowing the receiver buffer, the CWND is used to prevent the sender from overloading the network.

In the most common versions of TCP, four mechanisms are typically used to control congestion, namely the slow start, congestion avoidance, fast retransmit, and fast recovery. The slow start algorithm is used at the beginning of a new connection to probe the network bandwidth; it increases the CWND by one maximum segment size (Msiz) when an acknowledgment is received; it stops executing when its CWND is greater than a value called the *slow start threshold*. After that, congestion avoidance scheme starts controlling the flow; it increases CWND by one Msiz per round trip time (RTT). Fast retransmit algorithm is triggered when a fixed number of duplicate acknowledgments (usually three) are received. TCP retransmits the potentially lost packet indicated by the acknowledgment and reduces its CWND to half of its value. Then, it increases its CWND by 1 Msiz when a duplicate acknowledgment is received. If there is one and only one packet lost in a single window, the CWND can be increased to the original CWND before the loss after about half RTT. After that, TCP can send a new packet when each duplicate acknowledgment is received, if allowed by the RWND. Finally it will send half a window new packets when it receives the first non-duplicate acknowledgment.

Error control includes two mechanisms, the error detection and the error recovery. TCP uses acknowledgment packets, timer, and retransmission to achieve error control. It uses cumulative acknowledgment, which means that when a packet gets lost, it prevents the acknowledgment from being sent and the window from sliding until the lost packet is recovered. Since it links the flow control, congestion control, and error control together, the sliding window mechanism becomes vulnerable when there are congestion loss and packet corruption in the network.

The use of TCP in satellite networks experiences several drawbacks including the following:

■ *Long propagation delay*: A typical round trip time in a GEO satellite network takes more than 500ms. In addition, it can be easily shown that the time t_{slow} taken by TCP slow start to reach the satellite bandwidth (SatBW) is estimated by the following expression, when every TCP segment is acknowledged,

$$t_{slow} = \text{RTT} \times \log_2(\text{SatBW} \times \text{RTT}).$$

For a connection with large RTT, TCP wastes a long time in slow start before reaching the available bandwidth. Consequently, short connections do not use the bandwidth efficiently. To overcome this limit, some experts have suggested using larger initial window (up to 4K bytes) rather than 1 Msiz for slow start. Some other experts have proposed to remove the delayed acknowledgment mechanism in the slow start so that every packet is acknowledged and the sender can increase his CWND more quickly. On the other hand, it appears that TCP connections with larger RTTs do not get their fair share of the bandwidth when they compete with the connections with smaller RTTs.

■ *Channel error*: Satellite channel is noisy compared to optical fiber channel. Bit error rates (BER) of the order of 10^{-6} are often observed. For mobile users, the bit error rates can be even higher because of channel fading and mobility. Because the traditional TCP Reno treats all losses as congestion in the network, this kind of link layer corruption can cause TCP to drop its window to a small size and lead to poor performance. TCP can convey information about non-contiguous segments received by the destination in the acknowledgments (ACKs) so that the sender can recover errors faster than TCP Reno, which can recover only one loss per RTT. Forward error correction (FEC) coding is usually used in satellite communication to reduce the bit error rate. However, the FEC consumes additional bandwidth by sending redundant information together with the data and transforms the original random error nature to one with bursty errors.

Attacks targeting the TCP usage on a satellite network include launching very short jamming at randomly selected moments. The effect of such attacks is to create errors on the transported packets. These errors will typically be detected but would not be classified as attacks.

■ *Reverse channel congestion*: In general, the forward channel bandwidth from the satellite to the earth terminals is much larger than the reverse channel bandwidth. When the reverse channel traffic load is greater than its bandwidth, congestion could happen. The congestion may cause poor performance in the forward channel because TCP uses ACKs to clock out data. In the best case, the ACKs are not lost, but queued, waiting for available bandwidth. This has an immediate consequence on the retransmission timer and slows down the dynamics of TCP window. To alleviate this problem, ACK filtering was proposed to drop the ACKs in the front of the IP queue by taking advantage of the cumulative acknowledgment strategy in TCP. The situation is even worse for two-way transfers. When the users are sending data and browsing the Web at the same time, a lot of data packets could be queued in front of ACKs in a FIFO queue, which increases the ACKs delay dramatically. In this case, a priority queue can be used to schedule the transmission of ACKs.

12.3.2 Enhancing TCP Protocol

The approach to modify TCP can be host-based or network-based. Several end host based approaches have been developed to enhance the TCP characteristics to provide better quality of transmission over satellite networks. These major end host based schemes can be classified into three categories: (a) window based enhanced TCP; (b) rate-based enhanced TCP; and (c) hybrid approach. Three end host based solutions can be distinguished for satellite communications:

1. **TCP Vegas:** Compared to TCP Reno, TCP Vegas addresses the congestion control by using the transmission rate as a congestion signal rather than packet loss (Brakmo, 1995). Every round trip time, the sender calculates its transmission rate based on the sending window and the measured RTT. This rate is compared to the expected rate, which is equal to the sending window divided by the Base RTT, knowing that the Base RTT is the smallest RTT measured before comparison. Low and high thresholds are used to trigger window additive increase or decrease, depending on whether the channel is under-utilized or over-utilized. Unlike TCP Reno, TCP Vegas can decrease its window in congestion avoidance. However, it has been shown that TCP Vegas is not stable when the number of connections becomes very large. In addition, the RTT value, as measured by the sender, may be affected by the congestion in the reverse channel rather than the forward channel. Consequently, TCP Vegas does not perform well in the asymmetric channel communications.

2. **TCP Westwood:** TCP Westwood is a sender-side enhancement of the TCP fast recovery algorithm that aims at improving the TCP performance in wireless and satellite networks (Mascolo, 2001). TCP Westwood estimates the bandwidth used by a connection by monitoring the inter-arrival time of the returning ACKs. Then, it uses the estimated bandwidth to compute the congestion window and the slow start threshold after congestion is detected (typically, the detection is performed after the occurrence of three duplicate ACKs or after a timeout). When a congestion is experienced, TCP Westwood

attempts to choose a slow start threshold and a congestion window in conformance with the effective bandwidth. While under same circumstance, TCP Reno blindly halves the congestion window. TCP Westwood outperforms TCP Reno. However, since the rate of returning ACKs depends on the reverse channel, the access scheme and the congestion status, the inter-arrival time may not reflect effectively the effect bandwidth in the forward channel.

3. **SCPS-TP:** Space communication protocol standards-transport protocol (SCPS-TP) is a set of TCP extensions for space communications (Durst, 1996). This protocol adopts the timestamps and window scaling options in RFC1323. It also uses TCP Vegas low-loss congestion avoidance mechanism. A receiver in SCPS-TP does not acknowledge every data packet. Instead, the acknowledgments are sent periodically based on the RTT. The traffic demand for the reverse channel is much lower than in the traditional TCP. However, it is difficult to determine the optimal acknowledgment rate and the receiver may not respond properly to congestion in the reverse channel. It does not use acknowledgments to clock out the data; rather, it uses an open-loop rate control mechanism to meter out data smoothly. The SCPS-TP uses selective negative acknowledgment (SNACK) for error recovery. A SNACK is a negative acknowledgment that is able to specify a large number of holes in a bit-efficient manner.

On the other hand, two strategies have been proposed by the network-based solutions to prevent the channel errors from the sender side so that they cannot be misinterpreted as congestion. The first strategy is to recover the errors at the link layer. This strategy has been used in SNOOP and AIRMAIL protocols. The second strategy is to split the end-to-end TCP connection at the wired/wireless network gateway. This strategy is used by several major extensions: I-TCP, M-TCP, Super TCPs and STP adopts this strategy. TCP connection splitting cannot only reduce the channel errors from the source TCP, but also shield the (possibly) long and variable delay from the source. Another approach requiring modifications to both end hosts and network router software is proposed (Katabi, 2002). Network-based solutions can be classified into three classes: the network layer active queue management, the link layer error recovery, and the TCP Connection Splitting.

Let us now describe six among the most important network-based solutions for enhancing TCP. Two solutions are selected from each class:

■ *RED*: A gateway implementing RED monitors the average queue size with a low pass filter. The decision to accept an incoming packet is based on comparing the average queue size $\overline{q}$ to two predefined thresholds λ and μ ($\lambda < \mu$). If the average queue size $\overline{q}$ is lower than λ, the arriving packet is accepted. If the average queue size is between λ and μ, the arriving packet is dropped or marked with a probability proportional to the average queue size. Finally, if $\mu \leq \overline{q}$ the arriving packet is dropped or marked with probability value equal to 1 (Floyd, 1993). Some form of per flow fairness is ensured in RED, since the packet drops of a specific connection during congestion is almost proportional to its arrival rate. RED attempts to maintain a low average queue size while admitting occasional bursts of packets. However, the random drop of incoming packet in RED could cause the TCP source congestion window to be halved even if the connection does not exceed its fair share. Therefore, one can say that RED may suffer from short-term unfairness.

■ *BLUE*: This extension attempts to reduce the packet loss rate experienced with RED (Feng, 2002). It uses packet loss and link idle events rather than queue length to manage congestion. The packet marking probability is increased when there are packet drops due to buffer overflow; conversely, when the queue is empty or the link is idle, the marking probability is decreased. By decoupling congestion management from instantaneous or average queue length, BLUE has been shown to perform significantly better than RED in term of packet loss and buffer size requirements in the routers.

■ *SNOOP*: This transport protocol essentially uses the TCP acknowledgments to trigger the link layer retransmission at the base station and suppresses the duplicate ACKs from propagating to the TCP sender (Balakrishnan, 1995). Therefore, it can protect the channel from errors and does not lead the TCP sender to reduce its window to half as end to end TCP does. Although SNOOP does not have any TCP layer code running at the base station, it needs to access the TCP header to get the sequence number and acknowledgment number. In addition, it does not work when IPsec protocol is used. SNOOP preserves the end-to-end semantics of TCP, but it cannot be used for satellite networks because of the long propagation delay of the satellite link, which could cause fairness problem if the base station keeps the ACKs to transmit end to end.

■ *AIRMAIL*: AIRMAIL is a reliable link layer protocol developed for indoor and outdoor wireless networks. By combining link level ARQ and adaptive Forward Error Correction (FEC) coding, it recovers the channel errors locally and can obtain better throughput and latency performance (Ayanoglu, 1995). However, there exists a complex interaction between the reliable link layer and end-to-end TCP. It is possible for the error to trigger the link layer retransmission while at the same time the duplicate ACKs of TCP propagate to the TCP source and cause TCP to halve its window and retransmit the same packet. Another problem is that not all the up layers need reliable link layer service, e.g., real-time traffic using UDP does not need reliable data transmission.

■ *Indirect TCP*: The basic idea of indirect TCP (I-TCP) is that the end-to-end TCP connection is divided into two connections: one is built from the server to the base station and another one is set up between the base station and the mobile users. The base station sends premature acknowledgments to the server and takes responsibility to relay the data to the mobile host reliably

(Bakre, 1997). The advantages of I-TCP include the separation of the flow control and congestion control of wireless and wired network, and a faster reaction to channel errors. However, this scheme violates the end-to-end semantics of TCP. In fact, the sender can receive an acknowledgment of a data packet while the data packet has not reached the destination. Some experts have argued, however, that using I-TCP for applications, that rely on application layer acknowledgments in addition to end-to-end TCP acknowledgments, does not comprise end-to-end reliability. Such applications include the FTP and HTTP.

■ *Super TCP*: Because GEO satellite channel is a FIFO channel, there is no out-of-order routing. In addition, congestion over the satellite link is unfeasible if the packets are sent at the rate of the satellite bandwidth. A connection splitting based solution is proposed to use one duplicate ACK to trigger the fast retransmission at the upstream proxy and to use a fixed window size for the satellite TCP connection. If there is only one connection in the system, the fixed window can be set to the satellite bandwidth delay product. However, multiple connections with different terrestrial round trip times and different traffic arrival patterns have not been addressed. Super TCP proposes a new sender algorithm using the same idea as in TCP new Reno. It uses partial ACKs to calculate the bursty loss gap and sends all the potentially lost packets beginning from the partial acknowledgment number. Although it is possible that the sender could retransmit packets that have already been correctly received by the receiver, it is shown that this algorithm performs better than TCP SACK in recovering bursty errors.

12.4 Packet Routing in Non-GEO Networks

In a non-GEO satellite network, each satellite has several neighboring satellites and can communicate with the other satellites via bidirectional *inter-satellite links* (*ISLs*). An ISL is nothing but a channel allowing two satellites to communicate when they see each other. The links between satellites are called *intra-plane ISLs* if the satellites are in the same orbital plane. They are called *inter-plane ISLs* when the two satellites belong to different planes. These ISLs are characterized by the following five features: (a) the intra-plane ISLs are maintained at all times between the satellites; (b) all the satellites move in the same circular direction within the same plane; (c) the propagation delays on the intra-plane ISLs are always fixed; (d) the propagation delays are highly variable for inter-plane ISLs; and (e) the inter-plane ISLs are operated only outside the polar regions. Figure 12.4 depicts a general satellite network architecture allowing building a route between two mobile terminals through three satellites.

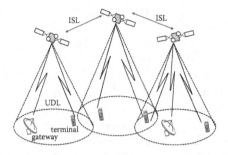

Figure 12.4   General satellite network architecture.

Current terrestrial routing protocols are not capable of providing QoS guarantees in LEO satellite networks with satellite-fixed cells, due to the inherent time variance of the user traffic on the inter-satellite links. In particular, the link capacities may be sufficient to accommodate a specific session at the session setup stage on the determined route; but the same route may not be able to maintain the required QoS for the entire duration of the session because the loading on the links change in time as the satellites move along their orbits. Hence, new routing protocols that take into consideration the changes in the loading of the inter satellite links due to the motion of the satellites were needed. Various connection-oriented routing protocols through non-GEO-satellite networks have been proposed for the transport of real-time multimedia applications that are sensitive to delay variations and impose strict delay bounds.

On the other hand, it appears that connection-oriented routing protocols through a satellite network should consider the link handover and minimize its impact on each individual connection. They may also consider the dynamic network topology as a periodically repeated series of K topology snapshots. In addition, they should maximize the total number of sessions that are served by the network with respect to the QoS requirements that are required by each of these sessions.

A simple routing solution may aim at finding a new route for the session, whenever the current route fails due to bottlenecks on the links or due to link failure. Even though such a solution is feasible in a terrestrial wireless network, where handovers are infrequent and random, it is not an optimal solution for LEO-satellite networks. In fact, defining a complete new route may cause high messaging traffic and processing load on the LEO network. On the other hand, predictive routing protocols should provide guaranteed QoS in the satellite networks. They make use of the deterministic nature of the LEO satellite topology to foresee the traffic load on the ISLs up to a short period of time in the future. The optimal path is selected from the path set to reduce the link changes and balance the user traffic. This protocol, however, does not consider inter-satellite handovers. In addition, the computation overhead it generates grows noticeably as the value K increases.

12.4.1 Predictive Routing Protocols

The drawbacks discussed in the beginning of the ongoing section can be overcome by exploiting the *predictability* of the non-GEO satellite network topology. Let us, for the sake of clarity, consider the case of the LEO satellite networks and present the protocol described in Erçetin (2000) for predictive routing.

Let us assume that the topology at a given time t can be determined from the information on the connectivity of the constellation and the satellites' speeds. The users in the footprint of each satellite can also be determined by using the location information of these users at any particular instant. Therefore, the total traffic that needs to be routed by each satellite at a particular time in the near future can be predicted. This fact can be used, at the call setup, to determine multiple routes for the same call to help avoiding predicted bottlenecks on the links. Consequently, the predictive routing approaches present two advantages: First, the processing delays and the messaging overhead it generates due to handover route recomputation, for an ongoing call, can be avoided. Second, the bandwidth utilization is improved since only the required bandwidth is reserved.

The predictive routing method presented in Erçetin (2000) assumes that, when a user sends a request for a new connection, the request is sent to the gateway. The request message reports the locations of the source and destination users, and a requested bound for the delay. In determining a route between a pair of end-users in a LEO satellite network, the following facts should be taken into consideration in addition to the limitation of network resources:

- There may be a very large number of ongoing sessions with different QoS requirements.
- If a session is accepted to the system, its QoS requirements should be satisfied during its lifetime.
- The network resources should be used efficiently. In particular, the messaging and signaling overhead of the outing protocol should be minimized.
- The amount of information stored for each session at any satellite should be minimized.

Initially, the predictive protocol assumes that the gateway determines a route for a given call according to the available satellite link bandwidths available at the instant where the session request arrives, namely t_0. Due to satellite mobility, at time $t > t_0$, the satellite may serve users who may be able to require to use the same links as those used by the call under observation. This may result in an increase in the load (or worse, a congestion might occur) on these inter-satellite links. Any new accepted call should not degrade the QoS of the on-going calls. This means that the sessions that are in progress should have a certain priority over the new sessions. To ensure that the quality of service of on-going sessions is not degraded, the route chosen for a new session should not cause congestion.

To predict future loads on the inter-satellite links, the link state information (including the available resources on the links for serviced sessions and the routing table) is collected from the satellites and used by the ground gateways. The remaining resources available on the links along a route are checked to ensure that there is always sufficient resources for the session, for all time t, such that $t_0 < t < t_0 + TS$. If the route cannot accommodate the session needs at some time t_1, $t_0 < t_1 < t_0 + TS$, because the available bandwidth on the route is less than the required amount, for example, a new route for the session is determined. If this second path is also infeasible at some time t_2, $t_0 < t_1 < t_2 < t_0 + TS$, then another route for the same session will be determined. The procedure is repeated until a feasible path for all t, satisfying $t_0 < t < t_0 + TS$, is determined. If no feasible path can be found for any period of time between t_0 and $t_0 + TS$, then the session is blocked.

In conclusion, one can say that the protocol determines a set of paths

$$\{p_0(t_0), \ldots, p_n(t_n)\},$$

where t_i denotes the time at which the system starts using the path p_i to route packets from source satellite to the destination satellite $t_0 < t_1 < \ldots < t_n < t_n + TS$, and t_i. To converge, this process needs to provide that, for all j, $t_{j+1} - t_j > a$, for a given positive constant a. In that case, the number of paths is lower than $TS/a + 1$. This constraint can be easily achieved in a LEO network.

12.4.2 QoS-Based Routing in LEO Satellite Networks

The objective of a QoS-based routing algorithm is to select routes that are able to comply with the requirements of communicating users. They should be, for example, capable of reducing the delay jitter while guaranteeing the bandwidth required by the user. Some QoS-based protocols use the location information of satellites and ground stations to predict the availability of satellite links to build stable paths, for connection requests, and to reduce the probability of link handovers during connection duration. The routing process allows the establishment of a requested connection by allocating the required resources (such as bandwidth) along some path between the source and the destination nodes. The allocated resources are released when the connection terminates.

Various QoS-based routing schemes have being provided in the literature for non-GEO satellite networks. Among these schemes, we describe in the following one, as discussed in Chen (2003). The scheme assumes that, when a ground station issues a connection request, it describes several parameters including the location of source s (using GPS service, for example); the ID of the destination ground station d; the expected connection duration; and the requested bandwidth (bw). When the traffic has a constant bit rate, the requested bandwidth, say bw, is fixed through the connection lifetime. When the traffic is variable, the requested bandwidth bw can be specified, like in ATM networks, by the maximum bandwidth and sustained bandwidth. The protocol includes three steps: deterministic UDL routing, probabilistic ISL routing, and handover rerouting.

- *Deterministic UDL routing*: As soon as the connection request is received, the source ground station selects the ingress and the egress satellites (we say in that case that an UDL routing is performed) for the path between the source and the destination. The satellite with sufficient bandwidth and maximum remaining coverage time to the ground station is selected. By doing this, the probability of inter-satellite handover is reduced. The computation of the remaining coverage time of satellites can be done with the knowledge of location information.
- *Probabilistic ISL routing*: The ISL routing is executed after selecting the ingress and the egress satellites. Two major parameters are managed for each link j: the propagation delay d_j and the probability p_j that measures the probability that the ISL link will not be shut down either before the connection ends or an inter-satellite handover occurs. The value of d_j of satellite link j can be easily deduced from the satellite trajectory information, at any time. Because it is easy to know the exact time when an inter-satellite handover would occur is after the deterministic UDL routing, one can predict the ISL handover time of link j and deduce the probability value p_j. The selection of a path is based on the Dijkstra's algorithm. The latter is used with the following cost function for link j:

$$C_j = \begin{cases} d_j\left(-\log(p_j)\right), & \text{if } Av\_B \geq bw \\ \infty, & \text{if } Av\_B < bw \end{cases},$$

where Av_B is the available bandwidth. Finally, once an ISL path is found, the required bandwidth is allocated along the path. Notice that $C_j = d_j$ when $p_j = 1$ and that the cost C_j tends to become very large when p_j is close to 0.

- *Handover rerouting*: Assume that, at a given time t, the ground station moves out of the footprint of its access satellite S_{old}, a new satellite S_{new} with the maximum coverage time should be selected as the new access satellite. In that case, the rerouting step can perform the following, instead of computing a new ISL path immediately:
  - If the satellite S_{new} is already on the current ISL path, the subpath of the current path from S_{old} up to S_{new} is deleted and the reserved bandwidth is released. The new ISL path starts with S_{new}.
  - If S_{new} is not on the current ISL path, a direct link with sufficient bandwidth to one of the satellites on that path is determined (starting from the other end of the path).

12.5 Mobility and Location Management in Satellite Networks

Mobility management allows locating roaming mobile terminals (MTs) at any time to deliver services and to maintain connections as the MT moves from one service area to another. Mobility management performs two major processes: The first process is the location management. It aims at locating the MTs with the main needs to deliver incoming sessions to them at a reasonable cost. Location management is a two-step procedure composed by the location registration and the call delivery. In the location registration (or location update) step, the MT notifies, from time to time, the network about its current access point, and the network database (DB) stores the position or registration area, if it has changed. In the call delivery stage, the network DB is queried and the current cell (called also spotbeam) where the MT is roaming is found.

The second process is the handoff management. It is responsible for the transfer of ongoing sessions to adjacent cells as the MT moves from one cell in the network to another, while attempting to guarantee the continuity of the active sessions. The handoff management enables the network to maintain user's connections as the MT changes its access point to the network. It is composed of three tasks: the initiation, new connection generation, and data-flow control. During initiation, the user (or the network) detects a change in the network ongoing connections identifying the need for handoff (such as a deterioration of the signal strength of the ongoing connection). New connection generation requires locating new resources for the handoff and performing additional routing operations. Finally, during the data-flow control stage, the data from the old connection path is delivered to the new connection path according to agreed-upon service guarantees.

Location and handoff management in satellite networks highlights various concerns that need to be addressed: (a) optimizing the route for each connection; (b) minimizing the signaling load on the network; (c) reassigning efficiently the bandwidth; (d) processing efficiently packets; and (e) securing all the operations and exchanged data. Resources management is also an important issue. In the following subsections, the description of these two processes will be detailed.

12.5.1 Location Management in Satellite Networks

In location management, four major functions can be distinguished: Location registration (LR), location interrogation (LI), location update (LU), and paging (PG). Functions LU and LR are triggered by the *mobile terminal* from time to time, periodically, or after crossing borders of a cell. The functions IG and PG are triggered by an incoming call to the MT.

In current terrestrial cellular systems (such as GSM), the coverage area is partitioned into a number of Location Areas (LA's). Each LA consists of a group of neighboring cells. Each MT triggers a location update when an MT enters into a new LA. It also triggers the location update periodically. In addition, when an incoming call arrives to a MT the LI function is executed. After this action, the network pages all cells within the LA *where* the MT is roaming. The LU scheme experiences inefficiencies. For example, when a mobile terminal moves back and forth various times across the border of two LAs, an excessive signaling cost may be generated. In addition, to minimize location management cost, the LAs must be designed carefully and should have the same size and shape.

Various schemes have been proposed for the location management in satellite networks. They mainly assume that the covered area of the Earth surface is divided into fixed LAs. Each LA is associated with a fixed earth station (FES). The size of a location area is bounded by a threshold that depends on the dynamics of the satellite network. The sum of an FES's instantaneous coverage (i.e., the surface covered by at least one of the location areas for a given instant) is changing as the satellites move relative to the Earth. However, for each FES it is possible to define a guaranteed coverage area (GCA), which is the geographic area over which the FES can provide service all the time. Obviously, one can see that the size of the GCAs also depends on the constellation of satellites.

Having a fixed LA associated with a FES would present some advantages. In particular, an incoming call can be simply routed by the network to the FES responsible of the LA where the MT is known to be located. From the network viewpoint, the location of the MT is "somewhere within reach of the FES." Using the same scheme as for terrestrial cellular networks, an MT will location update only if it has lost the FES's location area broadcast channel. To prevent ping-pong location updates, the spotbeams covering the boundary between two LAs will have to broadcast the identity of both LAs. The FES may provide a way of intelligently reducing the area over which it pages in the event of an incoming call. Two schemes for efficiently paging the MT within the LA are presented later.

In the case of Globalstar, the LA can be equal to the GCA or it can be only part of it. A FES can be assigned several LAs within its GCA. The GCAs can be large enough so that the Earth can be covered using a reasonable number of FESs. They also can have the maximum possible size for an FES at a given arbitrary location to guarantee certain minimum elevation angles and satellite diversity conditions for the mobile terminal. The GCA concept is also applicable to Iridium and Inmarsat thanks to the inter-satellite links. In these cases, the guaranteed coverage area of a FES can be as large as desired.

Location management can be performed following two approaches:

1. *Using the last spot beam position.* With location update, the FES identifies and records the instantaneous size, shape, and location (latitude, longitude) of the spotbeam in which the MT last made contact. For an incoming call, the FES would first page only those spotbeams required to completely cover

the recorded area. However, in case the mobile does not respond, the paging is repeated over the whole LA.

2. *Using terminal position fixing.* Using the position fixing capability, the FES would record the position measured and transmitted by the MT using a location update message including the latitude and longitude of the MT and an uncertainty radius that determines the circular area (or *uncertainty area*) where the terminal can be found at any time. If the MT moves outside of its *uncertainty area*, it would perform a *position update*, which is a kind of location update between the MT and the FES. If at any time the MT loses the location area broadcast of the FES, then it would search for a new location area broadcast and location update to the new FES, invoking the location update function. On the other hand, when an incoming call occurs, the FES only needs to page the spotbeams that cover the MT's declared uncertainty area. Depending on the mobility of the MT and its users' incoming and outgoing call rates, the MT might vary the uncertainty area radius to minimize either the paging area or the number of position updates in an attempt to minimize the total spectrum and power resource consumed by this signaling.

Several attacks can be launched against the location management system in satellite networks. In particular, terminal position fixing messages can be replayed or forged.

12.5.2 Handover Management in LEO Networks

The constant movement of satellites causes variations of network connectivity and satellite link delays. The handover is required to maintain the active connections. Supporting continuous communication over a LEO satellite network may require replacing ISLs and changing the IP addresses of the communication endpoints. Therefore, both link layer and higher layer handovers may be required for satellite networking.

One can easily notice that the mobility LEO satellite networks are very similar to cellular radio systems. In fact, in both systems, the relative position between the cells and the mobile hosts changes continuously, requiring handover of the mobile hosts between adjacent cells. However, some differences can be seen. In particular, while the mobile hosts move through the cells in the cellular systems, the cells move through the mobile hosts in LEO systems. In addition, the cell size of LEO satellite systems is larger compared to cellular systems and the mobile host's speed can be neglected in LEO satellite systems since that speed is negligible compared to the LEO satellite's rotational speed.

Handovers in satellite networks can be roughly classified into two categories: the link layer handover and the network layer handover (Chowdhury, 2006). The link layer handover occurs when one or more links between the communication endpoints have to change due to dynamic connectivity characteristics. The network layer handover is needed when one of the communication endpoints (which can be a satellite) changes its IP address due to the change of coverage area of the satellite or mobility of the user terminal, a network or higher layer handover is needed to migrate the existing connections of higher level protocols (e.g., TCP, SCTP, etc.) to the new address.

12.5.2.1 Link Layer Handover

The link layer handover can be further classified into three subclasses: (a) spotbeam handover, which happens when the end point user crosses the boundary between the neighboring spotbeams of a satellite; (b) satellite handover; which occurs when the existing connection of one satellite with the end user's attachment point is transferred to another satellite; and (c) ISL handover: This type of handover occurs when inter-plane ISLs are temporarily switched off due to the change in distance and viewing angle between satellites in neighbor orbits (Chowdhury, 2006).

■ *Spotbeam handover*: The footprint of an individual satellite is a circular area on the Earth's surface divided into smaller cells (or spotbeams). A spotbeam handover involves the release of the communication link between the user and the satellite providing communication service on the current spotbeam and getting a new link in the next spotbeam to continue the communication via the same satellite. No other satellite can be involved in the handover process. Various channel allocation strategies can be used to assign a channel (a new link) to a call, including The Fixed Channel Allocation (FCA) based handover schemes; the Dynamic Channel Allocation (DCA) based handover schemes; and the Adaptive Dynamic Channel Allocation (ADCA) based handover schemes (Chowdhury, 2006).

In FCA schemes, a group of channels is permanently assigned to each cell, according to the frequency reuse metrics. A handover call can only be assigned a channel if there is one available in the group of the cell. If no channel is available, the call is blocked or dropped. The DCA based handover schemes use dynamic channel allocation, assuming that all channels are grouped together in a so-called *central pool*. Any cell requiring a channel will be allocated a channel from the pool with respect to the channel reuse metrics. An allocated channel for a session is removed from the pool during session lifetime. It is transferred back to the pool for future reuse when the call is terminated.

Finally the ADCA schemes use guard channels during handover. A handover protocol with guard channel technique has to make use of the tradeoff between the number of guard channels and the number of regular channels. A large number of guard channels will create new call blocking, while a small number of guard channels may block the handover calls. The ADCA scheme monitors the current traffic load and dynamically adapts the optimal number of guard channels according to user location.

■ *Satellite handover*: A satellite handover occurs when a satellite involved in the connection between two users cannot provide service to a user. Satellite handover schemes should aim for selecting the most suitable satellite depending on the probability of a new call being blocked during handover, the probability of a handover call being dropped during handover, and the quality of communication provided by the satellite. Various selection criteria of the next satellite have been used including the following four criteria: (a) select the satellite that offers maximum service period, thus minimizing the number of handovers and achieving low call failure; (b) select the satellite with maximum number of free channels, thus achieving uniform distribution of calls among the satellites; (c) select the closest satellite to avoid link failure; (d) select the satellite with the longest remaining mutual visibility time; and (e) select the satellite with the longest remaining mutual visibility time.

■ *ISL handover*: This type of handover is specific to satellite constellations that use ISLs among neighboring satellites for communication. An ISL handover may occur at different locations in the orbit, for some LEO networks (e.g., Globalstar) or at the polar region only (e.g., Iridium). The ISL handover schemes attempt to solve basic issues such as determining where the ISLs have to be switched off between neighboring satellites and ongoing connections handed over to different satellites; find optimized routes with minimum ISL handovers between satellite pairs; and reduce the number of rerouting attempts during ISL handover.

12.5.2.2 Network Layer Handover

LEO satellites are not stationary with respect to a fixed user on the Earth's surface. Therefore, the visibility period of a satellite in a cell is very small due to constant rotation of the LEO satellites. For this reason, a user terminal can be served by a number of satellites during a connection (Jamalipour, 2001).

Two scenarios can be considered when dealing with the network layer handover. In the first scenario, the satellites merely operate as router. In this case, the satellites do not have any onboard equipment to produce or consume data and, as they move, new communicating fixed/mobile hosts come under their footprints or spotbeams and require a network layer handover during the change of communication links from one satellite or spotbeam to another.

In the second scenario, the satellites act as mobile hosts with all the onboard equipments, which exchange data with ground stations. In that case, the satellites should maintain continuous connection with ground stations. Therefore, the IP address of a satellite has to be changed when a network layer handover to a new ground station takes place.

During the handover, three different phases are considered: the initiation, the decision, and the execution. The decision phase is realized by the handover controlling schemes and can be network controlled or mobile-controlled. In the first case, the network monitors the link quality and decides whether to initiate handover, while in the second, the mobile host is responsible for monitoring the link quality and initiating the handover. On the other hand, the execution phase of handover is a composition of connection establishment and connection transfer scheme. The network layer handover is performed using a hard handover scheme, a soft handover scheme, or a signaling diversity scheme (Efthymiou, 1998). In hard handover schemes, the current link is released before the next link is established, which may result in connection blocking during handover.

During soft handover, the current connection is not freed until the next connection is definitely established. The signaling diversity based scheme is quite similar to the soft handover, except for some differences. In fact, the signaling procedures in signaling diversity schemes are performed through both the new and old links, while user data is sent through the old link. In addition, no synchronization between links is needed as the old link is used for data and the new link is used for signaling data.

12.6 Attacks against Satellite Networks

Satellite networks reveal a number of vulnerabilities that can induce serious threats to the service they provide. Satellite networks have several points of failure that can be exploited by adversaries including, but not limited to, the following:

■ The wireless communication links, which are vulnerable to signal blockage and jamming;
■ The radio-navigation system, which can be a victim of signal blockage and jamming;
■ The onboard and asset-tracking units, which are vulnerable to tampering and unauthorized modifications;
■ The data protocols used for the transfer of monitoring data, which are vulnerable to unauthorized modifications;
■ The information systems of the monitoring center, which are vulnerable to attacks such as viruses and distributed denials of service.

In the following subsections we will discuss some important threats and highlight particularly the denial of service attacks.

12.6.1 Threats

In the following, we list some of the important security threats in the satellite networks. Some among these threats can be considered highly damaging for hybrid satellite networks including satellite segments (Roy-Chowdhury, 2005).

12.6.1.1 Insider Attacks

A malicious adversary can obtain access to a satellite network as a legitimate node, provided that there is weak access control, or that he has successfully got control of the password of some legitimate node in the network (by sniffing, for example). In such cases, the adversary can launch a large set of attacks. The size of this set will depend on the permission levels the adversary has gained. In particular, the attacker will be able to read confidential data. The countermeasures set up to provide source authentication, confidentiality, and privacy should be able to prevent insider attacks. Additional measures against insider attacks would require implementing intrusion detection mechanisms, which will be able to detect the improper actions attempted by the adversary.

12.6.1.2 Threats against Communication Privacy

In satellite networks, the important threats that can target information privacy include eavesdropping and unauthorized access to confidential data. This is made possible mainly because of the nature of broadcast links to satellites. Indeed, every entity under the coverage of a satellite can receive all satellite transmissions, provided that it is appropriately equipped. In particular, if the data is transmitted in the clear, the malicious adversaries listening to the transmission can collect all the information that is flowing near them. Consequently, the collected data can lead to the leakage of classified information for sensitive applications.

Security measures that can be taken to protect the privacy of communication should include data confidentiality, which can be accomplished by message encryption and necessitates coordination between the senders and the receivers so that they can synchronize the correct cryptographic keys they use to perform the encryption/decryption operations. Providing a security mechanism should, in addition, include the selection of suitable cryptographic algorithms, so that the needed actions are performed without affecting the overall performance of the satellite network and the coordination of keys between users.

12.6.1.3 Packet Modification Attack

When the traffic flows over an open network through a path, an adversary who is listening to the path may be able to intercept both control and data packets. He also can modify the flowing packets and send them to their destination, no matter whether the destination is a ground terminal, an end user, or another entity. The attacker does not need to masquerade as a legitimate node to perform a packet modification attack. When the corrupted message reaches the intended destination, the latter would believe that the modified packets are coming from the true source. Message modifications can lead to abnormal behavior of the nodes. They can be prevented by appending message integrity check information to every message.

The utilization of message authentication codes would require that the source and destination of a message share the same cryptographic key that is required to generate and validate the MACs. Key management and distribution represent, therefore, important issues to handle for satellite networks. The satellite network architecture and the node capabilities will impact the techniques and mechanisms that will be used to address these issues. On the other hand, the security requirements and policies imposed by the applications would command whether message authentication should be performed only at the communication end points, or whether intermediate nodes should be involved in the verification of the integrity of the flowing message.

12.6.1.4 Sending Forged Commands

Obviously, it is vital that the control of the satellites in a network be maintained by an appropriate control center that is able all the time to manage important functions such as signaling and satellite location. A malicious user that is appropriately equipped can send illegitimate control and command messages to the spacecraft, making the satellite perform operations different from their intended use. As a consequence of the forged commands, legitimate operations and communication connections can be disrupted. They also can lead to hijacking of sessions.

This attack can be prevented if the sources of the messages can be properly authenticated by every receiver. Such function would require relevant mechanisms for authentication, such as digital signatures. To this end, the digital signature of a message, to be sent, should be appended by the source to the message. The algorithm to be used for this depends on the network infrastructure and the node capabilities, among other factors. Additionally, the level of security required would induce the setup of some authentication policy. Such a policy can decide, for example, whether only the end users should authenticate each other, or whether authentication should be performed on a per-hop basis. The latter scenario can be needed in situations where the satellite should stop illegitimate messages. If a satellite is able to authenticate the source of every packet it receives, it will relay only those messages correctly authenticated. However, allowing the satellite to perform verification of authentication can lead to other attacks.

12.6.1.5 Traffic Analysis

In some network scenarios requiring high security levels, it might be necessary to make sure that no outsider can know which parties are involved in a particular communication. This would require that traffic analysis of the data flowing in the network be prevented from reading important information in the packet. However, traffic analysis attacks are difficult to prevent even if the network is secured for data confidentiality and data integrity and source authentication. An adversary only needs to read the packet headers for the source and destination information to do successful traffic analysis. Protection against address sniffing uses mechanisms (such as those using the security channels) that are able to hide the actual source/destination headers.

Finally, let us consider another type of attacks, called attacks from space, which can be performed by well-equipped attackers. The object of these attacks can be to send false information to the LEO satellites or the mobiles connected to them in order to block a handover or induce false routing decision. The false information is transported by replayed packets or modified commands involved in the signaling message exchange.

12.6.2 *Denial of Service in Satellite Networks*

Denial of service attacks aim at preventing legitimate users from accessing a service they are authorized to access, to get the required quality of service, or to use a resource they need. Assuming that the satellite networks and their components are properly configured and protected physically against unauthorized accesses, a DoS attack targets only the utilization of scarce, limited, or non-renewable resources in the satellite networks. In particular, DoS attacks such as destroying or altering the configuration information or physically destroying or altering network components do not apply in such situation.

Consequently, this section will focus on DoS attacks that target the victim's major resources such as the memory of a system, the disk space of the victim component or service, and the network bandwidth. In a typical DoS attack, the intruder may be able to utilize an important amount of resources at the victim component by sending some specific packets based on the resource allocation caused by the processing of each packet. Moreover, the intruder may also utilize the available bandwidth entirely by generating a large number of spurious packets. To overcome the rules that limit the number of requests issued by a given source for a given service or target, DoS attacks can be distributed or emulate several sources by setting false source addresses in the requests the intruder generates, or by coordinating several sources as distributed intruders (as explained in a Chapter 2).

To prevent simple DoS attacks, the satellite network can set up methods that filter requests with fake source identification. However, a straightforward setup of data origin authentication techniques does not solve the DoS problem all the time. This anti-blockage technique is unsuccessful in satellite networks because of two facts. First, the broadcast nature of satellite networks implies that all the terminal nodes would receive all traffic from the servers and the anti-congestion technique based on the message exchange would be ineffective because the packets issued by the server and intended for impersonated origins would be received by the impersonating entity, which is able to reply appropriately (with the expected weak authentication messages, if any). Second, the extra delay that would be caused by integration of an anti-blocking mechanism would be unacceptable for many

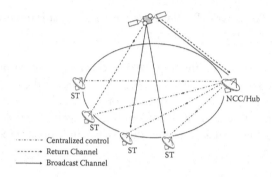

Figure 12.5   A satellite network.

applications, when added to the long delays experienced. An authentication scheme protecting a satellite network from DoS attacks should not only be secure against masquerade, but also be efficient in terms of resource utilization.

To overcome these drawbacks, several approaches have been developed. Let us now describe one of the proposed schemes (Önen, 2004). To do so, let us first give a brief description of the satellite network on which we will operate the proposed scheme. Figure 12.5 depicts the proposed scheme. The network is composed by Network Control Center (NCC/Hub), a GEO-satellite, and four satellite terminals (*STi*, *i* < 4). The network control and management are centralized in the NCC using management functions such as the address and key management functions. We assume that clock synchronization is provided between the NCC and the terminals. A satellite terminal *STi* has a unique identifier ID_i and shares a secret key K_i with the NCC, which is pre-determined by the NCC. We finally assume the time is slotted into fixed slots. During the jth slot, the NCC reliably broadcasts a random value, *Nj*, for the need of replay detection that is different from those sent in the previous slots.

The proposed protocol works as follows: A satellite terminal, *STi*, sends control messages to the NCC via the satellite, during slot *j*, using a data transfer protocol. The structure of a message control contains the following fields: the identity of the satellite terminal (ID_{ST}), the sequence number *Seq* of the message sent by *ST* (which should be equal to *Nj*), a payload *m*, the hash value $h(m)$ of *m*, and a message code identifier $MAC(K_i, Seq, h(m))$, where K_i is the key shared by *STi* and NCC.

Upon receipt of a request sent by *STi*, during the interval *Tj*, the NCC first verifies that the sequence number *Seq* is equal to the actual nonce *Nj*. Then, it computes the shared key K_i and determines the message code identifier using the identity ID_i and $h(m)$, as retrieved from the header of the request message. In order to verify the authenticity of the request, the NCC compares the value $MAC(K_{ST}, Nj, h(m))$ with the one retrieved from the header. If they are equal, the NCC further verifies the integrity of the message by comparing the received value of $h(m)$ with the hash value of its payload. This two-step verification process allows for fast discarding of the inconsistent packets. Once both verification steps of a request sent by a *STi* terminate successfully, the NCC processes the request and allocates the necessary resources for it.

Based on the secrecy of K_i shared by the corresponding *STi* and the NCC, an intruder cannot generate a valid request. Assuming that the intruder could not send replayed messages eavesdropped within the same time slot, one can say that replays will be easily detected by the NCC based on the difference between the current nonce value Nk and the value included in the header of the replayed request, Nj $(j < k)$.

Now let us discuss the assumption related to the impossibility of sending a replay in the same time slot. One can say that it is practically impossible to intercept any message from the up-link in a satellite network and that the intruder can only intercept a legitimate request from the down-link, that is, after an end-to-end latency for a GEO-satellite system. This states that, if the time slot duration is lower than the end-to-end latency, the assumption is reasonable.

Let us assume that the time slot duration is greater than the end-to-end latency. The approach can be improved in the following way: one can introduce a stateful verification mechanism that allows removing the synchronization requirement between the NCC and the terminals and asking the NCC to keep some state information about successful authentication attempts occurring in each time slot. The improved approach will work as follows:

The terminal *STi* increments the sequence number *Seq*, initially equal to the nonce Nj of slot j, for every new request within the valid slot. Therefore, when the NCC receive another request from a terminal that previously has been added to a special table, it will also verify that the sequence number sent by *STi* is greater than the one in the table, and replace the value with the new one if the message is authenticated. Because the intruder is not able to issue a valid request, it therefore will not be capable to launch a replay attack within a same interval where requests have been eavesdropped, because the sequence number changes with every message.

Let us finally notice that the new version presents also all the secure properties offered by its first version.

12.7 Securing Satellite Networks

Significant works have been done to secure wireless communication in general. Some among these works can be applied to satellite networks. More recently, several proposals have been developed specifically to secure satellite networks (Roy-Chowdhury, 2005). We briefly discuss in this section some among the most important.

12.7.1 IPsec-Like Solution

The first works have focused on using existing, standardized technology originally designed for terrestrial networks, to fix well-known security holes in satellite networks. Several proposals for data confidentiality and authentication have called for the use of IPsec, which has been widely adopted for security at the network layer.

As explained in Chapter 3, IPsec has two special components: the Authentication Header (AH), which provides integrity protection to data packets, and the Encapsulating Security Payload (ESP) that provides encryption and optional integrity protection. In addition, the use of IPsec requires the generation of a *Security Association* (SA) between the source and destination end points. The SA specifies the various security attributes for the particular session. IPsec allows strong security for data confidentiality and authentication, but it has a heavy byte overhead. In particular, it adds 10 bytes overhead to every data packet in the ESP mode. In addition, if authentication is used, ESP adds 16 bytes or more for the integrity check value, and another 8 bytes or more of initialization vector if the encryption algorithm uses an IV.

IPsec has been designed primarily to secure point-to-point communication; it appears that it is not well suited for group communication, because of the lack of the dynamic key-establishment procedure necessary to secure communication in groups where the membership changes with time. Furthermore, IPsec does not authorize the authentication of intermediate nodes, which might be of good support in some security environments.

Some other drawbacks can be mentioned for the IPsec suite. First, the set up of SAs using IKE can be complex and expensive. In particular, if the network entities do not have pre-shared secrets, then the IKE protocol would require the use of public key pairs, which arises the need for PKI. Second, one can notice that IPsec is unable to coexist with other protocols (e.g., HTTP), since the keys used for encryption in the IPsec ESP are known only to the two endpoints and therefore any intermediate node in the network cannot decrypt the traffic. In particular, the HTTP cannot function properly when the IPsec ESP is used. Because a HTML page is encrypted from end to end, a HTTP proxy cannot read the page in order to get the embedded objects. Thus, the use of IPsec leads to a severe degradation in performance for the transport protocol (e.g., TCP) and the HTTP proxy.

To mitigate IPsec problems, several proposals have been made. In particular the concept of breaking up IPsec encryption into multiple encryption regions or zones on a single packet has been proposed. In multilayer IP-security protocol (ML-IPsec), the approach adopted is to encrypt different regions of the IP packet using different keys (Zhang, 2004). The TCP payload is encrypted with a key K, which is shared only between the endpoints, while the original IP header and the TCP header are encrypted with a key K', which is shared between the end points and also with intermediate authorized nodes. Therefore, the TCP PEP can decrypt the header portion of the ESP packet with K' and read the TCP header. However, the PEP cannot read the TCP payload and thus cannot access the actual data.

The use of ML-IPsec, however, necessitates that the security attributes be distributed properly to all the related entities—this is done by defining a new type of security association called *Composite Security Association* (CSA). CSA is a set of SAs that together offer a multilayer security protection for the traffic stream. Also, ML-IPsec needs trust in third parties like the satellite gateways. Moreover, ML-IPsec, as envisioned currently, is strictly for point-to-point communication and has no support for groups. Another problem arises with application level optimizations that do not work in the presence of network level security.

To mitigate the reduction induced by the use of IPsec, several solutions been proposed. In particular, two solutions have been suggested in Olechna (2001) to this issue. The first method proposes placing the TCP gateways at the endpoints. The end-to-end TCP optimization can be performed on the traffic; then, the traffic can be encrypted using IPsec. This approach improves the performance, but increases congestion. In fact, if a packet is lost or experiences an error, the TCP is able to apply congestion avoidance and the transmission window can be reduced by half. The second method proposes to split the secure connection into two connections. The first connection is set up between the client and the satellite gateway, and the second connection is implemented between the gateway and the Internet server. This allows the satellite gateway to decrypt the IPsec packet, read the headers, and optimize the performance. The second approach, however, assumes that the satellite gateway is trustable, which might be unacceptable to mobile applications who want strong end-to-end security.

12.7.2 Securing HTTP Sessions over Satellite Networks

One practical method to secure HTTP over a satellite network is to split the end-to-end IPsec tunnel into a sequence of tunnels (Olechna, 2001). Three IPsec tunnels can be built. The first tunnel connects to the client and is terminated to the client proxy. The second tunnel is created by the client proxy to link it to the gateway TCP proxy. The third IPsec connection is created from the gateway TCP proxy to the Web server. In this method, the Web traffic can be read completely by the client proxy and the hub proxy. The two proxies are able to perform the TCP enhancements because they can read the TCP header.

This method, however, presents some drawbacks. In fact, the IPsec handshaking between the client and the server can spoofed by the client proxy on the client end. It also can be spoofed by the TCP gateway proxy on the server end.

In addition, the hub HTTP proxy can perform HTML caching and object prefetching from the server because it can read the base HTML page as it is returned to the client on a HTTP request. Thus, when the security requirements recommend that the traffic be unreadable to intermediate nodes, the above approach will not work. To overcome this limitation, the approach can be extended into a layered IPsec method in order to allow portions of the HTML content to be also accessible to the proxy servers. To this end, we suppose that two keys K and K' are set up. Key K is known only to the client and the server, while key K' is known to the client, the Web server, and the intermediate proxy servers at the client and the gateway. When the client generates HTTP requests, the requests are encrypted using K' and the client proxy server can read them and send local acknowledgments.

A major issue in the aforementioned method is the handshaking mechanism required to set up the layered IPsec tunnels. To maintain a reasonable level of security, one can assume that the connection is set up first between the client and the server, who negotiate both K and K', separately from other parameters of the security association. Then, the handshaking mechanism can provide securely the key K' to the client and the hub proxy servers. The client and the hub proxy servers are required to authenticate themselves correctly before they can receive the other key or access the IPsec traffic.

References

I. F. Akyildiz, G. Morabito, and S. Palazzo, TCP Peach: A New Congestion Control Scheme for Satellite IP Networks, *IEEE/ACM Trans. on Networking*, 9(3), June 2001.

E. Ayanoglu, S. Paul, T. F. DaPorta, E. K. Sabnani, and R. D. Gitlin, AIRMAIL: A Link-Layer Protocol for Wireless Networks, *ACM-Baltzer WirelessNetworks*, 5:47–67, 1995.

A. Bakre and B. R. Badrinath, Implementation and Performance Evaluation of Indirect TCP, *IEEE Transactions on Computers*, 46(3), March 1997.

H. Balakrishnan, S. Seshan, and K. H. Katz, Improving Reliable Transport Protocol and Handoff Performance in Cellular Wireless Networks, *ACM-Baltzer Wireless Networks*, pp. 469–481, December 1995.

L. Brakmo and L. Peterson, TCP Vegas: End to End Congestion Avoidance on a Global Internet, *IEEE J. Select. Areas Comm.*, 15:1465–1480, October 1995.

C. Chen, QoS-based Routing Algorithm in Multimedia Satellite Networks, in *Proc. of IEEE 58th Vehicular Technology Conference (VTC2003-Fall)*, vol. 4, (Orlando, FL), pp. 2703–2707, Oct. 2003.

P. K. Chowdhury, M. Atiquzzaman, and W. Ivancic, Handover Schemes in Satellite Networks: State-of-the-Art and Future Research Directions, *IEEE Communications Surveys and Tutorials*, Vol. 8, No. 4, August 2006.

R. C. Durst, G. Miller, and E. J. Travis, TCP Extensions for Space Communications, in *ACM MobiCom Conf.*, Nov. 1996.

N. Efthymiou, Y. E. Hu, R. E. Sheriff, and A. Properzi, Inter-segment Handover Algorithm for an Integrated Terrestrial/Satellite-UMTS Environment, *IEEE In. Symposium on Personal, Indoor and Mobile Radio Communications, PIMRC*, Boston, Massachusetts, USA, pp. 993–998, Sept. 8–11, 1998.

Ö. Erçetin, S. Krishnamurthy, S. Dao, and L. Tassiulas, A Predictive QoS Routing Scheme for Broadband Low Earth Orbit Satellite Networks, in the *11th IEEE Int. Symp. on Personal, Indoor and Mobile Radio Communications (PIMRC 2000)*, Sept. 18–21, 2000, London UK.

C. Faller, B. H. Juang, P. Kroon, H. L. Lou, S. A. Ramprashad, and C. E. W. Sundberg, Technical Advances in Digital Audio Radio Broadcasting, *Proc. IEEE*, 90(8):1303–1333, August 2002.

W. Feng, K. G. Shin, D. D. Kandlur, and D. Saha, The BLUE Active Queue Management Algorithms, *IEEE/ACM Transaction on Networking*, 10(4):513–528, August 2002.

S. Floyd and V. Jacobson, Random Early Detection Gateways for Congestion Avoidance, *IEEE/ACM Transaction on Networking*, 1(3):397–413, August 1993.

Y. Hu and V. O. K. Li, Satellite-Based Internet: A Tutorial, *IEEE Communications Magazine*, 39(3):154–162, 2001.

A. Jamalipour and T. Tung, The Role of Satellites in Global IT: Trends and Implications, *IEEE Personal Communications*, pp. 5–11, June 2001.

D. Katabi, M. Handley, and C. Rohrs, Congestion Control for High Bandwidth-Delay Product Networks, in *ACM SIGCOMM'02*, August 2002.

G. Maral, J. de Ridder, B. Evans, and M. Richharia, Low Earth Orbit Satellite Systems for Communications, *Int. Journal of Satellite Communications*, pp. 209–225, 1991.

S. Mascolo, C. Casetti, M. Gerla, M. Y. Sanadidi, and R. Wang, TCP Westwood: Bandwidth Estimation for Enhanced Transport over Wireless Links, *MobiCom '01: Pro. of the 7th Annual International Conf. on Mobile Computing and Networking*, Rome, Italy, 2001, pp. 287–297.

E. Olechna, P. Feighery, and S. Hryckiewicz, Virtual Private Network Issues Using Satellite Based Networks, *MILCOM 2001*, vol. 2, 2001, pp. 785–789.

M. Önen and R. Molva, Denial of Service Prevention in Satellite Networks, *IEEE Int. Conference on Communications, (ICC) 2004* , vol. 7, Jun. 2004, pp. 4387–4391, Paris, France.

A. Roy-Chowdhury, J. S. Baras, M. Hadjitheodosiou, and S. Papademetriou, Security Issues in Hybrid Networks with a Satellite Component, *IEEE Wireless Communications*, December 2005.

C. Wullems, O. Pozzobon, and K. Kubik, Signal Authentication and Integrity Schemes for Next Generation Global Navigation Satellite Systems, in *Proceedings of the European Navigation Conference (ENC-GNSS 2005)*, Munich, Germany, July 2005.

Y. Zhang, A. Multilayer, IP Security Protocol for TCP Performance Enhancement in Wireless Networks, *IEEE JSAC*, 22(4)(2004):767–776.

PROTECTION TECHNIQUES FOR MOBILE APPLICATIONS

IV

Chapter 13

Security of Mobile Payments

13.1 Introduction

In the economic world, payment presents one of the main mechanisms motivating individuals and communities to share their products. In general, payment is based on the exchange of different amounts of payment means (e.g., money) for some required services or products. This exchange should be protected against the misbehavior of the customer and the merchant as well as against any external threat. For this purpose, mechanisms with different levels of security guarantees varying with the value of the transactions are employed. Further, trusted third parties are defined to control payment between involved entities. The history of money shows that the ways of representing value have become increasingly abstract over time. As money has evolved, so have the methods of performing payments. Nowadays, payment systems are either account-based or token-based. Payment means, which realize transfer between accounts, include checks, payment cards, and bank transfers. Token-based systems include cash and pre-paid cards to access specific services.

13.2 Issues of E-Payment

Electronic payment is an alternative to the traditional payment, which is basically performed physically. Traditional payment may be realized by presenting a credit card to a merchant and signing on a payment form as evidence of the payment transaction made to the merchant. Electronic payment (or e-payment) is commonly defined as the transfer of an electronic means of payment from the payer to the payee. Taking into consideration that transmitting and receiving the payment means can take place over the air, we can adopt the following definition for mobile payment (or m-payment): *A mobile payment is the transfer of an electronic means over a mobile network from the payer to the payee.* This definition is more precise than the one presented in Kruger (2001); in fact, the proposed definition covers a larger range of possible applications and networking technologies. In addition, no constraints can be placed on the nature of the device used to transmit the needed information over the wireless link. A mobile customer needs to connect to a wireless network to initiate a payment. The network could be GSM or any other cellular network.

13.2.1 Electronic Payment Basics

Existing electronic payment systems can be characterized by considering the way they organize money transfer. Two categories of payment systems can be considered: account-based and token-based payment systems. In an account-based payment system, money is represented by an account balance in bank accounts. It is transferred between accounts set up by engaging entities with their banks. Often, account-based payment systems can be of two types depending on the payment clearing periods: credit-based and debit-based systems. In a debit-based payment system, the account owner is allowed to spend the money up to the current balance; in a credit-based payment system, the owner is allowed to spend more than the current balance using a payment authorization (such as a credit card). It can be observed that the difference between credit-based and debit-based payment systems is characterized only by the billing periods.

In a token-based payment system, electronic money stands for physical money that the customer exchanges with electronic tokens that he uses to pay for products and services. A merchant collects the tokens and sends them appropriately to a trusted third party (the bank, for example) to redeem the money (by means of money transfer to the merchant's account, for example). In a token-based payment system, the owner is not required to have a payment authorization from the bank for every payment operation. Thus, token-based payment systems would have a lower operational cost compared to account-based systems. Two schemes are known to work under the token-based mode: electronic cash (Yu, 2001) and micropayment systems (Yen, 2001).

In addition, electronic payment systems can be evaluated based on a number of criteria, such as the security provided, the business roles, and the functional characteristics. Considering the second criterion (the business characteristics), one can say that a number of actors perform different roles in order to deploy a payment system. In the following, Figure 13.1 depicts a model in which five roles have been

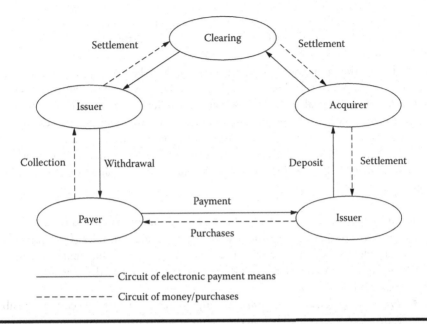

Figure 13.1   Electronic payment model.

defined. These roles have different relationships and interactions with each other and can be observed in an electronic payment system. Transactions can be executed between actors performing these roles. The following are the major roles as depicted by Figure 13.1, where the case of account-based payment is considered:

- *Payer*: The payer (e.g., customer) can withdraw electronic payment elements from the issuer and use them in a payment transaction in order to purchase physical and electronic products and services from the payee.
- *Payee*: The payee (e.g., merchant) delivers the products upon receipt of a payment document (or transcript) that one can define as a set of three objects: the payment request, the evidence on the payment claim (such as a signature), and the electronic payment means (or instrument) used to produce the evidence. Typically, the transcript is transmitted to the acquirer.
- *Issuer*: The issuer is the financial organization issuing valid electronic payment means. The issuer collects real money from the payer in return for the withdrawn electronic money.
- *Acquirer*: The acquirer verifies the validity of the deposited payment documents, and forwards them to the clearing. After the settlement (or transfer) of funds between the issuer and acquirer, the acquirer credits the payee's account with the amount of money stated by the deposited evidence.
- *Clearing*: The clearing receives the payment transcripts from the acquirers and verifies them. For each acquirer, the clearing computes the total amount

due from each issuer involved with the acquirer in a transaction. Then, the clearing initiates the settlement of funds between issuers and acquirers.

Although the model of Figure 13.1 is depicted in the context of account-based payments, it can easily be considered by any other payment system to represent the actors of the typical system. However, the content of the payment document and the order in which the operations of a payment transaction are processed may vary based on the payment characteristics, the payment element, and the nature of protection. The payment characteristics determine the actual interpretation of the overall model. They include, but are not limited to the following classification concepts:

- *Local versus central account*: An electronic payment system assumes that the payer always has an account that can be managed centrally (by a bank, for example), or kept locally with the payer while an account image is kept by the issuer. In particular, token-based payments require that the account be managed locally.
- *Pre-payment versus post-payment*: In a pre-payment mode, the collection sub-transaction is executed before the payment sub-transaction, meaning that the balance of the payer's account is credited with the initial amount of money before the purchase is made. The balance is decreased after the execution of every payment sub-transaction. If, in such a situation, there is an active connection between the payee and the issuer, and the issuer is managing a central account for the payer, then the pre-paid mode is called pay-now mode. In the post-payment mode, the collection sub-transaction is performed after the running of the payment sub-transaction. The balance of the payee's central account is credited with the amount of purchase after the purchase has been made. While the post-payment mode is suitable for credit-based payment, the pre-payment mode fits better the debit-based electronic payment systems.
- *On-line versus off-line*: In the on-line payment mode (also called instant-deposit), the payee has to deposit each payment document with an acquirer at the moment of the current transaction. On-line payments require a network connection to a third party. They often involve a high cost per transaction. In addition, one can see that they are efficient when the amount of the payment is very high. On the other hand, the payer needs only to forward the transcripts after the transactions were performed in an off-line payment mode (or later-deposit), where no communication with a third party is required. The off-line payment is thought to be suitable in the case of small size payment amounts.
- *Payment per event versus subscription*: Payment per event states that every use of service is paid separately. Subscription means that a right for an unlimited use of a service during a certain period of time is paid once. When using the service, the payer forwards the subscription right to the payee, as an equivalent

of a payment transaction. Subscription is typically an off-line payment mode, meaning that the payee does not need to go for an on-line verification of the subscription right forwarded by the payer.

13.2.2 Security Requirements

Security and privacy are two major concerns in electronic payments. Indeed, it has been shown that two crucial issues can be distinguished in e-commerce: They are the credit card security and the disclosure of personal information. With respect to security, it appears that payers, payees, and issuers present very constraining security requirements that we summarize in the following:

- *Payer requirements*: The security requirements of the payer are related to the collection, withdrawal, and payment sub-transactions. The balance of the payer should be increased according to the exact amount of money paid to the issuer during the collection sub-transaction. Likewise, the payer should get the necessary electronic payment elements during the withdrawal. The balance of the payer's account should be decreased only with the exact amounts specified in the payment sub-transaction. No entity other than the payer should be able to perform valid payment transactions with the electronic payment elements. In the case of token-based payment, no entity should be able to impersonate the payer, or fabricate claiming that an element (or a token) was already used in other payment transaction.
- *Payee requirements*: The security requirements of the payee are related to the payment and deposit sub-transactions. The payee should receive all the amounts specified during valid payment transactions he has been involved in. The payer should perform the payment sub-transaction with the amount specified by the payee.
- *Issuer requirements*: The security requirements of the issuer and acquirer are related to the collection, withdrawal, and deposit transactions. The balance of the payer's account should be increased unless the payer explicitly transfers money to the issuer during collection. No entity, except for the issuer, should be able to produce valid electronic payment elements on behalf of the issuer. Payers should only obtain valid payment elements if they are involved in a withdrawal sub-transaction, meaning that electronic payment must be unforgeable. The payee should not be able to deposit more than once a payment transcript of a valid payment sub-transaction. A duplicate deposit detection mechanism should be made available.

On the one hand, while the lack of privacy is a major concern for individual users, the need to provide an unrestricted anonymity is another concern for anonymous payment procedure and entities such as the law enforcement agencies and governmental administrations. On the payer side, various privacy-related requirements can be satisfied, including the following:

■ *Unobservability of information*: This requirement refers to the impossibility for any entity not directly involved in an ongoing transaction to collect useful information about that transaction.

■ *Unlinkability of payments*: This requirement assumes that an issuer is not able to link two different payment transactions to the same payer. If anonymous payments can be linked, revealing the identity of the payer in only one payment transaction leads to loss of anonymity in all the other linked transactions.

■ *Untraceability of payment elements*: This requirement imposes that the payment means, used in a payment transaction, are not linkable to withdrawal transactions. Untraceability guarantees that payers remain anonymous during payment toward payee and issuer. Payer's identity remains linked only to the collection and withdrawal sub-transactions.

13.3 Overview of Electronic Payment Systems

A classification of electronic payment systems can be made based on a classification into four categories: credit card–based payments, electronic cash, electronic checks, and account transfer.

13.3.1 Credit Card-Based Payments

The most common way of using credit card payments consists of just sending credit card information (such as the card number and expiry date) to the payee over a secure channel such as the SSL or the Secure Electronic Transaction (SET), which was designed to eliminate the security vulnerabilities of the SSL. Although the credit card information is securely sent to the payee, some important problems may occur. The payers have to disclose some information about their credit cards. However, this is conflicting with the fact that the credit card number is actually the secret on which the whole system is based.

13.3.1.1 SSL-Based Payment Systems

SSL is currently the most widely used protocol for providing security for the payer/payee Internet link. To support service provision, SSL provides two layers, the handshake protocol and the record layer. The handshake protocol is responsible for initializing and synchronizing cryptographic channel between the payer and the payee, while the record layer provides confidentiality and authentication of the payment and card-related information as well as protection against replay attacks. Typically, to set up a channel, SSL performs five typical steps:

1. The payer (or customer) first sends a *ClientHello* to the payee's site. *ClientHello* includes information such as the SSL version, data compression method to use, session ID, and a random number that is used to properly identify the channel to be started.
2. The payee's server responds by a *ServerHello* message. Then, it sends a *ServerKeyExchange* message containing the server's public key. Finally, it sends a *ServerHelloDone* message to indicate that it has finished the initial negotiation of the channel setup.
3. The payer sends its certificate, if requested by the payee's server, along with a *ClientKeyExchange* message containing the key information that will be used (between the payer and the payee's server) to generate a master secret key and the keys that will be subsequently used for encryption of the information related to the credit card and payments. The payer also sends a *CertificateVerify* message to prove that he/she has the private key corresponding to the key occurring in the certificate.
4. The client sends a *ChangeCipherSpec* message to indicate the starting point of a protected channel. Then he/she sends a *ClientFinish* message containing a hash of the handshake messages exchanged. The message is encrypted and authenticated.
5. The payee's server sends back another *ChangeCipherSpec* message for the generation of similar keys. It also sends a *ServerFinish* message to finish the setup of security features.

It appears clearly from the aforementioned steps of the SSL function that a credit card-based payment system using SSL will have the following features:

■ SSL protects the confidentiality of the payment transaction using symmetric encryption. It also guarantees the confidentiality of transmitted data against interception attacks and ensures integrity protection for the transferred data.
■ SSL uses the payee's server certificate as the basis for payee's authentication. To this end, the payer can check the server authentication by verifying its ability to decrypt information encrypted using the server's public key. In addition, SSL can provide customer authentication if the customer has a public key signed using a certificate issued by a CA trusted by the server. SSL provides protection against the third party replay attacks by using a random number during handshake.
■ SSL provides no non-repudiation services; that is, neither the customer nor the merchant has any cryptographic evidence, to the third party, that a transaction has taken place.

13.3.1.2 SET-Based Payments

Secure electronic transaction is an open encryption and security specification designed to protect credit card transactions on the Internet (SET, 1997). SET is not a payment system itself. It is rather a set of security protocols and formats enabling users to employ the existing credit card payment infrastructure on a network, such as the Internet, and in a secure manner. Basically, SET offers three basic services:

1. It provides a secure communication channel for all entities involved in a payment transaction.
2. It provides a high level of trust between payers and payees by the use of X.509 v3 digital certificates issued by trusted certification authorities (see Chapter 2).
3. It ensures privacy by allowing the transported information (related to payment and credit card) to be only available to the different parties in a transaction when and where it is necessary.

The SET protocol utilizes cryptography to provide confidentiality of information. It ensures payment integrity and support to payer and payee identity authentication. For authentication purposes, cardholders, merchants, and service finance providers (SFP) will be issued digital certificates by their supporting certification authorities. It also uses *dual signature*, which hides the customer's credit card information from payees and also hides the order information from SFPs. The steps performed to protect the privacy of a payment transaction are listed by the following 6-step procedure and depicted by Figure 13.2.

1. *The customer opens an account.* The customer obtains a credit card account, such as MasterCard or Visa, with a service finance provider that supports electronic payment based on the SET.
2. *The customer receives a certificate.* After verification of identity, the customer (or future payer) receives an X.509v3 digital certificate, which is signed by the bank. The certificate contains the customer's RSA public key and its expiration date. It also establishes a relationship, guaranteed by the bank (or the related certification authority), between the customer's key pair and his/her credit card.
3. *The payer places an order.* This is a step involves the payer first browsing through the merchant's web site to select items and determine their prices. The payer then sends the list of the items to be purchased from the payee (or merchant), who returns an order form containing the list of items, their individual prices, a total price, and an order number. In addition to the order form, the payee sends a copy of his certificate so that the payer can authenticate the payee.
4. *The payer sends the order and payment.* The payer sends both an order and payment information to the payee, along with the payer's certificate. The order confirms the purchase of the items in the order form. The payment contains

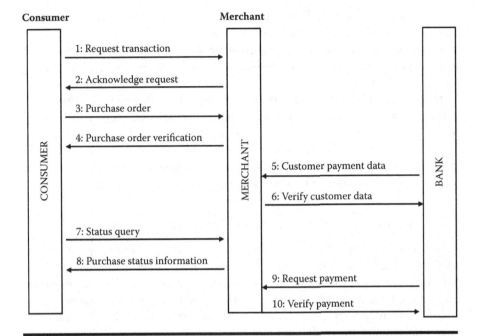

Figure 13.2 Steps performed to protect the privacy of a payment transaction using the SET protocol.

credit card details. The payment information is encrypted in such a way that it cannot be read by the payee. The payer's certificate enables him to authenticate the customer, the order received, and the payment information.

5. *The payee requests payment authorization.* The payee sends the payment information to his finance service provider requesting authorization that the customer's available credit is sufficient for this purchase.

6. *The payee confirms the order.* The payment elections containing the account information are forwarded to the payer. Then, the payee sends confirmation of the order to the customer. The merchant provides the service (or ships the products) to the customer and requests payment. This request is sent to the payment gateway, which handles all of the payment processing.

In conclusion, it appears that the principal security features can be provided using public-key cryptography:

- X 509 certificate holders are authenticated as legitimate. The payee can be assured that the transaction is from a registered certificate holder.
- Dual signatures are used to link a payment order sent to the payee with the account information forwarded to the payer. This means that the credit card number is cryptographically bound to the transaction by a digital signature.
- The integrity of the entire payment transaction is guaranteed.

Concerning security, SET-based payment systems provide better protection compared to those using SSL/TLS systems. However, it has not become popular, mainly because of its complexity, expensiveness, and the obligation for the customers to have a public/private key pair. Another disadvantage can be observed with SET, since its liability is questionable. Indeed, legal responsibility could be moved from the payee to the payer, as SET transactions are digitally signed by the payer. Therefore, consumers cannot repudiate transactions. In addition, the concepts of responsibility and non-repudiation represent special subjects of discussion. Attacks using Trojan horses can thieve payer's private keys, for example, and use them. Moreover, the payer can always deny a transaction suspecting the issuer of having performed it, when the payer's private key is operated and maintained by the issuer.

13.3.1.3 Alternatives to SET-Based Payment

A large number of merchants are disinclined to implement secure e-payment systems to prevent fraud, thinking that they may prevent rather than encourage consumers. This has probably limited the development of SET-based payment systems and led the industry to look for simpler credit card-based payment systems. SET-based payment systems are still able to provide authentication of the cardholder (e.g., protection of the credit card number) and reduce the risk of the merchants as opposed to imposing it to the payers. Different schemes have been proposed to comply with this point of view. They all allow the issuer to authenticate its cardholders using various techniques. Depending on the technique, authentication can just be based on a simple password, which has the advantage of not requiring additional software or hardware to implement at the payer's site, or it uses a digital signature with smart card. Therefore, authentication provided by the aforementioned systems is intimately linked to the payment transaction with the payee, and authorization message communicated by the issuer to the payer. In general, this is done by the use of SSL only.

An alternative solution to the SET is the use of credit card numbers that are valid only once or for a short period of time. In this context, a customer can request from the issuer a temporary credit card number that can be added to the official number of the credit card. Upon receiving the temporary number, the payer can use it instead of the official card number over SSL (or TLS). Since the temporary number is valid only for a short period of time (or for only one payment), it can be used by an adversary after the period has expired.

13.3.2 Electronic Checks, Transfer, and Cash

Payment systems using electronic checks or account transfer have the following three properties: (a) payers have central accounts (typically different from credit-card accounts); (b) during the payment phase, the payer authenticates himself to a central server (or financial service provider managing the account) and the correct amount of money is deduced from his account; and (c) the central server confirms the payment acceptance to the payer and sends him the correct amount of money. Various systems have been developed and are available in the literature. The types of account they use all fall into four classes: pre-paid account, account linked to a credit card or a bank account, direct bank account, and third-part account.

With pre-paid account, a customer can obtain some pre-paid money value that is associated with a unique number. The unique number and the related value are stored on a central server (the issuer) and constitute an account, which is protected with a PIN code or password. Typically, a payment using a pre-paid account would work following a three-fold process: (a) the payer receives from the payee a transaction identifier and the amount of money to be paid; (b) the payee sends to the issuer his unique account number and PIN code related to the amount, together with the transaction identifier and the exact amount of money; and (c) the issuer confirms to the payee that the payment related to that transaction identifier has been successfully performed, and debits the payee's account. Therefore, it appears that electronic payment systems based on pre-paid accounts are technically simple to build, suited for relatively small payments (such as phone cards), and that they constitute a good means for anonymous payments. An alternative to having a pre-paid account, the central account can be linked to a real credit card or a bank account. Payments in this context are credited to the real account and are settled periodically. Compared to a credit-card payment using SSL, this approach offers the following advantages: (a) the payer does not have to send sensitive information directly to the payee; and (b) payment transactions require payer authentication before they are executed.

Regular bank accounts can be used in e-payment systems. This is achieved through account transfers initiated via electronic banking, for example. In this context, mobile operators can be used to manage the payment server and customers can register with the server. The authentication of the payer relies on the authentication of the SIM card. A typical payment transaction in this system is described as follows: the payer can use a random code occurring in an SMS message received from the central server. This random number is then introduced appropriately to complete the transaction. Another scheme only requires of the payer to call a unique phone number of the payee at the time of payment.

On the other hand, electronic cash is a payment protocol that does not involve transfer between accounts, but that involves the transfer of electronic tokens, which may be authenticated independently of the issuer. This is commonly achieved through the application of self-authenticating tokens or tamper proof hardware. The system provides nothing but a secure tunnel between the payer (or his card) and the payee (or his secure payment module). The channel is utilized to exchange all messages needed for the payment transaction. Many electronic cash protocol have been proposed in the literature. They provide a large spectrum of security features such as privacy. Very few protocols have offered anonymity. An example is given by Ecash (Schoenmakers, 1998), which represents an on-line payment solution that adds anonymity to the prevention of double-spending by storing the trace of all tokens paid. Ecash consults the issuer for each payment.

13.4 Privacy and Anonymity in Electronic Payment

Whenever a customer runs a payment transaction, some information that can be considered private or sensitive by the customer may be stored into a database in some place. Furthermore, all these records can be linked so that they constitute, in fact, a single record of the user's profile. Organizations link records from different sources for their own protection. Unquestionably, it is in the interest of a bank looking for a particular decision (a credit application, for example) to know that the user has committed defaults several times in the past. However, that information, in malicious hands, provides neither protection for businesses nor better service for consumers. In addition, malicious entities can use a stolen credit card number to trade on their victims' good payment records.

Privacy and anonymity are two issues that should be addressed to provide resistant solutions to the aforementioned attacks. But, before discussing the major mechanisms for privacy and anonymity, let us consider in the following subsection some generic definitions and requirements.

13.4.1 Privacy and Anonymity Basics

Digital identity, privacy, and anonymity can be defined as follows. Privacy is related to the personal information of an individual such as his digital identity, his interests, his behavior, etc. Privacy of personal data relates to the individual's interest to restrain other individuals and organizations from accessing his personal data. Individuals must at least be able to use complete control over their data, as well as over the visibility and the use of their data.

As for anonymity, it characterizes the state of being not identifiable within a set of entities, individuals, or objects (Pfitzmann, 2001). Therefore, anonymity focuses on hiding the identity of an individual relative to a certain set of subjects. It can be provided and checked at different levels: the payment application, network, and third party levels. The degree of anonymity that a payment system can provide depends on the mechanisms used and the number of entities using the system. Anonymity can be seen as a method of privacy protection. To provide it, a pseudonym is generally used instead of the real identity. Despite the linkability of actions related to a pseudonym, two situations may occur: (a) the pseudonym can be linked to real user identifiers; this is referred to as pseudo-anonymity; and (b) the pseudonym cannot be linked; this is called full anonymity.

The major requirements for electronic systems with respect to the anonymity of the payer are the following:

- Tracing the identity of a payer through the payment means should be impossible to achieve, meaning that the identity of the payer should only be linked to the collection and withdrawal sub-transactions. This will guarantee that the issuer will not be able to trace payments performed by a given payer.
- Tracing the payment transactions a given payer has conducted should not be possible to perform. This indicates that no party that is not directly involved in an ongoing payment transaction should be able to obtain useful information about the transaction.
- Different payments from the same payer should be unlinkable, even if the identity of the payer remains hidden. This states that the issuer and the payee should not be able to link two different payment transactions to the same payer.

Most of the commercial e-payment systems available do not provide anonymity, while payer authentication is considered as an important feature to have. For logical reasons, the credit card number, in credit card-based payments, is known by the issuer and can be linked to the real identity of the payer. Therefore, credit-based payment systems cannot protect the user's privacy, as the issuer knows every purchase the payer can make. One can notice, however, that one-time credit card–payment systems offer some privacy toward the payee, but not toward the issuer. Additionally, electronic account-based payments do not provide anonymity with respect to the issuer, since the account can be linked to the payer's identity. However, because the account is identified by a pseudonym, it can be assumed that the public cannot associate the pseudonym with the corresponding real identity. Moreover, electronic cash-based systems that can be used on-line do not provide anonymity toward the issuer, but may, to some extent, provide anonymity toward the payee as the link between the payer's real identity and the payer's purse identifier is not known by the payee.

13.4.2 Mechanisms for Unconditional Privacy and Anonymity

Physical cash is a traditional way of anonymity. In cash-based payments, digital cash equivalent of physical cash is based on the concept of blind signature, which allows a user to receive a message signed by a signer without revealing the contents of the message to the signer. A standard scheme for this mode of payment would work as follows: The tokens are random strings signed by the bank. Thus, they cannot link the identity of the user (which can be known during the withdrawal of the token) to the final obtained token (and thus to the transaction to which the token is related). Nonetheless, this simple scheme allows the duplication of tokens. As copying cannot be avoided, so double-spending should be detected. To this need, the banks should keep every spent token within a database and the payee should go on-line, on any receipt of a token, to check whether the token has already been spent. If that is the case, the payment transaction is rejected.

An alternative solution that can supplement the blind signature assumes that every user of the payment system has two accounts: a personal account and an anonymous account (Camenish,1994). The anonymous account is established so that its number cannot be associated with the user. In addition, the customer has a secret key K_u to use with the anonymous account. The customer can transfer money from his personal account to his anonymous one using the blind signature approach. This signifies that anonymous tokens are withdrawn from the personal account; the account number of the anonymous account is embedded into the anonymous tokens, to make sure that the token can be paid only to the anonymous account. The payment transaction is performed by the transfer (authenticated by K_u) from the anonymous account to the payee's account. A payment system implementing this approach would have the following features: (a) the payments conducted from the same anonymous account are linkable; (b) the database needed for the anonymous accounts is relatively small; and (c) the customer can have a large number of anonymous accounts.

Motivated by the fact that going on-line for the verification of each received token would generate a large cost, various off-line solutions have been developed to detect double spending (after payment). A generic scheme adopted by these solutions works as follows: The structure of a token is not chosen randomly. Instead, it contains two parts, a public and a private part. The public part of the token is blindly signed by the issuer during withdrawal (assuming that the identity of the individual that withdraws the token is encoded into the public part). To complete the payment transaction, the payee challenges the payer, who responds by sending the public part along with a response to the challenge. The token, challenge, and response are later provided to the issuer for storage in the database of the spent tokens. The issuer verifies the received information and checks whether the token is already in the database. If double spending is noticed, the issuer is able to link the operations to the identity of the individual who withdraws the tokens.

Let us now notice that the anonymity property of a payment system is closely related to connection anonymity. It is commonly known that, if the connection used for payment is not anonymous, the tokens can be traced back to their source, for example. Anonymous payment systems that have been developed so far provide communication anonymity in an integrated manner, meaning that data and communication anonymity are provided at the same time. Despite the advantage provided by the integrated approach, it appears that building an anonymous payment system on top of an anonymous communication protocol is somehow attractive.

13.4.3 Conditional Anonymity in Payment Systems

The preceding subsection discussed electronic payments that aim at providing anonymity to the payers. Anonymity of electronic money can be misused by many types of malicious customers to carry out attacks such as overspending, illegal purchase, blindfolding (i.e., attacks engaging banks in non-standard protocols for withdrawal), and performing a high number of micropayments in a very short period of time. Therefore, to make anonymity acceptable by the actors involved in payment transactions (i.e., payers, payees, issuers, and governments), mechanisms for conditional (or controlled) anonymity are needed. Typical mechanisms include (but are not limited to) the following:

- *Traceability mechanisms*: These mechanisms, also called revocability mechanisms, allow the tracing of anonymous tokens, original owners of tokens. Such mechanisms can be triggered by a customer needing to trace his/her tokens.
- *Limitation of payment amounts*: These mechanisms can be used to limit the amount of anonymous payments that a customer can make.
- *Double/over spending transactions*: Such control mechanisms allow the detection of double spending and over spending transactions. Integrated to on-line systems, they can reject the incorrect transactions.
- *Transferability mechanisms*: Some cash systems allow the receiver to spend the tokens received during a payment without interaction with the issuer. A transferability control mechanism will provide a complete control on the transfer of electronic cash. It can also prevent any transfer or impose a specific policy based on delegation.

Mechanisms for revoking anonymity under well specified conditions should meet various requirements, including (a) the anonymity should be revocable on a per transaction manner; (b) the anonymity of a transaction can be revoked only by a trustee. It is revoked only when it is necessary; (c) the trustee revoking the anonymity should not have any capability other than tracing; and (d) the issuer should not be able to double spend money on behalf of the payer. To achieve the aforementioned objectives of revocation, three schemes of mechanisms have been developed: the *unlinkable revocable*, *linkable revocable*, and *trustee linkable* schemes.

Revocability, in unlinkable revocable anonymous systems, is enabled during the establishment of a customer account. The customer, the issuer, and the trustees (assumed to be two) set up a cooperative scheme that allows the user to give the trustees the information that will permit owner tracing. The scheme uses different types of blind signatures to provide linking a message-signature pair to the corresponding sender. It can also use the concept of dual verification signatures, assuming that both the issuer and trustee sign tokens. Thus, verification of the signatures can be done without the involvement of the trustee or the bank (Jacobson, 1996).

Using linkable anonymous payment systems, the customers should set up two accounts, say the personal and anonymous accounts. An anonymous payment is defined as a transaction from an anonymous account to a payee's account. The scheme assumes an efficient method for transferring money from a personal to an anonymous account without revealing the relationship between them. On the other hand, total unlinkability of personal and anonymous accounts is achieved using a blind signature scheme making the following properties come true: (a) the trustee knows the correspondence between the two accounts; and (b) tokens withdrawn from a personal account can only be deposited into the corresponding anonymous account. Therefore, even though the customer's identity is not revealed, the issuer can link different payment transactions when the same anonymous account is used. The last scheme allows owner tracing as the trustee can at any time find the source of a transfer, given the anonymous account number. The scheme also allows token tracing as the trustee can find the destination of a transfer, given the personal account number.

Let us now mention that several trustee linkable revocable anonymous payment systems have been developed. They all use a specific technique to involve the trustee in blinding the tokens. The system provided in M'Raihi (1996), for example, allows a customer (having an account) to agree on a shared secret with the issuer. The issuer certifies the shared secret so that the customer can anonymously associate pseudonym with the trustee, who then acquires knowledge of the link relationship between the shared secret and the pseudonym. During withdrawal, the blinding of the tokens is delegated to the trustee, unblinded tokens are signed by the issuer, and the blinded tokens are signed by the trustee. This enables the trustee to associate tokens to the secret-pseudonym pair.

Finally, it is worth noticing that the mechanisms for controlling transferability and amount-limitedness are easier to implement within an anonymous payment system. To provide amount limitedness, the flow of tokens that a customer can spend is controlled, for example, by a mechanism that only allow customers to withdraw a limited amount of electronic cash in a specific period of time. Additionally, anonymous cash can explicitly be designed as non-transferable.

13.5 Mobile Payment Systems

Lately, the necessity to perform electronic payments on the move has emerged. The development of wireless communication technology has provided the ability to access the Internet and perform e-payment transactions using mobile devices such as cellular phones, portable computers, or personal digital assistants. Performing an e-payment transaction where at least one involved party is a mobile user is called mobile payment. An m-payment can be characterized by the use of multiple attributes including (a) the transaction environment, which can be a remote, local, or personal environment; (b) the transaction volume, which represents the amount of money transferred over the mobile network from the payer to the payee. Micropayments typically consider the payment of amounts lower than $10. Amounts exceeding this range are commonly called macropayments; and (c) the time when the payment transaction is performed. Three categories of payments can be distinguished: pre-payment, concurrent payment, and post-payment. A concurrent payment would

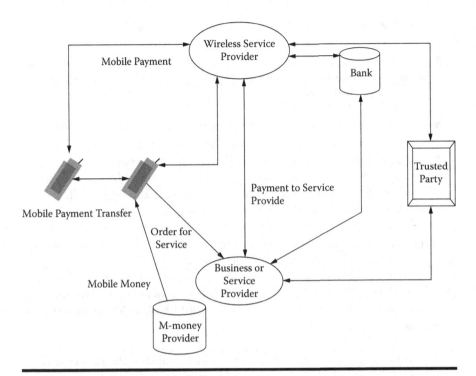

Figure 13.3   Mobile payment scenario.

be attributed to a debit card that allows purchases immediately after confirmation through entry of the PIN and password.

Currently, several mobile payment scenarios can be used. Figure 13.3 depicts an overview of a mobile payment scenario and shows the major actors. Five actors can be involved in m-payment systems. They are the mobile customer, the mobile service provider, the bank, the business provider, the trusted party, and the m-money provider. As shown in Figure 13.3, a mobile user could order products and services from one or more service providers, who will then contact either a trusted third party, wireless service provider, or a financial institution for verification involving the customer and the amount of purchase. The payments can be made to the business or service providers through a bank, a wireless service provider, or another payment party. These can then be subtracted from the customer's bank account, withdrawn from his/her m-wallet, or added to his/her wireless phone bill. It is also possible for the user to pay for products and services using "mobile money" that has been provided for him by another mobile customer or third party mobile money provider using a pre-paid or post-paid service.

One can notice that the scenario depicted by Figure 13.3 could be used for different types of mobile commerce involving both micro and macro payments. These include P2P, B2C, and B2B mobile commerce.

The role of the major actors is therefore summarized as follows:

- *Financial service provider (FSP)*: The FSP provides the back-end system for the payment settlement. There are three types of FSPs in relation to m-payments. They are the banks, the mobile operators, and the provider of electronic cash systems.
- *Mobile/wireless service provider (MSP)*: A MSP provides products and/or services to the end-users. A MSP can be a simple merchant.
- *Payment service providers (PSP)*: PSPs provide the m-payment system and channel for the financial service provider. A PSP establishes transactions between the FSPs, end-users, and merchants basically by providing them with software and interfaces between these actors.

Typically, two major reasons explain why the security of mobile payment systems is not completely achieved: the limitations of wireless environment and the security of the mobile system itself. Performing payment transactions in wireless environments basically suffers from resource limitations of mobile devices and from the characteristics of wireless networks (Wong, 2001). Mobile devices have the following limitations:

- Computational capability of the processors included in the devices is comparatively lower than what it provided by personal computer.
- Network connections set up over wireless links are less reliable, since packet losses occur more frequently than on fixed networks; the lost packets need to be retransmitted. This may induce high latency.
- Connection cost of wireless networks is higher compared to that of fixed networks.
- Data transmitted over wireless networks is easily eavesdropped.

From the aforementioned limitations, it comes that performing payment transactions over mobile networks is time consuming. Moreover, performing mobile payment transactions on low computational capability mobile devices lasts for a larger time to complete all running payment transactions. In addition, performing payment transactions over wireless networks using such limited devices will charge customers larger costs. Moreover, transmitting transactions over radio links can be easily eavesdropped and using cryptographic techniques to protect against eavesdropping requires high computational capability mobile devices. Thus, performing mobile payment transactions over mobile networks brings up concerns about the security of the underlying payment systems. Another concern is that mobile payment applications should be compatible with the traditional e-payment applications so that the existing infrastructure can continue to operate.

It appears from the previous discussion that e-payment systems, which are originally designed for fixed wired networks, cannot be mechanically applied to mobile environments. To enable mobile payment and overcome these limitations, a number of useful frameworks have been built based on the migration of the existing wired network-based payment solutions or on building a system purposely for mobile networks. Solutions provided through migration are classified into proxy-based and agent-based solutions, whereas those provided through building a new system are non-proxy-based ones.

The following subsections will discuss the major features of the available solutions and study some examples.

13.5.1 Proxy-Based Solutions

A proxy-based mobile payment solution authorizes a customer to perform a payment transaction using an existing wired network payment scheme through a proxy server operating on behalf of the customer involved in a transaction and connected to the fixed network, assuming that the customer, who already has the existing payment infrastructure, does not want to upgrade the infrastructure. The set up proxy server will serve as a medium between the mobile devices and the payment infrastructure. To perform a payment transaction to a payee, the payer sends a request from his mobile device to activate the proxy server to perform the transaction with the payee on behalf of the payer. The only thing to provide therefore is to guarantee the authentication of the payee.

Two examples of proxy-based payment systems can be considered. The first system is based on the use of the SET protocol and the second utilizes the WAP protocol. They are called the *Three Domain SET approach* (Wrona, 2001) and the *Dai et al's Scheme* (Dai, 2003), respectively.

13.5.1.1 Three-Domain SET Approach

This is a payment performing SET protocol through the following six steps:

1. The payer informs the payee that he/she is starting a SET-based payment.
2. The payee notifies the payer that the payment session is about to be built.
3. The payer is redirected to the issuer's server, which contains all payer information, including those related to the credit card.
4. The issuer requests the payee to provide the authentication information to confirm the payment.
5. The payer provides the authentication information to the issuer. And the issuer completes the SET payment transaction on behalf of the payer.
6. After transaction completion, the payer is redirected back to the payee's site.

The Three Domain Set approach offers several advantages, including (a) the payer does not necessitate to store anything on his mobile device; (b) a few messages are transmitted over the wireless link connecting the payer; (c) very limited computation is performed by the payer; and (d) all updates can be done only at the issuer server, which provides larger flexibility. However, the approach presents the following disadvantages: (a) the payer has to fully trust the issuer acting on his behalf; (b) the payer authentication information is required to be stored on the issuer's server. This information can possibly be lost or used by malicious internal attackers; and (c) the issuer can trace the spending operation of a payer and build his profile. The latter disadvantage would affect the need for privacy.

13.5.1.2 Dai and Zhang's Scheme

This is a payment performing WAP protocol allowing the mobile payers to run payment transactions using mobile phones. The following steps are depicted in Figure 13.4 and describe the operations performed by this scheme, which involves the WAP gateway. Indeed, payers and payees using this approach are assumed to have WAP-enabled equipments.

- The payer sends a request to the payee via the WAP gateway.
- The payee sends back the payment type (i.e., credit or debit) to the payee via the WAP gateway.
- The payer selects the payment type to the WAP gateway.
- The WAP gateway sends a contract to the payer via the short message service center (SMSC).
- The payer signs the contract and sends it back with the signature to the WAP gateway via the SMSC using the standard PKCS#1.
- The WAP gateway verifies the signature and performs the transaction to the payee, on behalf of the payer.
- The payee sends the receipt to the payer via the WAP gateway.

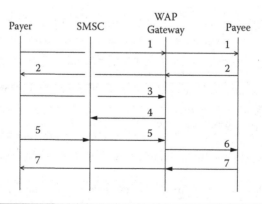

Figure 13.4   A WAP-based payment system.

The Dai and Zhang approach guarantees that the payee's authentication is made by the WAP gateway using his digital signature. It also allow the security of the payer to rely on the security and trustworthiness of the WAP gateway, since the gateway is able to impersonate the payer because it has the payer's private key.

13.5.2 Agent-Based Solutions

An agent-based m-payment solution makes use of a mobile agent technology to allow mobile users to perform payment transactions on an existing payment system deployed on a fixed network. The basic idea supporting the agent-based approach is to allow a mobile payer to send an agent (a transported code) containing payment-related information and acting on behalf of the payer to run the transaction in the payee's fixed environment. This induces two important benefits: the reduction of connection cost, since the payer is required to remain connected for very short periods, and the reduction in the computational load at the payer's mobile device, since the agent created and sent by the payer is executed away from the payee.

Various agent-based solutions have been developed. SET/A has been a common solution that integrates the SET protocol (Romao, 1998). SET/A allows a mobile agent to execute a payment transaction on behalf of a customer out of his mobile device. The basic operations of SET/A are performed as follows (Figure 13.5):

1. The payer creates an agent, referred to as *Agent(Payer, tr)*, and sends it to the payee requesting a SET initialization. The agent has knowledge of the payment related information along with a *SET wallet*. The agent stays at the payee's server to run the payment transaction *tr*.
2. At the payee's site, the user's payment request, denoted by *PReq*, is generated and sent to the payee. PReq contains the order information and the payment information formatted as required by SET protocol.

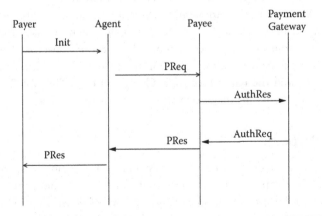

Figure 13.5 SET/A payment system.

3. The payee performs the transactions using the protocol SET until receiving the authorization response from the payment gateway. Then, the payee generates the payment response, *Pres*, and passes it to the agent acting on behalf of the payer.
4. After receiving the response Pres, the agent returns the result of the transaction to the payer.

One can notice that, using SET/A, the payer has to be connected to the network only during two limited periods of time. In the first period, an agent containing the user's request for payment transaction is sent to the payee. During the second one, the response is received along with the result of the transaction execution. Thus, one can be persuaded of the following two facts: The cost of connection is reduced and the computation load at the user's site is also reduced. Unfortunately, it comes into view that executing the agent at the payee's site presents some security drawbacks. The operations performed by the agent can be vulnerable to attacks. In particular, the actions executed by the agent may contain the random generation of a secret key that is used for the encryption of the payment information (to be transmitted to the payment gateway). This key can be retrieved by the payee during generation.

To overcome SET/A limits, a modified agent-based SET payment system, referred to as SET/A+ (Wang, 1999), is provided. It is in charge of performing the generation of PReq at the user's site and considering the payment together with a brokering and negotiation agent, which is sent by the user to collect information about products. Thus, SET/A+ solves the problem of the key compromise occurring with SET/A; however, it adds a high computation load at the customer's site.

13.5.3 Non-Proxy-Based Solutions

A non-proxy-based mobile payment system does not require a proxy server. Instead, it integrates a lightweight cryptographic technique to reduce the computational and the communications loads of the customer's mobile system. It also provides better security. Many solutions have been proposed to secure non-proxy-based mobile payments. Among these solutions, one can distinguish two solutions, the Paybox (Paybox, 2001) and the Kim's Electronic Cash solutions (Kim, 2002). In the following, we will discuss the major operations and features of these schemes.

13.5.3.1 Paybox

This is an easy way to develop a payment system based on a mobile network system, in the sense that the payer and the payee are required to use mobile terminals with properly identifying phone numbers. Paybox involves a third entity, the Paybox server. To execute a payment transaction, the payer and the payee need to have a bank account set up. The main steps in performing a payment with Paybox are described as follows:

- After the agreement of the payee to pay, the latter connects the Paybox server via the network and passes the requested amount and the payer's mobile phone number.
- The payer is called back to deliver the payment authorization by simply providing a predefined PIN number.
- Upon receiving the payer's authorization, the Paybox server transfers the account of money from the payer's account to the payee's.

It is easy to see that the security of Paybox relies greatly on the security features supplied by the communication network interested in the transaction execution. However, since the payer authentication is obtained by introducing the PIN number into the Paybox system, an attacker can copy it and utilize it in a future attack.

13.5.3.2 The Kim's Electronic Cash

This protocol reduces the computation load of the mobile customers by deploying only hash computations and digital signatures. Three entities are involved in the execution of a payment transaction: the payer, the payee, and the bank. They cooperate to the achievement of the subsequent steps:

1. The payee sends the hash value of the amount of money Am concerned by the transaction along with the transaction code, say Tc, to the payer. The message sent has the form $<h(Tc, Am), Am>$.
2. Upon receiving the message, the payer sends a request to the bank and asks for a payment token. The request $PReq$ has the following form: $PReq = <IDp, Pwd, h(Tc,Am), Am>$, where Pwd is a password shared with the bank and IDp is the identity information of the payer.
3. The bank checks whether the password is valid and issues a payment token, Tk, to the payer. The token has the form $Tk = E(K, h(Tc, Am))$, where $E(K, -)$ is the encryption operation using the private key K of the bank.
4. The token is then delivered to the payee by the payer along with Tc and the identity information of the bank IDb.
5. Upon receiving the token, the payee deposits the amount by sending back the token along with his identity information.

A study of this protocol shows that the Kim's Electronic Cash does not put a high computation on the mobile devices. It presents, however, many drawbacks, including the password is provided in cleartext; the payment token provided to the bank is not linked to the payee; the payer and the bank are unable to verify the validity of $h(Tc, Am)$; and the protocol lacks non-repudiation property for all messages (Kungpisdan, 2005).

13.6 Analysis of Existing Mobile Payment Systems

Existing mobile payment systems can be analyzed and compared based on several dimensions including the following: the trust relationship, the constraints of wireless links, the security against attacks, the nature of the token used, and the cryptographic operations.

13.6.1 Analysis Parameters

Analysis of mobile payment systems includes the following parameters.

■ *Trust relationship*: Several works that examined trust in business relationships have identified the trust to be a key factor for successful long term partner relationships. Trust in a mobile payment environment is an essential requirement to the development of mobile services. The payer, in a proxy-based mobile payment, needs to fully trust the proxy server, since his/her sensitive information is stored there. On the other hand, the payer in agent-based and non-proxy-based protocols does not have to trust any of the other entities operating in the payment environment. Trust based mechanisms are the protecting measures of mobile payment. Combined with security mechanisms, they ensure timely, accurate, and complete transmission, receipt, and safe execution of payment transactions. Trust based mechanisms comprise the access control to payment servers, action authorization, delegation control, and clear security policy definition.

■ *Constraints of the radio link*: The main goal of the existing mobile payment systems is to overcome the constraints of the radio links. To this end, they put mechanisms that reduce the amount of data transmitted over the radio link. With this respect, it can be observed that proxy-based mobile payment systems represent efficient solutions, since they allow the smallest amount of data to be transmitted over the radio link, compared to the non-proxy-based mobile payment and the agent-based mobile payment systems. In proxy-based protocols, a payer sends only a request to trigger the execution of a payment transaction at the proxy server (on his behalf). However, there is a cost to pay for this; the payer needs to stay online during the transaction duration. To alleviate this, agent-based protocols focus on the reduction of the connection duration.

It is commonly believed that proxy-based and agent-based mobile payment systems are expected to be applied to secure account-based mobile payment systems, which necessitates high communication and computational load. But, given that operational costs of token-based payment systems may be high, the proxy-based framework may not be appropriate for this mode of mobile payments. Moreover, let us notice that proxy-based and agent-based payment system protocols are inclined to enable mobile payment transactions extending fixed network payment frameworks while the non-proxy-based payment solutions are likely to design new mobile payment frameworks based on existing wireless protocols (such as the WAP and WTLS), and deploy simple cryptographic tools to reduce communication and computational loads.

■ *Security against attacks*: The security of the payment information transfer over a wireless network relies on the security technique applied to protect it. This means that the messages exchanged for the need of mobile payment achievement between the payer, the payee, and the issuer (and the trustee, for some solutions) should be secured by implementing security mechanisms at the application and transport layers. This is the case for the mobile payment SET/A that uses the protocol SET. In addition, actions performed at the payee's server or the bank site on behalf of the payer should be protected against various attacks, including double and over spending, token modification, key copying, and payer masquerading. While some mechanisms are integrated in the payment frameworks, various payment solutions do not provide a large protection.

■ *Secret nature*: Shared secrets can be used in mobile payments to authenticate the payer and the message exchanged for the need of transaction commitment. Shared secrets stand for a large spectrum of forms varying from a cleartext password delivered via a protected channel to a digital signature made using a private key linked to a X.509 certificate generated by a legitimate certification authority. The payer, for example, can establish a secure communication channel by executing a key exchange protocol and, then, providing his/her password to be granted an access. In a credit card payment, the payer sends the payee his credit card number (which represents the secret shared with the card issuer) along with payment related information via a secure channel, such the SSL channel, to request a payment to the payee. Using SET protocol, the mobile payment allows the payer to submit his/her credit card information as an authentication token that is encrypted with the payment gateways' public key and signed with the payer's private key. When the token is transferred to the issuer, the issuer can determine whether the request has been originated by the payer and whether the token is valid.

Another type of sensitive information in payment systems that needs a real protection is related to the payer's account. Several security issues are related to this information. The fact that the SSL-based credit card payment system guarantees that this information is securely transferred, for example, does provide a complete protection of this information, since it is still revealed to the payee who is often an untrusted entity.

13.6.2 Case Study: Analysis of a GSM-Based Payment System

Various GSM-based payment systems have been developed. We consider in this subsection a payment system described in Claessens (2002) and involving payment related information exchanged via GSM. The payment system is context-general and independent of the wireless system. Using more advanced wireless techniques, such as UMTS, a more secure scheme would be set up following the same system architecture. The actors of the payment system are the customer (or payer), the merchant, and the deliverer, that is, the local representative of the merchant with respect to the payer. The network operator plays the role of the bank. To play this role, the network operator subtracts the necessary amount of money from the payer's account and adds this amount to the merchant account. The accounts can be credit or pre-paid based.

13.6.2.1 System Architecture

The payment protocol allows a customer to initiate a payment transaction over the GSM and receive a payment receipt. It performs five main payment operations executed after finishing the purchase request, the purchase confirmation, and the verification of the order. The payment operations are (a) the verification by the payer; (b) the debit operation; (c) the intra-GSM operation; (d) the verification by the merchant; and (f) the payment acknowledgment. Let us see how these operations are handled.

- *Payer verification*: Upon the receipt of a signed purchase confirmation message from the merchant, the payer starts the verification operation. The purchase message should contain the transaction identity TId, the merchant identity MId, the GSM operator identity GSM-O(m), the description of the products, and the amount of money to pay. The payer verifies whether all the ordered products are listed in the message, and whether the amount and the digital signature are correct. The authentication of the merchant relies on the GSM–related information (the payer knows the merchant's GSM number), on SSL, and the public key used to check the signature. This, however, requires that the mobile phone and the computer used in SSL should be connected.
- *Debit operation*: The payment application is installed on the SIM card, which is invoked on the receipt of the purchase confirmation message. During debit operation, the payment application asks the payer a confirmation for sending a debit account message to the GSM operator, say (GSM-O(c)), including the amount of money to be paid, the TId, MId, and GSM-O(m). The authentication of the payer relies on his/her GSM connection. In addition the TId permits further verification by the merchant.
- *Intra-GSM operation*: The GSM operator GSM-O(c) deducts the exact amount of money from the customer's account and forwards the debit account

message to the GSM denoted by GSM-O(m), which is in charge of adding the amount of money to the merchant's account. The GSM-O(m) sends then a signed delivery confirmation to the merchant. This message contains the amount of money and the TId.

■ *Merchant verification*: The merchant checks whether the delivery confirmation message is originated from GSM-O(m). The digital signature of GSM-O(m) is also verified. Then the merchant checks whether the amount of money included in the delivery confirmation is the one included in the corresponding purchase confirmation message, using TId based on the TId.

■ *Payment acknowledgment*: The merchant sends an acknowledgment to the payer via SMS. The acknowledgment contains a hashed description of the products, the TId, a timestamp, information on the delivery, and information related to the payer. The acknowledgment is digitally signed by the merchant. The occurrence of the TId and timestamp in the acknowledgment guarantees that it cannot be replayed by the customer.

13.6.2.2 Security Analysis of the Application

The security characteristics provided by SSL and GSM constitute together a basis for the security of the aforementioned mobile system. Without a doubt, the payer can securely initiate a payment via SSL. The payer receives a confirmation via SSL and SMS. This ensures the authentication of the merchant and the confirmation of the purchase. In addition, the payer is unable to modify the payment related information, when he informs GSM-O(c). In fact, the merchant can notice all the modifications the payer can make. The presence of digital signatures applied to the major messages would increase the protection of the system and reduce the requirement for a trust relationship between the payer and the merchant. However, this needs the issuance of digital certificates by mutually trusted certification authorities.

This payment system works properly if one can assume that the acknowledgment contains a unique description of the products and that the GSM operator is trusted to transfer the exact amount of money from the payer's account to the merchant's account. In addition, the payer's mobile phone should be considered secure, since the force of the proposed mobile payment system relies on this terminal. Therefore, a special care should be taken to protect physically the terminal from being stolen. The security of the payment system relies also on the quality of protection provided by the SSL and GSM. Although SSL protocol is expected theoretically to provide a high level of security, implementation flaws allow various attacks to be launched, including Web spoofing and false server certificates. The GSM system unfortunately provides a low level of security (as shown in Chapter 5). Furthermore, because the SMS messages are sent through the SMS centers, these centers have to authenticate the source of the messages based on GSM authentication. If this is not carried out, the SMS messages can be sent with a spoofed source.

Moreover, the SMS message forwarded to the SMS center of the addressee cannot be checked for source authentication.

All these comments show three facts: (a) the underlying technologies used with the proposed mobile payment system have many security weaknesses. By and large, the security of the mobile payment system will improve if the security weaknesses of SMS, SSL, and GSM are overcome; (b) the proposed mobile payment system does not provide a high degree of privacy, since the merchants know at least the mobile phone number of their clients. However, this number does not necessarily reveal the real client identity, particularly when a mobile phone with pre-paid card is used. The capability of hiding phone numbers would undoubtedly improve the payment system; (c) privacy problems occur with the GSM operator, because the latter knows exactly which clients are buying which products from which merchant and for what price. Privacy limitations may be used for client profiling and merchant damaging.

13.7 Mobile Agents and Mobile Payment

A mobile agent is a generic term denoting a software procedure that can operate autonomously to achieve the execution of a given task in a remote site. A mobile agent can communicate, travel from one host to another, can be sent by a customer, and will finally get back to report on their results. Despite the advantages of mobile agents, some security issues need to be undertaken before an agent concept can be largely used in mobile commerce. Indeed, protecting the secret information/key, by producing a signed transaction, is very difficult to achieve in an agent-based environment.

13.7.1 Securing Mobile Agents against Malicious Hosts

Mobile agents should be protected while they are in transit from one host to another. The communication between the agents and the customers and between the agents themselves should also be protected. The communication channels involving agents should be cryptographically secured. Agents, customers, and hosts can eavesdrop or tamper with communication. They can also impersonate participating entities in a payment transaction. Typical solutions for this need include the SSL, SET, and IPsec platforms. The protection of the cryptographic keys that are used to secure the communication of an agent is clearly dependent on the other mobile agent involved in the communication. If the two agents transport their own keys, they may be able to secure the keys against the malicious hosts they can visit. Three types of malicious entities should be considered: the malicious agents; the malicious hosts; and the malicious clients.

Agent authentication allows a host to identify the agent (or its signer). Agents should run in an environment where they do have limited privileges, while they are protected from each other properly. To achieve specific tasks, however, agents can request extra privileges and can be granted, if they are allowed to. Additionally, a resource allocation should prevent the agents from overusing the resources (such as flooding the host). Some programming languages, such as Java and authorization languages, can provide methods to comply with such requirements and protect against malicious agents. With an authorization language, for example, a security policy can be implemented in a host and specify which agent is allowed to execute which operations.

Protection against malicious hosts constitutes a hard problem to deal with. A malicious host can explore the agent code, the data carried by agents, and the control flow. It can also deny the execution of an agent and discover the interaction an agent has with other agents returning errors to the system calls generated by an agent. It is commonly believed that providing a perfect protection against malicious hosts is impossible. However, security mechanisms consider prevention, protecting data, execution integrity, and execution privacy. This will be detailed in the following subsection.

On the other hand, malicious users can exhibit unauthorized actions such as tampering with a communication, sending malicious agents, or modifying transactions. A malicious user can decline to pay for a purchase made by an agent, or reject a message. To protect against malicious actions, non-repudiation and auditing services may be needed. Double signing an agent may also be needed with the aim to identify who the generator of the agent that is intended to be send to a host is. A legal framework should, however, be available to fix the responsibility of the user for the tasks he/she assigned to mobile agents.

Solutions dealing with malicious hosts form three categories: (a) the solutions avoiding malicious hosts, (b) the solutions for agent execution integrity, and (c) the solutions protecting the data transported by the agent. With the first category, mobile agents can only visit a reduced number of trusted and authenticated hosts occurring in a list that is updated using various techniques. A mobile agent performs a payment transaction only at the host it trusts. The trusted host can be determined when a trusted hardware is attached to the host. Sensitive code of the agent can be sent encrypted to the trusted hardware. The execution environments (including the cryptographic keys they contain) cannot be controlled or modified by the hosts. An example of trusted hardware can be given by a smart card.

When a mobile agent visits a host to execute a payment transaction, the data it carries should be protected from the malicious host. A set of security properties that should be achieved to protect the carried data include (a) data confidentiality, which states that only the originator is able to extract the information; (b) forward privacy, which states that the identity of the previous host visited by the agent is not revealed and that no collected information (from the visited hosts) can be modified in the future; and (c) insertion avoidance, which assumes that data cannot be inserted unless explicitly declared. A number of solutions for the provision of these requirements have been developed. Typically, these solutions are based on cryptographic actions such as (a) the hosts digitally sign the data they may give to an agent; (b) the data is encrypted with the public key of the agent originator; and (c) the hash value of useful information is inserted before signing (Volker, 2001). Let us notice finally that the solutions provided do not cover all the properties and that some flaws have been discovered with some of these solutions.

Execution integrity ensures the correct execution of a mobile payment agent. The object of execution integrity is the protection of the agent code and state from outside and internal attacks. This can be made by digitally signing the code and state and encrypting them with the public key of the recipient host. Solutions to achieve the protection of code and state employ different techniques. The major ideas behind these solutions vary from the definition of invariants (that should be verified to prove the integrity of the state) to the cryptographic traces, which are defined as digitally signed logs of the operations executed by a mobile agent. The traces allow the agent originator (after execution terminates) to check the correctness of the execution history or whether hosts have hacked the mobile agent code. An alternative technique based on intrusion tolerance assumes that multiple agents, generated with the same functionality, are sent to different hosts where they can be executed. A voting process can be organized to survive malicious hosts and provide execution integrity.

Finally, let us notice that while execution integrity provides protection of the agent's code and state, it unfortunately does not necessarily keep the code and state private. Some measures for mobile agent privacy have been developed. They include the environment key generation technique, function hiding, and code hiding. The environment key generation states that an agent can be encrypted with a key (generated when certain environmental conditions are valid) in order that it can be decrypted and executed (in a host) when the environmental conditions are satisfied (Riordan, 1998). Function hiding is a cryptographic way that takes a function, transforms it, sends the encrypted function for execution, and transforms back the computed result when the agent returns it (Hohl, 1998). Code hiding can be achieved by adding a mix-up code or by distributing the agent's code among a group of interoperating mobile agents.

13.7.2 Securing Payment Transactions in Untrusted Hosts

The previous subsection considered the major features of solutions developed to protect mobile agents against malicious hosts. The goal of this subsection is to investigate whether the aforementioned provide a minimal level of trust for the electronic performed using mobile agents and to discuss some explicit agent-based solutions for secure mobile payment.

It is commonly agreed that avoiding the problem of malicious hosts is a good approach to allow mobile agents carry and execute payment transactions. These solutions, however, show some drawbacks involving cost, complexity, load, and legal issues. While the solutions for protecting mobile agent's data are efficient for protecting the context of a payment transaction, they are not effective for the security of the transaction itself. In fact, confidentiality can be needed by the payment transaction, meaning that only the originator can decrypt the information and not the mobile agent itself. However, for these transactions the agent would need to access, for instance, the private key. In this case, the protection of data is not sufficient.

Execution integrity, on the other hand, seems to be of low importance to payment transactions as it does not provide confidentiality. Even if the agent's private information is not protected from the malicious host attacks, it can be possible to forbid illegitimate use of it. The concept of agent replication can be well adapted to define the tradition *k-out-of-n* scheme and protect the payment transaction. The scheme typically works as follows: the transaction secret is dispersed over *n* mobile agents that are sent to different hosts. Confidentiality of the private information is achieved since none of the agents has the entire secret. The scheme supposes that less than *k* agents are running on malicious hosts and that at least *k* out of *n* agents should cooperate to complete the payment transaction.

Solutions provided for the execution privacy of mobile agents are important for the security of payment transactions. The environment key generation solution is difficult to use for transaction protection and should hide which environmental conditions to check. A malicious host can provide the agent with misleading information and make it unable to complete its task. The function hiding of a payment transaction can be a digital signature function with hard coded private key that the untrusted host cannot abuse. Finally, code hiding is useful for functions that need to be hidden for a limited period of time. To apply this approach to payment transactions, the secret data in the transactions need to be valid only for a short period.

To protect an agent-based payment system, solutions that avoid sending the agent to malicious hosts are difficult to deploy since they need to know where exactly the malicious hosts are located, which is difficult to achieve in open environments. To reduce the risk of payment tampering, a solution can attempt to avoid malicious hosts using the concept of master agent/multiple slave agents. The master agent can be static and the slaves can be mobile but having payment capabilities. A slave only travels to a host that may be malicious, negotiates a payment contract signed by the host, and returns with it to the master agent. The master agent has then the responsibility for evaluating the contracts and presenting the results to the payer. While this solution reduces tampering the slaves, it does not completely protect them since they are still vulnerable to malicious hosts providing false information (Kotzanikolaou,1999).

Two alternative solutions can be distinguished. The first solution considers that instead of giving the mobile agent direct access to the user's private digital signature key, a new key pair is generated for the mobile agent (Romao, 1999). The key pair is certified by the user, by binding the user to that key pair and to the transactions that the mobile agent will perform, using a certificate. The lifetime of the certificate can be short so that revocation is not needed. Therefore, it should be difficult for a malicious host to discover the private key before the certificate expires. In addition, the certificate can contain constraints that prevent the private key from being used for arbitrary transactions. The second solution allows the private signature key to be blinded (Ferreira, 2002). A blinded signature can be produced using this key in such a way that only the resulting signature can be unblinded (and not the key). Mobile agents carry the blinded signature key and a signed policy that defines the restrictions under which the signature key may be used. The blinding factor can be given to a third party or to the mobile agent. With the third party, the private key is cryptographically protected, as opposed to just being hidden or distributed over multiple agents.

13.8 Multiparty Mobile Micropayment Systems

Using mobile communication systems, a customer may move under the coverage of a new network, place calls or send a service request, and use the services of local and remote value-added service providers (VASP). A basic model, network independent, in which multi-party payments are needed, is described as follows: A *customer* attaches to the network through an access network operator over a mobile wireless link. The customer makes calls to another user, accesses a value added service, or sends packets through the access network operator that he pays in real-time. The connection may involve different network operators before reaching the destination user, a service provider, or VASP. The customer releases a stream of micropayment tokens into the network to pay all the SPs as the call proceeds. The mobile sends a payment token to the local network operator who forwards a copy to all the downstream entities. The payment token is worth a different amount to each entity, and this amount is fixed at call setup. Tokens are based on hash chain constructions. They are purchased by the customer from one of several online brokers and are spent via a designated specific SP, called the *enforcer*, who prevents corrupted actions that the customer or the SPs can launch. After the call, payment tokens can be efficiently redeemed by each SP at its broker. One can envision a broker per area or region who will redeem for multiple SPs in that area (Broker1, ..., Broker4). Only the final token received needs to be redeemed as the other tokens can be derived from this. In addition, the unspent tokens can be spent on a different call to a different destination, with the same enforcer, or they may be refunded.

Figure 13.6 depicts the multi-party payment basic model and illustrates a call placed involving two network operators (SP_1 and SP_2) and one service provider (VASP, SP_3). It assumes that the m-payment is made in a real time way, because of three reasons: First, the total usage is not known in advance. Second, the customer can be a roaming mobile and cannot be trusted to pay the full amount after the call (or service) is completed. Finally, the tariffs might be set when the connections are established and presented to the customer's mobile device for agreement before the call begins.

To pay all involved SPs in multi-party micro-payment, the mobile customer releases a stream of prepaid tokens. Each party providing a service takes exactly its share of the payment. It can be seen that the multi-party payment scheme described

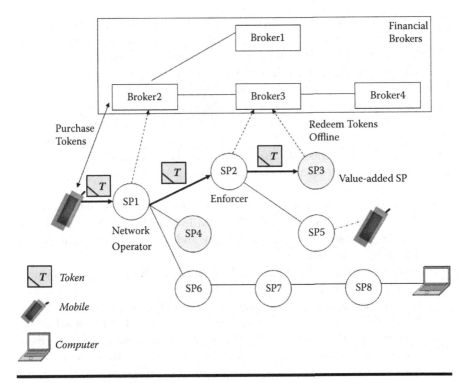

Figure 13.6 Multi-party payment model.

with that example needs the ability to handle repeated payments of small amounts. In addition, when mobiles pay in real-time the need to settle payment through a distant home operator may not be appropriate. Moreover, the need for several interconnection agreements between network operators and service providers may be difficult to handle and can be removed. To overcome some of these limits, the mobile customer creates tokens by repeatedly applying a one-way hash function to a root value *PN* to generate a *payment hash chain*, and he can nominate any specific SP, called the *enforcer*, through which the tokens will be spent.

13.8.1 Micro-Payment Requirements

The multiparty payment involves four major operations: the call setup initiated by the roaming customer; the assembly of pricing contracts done via unsigned messages exchanged between SP_1, SP_2, and SP_3; verification and signing of contracts; and the distribution of fully signed contracts.

Real-time payments reduce the large spectrum of trust assumptions, security risks, and overheads of payment transactions. They allow the service providers, involved in the request of the customer, to provide service safely, provided that he can perform a multi-party payment for services, no matter who the user is, what his credit rating might be, or where its home network is located. Real-time payments for networked services will enhance the use of continuous micropayments and justify the need for a real-time multi-party micropayment system, allowing customers to pay every entity involved in a particular scenario of service provision. Such payment schemes bring with them increased flexibility of usage, and opportunity for the development of new services. They also carry some security concerns, particularly when they are done over a wireless network. This imposes several requirements to be efficient and secure. Depending on the nature of services the following non-exhaustive list describes the major requirements to overcome wireless vulnerabilities and provide an acceptable level of security.

■ *Real-time payment recipients and location requirement*: A mobile customer should be able to pay all parties involved in providing multiple services in real-time, despite his/her current location and lack of on-line contact with a distant home location.

■ *Payment verification requirement*: Any entity accepting a (micro) payment should be able to check its validity, without the need to contact a third entity for this. Each payee should be guaranteed the ability to redeem a valid token. Redeeming should be made with a token issuer or a broker acting in behalf.

■ *Customer signatures requirement*: The use of digital signatures should imply the existence of public key infrastructures and requires an important effort of cross certification to allow a maximum of customers to sign documents while roaming.

■ *Payment flexibility requirement*: The customers should be able to pay for a service using tokens specific to any entity that appears in the service scenario requested by the customers. The major current solutions require the appropriate tokens or pre-paid cards to let customers be able to use and pay the local provider.

■ *User trust requirement*: The customers might be the least trusted entities within the system. Strong on-line authentication must be performed at the start of a payment session. Extensive blacklists of stolen identities and equipment must be maintained to restrain fraud.

■ *Requirement on identified payees*: A payment token should only be redeemable by the intended payees. This should prevent tokens from being used by malicious entities and eavesdroppers.

The aforementioned list of requirements shows that multi-party systems require more security mechanisms than a simple mobile payment system.

13.8.2 Chain-Based Micro-Payment

The typical multi-party payment scheme involves brokers and uses the concept of payment commitments. It works according to the following four-step process.

13.8.2.1 Payment Chain Purchase

A mobile user buys a prepaid value from a broker (generally through his mobile terminal). He creates tokens by repeatedly applying the hash function to the root value, denoted by P_{rt}. To obtain the commitment, the mobile user sends the broker the final hash ($P_0 = H^n(P_{rt})$), the length n of the chain, the desired total value associated with chain V, the enforcer identity, which is a service provider selected by the user to spend the payment hash, and the macropayment details, all encrypted by the broker's public key. The broker commits to the hash chain by sending the message *CommP* to the user. We have

$$CommP = <P_0, n, V, Enforcer>_{SignB},$$

where $<->_{SignB}$ represents the signature of the broker. The payment hash value is defined by V/n. The broker commits to the hash chain, or promises to pay its value, by digitally signing the *payment chain commitment CommP*, consisting of the chain details sent by the mobile user. The commitment shows that each *payment hash* from the chain represents pre-paid value, redeemable at the broker. Therefore, the role of a commitment is to fix the payment hash value. The commitment shows that each payment value from the chain represents a pre-paid value redeemable at the broker.

The value of a single payment hash is fixed later by the enforcer, on a per call basis. This allows the same hash value to be used to pay all parties. By fixing the enforcer in the commitment, the mobile cannot spend payment hashes more than once and attempt to spend again at other providers.

13.8.2.2 Assembling a Pricing Contract

To apply for a service, the customer informs the enforcer that he is about to use the service and sends it a payment chain commitment. A *signed pricing contract* is generated by the service providers involved in the service provision. The contract is used to exchange payment chain commitments and fix the starting hash for the service. Each entity involved in the service provision will use a different payment chain to pay the downstream entities. On receiving the user's request, each service provider digitally signs the assembled pricing contract and adds the price it requires for its part of service. The signing starts with the final service provider(s) in the service provision and the partially signed contract flows back to the enforcer. The signed pricing contract, depicted in Figure 13.7, allows verifiable tariffs; decides on the value per payment hash; fixes the starting hash in the payment chain; links the single commitment to all the SPs; and creates a record of the requested service.

The major fields in the pricing contract are the Transaction identifier for the contract (TID), the identity of each network operator and VASP involved, the charging mechanism and individual tariff rate for each SP, the starting payment hash

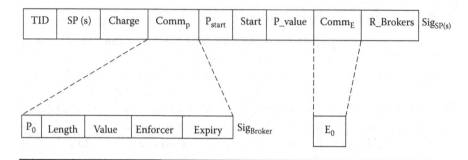

Figure 13.7 Structure of a pricing contract.

from the payment chain for the service (Charge), the value per payment hash for the duration of the call, fixed by the enforcer (P_value), and the expiry period that limits the contract to prevent double redeeming. The charging mechanism and call tariff vary according to different parameters including the service requested, the current network load, and the time of day of the service call. The pricing contract also describes the tariff rate and the charging mechanism (such as per second, for phone calls).

After the enforcer has signed, the finished document is sent to the customer and to all the other service providers. The signatures prove that each service provider has taken part in the service and is due payment. The enforcer is responsible for ensuring that the pricing contract is constructed correctly using a three-step protocol. During the first step, each SP involved in service provision adds its charging details to the contract. With the second step, each SP digitally signs a hash of the fully assembled pricing contract, checking that their input has not been altered. In step three, the finished contract is forwarded to each SP involved.

13.8.2.3 Making Payments

Every period of time, the customer releases a payment hash, in this case starting with SP1 (the enforcer) from a new payment chain. The enforcer verifies that the payment is valid by performing one hash function on it to obtain the previous payment hash. The enforcer forwards the payment hash and his own endorsement hash to the other SPs. Each SP independently verifies both the payment hash and the endorsement hash. Since the hash function is one way, payment hashes cannot be forged, and knowledge of the payment hash is a proof of payment the user pays the total charge to the enforcer, which subtracts the amount it is owed and pays the downstream service provider(s) the remaining amount.

To explain how this works, let us consider that the multi-party payment is done in a way that the customer begins taking the service by releasing a payment hash equal to the total amount due per charging unit. For this, let us consider the case depicted by Figure 13.6 and assume that the α, β, and γ are the charges per unit required by SP_1, SP_2, and SP_3 respectively. Let P_x, Q_x, and R_x be the payment chain acquired by the customer, SP_1, SP_2, and SP_3 respectively. The payment flow generated every time unit by the user, SP1 and SP2 is given respectively by following sequences:

$$\left(P_{(\alpha+\beta+\gamma)}, P_{2(\alpha+\beta+\gamma)}, \ldots, P_{i(\alpha+\beta+\gamma)}, \ldots \right),$$

$$\left(Q_{(\beta+\gamma)}, P_{2(\beta+\gamma)}, \ldots, P_{i(\beta+\gamma)}, \ldots \right), \text{ and}$$

$$\left(R_{\gamma}, R_{2\gamma}, \ldots, R_{i\gamma}, \ldots \right).$$

On receiving $(P_{i(\alpha+\beta+\gamma)})$, SP_1 checks whether $P_{(i-1)(\alpha+\beta+\gamma)} = H^{\alpha+\beta+\gamma}(P_{i(\alpha+\beta+\gamma)})$. Similarly, on receiving $(Q_{i(\beta+\gamma)})$, SP_2 verifies whether $Q_{(i-1)(\beta+\gamma)} = H^{\beta+\gamma}(Q_{i(\beta+\gamma)})$. And the flow of payments and verifications continues downstream until the last SP is paid (SP_3).

13.8.2.4 Redeeming Tokens

At the end of a multi-party payment, the service provider sends a signed message to the broker containing the pricing contract, the highest spent payment, and the position of that hash from the final hash (P_0). The message has the form (*Contract*, P_i, i)$_{SignSP}$. The broker verifies that the chain is valid by performing i hashes to P_i and comparing the result with P_0 in the commitment. Then, it validates its signature on the commitment and checks whether this part of the chain has already been redeemed by the SP. After verifying that the chain is valid, the broker pays the SP the total amount requested.

The security analysis of the multi-party micropayment scheme shows that it meets the aforementioned security requirements. In fact, the scheme satisfies the following statements:

■ *A hacker cannot get hold of payment information during a payment chain processing*: The information contained in a purchase request cannot be seen or modified as it is encrypted with the broker's public key. The broker-signed commitment, containing the anchor of the chain, can be eavesdropped. But, without knowledge of the chain hashes, the commitment values cannot be spent. In addition, the secret chain values never leave the customer until they are spent. Although a hacker can prevent messages from reaching their destination, a reliable transport protocol with re-transmission can be set to carry this out to some extent.

■ *A hacker cannot redeem a money value, even if all payment messages related to it are visible*: A customer will not release payment hashes until a valid enforcer-signed pricing contract has been received. The pricing contract specifies the service providers that can redeem hashes and also specifies how much each hash will cost. Redeeming service providers are assumed to authenticate themselves to the broker, using a signature.

■ *A hacker cannot impersonate a value holder to obtain free service*: The identity of a user who holds a valid payment chain is not needed for payment. Service providers will only accept valid payment hashes for providing services, and are not concerned with user identity.

■ *A hacker cannot impersonate a valid SP during a service session*: A valid service provider holds a certificate, issued by a trusted Certificate Authority and a valid signature key. Even if a hacker does legitimately obtain such a certificate, he will be identified in the pricing contract and will be detected by some service providers who know the neighbors they interconnect with. They can also be detected by the payer.

■ *A user cannot overspend the total value of a payment chain*: The total value of the payment chain is specified in the broker-signed payment commitment. The signature prevents this value from being modified. The value must be spent through the enforcer, which is identified in the payment commitment. The chain is therefore a means of authenticating a temporary account at the enforcer. Without a valid endorsement hash matching a specific payment hash, the payment hash cannot be redeemed.

■ *A user cannot double spend payment hashes*: Payment hashes must be spent through the enforcer. The enforcer will record the highest spent payment hash, in order to verify subsequent payments and redeem hashes. Therefore, if an already spent payment hash is sent to the enforcer, it will be detected as not belonging to part of the chain above the highest received hash, and will therefore be rejected.

■ *Payment chain overspending by the enforcer can be proved to an independent third party*: The broker stores the total amount redeemed for each payment chain. When more than the total value is attempted for spending, it is immediately detected by the broker when redeemed.

On the other hand, several drawbacks can be discerned. We describe these disadvantages in the following. First, a denial of service attack can be easily launched. The attacker can eavesdrop the payment commitment, either during withdrawal or during service establishment. The eavesdropped commitment can then be replayed to the enforcer causing resources to be consumed as the contract is signed by the service providers. Second, the enforcer can take advantage of a number of situations. Customer payment hashes can be obtained in a call with a valid pricing contract. It might then be double spent by the enforcer. The enforcer can also create another pricing contract, paying itself more than it is allowed to. Third, user anonymity to the broker is limited because it depends on the macro-payment under use. While no identity information is required to be added in a payment commitment, all transactions on the same chain are linkable from the pricing contracts.

References

J. Camenish, J. M. Piveteau, and M. Stadler, An Efficient Electronic Payment System Protecting Privacy, in *Proc. of the 3rd European Symposium on Research in Computer Security (ESORICS)*, Gollemin (Ed), LNCS 875(1994):2 07–215, Springer Verlag.

J. Claessens, Analysis and Design of an Advance Infrastructure for Secure and Anonymous Electronic Payment Systems on the Internet, PhD thesis, Catholic University of Leuven, Dec. 2002.

Y. Dai and L. Zhang, A Security Payment Scheme of Mobile e-Commerce, in *Proc. of the Int. Conference on Communication Technology 2003*, vol. 2, pp. 949–952, Apr. 9–11, 2003.

L. C. Ferreira and R. Dahab, Blinded-Key Signatures: Securing Private Keys Embedded in Mobile Agents, *Proc. of the Special Track on Agents, Interactions, Mobility, and Systems (AIMS) at the ACM Symposium on Applied Computing (SAC'2002)*, pp. 82–86, Madrid, Dec. 2002.

F. Hohl, Time Limited Blackbox Security: Protecting Mobile Agents from Malicious Hosts, in *Mobile Agents and Security*, Vigna (Ed), Lecture notes in computer science LNCS 1419(1998): 92–113, Springer Verlag.

M. Jacobson and M. Yung, Revocable and Versatile Electronic Money, in *Proc. of the 3rd ACM Conf. on Computer and Communication Security*, pp. 76–87, March 1996.

M. A. Kim, H. K. Lee, S. W. Kim, W. H. Lee, and E. K. Kang, Implementation of Anonymity-Based e-Payment System for m-Commerce, in *Proc. of the Int. Conf. on Communications, Circuits and Systems*, West Sino Exposition 2002 (ICCCAS & WESINO EXPO'02), pp. 363–366, Chengdu, China, 29 June–1 July 2002.

P. Kotzanikolaou, G. Katsirelos, and V. Chrissikipoulos, Mobile Agents for Secure Electronic Transactions, in *Recent Advances in Signal Processing and Communications*, pp. 363–368, World Scientific Engineering Society, 1999.

M. Kruger, The Future of m-Payments—Business Options and Policy Issues, Background paper No. 2, electronic payment observatory (available at www.e-pso.net), 2001.

S. Kungpisdan, Modelling, Design, and Analysis of Secure Mobile Payment Systems, PhD Thesis, Monach Unversity, Australia, http://beast.csse.monash.edu.au/~srini/theses/keng.pdf.

M'Raihi, Cost-Effective Payment Schemes with Privacy Regulation, in *Advances in cryptology (ASIACRYPT'96)*, LNCS 1163(1996): 266–275, Springer Verlag.

PayBox, Business and Technical Information Regarding the Security at Paybox, Version 1.4, security white paper, Nov. 2001, http://www.paybox.net.

A. Pfitzmann and M. Kohntopp, Anonymity, Unobservability and Pseudonimity—A Proposal for Terminology, in *Designing Privacy Enhancing Technologies*, federrath (Ed.), LNCS 2009(2001): 1–9, Springer Verlag.

J. Riordan and B. Schneier, Environmental Key Generation Towards Clueless Agents, in Mobile Agents and Security, Vigna (Ed.), Lecture notes in computer science LNCS 1419(1998): 15–24, Springer Verlag.

A. Romao and M. Mirada Silva, Proxy Certificates: A Mechanism for Delegating Digital Signature Power to Mobile Agents, in *Proc. of the Workshop on Agents in Electronic Commerce*, Ye and Liu (Eds.), pp. 131–140, Dec. 1999.

A. Romao and M. da Silva, An Agent-Based Secure Internet Payment System, *Lecture Notes in Computer Science*, 1402(1998):80–93.

B. Schoenmakers, Basic Security of the ecash Payment System, in *Computer and Industrial Cryptography: State of the Art and Evolution*, Prencel and Rijmen (Eds.), Lecture Notes in Computer Science 1528, Springer, June 1998.

SET Secure Electronic Transaction Specification, Book 1: Business Description, version 1.0, May 31, 1997 (available at: http://www.win.tue.nl/~ecss/2IF30/set_bk1.pdf)

R. Volker, On the Robustness of Some Cryptographic Protocols for Mobile Agent Protection, in *Proc. of the 5th Int. Conf. on Mobile Agents*, Lecture notes in computer science, 2240(2001): 1–14.

X. F. Wang, K. Y. Lam, and X. Yi, Secure Agent-Mediated Mobile Payment, Lecture notes in Artificial Intelligence, 1599(1999): 162–173.

D. C. Wong and A. H. Chan., Efficient and Mutually Authentication Key Exchange for Low Power Computing Devices, *Lecture Notes in Computer Science* 2248(2001): 272–289.

K. Wrona, M. Schuba, and G. Zavagli, Mobile Payments—State of the Art and Open Problems. *Lecture Notes in Computer Science*, 2232(2001):88–100.

S. M. Yen, PayFair: A Prepaid Internet Micropayment Scheme Ensuring Customer Fairness, *IEE Proc. on Computers and Digital Techniques 2001*, Vol. 148, Part 6, pp. 207–214.

P. L. Yu and C. L. Lei, A User Efficient Fair e-cash Scheme with Anonymous Certifcates, in *Proc. of the 12th IEEE International Conference on Electronic Technology 2001*, Vol. 1, pp. 74–77, Aug. 19–22, 2001.

Chapter 14

Security of Mobile Voice Communications

14.1 Introduction

Voice communication has been the central service of the mobile communication networks for the last decades. As communication networks are nowadays moving toward packet switched networks, where all types of streams and information are transported by packets and delivered to the mobile users, it comes into vision that providing circuit-switched services can no longer be justified since it is inadvisable to have dedicated infrastructures to implement voice services, when their integration can be made on a packet-switched network in a scalable and cost-effective manner. In addition, the design, development, deployment, and operation of mobile multimedia services, in general, and voice services, in particular, present several advantages when they are transported on a packet-switched networks, as the networks experience less unnecessarily reservation during service access and their resources are better utilized. In addition, many new services will require more interaction with the user, close synchronization, and better provision of the quality of service.

Recently, a variety of voice-oriented services, where voice is essentially coded into streams of IP packets, have been offered. The technologies used to deploy these services over communication networks are often referred to as *voice over IP* (or VoIP). A typical VoIP system is depicted in Figure 14.1 and works as follows: At the sender side, the voice sound is first sampled using a microphone. Then, the sampled

527

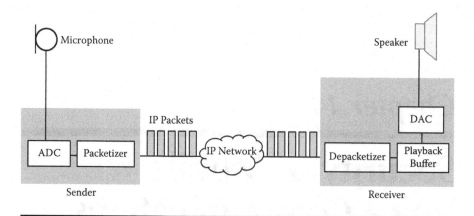

Figure 14.1 Voice over IP system.

traffic is translated into a digital representation by an analog-to-digital converter (ADC) to obtain a bit stream. The stream is packed into IP packets and sent over an IP network. At the receiver side the samples are first removed from the IP packets. Then, they are put in a playback buffer. This buffer is needed to compensate for the variation of the jitter generated between packets over the network. Finally, a digital-to-analog converter (DAC) converts the bit stream back into an analogue signal. To provide a two-way communication, VoIP replicates the system depicted in Figure 14.1 reversely between the receiver and the sender.

A typical VoIP implementation requires appropriate signaling protocols that are able to allow a subscriber to find other VoIP subscribers, provide for the registration of customers, and set up, modify, and break down a phone connection. To perform a call setup procedure to mobile phone, the location of the communicating VoIP phones has to be known. This means that the IP addresses and port numbers of the mobile terminals have to be known. This can be provided by a database maintaining all registered users along with their IP addresses and port numbers. When a user moves to a new location, only the database needs to be updated with the new IP address so that the user can be reachable on the new location. The servers that implement the location databases are globally locatable and play an intermediate role in the call setup. On the other hand, a VoIP solution should include also an appropriate transport protocol to carry real-time data and give feedback to other transport session users on the QoS parameters of the connection they are using.

Two signaling protocols are widely used in the VoIP solutions. They are the SIP (Session Initiation Protocol) protocol, which is recommended by the Internet Engineering Task Force (Handley, 1999), and the H.323, a protocol recommended by the International Telecommunications Union (ITU, 1998). Today, the two protocols are not compatible. The SIP is simple, scalable, and extensible. It requires four packets to establish a call, whereas the H323 requires longer setup time, needs 12 packets for the call setup, and provides a control within a session providing a way for conferences to coordinate input to the produced media. Two major advantages of SIP can be noticed with respect to H.323. First, the little number of packets needed to process a call setup and the fact that SIP runs on UDP, while H.323 uses both TCP and UDP during the call setup. The TCP is a reliable way to send data, because the data is acknowledged, ordered, and resent if not correctly received. UDP is just used to send data packets with a minimum of protocol overhead, no matter whether the packets arrive or not. When using TCP, a packet should be resent if it is not received correctly. This retransmission consumes much time and increases delay unbelievably, because large overhead is generated. The real-time protocol (RTP) is provided for this purpose (Schulzrinne, 2003). Various VoIP applications have been made available based on H.323 and RTP (e.g., Microsoft NetMeeting, Skype, and MediaRing Talk).

Mobile VoIP can be defined as the set of technologies, communication protocols, and voice-oriented services allowing voice to be transmitted over IP networks and services to be accessed by mobile stations. Generally, mobile VoIP assumes that the mobile stations involved in the service delivery have limited resources, implement an operating system capable of running add-on applications, and allow ad hoc protocols to transport real-time data. Mobile VoIP offers several advantages compared to the traditional public switched telephone Network (PSTN) and the cellular networks that render it a promising technology for businesses and consumers. In particular, the use of mobile VoIP allows the transmission of all types of streams, real-time data, and non-real-time data over the same network, enabling the emergence of a wide range of attractive services including voice-enabled electronic commerce and interacting multimedia services. Nevertheless, three major problems have to be addressed:

- the limited bandwidth and the constraints on other network resources present in the wireless environment may reduce the performance of mobile VoIP;
- the addition of security mechanisms to mobile VoIP systems in order to provide secure communications puts more limitations on the performance of the mobile devices and can even reduce the system flexibility of mobile VoIP; and
- the management of IP addresses needs to change the IP address of a mobile subscriber as he moves to a new area. Currently, such situation forces the termination of the VoIP call made by the moving user, since the current VoIP solutions require the hosts involved in a call to have fixed IP addresses.

The mobile network infrastructure providing voice communication over IP should enable mobile entities to roam within and between wireless access networks, while maintaining an ongoing phone call and providing little impact on the voice call quality. A typical architecture of such mobile networks is depicted in Figure 14.2, where the caller and the called entities are assumed to own an access account to the Internet via their Internet service providers (ISP). It is assumed that

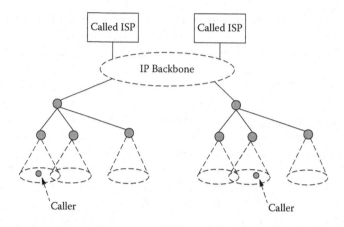

Figure 14.2 Typical architecture for mobile VoIP.

the caller's and callee's ISPs are able to provide other mobile IP services, in addition to the authentication, authorization, and accounting (AAA) functions. Therefore, to enable mobile IP telephony, the caller may use multiple service providers and identities. In the sequel, we briefly describe the properties of the architecture presented in Figure 14.2.

The caller connects to the Internet via an access network through an access node (AP, for wireless LAN) or a base station (BS, in a cellular network) in his surrounding area. But, because the caller may move out of range of that access node, he may need to perform multiple handovers between access nodes (ANs) during the duration of his ongoing call and, therefore, may interrupt his Internet access. In addition, before the mobile caller is granted any access to the VoIP service, he should perform a complete authentication and authorization procedure. From the user's perspective, the exchanges should be the same, whether the mobile user is connecting to an access node of his own provider or to an AN operated by a foreign provider, while the mobile is roaming.

For the caller, the procedure to acquire remote access, while he is leaving the coverage of first access node, denoted by AN_1, can be divided into three major steps:

- *Communication step*: The caller will need to scan for a new access node, say AN_2, and execute the appropriate handover.
- *AAA step*: The caller will authenticate to his ISP's AAA server via the visited network. The home AAA server, then, informs the visited network that the caller is authorized to be granted network access to VoIP service. A specific protocol is used to negotiate an authentication scheme between the caller and his AAA server and deliver these authentication messages.
- *Securing the wireless link step*: The caller and his AAA server perform mutual authentication and establish the needed keys to protect the voice data traffic

between them. To enforce access control, the communication over the wireless link may need to be authenticated and encrypted.

If the mobile caller is allowed to connect to the visited access network, the visited network will start to forward packets to and receive packets from the caller. The mobile caller can then obtain an IP address associated with the new IP subnetwork and access the Internet. Upon obtaining an IP address, the data traffic of ongoing sessions is redirected to the new location. To avoid losing an ongoing voice call, two major protocols can be used: (a) the SIP mobility, which provides mobility at the application layer (Wedlund, 1999) or (b) the mobile IP (MIP), which provides mobility at the IP layer (Johnson, 2004). Both protocols support route optimization in order to reduce delays for the voice transmission.

There are various challenging security concerns in mobile VoIP. The most obvious threat is impersonation, where a malicious user could masquerade as the mobile subscriber and send update messages that would result in the current call, made by the subscriber, being redirected to an unintended location. An alternative attack allows the attacker to intercept update messages, modify them, and resend them to redirect the call to an illegitimate location. The attacker could also eavesdrop on the update messages sent to a non-mobile host from the mobile host, and use these to track the movement of given mobile user. This would compromise the mobile user privacy.

The scenario that is typically used to build a VoIP call integrates a signaling plane, a media transport plane, and various telephony components. The signaling plane is used for transporting the necessary signaling information between the entities involved in the call. The media transport plane is used to carry the voice data between the components. For a security assessment, two related facts have to be considered:

■ The network infrastructure is not controlled by a single authority or a set of trusted providers. Signaling and media plane might be performed by untrusted network parts, components, and operators. Security, in this case, is hard to address.

■ The IP network, which is used by the signaling and media planes, is used by other services and both end system and infrastructure components are able to fulfill many other tasks, in parallel. Therefore, the resulting security problem domain is considerably larger compared to a standard PSTN based telephony system.

In this chapter we describe the mechanisms developed to handle voice over IP in mobile communications and secure its services. A particular attention will be paid to the mechanisms handling the loss and delay patterns related to terminal mobility, as well as attacks targeting the VoIP service and specific security mechanisms to protect against. Currently, it appears that a large number of light-weight implementations of mobile VoIP are emerging (e.g., Skype proprietary solutions). In addition, virtual VoIP operators, using SIP-based services through special handsets (e.g., Nokia E-Series handsets), are appearing on the market. Their emergence is quicker, as they do not need heavy infrastructure.

14.2 Basics on VoIP

This section provides an introduction to the main features of the SIP, H.323, and Real-Time Protocol (RTP), since they are widely used in VoIP technology. SIP is the most commonly used protocol to create and manage the VoIP media sessions, while RTP is the transport protocol in charge of the transmission of the data in a VoIP session.

14.2.1 Basic Signaling Protocols

VoIP signaling is the communication function useful for the different tasks including finding other VoIP phones on the IP network and controlling the procedure making a phone call. Both SIP and H.323 signaling protocols provide mechanisms for call establishment and teardown, call control, and other services. We describe in the following these two protocols and discuss their main functionalities.

14.2.1.1 SIP Overview

SIP is an application layer control protocol that can establish, modify, and terminate calls. Two main architectural elements can be distinguished in SIP: the user agent (UA) and the network server. The UA is implemented at the SIP end stations. The SIP user agent has two basic functions: receiving incoming SIP messages and sending SIP messages upon user actions or incoming messages. It contains a user agent client (UAC) that is responsible for issuing SIP requests and a user agent server (UAS) that is in charge of responding to SIP requests. The SIP UA also starts appropriate applications according to the session that has been established.

Three network server types can be found in SIP: the proxy server, the redirect server, and the registrar. However, a basic SIP call does not use servers. The SIP proxy server relays SIP messages, so that it is possible to use a domain name to find a user, rather than knowing the IP address or the name of the host. In that way, a SIP proxy can be used to hide the location of the user. A redirect server returns the location of the host; unlike the SIP proxy, it only has to send back a response with the correct location instead of participating in the whole transaction. Both the proxy server and redirect accept registrations from users, in which the current location of the user is given. The location can be stored either locally at the SIP server or in a dedicated location server. Another important function regarding SIP is the registration of the users with their provider's servers. When a SIP-based device (or UA) gets online, it first must get registered with a SIP Registration Server, called the *Registrar*. This process is handled by sending a REGISTER message. Registrations are not permanent; they bind the mobile user's ID with an IP address where the user can be reached.

The typical SIP operation involves a SIP UAC issuing a request, a SIP proxy server acting as end-user location discovery agent, or a SIP UAS accepting the call. A successful SIP invitation consists of two requests: the INVITE and the ACK requests. The INVITE message contains session specification that informs the called entity what type of media the caller can accept and where he wishes the media data to be sent. SIP addresses are referred to as SIP Uniform Resource Locators (SIP-URLs) and the SIP message format is based on the Hyper Text Transport Protocol (HTTP) message format, which uses a human-readable, text-based encoding.

Redirect servers process an INVITE message by sending back the SIP-URL, where the callee can be reached. The proxy servers implement an application layer routing for the SIP requests and responses. A stateful proxy holds information about the call during the entire time the call is active, while a stateless proxy processes a message without keeping the information it carries. The Registrar server is used to store the SIP address along with the associated IP address, so that when an INVITE request arrives for the SIP URL used in the REGISTER message, the proxy (or the redirect server) forwards the request appropriately. The INVITE message contains a session description expressed in SDP (Session Description Protocol), which is a format for describing multimedia session parameters for the purpose of session announcement. The message is received by a redirect server, which consults a location server to find out where to redirect the invitation. The function of the location server is not specified, but can be any object that can return a next hop address in the chain of finding the callee.

An example of SIP setup is depicted by Figure 14.3, where the caller calls another user (referred to as the answerer or callee) using his SIP identity, a type of *Uniform Resource Identifier* (URI), called SIP URI, which is similar to an email address and contains the user name and the host identifier (for example, caller-ID@host-ID). The caller sends a request called INVITE to the answerer provider SIP server proxy (SPA) via his provider's SIP server. If the answerer accepts the call, the media session is established.

14.2.1.2 H.323 Overview

The H.323 Standard defines the mechanism by which real-time information can be transmitted over packet-based networks that do not provide a guaranteed quality of service. It defines four major components for a network-based communication system: the Terminals, the Gateways, the Gatekeepers, and the Multipoint Control Units. The terminals are client endpoints attached to IP-based networks that provide real-time, two-way communications with other H.323 entities. The H.323 terminals implement the following functions:

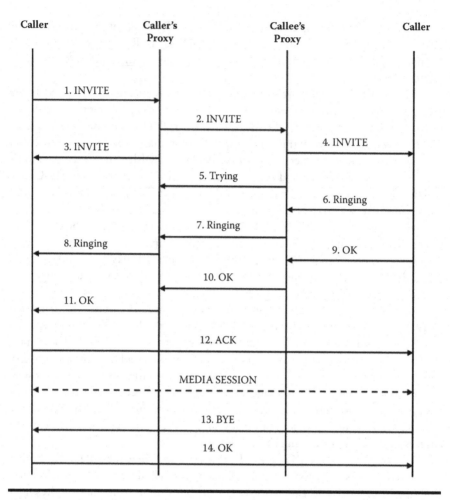

Figure 14.3 SIP setup.

- *Signaling and Control*: The terminals implement a standard for channel usage and capabilities, in addition to a protocol for call signaling, call establishment, and registration/administration/status (RAS) for communication with gatekeepers.
- *Real-time communication*: H.323 terminals implement RTP as a protocol for sequencing audio and video packets.
- *Codecs*: H323 implements pieces of software, referred to as Codecs, to compress audio/video before transmission. The decompression is operated to get the compressed packets back immediately after their reception.

The gateways provide the inter-connection between the packet-switched network and the switched circuit network (SCN). The gateway is not required when there is no connection to other networks. The gateway performs call setup and control on both the packet-switched network and the SCN. It has the responsibility of translating transmission formats and communication procedures. The gatekeepers perform at least four compulsory functions: the address translation (typical translations transform the phone numbers into the transport addresses), admission control, bandwidth control, and zone management. Gatekeepers can also support four optional functions: call control signaling, call authorization, call management, and bandwidth management.

Finally, the multipoint control units (MCU) support conferencing using a set of endpoints. Typically, a MCU consists of a multipoint controller (MC) and zero or more multipoint processors (MP). The MC provides control functions such as the negotiation between terminals and the determination of common capabilities for processing audio and video. The MP performs the necessary processing on the media streams for a conference involving audio mixing and audio/video switching.

H.323 supports five types of information streams between endpoints: Audio, Video, Data, communications control data, and call control data. Audio and Video streams are processed using audio and video *codecs*; they are transmitted and controlled using the Real Time Transport Protocol and the Real Time Control Protocol (RTCP) operating over an unreliable transport such as UDP. H.323 uses the concept of channel to structure the information exchange between communicating entities. A channel is a transport-layer connection that can be unidirectional or bi-directional. In particular, the following channels are distinguished in H.323:

- *RAS channel*: This channel provides a mechanism allowing an endpoint to communicate with its gatekeeper. Through the RAS channel, an endpoint registers with the gatekeeper and requests permission to place a call to another endpoint. If the permission is granted, the gatekeeper returns the transport address for the call signaling channel of the called endpoint.
- *H.245 control channel*: This channel carries the H.245 protocol messages for media control with capability exchange support. After the call participants exchange their capabilities, logical channels for media are opened using the H.245 control channel.
- *Call signaling channel*: This channel transports information for call control and supplementary service control. When the call is established, the transport address for the H.245 control channel is indicated on this channel.
- *Logical channel for media*: These channels carry the audio, video, and other media information. Each media type is carried in a separate pair of unidirectional channels, one for each direction, using RTP, for example.

H.323 specifies that the registration/administration/status channel and the logical channels for media transport are carried over an unreliable transport protocol, such as UDP. However, the H.245 control channel is specified to be carried over a reliable transport protocol, such as TCP.

14.2.2 Comparing the Basic Signaling Protocols

Let us address in this subsection a comparison between H.323 and SIP in terms of functionality, quality of service, and security features (Dalgic, 1999; Glassman, 2003). For the sake of clarity, we consider similar scenarios for both protocols involving, in particular, the SIP User Agent is equivalent to a H.323 terminal (or the packet-network side of a gateway) and the SIP network server is equivalent, from the operational point of view, to a H.323 gatekeeper.

14.2.2.1 Comparing the Functionality

In addition to the basic call service management, both SIP and H.323 support several call control services, advanced features, and capability exchange. The services they provide are similar, but they are performed with different approaches. In the following, we will discuss some of the signaling procedures for the provided services and compare how they are handled by the two protocols.

14.2.2.1.1 Call Setup and Tear Down

H.323 v2 call setup is based on RTP. The call setup needs a two-phase connection: the TCP connection and call connection. The H.323 v3 supports both TCP and UDP, which simplifies the call setup procedure. SIP call setup procedure is similar to H.323 v3. The tear down procedure is a reverse operation of the call setup. Either caller or callee entity can terminate a call by RELEASE COMPLETE (in H.323) or BYE (in SIP) messages.

14.2.2.1.2 Call Control Services

SIP and H.323 support call hold, call transfer, call forwarding, and call waiting. Call hold is defined as one call party disconnecting the voice communication without terminating the call, with the ability to reestablish the voice communication at a later time. H.323 defines two scenarios in the call hold service:

1. *Near-end call hold*: The hold is invoked at the holding endpoint as a local procedure.
2. *Remote-end call hold*: The holding endpoint sends a hold request to the remote endpoint requiring the held endpoint to provide Music on Hold (MOH) to the held user.

On the other hand, SIP uses a simpler approach to achieve the same call hold functionality. For a near-end call hold, no protocol assistance is provided. The client just continually receives the media stream from a server, but it does not generate any response. To achieve remote-end call hold, the holding side needs to send an

INVITE message to the other side, indicating a NULL set of receiving capability for any kind of media.

Call transfer enables a user to transfer an established call to a third party. Both H.323 and SIP support three types of call transfer: the blind transfer, the alternative transfer, and the operator-assisted transfer. Blind transfer works by applying the following steps:

1. Originator A connects with B,
2. Originator A asks B to connect with C,
3. A simply disconnects with B without any acknowledgment of connection between B and C.

The operator-assisted transfer works as follows:

1. First, the originator B sets up a connection with the operator A;
2. Originator A puts B on HOLD and sets up another connection with C. B and C set up the connection between them;
3. Originator A releases the connection with B and releases the connection with C.

The procedure of operator-assisted call transfer in SIP is very similar to the one in H.323, except that the equivalent SIP messages are sent out.

14.2.2.1.3 Call Forwarding

Call forwarding permits the called party to forward particular pre-selected calls to other addresses. Call forwarding services provided by SIP are usually instantiated with the LOCATION header fields, which contain the forwarding destination. SIP supports call forwarding busy, call forwarding no response, and selective call forwarding.

14.2.2.1.4 Call Waiting

Call waiting allows the called party to receive a notification that a new party is trying to reach it while it is busy communicating to another party. Consider that party C calls the party B while it is in another call with A, the call waiting procedure provided by H.323 works as follows:

- *Actions at the endpoint B*: On receipt of the request, B returns an ALERTING message to C and optionally starts a timer, and locally provides a call indication to the user. If the served user B likes to accept the waiting call, B stops the timer, and sends a CONNECT message to the calling point.
- *Action at the endpoint C*: On receipt of an ALERTING message, the calling endpoint may indicate call waiting to the calling user. Then the calling

user may wait until the waiting call gets accepted. He can release the call or choose other supplementary services.

SIP can provide call waiting service using the Call-Disposition header field, which allows the UAC to indicate how the server is to handle the call. The following is an example of call waiting service provided by SIP. The called party B is temporarily unreachable (e.g., it is in another call). The caller indicates that it wants to have its call queued rather than rejected immediately via a "Call-Disposition Queue" header field. If the call is queued, the server returns "Queued." When the callee becomes available, it will return the appropriate status response. A pending call can be terminated by a SIP BYE request.

14.2.2.2 Quality of Service

The relevant QoS parameters for VoIP flows are the bandwidth, latency, the delay jitter, the packet loss, and the call setup delay (Kuhn, 2005). Jitter refers to nonuniform packet delays. It is often caused by low bandwidth situations in VoIP. Jitter can cause packets to arrive and be processed out of sequence. Jitter can also be controlled throughout the VoIP network by using routers, firewalls, and other network elements that support QoS. Packet loss can result from excess latency, where a group of packets arrives late and must be discarded in favor of newer ones. It can also be the result of jitter, that is, when a packet arrives after its surrounding packets have been flushed from the buffer, making the received packet useless. The real-time constraints imposed to VoIP do not allow for a reliable protocol, such as TCP, to be utilized to mitigate packet losses. By the time a VoIP packet is reported missing, retransmitted, and received, the time constraints for QoS would be exceeded.

Signaling protocols should provide support for the communication of the required QoS parameters with the goal of satisfying the required QoS values, or rely on cooperating protocols. In the following, we consider the QoS support offered by the signaling protocols for the multimedia flows. Then, we examine the case of the call setup delay.

14.2.2.2.1 QoS Support for VoIP Flows

The gatekeepers, in H.323, provide a wide range of control and management functions, including address translation, admission control, bandwidth control, and zone management. Some optional functions include call control signaling, call authorization, bandwidth management, and call management. By contrast, the SIP does not supply the management or control functions by itself. Instead, it relies on other protocols.

Admission control checks whether the network has sufficient resources to support the QoS required for a call and accepts or rejects the call according to the availability of resources. To this end, the protocol must handle bandwidth management, call management, and bandwidth control. These activities are supported by H.323 but not by SIP. However, neither H.323 nor SIP supports resource reservation. Instead, they both rely on external means for resource reservation, such as the Resource Reservation Protocol (RSVP). H.323 v3 can offer some differentiated services based on QoS parameter negotiation (e.g., bit rate, delay and jitter). Upon initiation of a call, a terminal may request one of three service classes defined: "Guaranteed Service," "Controlled Service," and "Unspecified Service." SIP does not support a similar functionality.

14.2.2.2.2 Call Setup Delay

The call setup delay can be defined as the number of round trips needed to establish a call between the caller and the called entities. Call setup delay is very large in H.323 v1. This delay has been reduced significantly with the fast call setup procedure in H.323 v2. Besides, SIP and H.323 v3 provide significant short delays.

The fast call setup method is an option that reduces the delay involved in establishing a call to a number smaller than three roundtrips by including logical channel information in the SETUP and CONNECT messages.

14.2.3 *Voice Transport*

The RTP is an IETF standard specified in RFC 1889 (Schulzrinne, 1996). RTP is a transport protocol for real-time applications that provides end-to-end network functions and services suitable for transmitting real-time data, such as audio, video, or simulation data, over unicast or multicast network services. RTP runs on top of a non-reliable transport protocol, such as UDP, to make use of the underlying multiplexing and checksum services. RTP also provides a control protocol called *RTP Control Protocol* (RTCP) that is utilized to monitor data delivery and provide minimal control and identification functions.

RTP provides for the real-time data delivery specific services including sequence numbering, payload type identification (a type can be an audio sample or compressed video data, for example), timestamping, and monitoring the delivery. Security services for RTP and RTCP are provided by additional solutions, including IPsec encapsulation and Virtual Private Networks. The RTP standard also presents some mechanisms to provide this security. However, an alternative solution to secure is the RTP profile *Secure Real-Time Protocol* (SRTP). RTP security issues and solutions to enforce securing the RTP and RTCP traffic will be discussed in the sequel.

The RTP packet consists of a fixed header, a possibly empty list of contributing sources (source of a stream of RTP packets that has contributed to the combined stream produced by the RTP), and a payload. The payload contains the real-time application data. Detailed information about the payload types is given in the RTP

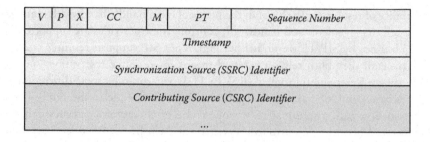

V	P	X	CC	M	PT	Sequence Number

Timestamp

Synchronization Source (SSRC) Identifier

Contributing Source (CSRC) Identifier

...

Figure 14.4 RTP header format.

standard (RFC 1889). The RTP header is depicted in Figure 14.4. The major fields are the following:

- *Version, V (2 bits)*: This field identifies the version of RTP. By default it is set to the value 2 for the RFC 1889 RTP specification.
- *Padding, P (1 bit)*: This field is set to 1 if padding has been applied to this packet.
- *Extension, X (1 bit)*: If this field is set to 1, the header is followed by exactly one extension field.
- *CSRC count, CC (4 bits)*: This field contains the number of CSRC identifiers that follow the RTP header.
- *Marker, M (1 bit)*: The interpretation of this field is defined by a RTP profile.
- *Payload Type, PT (7 bits)*: This field identifies the format of the RTP payload and determines its interpretation by the real-time application.
- *Sequence Number (16 bits)*: This field increments by one for each RTP packet sent. It may be used by the receiver to detect packet loss. The initial value of this field is random.
- *Timestamp (32 bits)*: This value reflects the sampling instant of the first octet in the RTP packet. As for the sequence number, the initial value is random.
- *SSRC (32 bits)*: This field identifies the synchronization source. The CSRC list (zero to fifteen items, each 32 bits in length) identifies all the contributing sources for the payload of the packet.

As noted above, the CC field in the fixed header contains the number of sources identified.

14.3 Security Issues in VoIP

Adding security constraints increases significantly the bandwidth usage, generating more latency and jitter, and thus reducing the overall QoS of the network. In addition, these requirements do not explicitly take into account the heterogeneous data flow over the network. Since voice and data streams are sharing the same limited bandwidth, significant amounts of data can generate a congestion in the network and prevent VoIP traffic from reaching its destination in conformance with time constraints imposed in the QoS it needs. However, the following potential risks have been identified for H323 and SIP signaling:

■ The signaling and voice data transport plane can be targeted by attacks that aim at breaking the integrity, confidentiality, authentication, or non-repudiation of the transported data.
■ The audio payload data and the signaling information, exchanged during a call, are sensitive to eavesdropping, jamming, and even active modification. The challenges become even more evident in an open environment where finding, choosing, and using services are subject to competition between service providers.
■ Compromising the identity of an end system or infrastructure component leads to additional risks even when using standard and non-compromised signaling mechanisms. If a malicious user can register with a H.323 Gatekeeper (or SIP server/registrar), he can potentially gain the identity of the victim. This can cause a potential invasion of privacy (since incoming calls are routed to the attacker and may give him information about the callers) and the possibility of misusing services.

14.3.1 Security Provided by H323

The security service provided by the H323 suite (as given by H235v2) is characterized by the following supports: the support of elliptic curve cryptography, the support for the Advanced Encryption System (AES) standard, and the use of several security profiles to help product interoperability. The Baseline Security Profile relies on symmetric techniques. Shared secrets are used to provide authentication and/or message integrity. Three supported scenarios are available for this profile. They are the endpoint-to-gatekeeper, the gatekeeper-to-gatekeeper, and the endpoint-to-endpoint.

The Signature Security Profile relies on asymmetric techniques. Certificates and digital signatures are used to provide authentication and message integrity. The signature security profile mandates the gatekeeper-routed model. Since this profile relies on a public key infrastructure rather than on pre-established shared secrets, it scales for larger global environments. In addition, to the Baseline Security Profile, it provides non-repudiation.

On the hand, one can notice the following features: (a) the baseline security profile is easy to implement but it is not really scalable for IP telephony due to the restricted key management; and (b) the signature security profile may have a critical impact on overall performance. This is due to the use of digital signatures for every message, requiring signature generation and verification on the sender's and the receiver's sides.

The voice encryption profile specifies the master key exchange during call signaling and the generation and distribution of media stream keys during call control. The following security mechanisms are described within the voice encryption security profile: the encryption of RTP packets with an variety of algorithms and modes that can be used; the key management with key and security capability exchange; and the key update mechanism and synchronization. However, the encryption and key management for RTCP and the authentication and integrity for RTP and RTCP are not covered by this profile.

Finally, the hybrid security profile relies on asymmetric and symmetric techniques. It is a combination of the baseline and the signature security profiles. Certificates and digital signatures are used within this profile to provide authentication and message integrity for the first handshake between two entities. During this handshake a shared secret is established and will be used, in the sequel, in the same way described for the Baseline Security Profile. The hybrid security profile mandates the gatekeeper-routed model. This profile provides high security without relying on pre-established shared secrets.

Version 3 of H.235 replaces H.235 version 2 by improving error reporting and incorporating of key management supporting the Secure Real-time Transport Protocol (SRTP). SRTP provides confidentiality, message authentication, and replay protection to the RTP/RTCP traffic. The RTP standard provides the flexibility to adapt to application specific requirements with the possibility to define profiles. However, SRTP does not define the key management protocol by itself. It rather uses a set of negotiated parameters from which session the keys for encryption, authentication, and integrity protection are derived.

Multimedia Internet Keying (MIKEY) describes a key management scheme that addresses real-time multimedia scenarios (e.g., SIP calls, RTSP sessions, streaming, multicast, etc.). The focus lies on the setup of a security association to secure multimedia sessions including key management and update, security policy data (Arkko, 2004). MIKEY defines three options for the user authentication and negotiation of the master keys; all of them are using the 2-way-handshake paradigm. They are the symmetric key distribution (pre-shared keys, MAC for integrity protection), the asymmetric key distribution, and the Diffie-Hellman key agreement protected by digital signatures.

14.3.2 Security Provided by SIP

Various security services have been made available for SIP including a digest authentication scheme that is based on a simple challenge-response paradigm. Using the RTP encryption, it provides confidentiality for media data. RFC 3261 commands the use of TLS for proxy servers, redirect servers, and registrars to protect SIP signaling (see Chapter 3 for details on TLS). In fact, TLS is able to protect the SIP signaling messages against loss of integrity, attacks against confidentiality, and replay attacks. It provides integrated key-management, mutual authentication, and secure key distribution. TLS is applicable on a hop-by-hop fashion between UAs and proxies or between proxies. However, a disadvantage of the TLS use in SIP scenarios can be noticed: TLS requires a reliable transport stack and cannot be applied to UDP-based SIP signaling.

IPsec may be utilized to secure, at the network layer, the SIP signaling. This security service is well suited to protect communications between SIP hosts in a SIP VPN scenario (e.g., between user agents and proxies) or between administrative SIP domains. IPsec applies to various transport protocols such as UDP, TCP, and SCTP-based SIP signaling. It may be used to provide authentication, integrity, and confidentiality of the transmitted data. It also supports end-to-end as well as hop-by-hop scenarios. The IKE protocol provides an automated cryptographic key exchange and management for IPsec. It is used to negotiate security associations for its own key management exchanges and for other IPsec-based services.

On the one hand, it is worth noting that several Internet standards have been developed to provide security enhancements for the basic SIP scenarios. Among the most important standards, one can mention (a) the SIP Authenticated Identity Body, which defines a generic SIP authentication token. The token is provided by adding an S/MIME body to a SIP request or response in order to provide reference integrity over its headers; (b) the SIP Authenticated Identity Management, which permits an administrative domain to securely verify the identity of the originator of a request; and (c) several Security Mechanism Agreements for SIP, such as HTTP authentication.

14.3.3 Key Management

The key exchange is a fundamental security mechanism for VoIP. It is essential to specify what security guarantees it provides, because a difference between the expectations of the transport-layer protocol and the security properties, actually ensured by the key exchange protocol, can generate important vulnerabilities.

14.3.3.1 SDES

SDP's Security DEscriptions for Media Streams (SDES) is the key transport extension of the SDP protocol, which is a format for describing multimedia session parameters for the purpose of session announcement, session invitation. SDP is purely a format specification. It is independent of the transport layer and may be carried, for example, by SIP (Handley, 1998). SDES provides a way to signal and negotiate cryptographic key(s) and other session parameters for media streams in general, and for SRTP in particular (Andreasen, 2006). The only method supported for key exchange specifies that the key itself must be included in plaintext. In other words, the key is embedded directly in the SDP attachment of a SIP message. Therefore, the protection of the key depends exclusively on SIP and can be provided by TLS, if the transport layer is TCP, or by S/MIME. The use of TLS is critical because it does not provide end-to-end protection over a chain of proxies and assumes that the hops in a SIP proxy chain can be trusted. On the opposite, S/MIME provides end-to-end confidentiality and authentication for SDP payload encoded as MIME. In addition, S/MIME does *not* provide any replay protection, and the application must provide a separate defense against replay attacks. In general, most applications have limited replay protection because it requires state maintenance and/or loose clock synchronization.

In the following, we will discuss how an attacker can exploit the lack of replay protection in S/MIME protected SDES to break the security of an SRTP session.

14.3.3.2 ZRTP

ZRTP specifies an extension header for RTP to establish a session key for SRTP sessions using authenticated Diffie-Hellman key exchange (Zimmermann, 2006). The main advantage of ZRTP appears in the fact that it does not require prior shared secrets or the existence of a public-key infrastructure. Because Diffie-Hellman (DH) key exchange does not provide protection against man-in-the-middle attacks, ZRTP uses a *Short Authentication String* (SAS), which is essentially a cryptographic hash of two Diffie-Hellman values, for key confirmation. The communicating parties confirm the established key verbally over the phone, by looking at their respective phone displays and reading the displayed SAS values to each other. After that, the two parties may rely on key chaining: the shared Diffie-Hellman secrets cached from the previous sessions are used to authenticate the current session. To explain this, Figure 14.5 depicts a ZRTP key exchange between two users that we denote by User A and User B.

The HELLO message shown in Figure 14.5 contains SRTP configuration options and a unique value, called ID, which is generated at installation time. This value can be used by the recipient to retrieve the cached shared secrets. The HELLO and HELLOACK are optional messages. The sender of the COMMIT message (User B, in Figure 14.5) is called the *initiator*, and User A is the *responder*. In the COMMIT message, hash, cipher, and pkt describe the hash, encryption, and public key algorithms, respectively. They are selected by User B from the intersection of the lists of algorithms in the sent and received HELLO messages. User B generates a random exponent x and computes the value g^x mod p, where the generator g of the Diffie-Hellman group and the prime number p are determined by the algorithm pkt. The value hvi, called the hash commitment, is the hash of the Diffie-Hellman value generated by User B, concatenated with hash, cipher, *SAS*, and pkt from the HELLO message send by User A.

Upon receipt of the COMMIT message, the responder User A generates his own Diffie-Hellman secret and computes the corresponding public value. Each secret already shared between User A and User B has an ID, which is the HMAC of the string "Responder" computed using this secret as the key. User A and User B

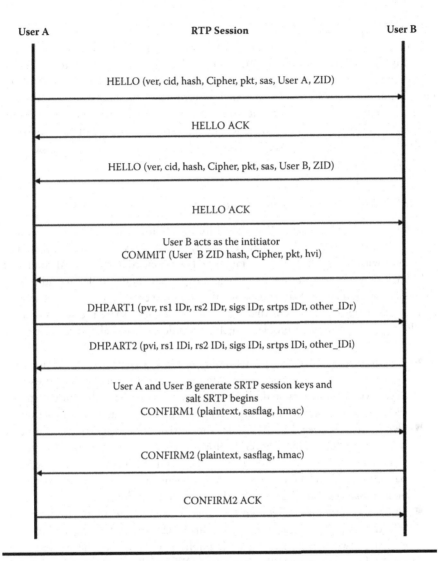

Figure 14.5 Establishment of SRTP session key using ZRTP.

retrieve their shared secrets rs1 and rs2 using their ID. The behavior of User B in response to User A's DHPART1 message is very similar.

Upon receipt of the DHPART2 message, User A checks that the User B's public DH value is not equal to 1 or p-1 (since the standards state that this check thwarts man-in-the-middle attacks). If the check succeeds, User A computes the hash value of the received value and checks whether it matches hvi received in the COMMIT message. If not, User A terminates the protocol. Otherwise, he stores the shared secret IDs received from the DHPART2 message as set A. User A then computes the set of shared secret IDs that he expects to receive from User B. For each secret, the ID is computed as HMAC of the string "Initiator," keyed with the secret itself. Let B be the set of the expected IDs. User A then computes the intersection of sets A and B. Secrets corresponding to the IDs in the intersection are stored as set D, sorted in the ascending order. The final session key is computed as the hash of the joint Diffie-Hellman secret concatenated with the set D of shared secrets. Finally, cached shared secrets rs1 and rs2 are updated as rs2:= rs1 and rs1:= HMAC (session key, "known plaintext") on both sides. The master key and the salt for the SRTP session are computed as HMAC of known plaintexts using the new session key (Zimmermann, 2006).

14.3.3.3 Multimedia Internet Keying

MIKEY is another key exchange protocol for SRTP. It can operate in three different modes: pre-shared key with key transport, public key with key transport, and public key with authenticated Diffie-Hellman (DH) key exchange. An advantage of MIKEY is that it allows the key to be negotiated as part of the SDP payload during the session setup phase in SIP. Thus, it requires no extra communication overhead. A clear disadvantage of MIKEY is that it requires either prior shared secrets or a separate PKI, with all attendant problems such as certificate dispersal and revocation.

- *Pre-Shared Key Transfer*. In this mode, the key is generated by the initiator and transferred to the responder. The message is integrity-protected using a keyed MAC and encrypted. The respective keys are derived from the shared secret and a random value using a cryptographically secure hash function.
- *Public Key Transfer*. In this mode, the initiator's message transfers one or more TGKs (i.e., values chosen uniformly at random by the initiator) to the responder using the responder's public key, and set of media session security parameters. Thus, the encryption and authentication keys are derived from an *envelope key* (Envk) chosen by the initiator at random.
- *Public Key with Diffie-Hellman Exchange*. Let *G* denote a large cyclic multiplicative group with generator g. In this mode, the initiator and responder exchange two messages using their public keys to derive the needed keys. In the first message, the initiator uses Diffie-Hellman scheme to start the set up of a secret key and random number is also included to provide security features such as the anti-replay and responder authentication. The second message is sent by the responder to complete key derivation and mutual authentication.

14.4 Mobility Issues

The major issues to support mobility are the handover management and the location. We discuss in this section an overview on SIP and Mobile IP handoff schemes before detailing a hybrid handoff algorithm. The latter scheme reduces the handoff delay for multimedia applications. It avoids the need for tunneling the packets throughout the handoff period. We also discuss some issues related to IP mobility support and location techniques.

14.4.1 Handover Issues

14.4.1.1 SIP Handoff

This is an application layer handoff. It offers a complete transparency regarding the protocols implemented at the lower layers. To show how the handoff takes place, let us assume that the mobile host is registered with some network (its home network), on which there is a SIP server, which receives registrations from the mobile host each time the mobile host changes its location. The mobile host does not need to have a statically allocated IP address on the home network. When the correspondent (static) node sends an INVITE to the mobile host, the SIP server has the information about the current mobile device's location and redirects the INVITE to that location.

When the mobile device moves during a communication session, it sends a new INVITE to the correspondent node using the same call identifier as in the original call setup. It inserts the new IP address (obtained from the DHCP server at the new location) in the Contact field of the SIP message. This informs the correspondent host where the INVITE issuer wants to receive the following SIP messages. To redirect the data traffic flow, it indicates the new address in the SDP field, where it specifies transport address. One can notice, therefore, that the SIP based handover is dependent on the rapidity with which the DHCP server processes the new IP address and the time taken for handling the re-INVITE message.

14.4.1.2 Mobile IP Handoff

Although a majority of the handovers that a mobile user will make during a session can be handled by link layer mobility mechanisms, it may be needed to support these handovers that occur across IP subnet borders. Two situations where a user may perform such a handover can be distinguished: First, the user's ISP may have segmented its wireless access networks into smaller subnetworks to limit the amount of multicast traffic, for example. When the mobile user roams between two access nodes attached to different subnetworks, he generally needs to acquire a new IP address. Second, when the mobile terminal is equipped with multiple interfaces, a handover may occur between the different interfaces. Generally, this task requires the use of a new IP address.

Mobile IP is a transparent solution that is able to process mobility at the network layer. This protocol was designed to solve the mobility problem by allowing the mobile node to use two IP addresses: a fixed home address and a care-of-address that changes at each new point of attachment. It is assumed that every mobile node has its home network and a statically allocated IP address on its home network. The support for mobility is provided by adding IP tunneling to IP routing. This solution is transparent because, when two nodes communicate with each other and one of them moves to a new subnet, then the other node is completely unaware of this mobility, and it continues its communication using the static address. The Mobile IP architecture mainly consists of four components: the mobile node (MN); the correspondent node (CN); the home agent (HA); and the foreign agent (FA). The CN participates in a communication with the MN and can be a fixed or a mobile node. The HA is a default router on the home network. It is responsible for storing the current locations of all the mobile nodes in its network. It intercepts the packets from CN and tunnels them to the current location of the mobile node.

The FA is the default router on the foreign network. It receives the tunneled packet from HA and forwards them to the MN in its network. Three major functions are performed by Mobile IP:

- *Mobile agent discovery*: When a mobile node is away from its home network, it needs to find agents to maintain access. The mobile agent discovery helps finding the closest agents based on specific criteria.
- *Registration with home agent*: The mobile node registers its care-of address with its home agent in order to guarantee the continuity of service delivery while moving.
- *Packet delivery using IP tunneling*: A tunneling function allows forwarding messages from the HA to the FA to finally deliver them to the mobile node.

The HA and the FA are responsible for broadcasting agent advertisements in their subnets. The roaming MN detects that it has moved to a new subnet when it hears such an advertisement from an agent that is not its HA. On hearing an advertisement, the HN sends a registration request to its HA through the FA with a care-of address (CoA). The HA then replies by accepting or denying the request. The CN will be completely unaware of this mobility and will keep transmitting its packets to the mobile node's home address. The HA will intercept these packets and performs the IP-in-IP encapsulation and tunnels them to the CoA of the FA. The FA then extracts the packets and forwards them to the mobile node.

14.4.1.3 Hybrid Handoff

Assume that the CN is involved in a communication with a MN established using SIP. If the MN moves to a new subnet, with the session being still active, it uses the Mobile IP scheme and receives the packets in the new subnet using a tunneling being achieved by its Home Mobility Agent (HMA). The tunneling is required only until the MN receives a new IP address. As soon as the MN gets the new IP address, it sends an SIP RE-INVITE message, with the same Call-ID, to the CN to inform it about the new IP address and the tunneling is stopped. Then, the CN sends its packets directly to the new location of the MN.

14.4.2 IP Mobility Support

When a mobile user moves to new subnet, several issues related to the IP layer should be addressed in addition to redirecting the data flows and avoiding breaking ongoing sessions. These issues include the following:

- *Availability of global IP addresses*: When IPv6 is used, the availability of global IPv6 addresses is practically unlimited. In IPv6 networks, one can assume that the MN is provided with a routable address on the network he visits. The limited availability of routable IPv4 addresses has lead to the use of NAT gateways and non-routable addresses in many access networks. A Mobile IPv4 user can either acquire an address of his own (a co-located CoA) on the visited subnetwork, or use the IP address of a FA. Providing the MN with a "routable" co-located CoA simplifies its ability to use optimal routing.
- *Optimal routing*: Some layer-3 mobility management protocols require that packets between two users, say A and B, are sent via their respective home networks. Although such an approach works well, we believe that optimal routing is critical to achieve good performance for VoIP communication between A and B. Unlike IPv6 networks, where different ways to achieve optimal routing to mobile hosts are provided, one can assume that the lack of mobility support in deployed end-systems and the use of ingress filtering routers limit User A to the use of SIP mobility to achieve optimal routing, in IPv4 networks.
- *Detecting layer-3 mobility*: Using information from the link layer to trigger layer 3 handover mechanisms is useful. The alternative would be to use dedicated layer 3 mechanisms to detect movement across IP subnets such as listening for periodic *Router Advertisements* from a FA or a common router. However, such frequent advertisements may burden the capacity of the link. Thus, one can assume that it is worth it for mobile users with ongoing real-time sessions to actively probe for routers after a WLAN handover; one can also assume that the mobile stations can utilize layer-2 triggers in their movement detection process.

When User A performs a handover between WLAN APs, he may not know whether he still resides on the same IP subnet or whether he has made a layer-3 handover. However, APs that logically reside on the same network are configured with the same *Extended Service Set Identifier* (ESSID), with IEEE 802.11, meaning that all APs serving the same LAN would announce the same ESSID. Therefore, the mobile would be able to notice the change of IP subnet by simply checking the changes in the ESSID announcements.

■ *Acquiring a CoA*: The CoA can be the address of a dedicated agent on the foreign network, i.e., the address of a FA. The alternative, for a MN, to acquire a CoA of his own (a co-located CoA) on the visited network using DHCP, for example. Using foreign agent CoAs allows multiple MNs to share the same CoA. This saves IP addresses compared to the co-located CoA approach handled by DHCP, where a pool of addresses on the foreign subnet must be reserved for visiting MNs. On the other hand, when using co-located CoAs, a MN will have more control on implementing its own routing policy and FAs may become bottlenecks as they have to process each packet to all served MN. Furthermore, for Mobile IPv6 and for SIP mobility the notion of FAs does not exist. Figure 14.6 describes the messages exchanged to acquire a CoA in the case of MIPv4.

■ *Duplicate address detection*: When co-located CoAs are used on multi-access networks, the mobile user or some other entity should verify, on its behalf, the uniqueness of the assigned address to avoid address conflicts. In fact, there is always a risk that some user takes an IP address without checking the uniqueness with the system administrators or without using for example DHCP. To cope with such situations, DHCP and IPv6 stateless autoconfiguration use a feature known as Duplicate Address Detection (DAD). To check whether an address is already in use, a node can transmit one or more *ICMP Echo Requests* and wait some reasonable amount of time to verify that no corresponding *ICMP Echo Reply* is received.

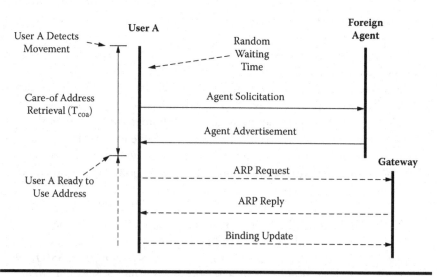

Figure 14.6 Acquire a CoA in IPv4 with FA.

14.5 The Security Threats to Mobile VoIP

A threat to VoIP can be a loss of availability of the VoIP service or a telephone fraud. Typically, four categories of important threats to VoIP can be distinguished. They are related to the abuse of access, denial of service, eavesdropping, and masquerading. In the following, we discuss some of the threats that are critical to VoIP provisioning and should be addressed by any VoIP service provider.

One can easily believe that the signaling and the media transport planes can be targeted by attacks targeting the integrity, confidentiality, authentication, or non-repudiation of the transported data. In addition, the signaling information exchanged between the components is sensitive to eavesdropping, jamming, and active modification. The challenges generated by these threats become even harder to face in an open environment, where finding, selecting, and running services are subject to competition between service providers.

The specific functions provided by the VoIP, such as the registration service as provided by a gatekeeper, can be targeted by different attacks. In addition, the components involved in the surrounding environment hosting specific VoIP functions (such as the management interfaces, used to configure the IP telephony) can be attacked. Moreover, functions that are not directly involved with the VoIP offer, but complement it, can be attacked. For instance, a VoIP enabled firewall might be weakened due to the fact that it supports IP communication and can temporarily open communication paths that could be used by attackers instead of regular voice conversation.

14.5.1 Basic Threats to Mobile VoIP

In this subsection, we discuss the major features of the four major categories of threats mentioned in the beginning of this section.

14.5.1.1 Abuse of Access

This is a category of threats where malicious mobile users (or programs acting on behalf of mobile users) misuse or abuse their access to the mobile VoIP systems. Abuse of access can have multiple forms. A first form of abuse of access is the so-called click-to-dial service, where an enterprise, setting up a specific service, calls back users on his demand, via the regular phone system. The service set up is typically offered through a Web page allowing any user to enter any phone number. An attacker can use this threat to cause some financial losses to the enterprise.

A second example of abuse of access can be provided by phone applications that present an API to other applications allowing them to initiate calls or insert objects in active calls. The inserted objects can be malicious. Examples of such applications include the Skype application. Generally, the presented API is protected by an access control mechanism asking the users whether they want to allow the phone application to control the API use. The Skype system, for example, authorizes another application to manage all the aspects of the Skype client.

14.5.1.2 Denial of Service

This is a category of threats aiming at conducting attacks to deliberately cause loss of availability of the mobile VoIP service. DoS threats against a mobile VoIP can be identified at several communication levels including the transport/network-level, server level, and signaling level threats.

- *The transport level DoS threats*: A DoS attack may be launched by flooding a target. An example is given by the ping of death or Smurf attack.
- *The server level DoS threats*: A server may be made unavailable by simply modifying some stored information in order to prevent authorized users from accessing the service, or by overwhelming the server with a large number of requests.
- *The signaling level DoS threats*: A DoS can make the SIP protocol unavailable to handle legitimate SIP messages by overloading the protocol server with too many messages. Unauthorized mobile users can also generate over-usage problems inducing the degradation of the QoS for the legitimate users. An example of an unwanted incident caused by a DoS attack is the disruption of network services by collapsing the entire signaling protocol.

Further DoS attacks that are specific to mobile VoIP include the following attacks that can be launched against SIP.

- A DoS attack on a call can be performed by sending spoofed bye messages to the user participating in a SIP call in order to close the call.
- The entities involved in the SIP signaling are vulnerable to DoS attacks. They can be simply flooded using Register or Invite messages.
- Illegal SIP messages can overload SIP, since SIP ensures that any entity receiving a SIP message must investigate the entire message before it can state its validity.

14.5.1.3 Eavesdropping Threats

This class of threats attempts to list signaling or data packets by copying legitimate messages between the communicating entities. Eavesdropping can be used against privacy, where an attacker can collect information in an unauthorized manner, obtain information about the origin and destination of a call, overhear a private conversation, or intercept personal information related to a mobile client's account, for example. The eavesdropping attacks in mobile VoIP constitute a real menace as there are packet-sniffers largely available that can be used for eavesdropping on VoIP traffic. In particular, this attack can be easily mounted in a WLAN.

Using the information obtained by eavesdropping on the signaling, an attacker can manipulate fields in the media stream and make fraudulent VoIP calls or inject their own data. This threat can be easily mounted in VoIP as SIP messages and media streams are always sent unencrypted, in practice, to allow interoperability or ease the execution of functions by the wireless network (even though they are encrypted).

14.5.1.4 Masquerading Threats

These threats allow an entity to pretend being another entity. Masquerading can lead to call charging fraud, violation of privacy, and breaking of integrity. A masquerading attack can be carried out by hijacking a link after the authentication process has been performed or by eavesdropping of authentication information and subsequently replaying it. An attacker can steal the identity of a legitimate user and obtain access by masquerading as the real user. He can gain unauthorized access to mobile VoIP services. By applying a replay attack, the attacker can capture the authentication credentials of an authorized user and replay the authentication message at a later time to obtain fraudulent access to a service.

The simplest form of masquerading in SIP is the reuse of username and password that can be obtained through interception. The authentication information can be obtained for the purpose of masquerade by reverse engineering of passwords, in the case of SIP digest authentication. To this end, the attacker may send several false challenges to the SIP user agent in the user's terminal to generate a list that can be used to break the cryptographic hash of the password as computed by hash functions, such as MD5. A Masquerading attack can then be combined with modification of data to obtain access to services to place an unauthorized call.

14.5.1.5 Obtaining Control of an End System

When an attacker obtains physical access to an IP telephone, he is able to reset it to its default configuration and can provide backdoors or weak initial passwords. Moreover, the remote management interface that the IP telephone allows it to use is vulnerable to attacks. In addition, since the administrator password is sent in plaintext, this makes the communication vulnerable to sniffing. Finally, the administration password can also be attacked with a series of automated brute-force trials, since it has a limited length and restricted alphabet as it has to be typed in via the telephones keypad. Once the initial access to the device is achieved, its administrative password for both local and remote WWW access can be set to a value known and becomes usable for further malicious operations.

The mobile device configuration is visible to an attacker and can be used for gaining additional information about the attached network and the mobile user. Information such as IP-phone address and IP address of the H.323 gatekeepers can be obtained and changed. Doing so would allow an attacker to change the phone capabilities in a way that it registers with a gatekeeper that is under the attacker's control. This gives the attacker the opportunity to read and even change the access to the VoIP signaling and to get information about communication relations, enabling QoS degradation and denial of communication services. In particular, the user can be unable to receive calls from or originate calls to specific phones. He can also access to the voice content for outgoing calls to participants, if they are routed via a H.323 to the PSTN gateway of the attacker.

14.5.1.6 Attacking User's Privacy

VoIP applications utilize RTP packets transmitted via the UDP protocol to carry audio data streams while using basic mechanisms for symmetrically encrypted audio payloads in RTP packets are described in an appropriate RTP profile (Schulzrinne, 1996). The eavesdropper is able to identify the data streams that form the audio connection(s) thanks to the public availability of VoIP protocol stacks and an in-detail description of protocol mechanisms, and despite the fact the ports used for these streams are typically negotiated in a dynamic manner.

14.5.2 Security Requirements for Mobile VoIP

In order for mobile VoIP to be widely adopted and generalized, a minimum set of security facilities must be provided. In particular, many experts argue that end-to-end authentication between the caller and the callee is not only possible, but it should also establish session keys, which can be used to protect the subsequent voice data stream. A minimal set of requirements to provide secure mobile VoIP call should contain the following rules (Vatn, 2005):

- *A call should only be established with the callee that the caller expects*: Securing the registration messages will defeat some of the simple redirection attacks. To ensure that the callee is really who the caller is communicating with, end-to-end authentication can be used.
- *The voice stream should be protected against eavesdropping*: The caller, say A, initiating a secure call to B should expect to be able to request privacy during call establishment and voice data transmission. The need for this service is probably greater for mobile VoIP than for VoIP and regular telephony, because the possibility of eavesdropping of an IP call is greater, in particular since many tools to do this are readily available. A solution to solve this can be based on session keys to encrypt and guarantee the integrity of the audio streams that are generated during a call.
- *Undesired calls should be blocked and VoIP spamming avoided*: Spamming and unwanted calls are mainly caused by the cost of VoIP, which will experience a similar situation as it is currently the case for email spams and unwanted mails.

An authentication handshake at call establishment can be used to reject a call automatically based on user preferences.

■ *Charging the call must be done correctly to the caller*: If charging calls is needed, its correctness is essential. Typically, however, one can assume that flat rate will be used for Internet calls (fixed monthly cost or free).

■ *The information about caller's identity and who the caller is calling should not be revealed by eavesdropping*: This requires that the communication system must protect against eavesdropping. This can be achieved by encrypting the call setup messages and using TLS transport. The requirement related to callee's identity is hard to meet since, even if the system is able to protect all signaling activity from revealing the identity, there may be many other ways for an attacker to collect this information, by observing other traffic that the user sends and receives, for example.

■ *An anonymous call service must be provided*: The caller may require that his identity should be hidden from the callee. The system should allow a caller to be anonymous. However, the callee should be able to reject such calls. In fact, introducing an initial authentication handshake does not exclude the possibility for the caller to remain anonymous.

The non-exhaustive list of requirements can be addressed by enabling the caller and the callee to authenticate each other during call establishment. Authentication can be seen as part of a *keying protocol* used to establish a security association between the caller and the callee and to negotiate what cipher suite to use. A solution based on IPsec and the secure real-time transport protocol (SRTP) to protect real-time audio data can be used (Baugher, 2004). To address the security concerns raised by the need for privacy of the callee, the signaling messages may be needed to be secured between the related entities, in a hop-by-hop fashion (using TLS tunnels, for example).

14.6 Attacks on the Key Exchange

14.6.1 Attack on SDES/SRTP

Let us suppose that two legitimate users, User A and User B, have previously set out a successful VoIP session, and that an attacker was able to passively eavesdrop this session, without learning the session key and not being able to decrypt the data streams. Suppose also that User B was the initiator in this session, and SDES was used to transport SRTP key material. To provide confidentiality for the SDES message, the session has been using S/MIME to encrypt the payload. To describe that attack depicted in Figure 14.7, let us consider that the use of S/MIME was, in general, preferred to TLS for protecting SDP messages for the following reasons: (a) S/MIME provides end-to-end integrity and confidentiality protection, and

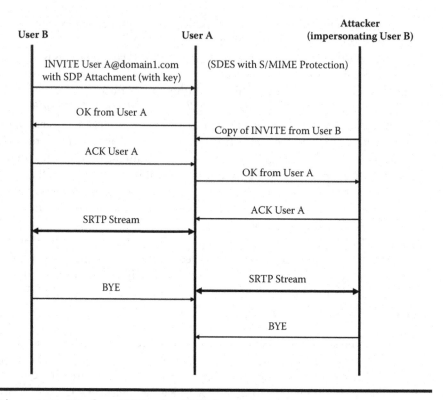

Figure 14.7 Attack on SRTP using SDES key exchange.

(b) S/MIME does not require the intermediate proxies to be trusted. However, S/MIME does not provide any anti-replay protection. Furthermore, let us notice that encryption in SRTP is simply the XOR of the data stream with the keystream.

After the original session has been closed, the attacker can replay User B's original INVITE message to User A, containing an S/MIME-encrypted SDP attachment with the SDES key transfer message. Figure 14.7 shows the sessions running concurrently; however, the attack need can suppose that one session is executed after the other. Since User A does not maintain any state for SDP, he will not be able to detect the replay. Using the old session's key material as his HMAC key, he will derive exactly the same master key and master salt as in the original session. Since the information is the same (including the sequence number), the generated session encryption key will be the same as in the previous session, and the keystream generated by applying AES to the (key, SSRC, SEQ) triple will be the same.

If User A now sends a packet in the new session that he thinks he is establishing with User B, the attacker can XOR the encrypted data stream with the data stream he eavesdropped in the original session. The keystream will cancel out, and the result will be the XOR of two data streams. Therefore, if the data streams contain enough redundancy or the attacker can guess parts of either stream, he will be able to completely or partially reconstruct the data of both streams.

In any case, encryption has been completely removed, and the attacker obtains a bitwise XOR of two payloads. Since VoIP packets are highly redundant, and the payloads at the initial packets are very predictable (i.e., whose digital encoding can be predicted), this should reveal a complete security breach. The attack is similar in spirit to the known attack on the protocol 802.11b WEP, which also allows the attacker to obtain an XOR of wireless packets.

The most important observation underlying this attack is that SRTP does not use any randomness on the responder side when the session keys are derived, even though the designers of SRTP were clearly aware of the danger of master key re-use. This highlights the need to use *automatic* key management mechanisms, since manual key management is more prone to result in key re-use. Among the automatic key management protocols compatible with SRTP are MIKEY, SDES, and ZRTP. Even though MIKEY was specifically designed as the key exchange protocol for SRTP, many VoIP implementations use SDES instead. While MIKEY contains built-in anti-replay protection and appears suitable for establishing SRTP keys, SDES is not suitable. Some SRTP implementations may take further measures to prevent key re-use and ensure freshness of the key material.

To prevent keystream re-use, SRTP responder should use its own fresh randomness as part of the key derivation process, e.g., as input to HMAC used in session key derivation. This randomness need not be secret. It can be publicly communicated to the sender as part of SRTP session setup to make certain that the sender derives the same set of session keys.

14.6.2 Attacks on ZRTP

14.6.2.1 Denial of Service

ZRTP is potentially vulnerable to denial of service attacks caused by attackers simply sending forged HELLO messages to end points. In response to each HELLO, a ZRTP endpoint creates a half-open connection, and keeps its parameters in memory. Sooner or later, the end-point will run out of storage or memory, and subsequent requests from legitimate clients will be refused.

14.6.2.2 Authentication

The main advantage of ZRTP is that it avoids the need for global trust associated with a public key infrastructure. ZRTP achieves this with the help of Short Authentication String (SAS), which is essentially a (keyed) cryptographic hash of Diffie-Hellman values along with other pre-shared secrets. After shared secrets have been used for authentication in one session, they are updated and kept by the participants for authentication in the next session. To authenticate the party on the other end of a VoIP session, the SAS value is read aloud over the voice connection. However, authentication based on SAS requires that some sort of display be available to the user. This is a severe problem for many secure VoIP devices. Therefore, we address the security of ZRTP in the situation where the user cannot explicitly verify SAS over the voice connection.

Authentication in ZRTP is based on the assumption that, in order to launch a successful man-in-the-middle attack on a pair of participants who already conducted several sessions, the attacker must be present on every session starting from the very first one. The approach works as follows: Each ZRTP user retains shared secrets rs1 and rs2 for users with whom he previously communicated. When initiating a new session, the user sends his ID, which is used by the recipient to retrieve the set of shared secrets associated with this ID. The session key is computed by hashing the joint Diffie-Hellman value concatenated with the shared secrets. Therefore, even if the Diffie-Hellman exchange is compromised, the attacker still cannot compute the session key because he does not know the shared secrets. Because the shared secrets are re-computed after each session, the attacker must be present in every session starting from the very first one, in which there was no shared secret.

Unfortunately, the fact that the IDs are not authenticated early in the protocol exchange makes the preceding reasoning incorrect. Consider a passive attacker who eavesdrops on a session between User A and User B and learns User B's ID. He then can set up a man-in-the-middle attack using the information exchanged in the previous session, as depicted in Figure 14.8. To do so, the attacker chooses random exponents x, y and computes $x = g^x \bmod p$ and $y = g^y \bmod p$, respectively. Then, he computes the hash of x concatenated with the set of algorithms chosen by

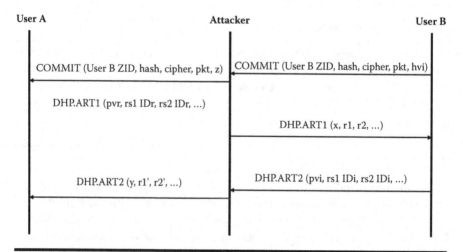

Figure 14.8 A man-in-the-middle attack on the ZRTP protocol.

User B for the ZRTP session. The attacker also replaces all shared-secret IDs with random numbers.

When User A receives the message from User B, he retrieves the set of secrets that he shares with User B and computes the set of expected IDs. Since the attacker has replaced all IDs with random numbers, they will not match. However, the protocol specification explicitly allows the set of shared secrets to be empty. In fact, the final shared secret, s0, can be calculated by hashing the concatenation of the Diffie-Hellman shared secret (DHSS) followed by the (possibly empty) set of shared secrets that are actually shared between the initiator and responder.

In addition, the specification does *not* require User A to stop the protocol, and instead it allows User A and the session key can be computed as the hash of the joint Diffie-Hellman value *alone* because User A believes that he doesn't have any shared secrets with User B anymore. Similarly, User B computes the same session key. The attacker knows both values. Therefore, he can compute SRTP master key and salt, and completely break SRTP encryption.

It is worth noting that, even if User A stops communicating with User B when the set of shared secrets is empty, this attack turns into a very effective denial of service, which allows the attacker to break off any session conducted between VoIP devices without displays. It is not clear whether the problem can be solved. In the absence of either the PKI, pre-shared secrets, or device support for out-of-band key confirmation by reading key hashes to each other, it is hard to see how the parties can carry out an authenticated key exchange. At the very least, the specification of ZRTP should not allow the key exchange to go forward when the set of shared secrets is empty.

14.6.3 Security of MIKEY

14.6.3.1 Secrecy

The goal of a cryptographically secure key exchange protocol is to establish a session key that is indistinguishable from a random bit string by anyone other than the participants Canetti (2001). It is easy to see that MIKEY does not satisfy this requirement when executed in the Diffie-Hellman mode. The shared key is derived as $g^{xi.xr} \bmod p$. where xi and xr state for the random value generated by the initiator and the responder, respectively. The joint Diffie-Hellman value is used directly as the key. In many Diffie-Hellman groups, e.g., in the group of squares modulo a large prime, testing group membership is not a computationally hard problem. Therefore, it is relatively easy to tell the difference between a random bit string and the key.

This does not necessarily lead to any exploitable weaknesses; however, it does prevent a rigorous proof of security from going through. Moreover, encryption schemes in which the derived key is intended to be used assume that the key be indistinguishable from a random value. A simple application of a deterministic hash function to the joint Diffie-Hellman value does *not* provably produce an output that is indistinguishable from random. By contrast, in protocols like TLS and IKE, the key is derived by hashing the Diffie-Hellman value together with some authenticated *random values* generated by one or both participants.

This use of randomness in key derivation, which is missing in MIKEY, is essential for the cryptographic proof of security to go through. To derive a key from the joint Diffie-Hellman value, MIKEY participants should be allowed to use an approach similar to the approach used in TLS and IKE, and use a universal hash function with public randomness generated by one or both participants.

Finally, one can witness also that using MIKEY with the pre-shared key mode does not satisfy perfect forward secrecy because the compromise of the pre-shared secret obviously leads to the compromise of all previous sessions.

14.6.3.2 Denial of Service

MIKEY offers very limited protection against denial of service attacks. In the public-key DH mode, the responder only performs modular exponentiation after verifying the message digest of the initiator's message. The attacker can still flood the responder with multiple copies of the same message, or he can send messages containing incorrect digest values. This will cause the responder to perform a large number of digest verifications.

14.7 Secure Real-Time Protocol

The Secure Real-Time Protocol (SRTP) provides message authentication, integrity, confidentiality, and anti-replay protection for RTP applications, by using a key management protocol that is able to provide the SRTP with a master key, which is used to derive the session keys needed to encrypt and authenticate the messages. SRTP provides a suitable protection by using transforms based on an additive stream cipher for encryption and a keyed-hash function for message authentication. It also provides an implicit index for sequencing and synchronizing.

MIKEY provides the SRTP with a master key, which is used to derive the session keys needed to encrypt and authenticate the messages. MIKEY would also take care of the cryptographic context initialization. This will be based on the RTP sequence number, for SRTP, and on the index number, for Secure RTCP. SRTP resides between the RTP application and the transport layer. For better clarity, we distinguish the SRTP sender and receiver sides. On the sender's side, SRTP intercepts an RTP packet, builds the corresponding SRTP packet, and sends it to the receiver. On the receiver's side SRTP intercepts the incoming packet (SRTP packet), extracts from it the RTP packet, and passes it up the stack.

14.7.1 SRTP Packet

The SRTP packet format is nearly the same as the RTP packet format. The SRTP packet header is identical to the RTP packet header, except that it adds two new optional fields: the *MKI* and the *authentication tag*. The new fields are placed at the end of the packet. The MKI field is used by the key-management protocol. It identifies the master key from which the session keys are derived. It may be used when re-keying for identifying a particular master-key within the cryptographic context. On the other hand, the authentication tag carries authentication data, if it is to be provided. The use of this field, however, could have an effect on the bandwidth consumption in cellular and wireless environments. It is computed at the sender site and verified upon the receipt, with the algorithm proposed in the cryptographic context. This feature provides authentication for the RTP header and payload, as well as indirectly providing replay-protection by authenticating the packet sequence number.

It is worth noting also that integrity protection is mandatory in SRTCP, since it must appear in the SRTCP packet. In addition, only the RTP payload will be encrypted along with possible padding, if needed. To provide RTP header confidentiality, end-to-end policies have to be considered.

SRTCP adds four new fields to the RTCP packet. These fields are the *SRTCP index*, an *encrypt flag*, the *authentication tag*, and the *MKI*. Since RTCP is a control protocol, the authentication of its messages must be ensured. The SRTCP index is a 31-bit counter explicitly included in the SRTCP packet. Its value has to be zero before the first packet is sent and increased by 1 modulo 231 after each sent packet. The E-flag indicates whether the current SRTCP packet is encrypted or not. The authentication tag is identical to the one present in the SRTP packet. Its length is variable (even though it is set by default to be 32 bits) and carries the message authentication data, as in the SRTP packet.

14.7.2 Message Authentication and Integrity

Message Authentication and Integrity are ensured by the computation and verification of the Authentication tag. For SRTP data, the sender computes the MAC of the Authenticated Portion (i.e., the RTP header and payload) concatenated with the roll-over counter (*ROC*) parameter and appends it to the packet. The receiver verifies this tag by performing a new Message Authentication and Integrity computation over the same parameters, and compares this to the one associated with the received packet. If both are equal, the message is valid, and if not, then the receiver must discard this packet, record the event, and an audit "AUTHENTICATION FAILURE" message is returned.

14.7.3 Key Derivation

Six different keys are needed to protect the RTP/RTCP session (i.e., SRTP and SRTCP encryption keys and salts, SRTP and SRTCP authentication keys). All six keys are derived from a single master key in a cryptographically secure way. Thus, the key administration protocol needs to exchange only one master key; SRTP then derives all the necessary session keys. SRTP will need, at least, one initial key derivation. But, the refreshment of these keys during the session and the associated master key lifetime need to be set up properly, since they are not considered.

14.7.4 Cryptographic Context

Clearly, each SRTP stream requires the sender and receiver to hold cryptographic state information, called cryptographic context, which include several common parameters such as the ROC, replay list, key derivation rate, and key lifetime. These parameters are independent of the encryption and authentication algorithm used; other parameters occur; they include the block size of ciphers, and the session keys are related to the specific security mechanism being used.

14.7.5 Packet Processing

The procedure used to create SRTP packets at the sender's side and extract from them the corresponding RTP packet at the receiver's side are performed as follows:

The sender behavior
1. Determine the cryptographic context to be used.
2. Derive the session keys from the master key from the key management protocol (MIKEY).
3. Encrypt the RTP payload.
4. If message authentication is required, compute the corresponding authentication tag and append it to the packet.
5. Send the SRTP packet to the socket.

The receiver behavior
1. Read the SRTP packet upon arrival.
2. Determine the cryptographic context to be used.
3. Determine the session keys from the master key from the key management protocol (MIKEY).
4. If message authentication and replay protection are provided, check for possible replay and check the authentication tag.
5. Decrypt the Encrypted Portion of the packet.
6. Remove the authentication tag from the packet and deliver the RTP packet.

14.7.6 Predefined Algorithms

A wide set of different algorithms for encrypting and authenticating the RTP messages are provided. Some of them are made as default mechanisms, also called *mandatory-to-implement* algorithms. In particular, the default cipher for encrypting the RTP payload is the AES. In addition, the NULL-cipher algorithm is provided, since it is mandatory-to-implement. On the other hand, the predefined algorithm to use is message authentication and integrity are HMAC-SHA1, which is based on a keyed-hash function and the *NULL Authenticator*.

14.8 Securing Mobile VoIP

For VoIP telephony to be commonly used one can consider that a large set of specific security services should be made available. Among the most useful security services, one can distinguish the end-to-end mutual authentication between the two entities involved in a VoIP session, session keys establishment during initial authentication handshake, and the protection of the subsequent voice stream between the parties involved in a VoIP session.

The major concern when supporting end-to-end security for a mobile VoIP call between two users is the placement of the security mechanisms. Three communication layers can be considered to address this concern. They are the network layer, the transport layer, and the session layer. A study of the current works shows the following: (a) to provide reliable data transfers, a large set of alternatives have used either IPsec (at the network layer) or TLS (transport/application layer); and (b) to provide real-time flow traffic the major alternative is represented by SRTP (at the transport/application layer). The security solutions offered by IPsec and SRTP offer the same services and can be compared based on the following three features:

14.8.1 Encapsulation Format

SRTP specifies the encapsulation format for the protected RTP packet and defines the fields in the RTP packet that are covered by the encryption and authentication algorithms. Two fields are added: the authentication tag and the master key index (MKI) fields. IPsec defines two IP headers (as described in Chapter 3): the *authentication header* and the *encapsulation security payload*. Only the ESP part of IPsec in *transport mode* is used. However, dissimilar to SRTP, the use of IPsec/ESP requires additional encapsulation.

14.8.2 Cryptographic Transforms

SRTP defines the protocols to use for encryption and sets the advanced encryption standard. AES enables the receiver to process the packets in random order, a feature that is desirable for real-time applications where packets are not delivered reliably and in right order. SRTP also defines protocols to use for packet authentication/integrity protection and sets the HMAC-SHA1 to be hash function by default. IPsec does not target any specific algorithm such as AES and HMAC-SHA1; however, it makes them available for ESP. It appears though that IPsec should be able to manage real-time voice traffic as efficiently as SRTP.

14.8.3 Session Key Generation Mechanism

SRTP requires a *master* key to be provided. Based on this master key SRTP can derive the session keys for its security transforms (e.g., encryption and authentication keys). Similarly, IPsec ESP relies on a keying mechanism to provide a security association for each direction in the communication between the caller and callee as well as performing user authentication, negotiation of security associations, and establishment of session keys. For this purpose the *IKE* has been provided.

Several countermeasures can be deployed in order to reduce the possibility for misuse of VoIP services by protecting the vulnerabilities that can be exploited. Basically, these countermeasures include authentication and authorization of user, integrity protection of signaling messages, and privacy protection of signaling and voice streams. More challenging, a means for providing denial of service protection needs to be identified. The eavesdropping threat applies to signaling data such as authentication information, information about the subscriber ID, or the phone number of the called party. More basically, the recommended countermeasure for protection of the VoIP various assets is the encryption of user communication and signaling.

To prevent an attacker from registering and using someone else's subscriber ID and authentication information in order to make free calls and malicious calls, sufficient protection mechanisms for the authentication information should be implemented. For example, in the 3G networks, the subscriber information and authentication keys are safely protected on the USIM card. On the other hand, the protection against loss of availability of the VoIP service is hard to provide. Some protection can be applied by introducing redundancy and service replication. As unauthorized use of resources can lead to loss of availability for VoIP users and DoS, some access control mechanisms should be implemented. Furthermore, techniques for incident prevention such as Intrusion Detection Systems can be used to detect incidents.

References

F. Andreasen, M. Baugher, and D. Wing, Session Description Protocol (SDP) Security Descriptions for Media Streams, *IETF RFC 4568*, July 2006.
J. Arkko, E. Carrara, F. Lindholm, M. Naslund, and K. Norrman, MIKEY: Multimedia Internet KEYing, August 2004, *RFC 3830*.

M. Baugher, D. McGrew, M. Naslund, E. Carrara, and K. Norrman, The Secure Real-time Transport Protocol (SRTP), March 2004, *RFC 3711*.

R. Canetti and H. Krawczyk, Analysis of Key-Exchange Protocols and Their Use for Building Secure Channels, in *Proc. Advances in Cryptology—EUROCRYPT 2001*, volume 2045 of *LNCS*, pp. 453–474, Springer, 2001.

I. Dalgic and H. Fang, Comparison of H.323 and SIP for IP Telephony Signaling, *Proc. of Photonics East SPIE'99*, Boston, Massachusetts, Sept. 1999.

M. Euchner, HMAC-Authenticated Diffie-Hellman for Multimedia Internet KEYing (MIKEY), *IETF RFC 4650*, September 2006.

J. Glassman, W. Kellerer, and H. Muller, Service Architectures in H.323 and SIP: A Comparison, *IEEE Communications Surveys*, fourth quarter 2003, Volume 5, No. 2, 2003.

M. Handley, H. Schulzrinne, E. Schooler, and J. Rosenberg, SIP: Session Initiation Protocol, *RFC 2543*, Mar. 1999.

M. Handley and V. Jacobson, SDP: Session Description Protocol, Request for Comments: 2327, available at http://www.ietf.org/rfc/rfc2327.txt.

ITU-T Recommendation H.323 (1998), Packet-Based Multimedia Communication Systems.

D. Johnson, C. Perkins, and J. Arkko, Mobility Support in IPv6, June 2004, *RFC 3775*.

R. Kuhn, T. J. Walsh, and S. Fries, Security Considerations for Voice Over IP Systems, NIST Special Publication 800-58, National Institute of Standards and Technology, 2005.

H. Schulzrinne, RTP Profile for Audio and Video Conferences with Minimal Control. *RFC 1890*, January 1996.

H. Schulzrinne, S. Casner, R. Frederick, and V. Jacobson, Rfc 3550: Real-Time Transport Protocol, July 2003, http://www.ietf.org/rfc/rfc3550.txt.

J.-O. Vatn, IP Telephony: Mobility and Security, Doctoral thesis, Department of Micro-electronics and Information Technology, KTH, Royal Institute of Technology Stockholm, Sweden, May 2005.

Elin Wedlund and H. Schulzrinne, Mobility Support Using SIP, in *2nd ACM/IEEE Int. Conf. on Wireless and Mobile Multimedia (WoWMoM)*, August 1999, Seattle, Washington.

P. Zimmermann, ZRTP: Extensions to RTP for Diffie-Hellman Key Agreement for SRTP, March 2006, available at: http://www1.tools.ietf.org/html/draft-zimmermann-avt-zrtp-01.

Chapter 15

Security of Multimedia Communications

15.1 Introduction

In the near past, all mobile services were simple communication-oriented network-based applications, such as circuit-switched voice calls or SMS messages. During the late 1990s, however, people started advertising content-oriented services and various multimedia solutions. In recent years, mobile multimedia communication services have grown rapidly due to the rich media content, broadband wireless networks, and flexible IP-based transport. Mobile multimedia can be defined as a set of protocols, standards, techniques, and mechanisms for multimedia information exchange over mobile networks. Therefore, a mobile multimedia system involves three actors, the owner of the multimedia data, the end-user of the multimedia data, and the mobile network that is in charge of delivering the multimedia data.

A mobile multimedia system manages and transmits multimedia data to provide the end-user with multimedia services, such as mobile information retrieval and mobile entertainment services. Multimedia information is composed and presented by more than one media type (including text, images, and sounds). It enriches the quality of the information and attempts a way to represent reality as sufficient as possible. Four concepts can be involved with mobile multimedia systems. They are mobility, multimedia communication, multimedia service provision, and security. Mobility involves three types:

- *End-user mobility*: The end-user (or customer) can move from one location to another while accessing a multimedia service. The operations performed during mobility, such as authentication, handover, and resource computation and reservation, should not affect the quality of media transmitted to the end-user, regardless of his positions.

- *Terminal mobility*: To receive and process multimedia data an end-user requires a device to fulfill these needs regardless of the user's location in a wireless environment. An end-user terminal can be a PDA, a notebook, a cell phone, or any other device that a mobile user utilizes to access remotely the multimedia service. The nature and technologies used on the end-user terminal should not affect the service provision.

- *Service mobility*: The multimedia service accessed by a mobile user may be provided by a mobile node. It can be used in different systems and move seamlessly among those systems.

On the other hand, multimedia communications can be classified into different (non-disjointed) categories based on four perspectives (Wu, 2006):

- *The content point of view*: Multimedia communications can be classified into *live content multimedia*, such as real time communication including voice over IP, video conferencing, and other methods, where multimedia content is generated on a real time manner; and *stored content multimedia* that is prepared ahead of time and stored in specific formats.

- *The delivery perspective*: Multimedia communications can be classified into *streaming multimedia*, where media content is played back while bit streams are being delivered to the receiver; and *download multimedia*, where media content is played back only after all the bits in the media streams have been delivered to the receiver.

- *The distribution point of view*: Multimedia communications can be classified as a *server/client multimedia* such as video on demand, where the media content is hosted-by or relayed through a central server; and *peer-to-peer* multimedia, such as end to end streaming, where entity is a content distributor and a content receiver with respect to other.

- From the interaction perspective, multimedia communications can be classified into *delay-tolerant* communications, for which the delay does not constitute a problem as long as the jitters are in a controlled range; and *interactive* communications, in which real time and fast response to user's interactions are a necessity.

To show the non-disjoint nature of the four classes, one can consider the VoIP application and show that it represents an example of a multimedia occurring in categories such as *live* content, *streaming* delivery, a *server-client* and *interactive* session categories.

The technical challenges by the mobile multimedia systems include several major issues: (a) to provide an error-robust efficient mobile media communication. This process would require error-strong media coding, error resistant transport, and efficient error suppression; (b) to adapt the mobile multimedia system to the bandwidth capacity changes and other terminal resources; (c) to achieve the miniaturization of mobile devices while providing more resources and power; and (d) to enhance the coverage of radio networks in terms of bandwidth and transport quality. The miniaturization is hard to achieve since it conflicts with other needs such as battery capacity and transmission range. On the other hand, the coverage of the radio technology allows radio networks of every size to provide a large set of applications and services, which should have location-independent service provision and enable high quality of service satisfaction.

The multimedia services provision involves four major players involved in the business with mobile multimedia: (a) the *network operator*, who is in charge of providing the mobile end-user with the infrastructure to access services mobile via wireless networks (e.g., 2G, 2.5G, 3G networks and ad hoc networks); (b) the *content provider*, who collects information and services to provide customers with convenient service collection adapted for mobile use and certify content and prepare it for end-users; (c) the *content creator*, who handles the computing infrastructure and content creation and provides the multimedia content via a transport scheme transparent to the mobile service provisioning; and (d) the application *developer and device manufacturer*, who are in charge of delivering hardware and software for mobile multimedia services and are not involved with any type of content creation and delivering.

Multimedia and network security issues are traditionally handled using cryptosystems. Cryptography provides confidentiality, authenticity, and integrity for a message transmitted through a public channel built often on an open architecture. It does not, however, protect against unauthorized copying after the message has been successfully transmitted. Digital watermarking is an effective way to protect copyright of multimedia data even after its transmission. Watermarking allows embedding a special pattern, referred to as watermark, into a multimedia unit so that a given piece of copyright information is permanently tied to the data. This information can later prove the ownership, identify a misappropriating person, trace the marked document's dissemination through the network, or simply can inform users about the rights-holder or the permitted use of the data.

Various watermarking approaches have been proposed during the recent years. However, most of these methods have focused on digital image watermarking, leaving a large number of challenges unsolved for video and audio watermarking, since various proposed video watermarking schemes have been based on the techniques of image watermarking and were not capable of appropriately protecting video data (Herrigel, 1998). In fact, applying a fixed image watermark, for example, to each frame in the video leads to some limitations in maintaining statistical and perceptual invisibility. Furthermore, such an approach is necessarily video independent as the watermark is fixed while the frame changes. In addition, using independent watermarks to each frame also presents a problem. Regions in each video frame with little or no motion remain unchanged frame after frame. Motionless regions may be statistically compared or averaged and used to remove independent watermarks (Swanson, 1997a).

Different requirements related to security and copyrights can be distinguished. The major requirements are listed in the following:

- The mobile communication networks should provide a secure environment which is able, at least, to provide the basic services (i.e., confidentiality, integrity, and privacy) for the multimedia transmitted. To this end, it is necessary to design ad hoc cryptosystems for multimedia data that cope better with its special characteristics and the real-time constraints it is submitted to.
- The protection schemes should be developed to protect owner's copyright. In particular, digital watermarking techniques require the development of efficient methods to resolve the rightful ownership of the invisible watermarking, while handling differently the data types involved in multimedia streams because of their special nature (including characteristics such as distortion and sensitivity).
- The mobile customer's rights should be protected. In particular, the rights of the authorized mobile customers to use a watermark must be addressed and should take into account the trust relationships between the owner and the customer.
- The security provided with multimedia multicasting to mobile users should be appropriately handled to allow sharing transmitted data between customers and protecting the rights and functions of all actors involved in the multicast.

Watermarking techniques have various applications. In addition to its use as a proof of ownership, a watermark can be used in different applications. In steganography, the watermark can be used to convey secret information. In data retrieving, it can bind semantic meaning to the host content, while it is useful in error recovery to communicate additional information to enable error control. Other important applications of watermarking include (a) the access control, to prevent unauthorized playback and copying of multimedia objects; (b) the fingerprinting, to identify the source of leakage in a content distribution network; and (c) the authentication of multimedia objects to ensure that they have not been tampered with during transmission to the mobile users.

As a conclusion, one can say that, to provide secure data transmission over mobile networks, it appears necessary to design schemes for multimedia because of the special characteristics including coding structure, volume of data processed, and the real time constraints they need to satisfy. The MPEG video encryption algorithms should, for example, provide an efficient real-time processing of streams so that they can cope with video delivery and security requirements. In particular, the encryption of MPEG video should provide efficient real-time processing and high security capability. On the other hand, the rights of the legitimate customers/ buyers have been addressed by a few works. Finally, multicasting security and copyright generate various challenges since they need differentiation of users and avoid sharing security objects.

Finally, it is worth it to notice that multimedia communications over ad hoc networks generates special challenges. Since ad hoc networks are characterized by highly dynamic connectivity, where nodes (or routers) can move and signal quality can change rapidly, the variation of resource availability is an important problem to solve for the delivery of high quality real-time multimedia streams. In ad hoc networks, the Quality of Service is hard to maintain during the whole session established to deliver a long duration stream, particularly when network-layer reservation mechanisms are used. Thus, multimedia applications should be able to cope with changes in network conditions, resource availability, and short-term and long-term QoS violations. Many experts believe that adaptive mechanisms at the application layer can constitute a complementary solution to those provided at the network layer QoS mechanisms. Typical examples of application adaptation mechanisms are the mechanisms that allow changing the frame-rate, frame size, and visual quality; and the mechanisms that allow switching to a different encoding scheme that consumes less bandwidth at the expense of lower quality.

15.2 Transmission Issues of Mobile Multimedia

Wireless links are characterized by a high packet loss and bit error rate. The bandwidth of these links is generally limited and also varies with changing of the environment. This is particularly true when mobile terminals, used for multimedia reception and display, are roaming from one network to another network and the network conditions may change dramatically. Common transport protocols, such as TCP, do not cope well with requirements to deliver multimedia contents. To overcome such shortcomings, several standards for delivery of multimedia data have been provided, including the real-time transport control protocol (RTP/ RTCP), the real-time streaming protocol (RTSP), session initiation protocol (SIP), and session description protocol (SDP).

In the following subsections, we provide an overview of the major issues addressed to provide transport error protection, congestion control for multimedia applications and service discovery, and transport security.

15.2.1 Transport Error Protection

In addition to the packet loss due to buffer overflow in the intermediate routers, the packets may also be lost because of errors in the wireless channels. Thus, efficient error protection schemes are essential for improving end-to-end multimedia quality. Traditional transport protocols use retransmission to recover packet losses.

This method, however, is questionable for applications that impose strict delay constraints, such as voice communications, particularly when attackers are targeting media stream by jamming them with short period noises. The forward error correction (FEC) can be used as better solution for error protection that is more suitable in real-time multimedia communication. FEC is a channel coding technique that protects the source data and appends redundant data during transmission, therefore adding transmission overhead.

Given the limited bandwidth of the radio links, the allocation of FEC protection-bits represents an interesting challenge. One approach to address it is to allocate the protection-bits unequally to different contents in multimedia applications. For example, stronger FEC protection should be applied to the base layer data than the higher layer parts (Zhang, 1999). In addition, since the wireless channel may dynamically change, it appears worthy to adjust the FEC protection level in response to the underlying changing network conditions. For example, GSM systems can dynamically distribute voice data and channel coding among the overall bandwidth to the possible best voice quality.

A more general form of the protection-bits problem is called the bit allocation problem, which jointly considers source coding and channel coding. The problems require users to make sure that the expected end-to-end distortion implies that the sum of the rates for the source coding and channel coding is smaller than the total available bandwidth, knowing that the expected end-to-end distortion consists of two components, source distortion DS and channel distortion DC. The source distortion is generated during the media source encoding, while channel distortion occurs when fragments of media stream are lost due to network congestion, or incorrectly received due to wireless channel noise. By resolving the aforementioned problem, one can see that both source coding and channel coding can affect overall media quality. Joint Source-Channel Coding schemes have been proposed to achieve an optimal end-to-end quality by adjusting the source and channel coding parameters, simultaneously. These parameters include quantization parameters, source rate, and the coding mode.

We notice, finally, that retransmission may still be applicable and play a complementary role to FEC in the case of media transmission where delay requirements are not so strict (e.g., delay-tolerant sessions). Several hybrid schemes involving FEC and retransmission have been proposed. They all made the choice that the delay bound can be achieved by limiting the number of retransmissions. Moreover, because high error rates are unavoidable in the wireless environment, energy efficient error-control is an important mechanism for mobile multimedia systems. This includes the energy spent in the physical radio transmission process, as well as the energy spent in computation, signal processing, and error control at the sender and the receiver. The total energy consumption per useful bit will depend both on the energy of transmission and the energy of redundancy computation. When error-correction mechanisms are implemented, the power consumption that is required to perform the error-correction mechanism can be considerable.

15.2.2 Congestion Control for Multimedia

Congestion control mechanisms are used to prevent the congestion failure in networks, to control traffic in the case of heavy environments, and to protect against denial of the transport service. A common method of congestion control is given by the transport control protocol (TCP), which employs an algorithm that increases the window by one packet, for example, when no loss is detected in a round trip and reduces the window in half, for example, when a loss is detected (see Chapter 14 for details). However, this congestion control algorithm presents two major disadvantages, from the transportation of multimedia perspective. First, the TCP congestion control algorithm introduces a large variation in the sending rate, which affects the quality of mobile multimedia delivery, by inducing unacceptable delays and jitters. Second, TCP relies on packet loss as the only indication of congestion. In fact, congestion may not be the only source of packet losses in wireless networks; wireless link errors may also cause packet losses. These reasons show that the TCP congestion control algorithm is not the appropriate tool to control congestion in the multimedia communications, as they are not able to utilize the wireless channel efficiently.

To overcome the shortcoming of TCP, two classes of congestion control algorithms have been developed: the window-based and the equation-based control (Rejaie, 1999; Padhye, 2000). The window-based congestion control approaches remains a congestion window as in TCP. However, the adjustment of the congestion window keeps a low fluctuation in the sending rate based on different parameters, such as an acknowledgment that is triggered at every incoming packet on the receiver side to measure packet loss and RTT, whereas the equation-based congestion control algorithms adjust the sending rate based on a throughput equation computed for TCP. The algorithms measure the current network parameters to determine the proper sending rate.

Many congestion control schemes have been proposed to improve the performance of transport over wireless networks by separating wireless links from the Internet control. These schemes are typically called *split-connection* algorithms. The main idea used in these schemes is to place a proxy or agent in the edge of the wired and wireless domains of the mobile network and differentiate between the congestion control in the wired and the wireless domains. In the wired domain, the aforementioned streaming protocol can be used, while in the wireless domain a special protocol is designed to handle the unreliability and unpredictability of the wireless link.

Many other approaches are proposed based on end-to-end mechanisms to initiate a collaborative service between the sender and the receiver using setup information. Typically, they develop heuristics to differentiate between the congestion packet losses and the random erroneous losses. While some methods have proposed a packet loss differentiation based on the use of packet delay information, other methods have used packet inter-arrival time to differentiate the cause of losses.

Other methods handled packet loss ambiguity by directly estimating the available bandwidth of the network path. TCP-Westwood uses the ACK streams to derive the so-called *Eligible Rate* with which the sender sets the congestion window after detecting a loss event, instead of dividing by 2 (Gerla, 2004).

Besides congestion there are many other factors that can affect the user-perceived QoS, which is a set of parameters that defines the end-user view on the performance of the multimedia application, including fading, mobility, multipath, and vertical handover. From the user perspective, some of these parameters are congestion related. Several other proposals focus on wireless networks (Mirhakkak, 2001). These proposal assume requirements that are very difficult to satisfy, particularly in multi-hop ad hoc networks. The most important tasks to achieve requirement satisfaction include minimizing application reconfiguration time in order to allow fast and flexible adaptation and applying progressive encryption to cope with QoS adaptation (Hamdi, 2008). Adaptation should be carried out to match resource availability (end-system and network) with subjective quality.

15.2.3 QoS Control of Mobile Multimedia

QoS control address the management of QoS parameters that can fall into four sets (as depicted in Figure 15.1) as follows. A process of QoS translation is typically needed to associate these set in the delivery of multimedia streams satisfying a specific QoS:

- *End-to-end perceivable QoS*: This set of parameters defines the mobile end-user perception of the multimedia application performance. Typically, such information allows the user to specify presentation features of the application as user's QoS preferences. Often, these parameters are not subject to negotiation as different users may have different understanding on what the words "good" and "bad" QoS mean. Typically, the translation of the perceivable QoS characteristics in more technical expressions is implemented by the multimedia application.

- *Application QoS*: Parameters involved in the multimedia application are used to describe end-to-end application performance (such as video frame rate, size, or visual quality) and are subject to negotiation with the other partner. A QoS contract can be used by the system to set up a multimedia session and enforce application of the required QoS. An adaptation process can be seen as a well-defined transition from one contract to another. For example, if a given QoS contract can no longer be fulfilled (due, for example, inability of coping with a bandwidth requirement due to handover), a second contract should be enforced. Thus, to provide continuous delivery of a multimedia stream, the multimedia application (helped by the transmission system) should derive various QoS contracts based on the available of system resources, network resources, and user QoS preferences. Intermediate nodes, on the delivery

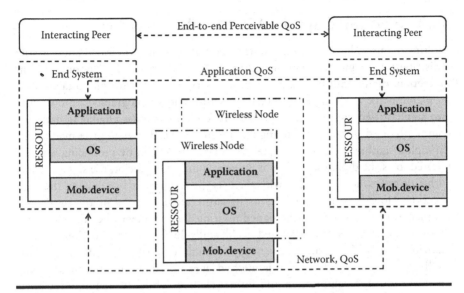

Figure 15.1 System model for distributed multimedia application.

path, may use these QoS contracts together with the available codecs to derive, negotiate, and enforce transport/network QoS parameters on behalf of those end-systems that are not able to explicitly or implicitly manage their streams during transmission.

■ *Transporting and routing QoS*: These parameters are used to specify the end-to-end requirements of the application with respect to network packet delivery. They parameters must be derived based on the media characteristics used as an input to the compression algorithm, the application QoS parameters, and the actual capabilities of the wireless network and the radio channels. The derivation may be easy to obtain for constant bit rate streams (such as audio streams). However, handling real time variable bit rate video codecs often requires traffic models to derive a set of parameters that may change based on the multimedia contents. The transport/routing QoS parameters are then specified to the transport layer entities for reservation, while the intermediate nodes may apply the provided QoS information to reserve network resources on behalf of the end systems.

■ *Mobile end-system QoS*: These parameters are used to specify local resource requirements (at the receiving node) and include information about memory, CPU, and battery capacity. The receiving system must be able to manage its resources and provide enough capacity to fulfill the application layer QoS requirements locally. Obviously, these parameters are not involved in any negotiation. Additional parameters include a common set of input/output configurations (e.g., addresses, ports, and codecs) for the multimedia streams in order to set up a valid end-to-end multimedia session.

15.3 Securing Copyright in Mobile Networks

Multimedia users have the ability to tamper with, produce copies of, and illegally redistribute the digital content of unprotected multimedia data. Without efficient protection, digital multimedia products and services will be unable to take-off in an e-commerce environment. Digital signature and cryptography are currently two standardized approaches to protect the digital contents. Digital signature is commonly used to authenticate digital transmissions. By passing the media unit through a signing process, a unique identifier is generated by producing a string referred to as the digital signature of the media unit. On receipt, the signed unit is authentic only when it matches with the decrypted signature by applying the hash function. However, the media and its signature are not bound in any significant manner. When transmitting the signed multimedia document, the document and its signature may become separated accidentally in transit or intentionally by a malicious entity. Thus, the receiver may not able to verify the authentic multimedia document. In addition, this method does not allow the multimedia document to undergo compression and format changes while still maintaining their authenticity.

The use of cryptographically secure license keys is another scheme to secure the digital intellectual property. The content of the documents are protected from manipulation and theft during delivery as the assessment of the document is only permitted to those who possess the appropriate key. However, the disadvantage of this solution is characterized by the fact that, after delivery of the document, the permitted recipient is able to reproduce perfectly the content and redistribute it. In addition, the intellectual property owner is not able to trace the responsibilities of pirating the properties.

15.3.1 Copyright Requirements

Multimedia protecting schemes aim at designing procedures to construct protecting objects for media data that are associated tightly with multimedia data so that it is difficult to remove the protecting object without damaging the multimedia data. These schemes have a wide variety of applications to digital rights management, including recognition of unauthorized copies, limitation on media copying operations, tracing of information leaks, and resolution of ownership disputes over digital content.

Multimedia protecting systems need to satisfy a large range of requirements including embedding effectiveness, fidelity, blind or informed detection, and robustness. A synthesis of the major requirements is given as follows:

- *Imperceptibility*: A protection system should not alter the cover image to the point of being useless. For this to be achieved, the protected object should be perceptually invisible. Invisibility requires that object embedded to the media contents, if any, remains unnoticeable when a user views the protected contents and does not degrade the quality of the content. Efforts have been made to provide protecting schemes that are capable of hiding the protecting component in such a way that they can be noticed when playing back the media. However, this requirement conflicts with some of the following requirements (tamper resistance and robustness, to name some).

■ *Robustness*: The object produced to protect a mobile multimedia document must be entirely resistant to any distortion that can be introduced during normal use either by an unintentional attack, a deliberate attempt to disable or remove the object presence, or malicious attack. Unintentional attacks involve transformations that are commonly applied to images during normal use, such as resizing. Robustness is the resistance of the protection against the removal of the object created for protection purposes by signal processing. The use of images and video signals in digital form commonly involves many types of distortions, such as lossy compression or, in the image case, filtering, resizing, contrast enhancement, and rotation. To be useful, the objects created to protect the media should be detectable even after such distortions have occurred.

■ *Capacity and speed*: A protecting scheme should allow for a useful amount of information to be embedded into the multimedia object. This can range from a single bit up to multiple paragraphs of text. Depending on the application at hand, the protection algorithm should allow a predefined number of bits to be hidden. General rules to define the size do not exist here; however, in the case of image protection case, the possibility of embedding into the image at least 300 bits should be granted (for the watermark, for example). In addition, the protecting algorithm should spread in a large area of the object to protect in order to prevent its deletion.

■ *Blind detection*: Blind detection refers to the ability to detect the objects created for the protection of a media document, without access to the original document. Due to the large size of uncompressed video files and the difficulty of indexing them to search for a specific frame, the blind detection is an important requirement in video watermarking.

■ *Low false positives and false negatives*: The protecting object related to a media object should be detected with high degree of reliability. Even in the absence of attacks or signal distortions, the probability of failing to detect the protection of a multimedia object and detecting a protection when, in fact, there is not one must be very low.

■ *Statistical imperceptibility*: This requirement necessitates that the protection scheme must modify the bits of multimedia object (e.g., image) in such a way that the statistics of the object are not modified in any manner.

■ *Security*: The protection scheme should be secure. This means that it is impossible to recover the changes, or to regenerate the protecting object after object alternations, even when the protection scheme and the protecting object itself is known.

■ *Real-time detector complexity*: For consumer-oriented watermarking applications, it is important that the complexity of the detection and extraction algorithms be low enough to execute within the specified real-time deadlines.

15.3.2 Watermarking

Digital watermarking has emerged as an effective technique for the protection of the author's rights. Informally speaking, a digital watermarking scheme is a procedure that embeds a "mark" in an object so that (a) it is hard to remove the mark without modifying (or damaging) the object; and (b) it can be detected or extracted later to make an assertion about the object. The object may be an image, audio, or a video. A simple example of a digital watermark would be a visible word placed over an image to identify the copyright.

Various watermarking approaches have been shown successful with the three major requirements mentioned above: capacity, imperceptibility, and robustness. Capacity refers to the amount of information that is being embedded in the watermark. Imperceptibility means that the marked data and the original data should be perceptually undistinguishable. Robustness refers to the fact that the embedded information should be reliably decodable after various alterations. In addition to these requirements, a watermark must be detectable or extractable to be useful.

Typically, a watermarking scheme consists of three components: the watermark, the encoder (or the insertion algorithm), and the decoder and comparator (or the extraction and verification algorithm). The verification algorithm authenticates the object by determining the real owner and proving the integrity of the object. The encoder, denoted by E, takes a media object I and a signature $S = s_1 \ldots s_n$ (or mark) and generates a new object, the watermarked object I'. This can be formally written as

$$I' = E(I,S).$$

The signature may be dependent on the object owner. The decoder function D takes an object I and an object J, which can be equal to the watermarked object I' or a corrupted object, whose ownership is to be determined, and generates a signature S'. Formally, this can be written as $S' = D(I,J)$. If the decoder and the encoder are associated then we should have

$$S' = D(I,I') = D\big(I, E(I,S)\big) = S.$$

The comparator function C compares the extracted mark S' with the owner mark S and output the value 1; if they match reasonably with respect to a threshold

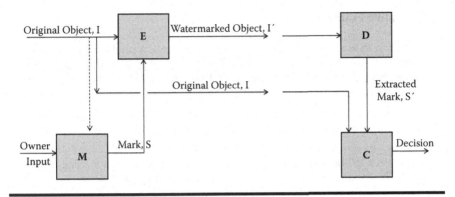

Figure 15.2 Typical watermarking scheme.

α this means that *C* computes the correlation *cor*(*S*,*S'*) between *S* and *S'* and outputs the value 1 if, and only if, *cor*(*S'*,*S*) ≤ α; otherwise, it outputs the value 0. Mathematically, the verification is written as follows:

$$C(\alpha, S, S') = 1, \ \text{if} \ cor(S', S) \leq \alpha; \ \text{else} \ 0.$$

This shows that, without loss of generality, a watermarking scheme can be defined as a triplet (*E*,*D*,*C*). Figure 15.2 depicts the three algorithms and shows a function M that is used to build the mark.

Depending on how the watermark is inserted and depending on the nature of the watermarking algorithm, the method used to extract it can involve various approaches. In particular, a watermark can be extracted in its exact form with some watermarking schemes implementing a procedure called watermark extraction. In other cases, the method can detect only whether a specific given watermarking signal is present in a media object by a procedure called watermark detection.

Four categories of watermarking techniques can be distinguished according to the type of document to be watermarked. They are: the image watermarking, the video watermarking, the audio watermarking, and the text watermarking. In addition, watermarking techniques can be classified, from the human perception perspective, into four different types as follows: visible watermarking, invisible-robust watermark, invisible-fragile watermark, and dual watermark. A visible mark is a secondary transparent overlay imposed into a primary image. This mark appears visible to a viewer on a thorough inspection. An invisible-robust watermark is embedded in such a way that any alteration made to a pixel value is perceptually not noticed and can be recovered only with a specific decoding mechanism. An invisible-fragile watermark is embedded in the object (e.g., image) such that any manipulation or modification of the object would alter or destroy the mark. Finally, a dual watermarking is a composition of a visible watermarking followed by an invisible watermarking (Mohanty, 1999).

Illustrative Example

Consider a RGB format image I having 512 × 512 pixels. A pixel is represented by a triplet $(i, j, p_{i,j})$, where (i, j) is the location of the pixel $p_{i,j}$ is its color $(p_{i,j} \leq 2^{24})$. Assume now that the owner of the image has a private key K. Then a signature made by the owner can be given by:

$$S(I) = \left(h\left(p_{1,1} p_{1,2} \cdots p_{512,512} \right) \right)_K$$

where h is a 192-bit hash function and $(-)_K$ is the encryption function using private key K. The encoding operation is achieved by performing two actions:

1. Decompose the signature $S(I)$ into a series of eight 24-bit segments, $S(I) = s_1...s_8$, and selecting randomly 8 pixels at positions $(i_1, j_1), ..., (i_8, j_8)$ such that $i_k \neq i_m$ and $j_k \neq j_m$ for all $k \neq m$.
2. Set the encoding $E(I, S(I)) = I'$ by taking the pixel $(i, j, q_{i,j})$ of I' equal to:

$$q_{i,j} = \begin{cases} p_{i,j} & \text{if } \forall k: (i, j) \neq (i_k, j_k) \\ p_{i,j} + s_k & \text{if } \exists k: (i, j) = (i_k, j_k) \end{cases}.$$

The decoding operation function D is the composition of two functions, $D = G \circ F$, where F and G are given by:

$$I = \left\{ (i, j, p_{i,j}): i, j \leq 512 \right\}, J = \left\{ (i, j, q_{i,j}): i, j \leq 512 \right\},$$

$$F(I, J) = r_{1,1}...r_{512,512}, r_{i,j} = p_{i,j} \oplus q_{i,j}$$

$$G(r_{1,1}...r_{512,512}) = s_1...s_k,$$

where the signature $s_1...s_k$ obtained by deleting all $r_{i,j}$ that are reduced to 24-bit $0...0$.

One can say that this watermarking does not distort the cover image since only eight pixels are modified over 2^{18} pixels. In addition this approach allows false positives and false negatives only when the number of segments computed at the output of $G \circ F$ is strictly smaller than 8, meaning that the encryption of the image digest contains a null 24-bit segment.

15.3.3 Digital Fingerprinting

The simplicity with which digital data can be perfectly reproduced has made multimedia hacking and piracy a growing threat for copyright holders. To address this issue a method of protecting copyrighted material, called digital fingerprinting, has been largely investigated. Digital fingerprinting is a method by which a copyright owner can uniquely embed a buyer-dependent, unremarkable serial number into every copy of digital data that is legally sold. The buyer of a legal copy is then dissuaded from distributing further copies, because the unique fingerprint can be used to trace back the origin of copy operation. In fact, all legally distributed copies of the same digital data are similar, with the exception of the unique buyer-dependent fingerprints. Thus, in contrast to active forms of security (such as encryption), fingerprinting is a passive form that is efficient after an attack has occurred.

Designing a fingerprinting scheme consists of designing a fingerprinting code and a watermarking scheme that is employed to embed the fingerprinting code into the multimedia data. The use of the fingerprinting scheme in a broadcast channel depends on the nature of the fingerprinting process and can be done in several ways. Formally speaking, one can describe the components of a fingerprinting scheme as follows: A fingerprinting code consists of a codebook and a tracing algorithm. A codebook is a finite set Δ of codewords of equal length, say n, over some finite alphabet A.

$$\Delta = \left\{\delta_1, \delta_2, \ldots, \delta_m\right\} \subseteq A^n, \delta_i = s_{i,j} \ldots s_{i,n}.$$

A tracing algorithm is a function Alg: $A^n \rightarrow P(A^n)\backslash\emptyset$ where $P(A^n)\backslash\emptyset$ is the power set of A^n, except the empty element.

A major problem facing fingerprinting should be addressed: A group of mobile hackers who possess distinctly fingerprinted copies of the same mobile multimedia data can exploit this multiplicity, compare the available copies, and detect and then render the fingerprints unreadable. Such an attack is referred to as *collusion*. Therefore, the first goal of a robust fingerprinting scheme is to ensure that some part of the fingerprint is capable of surviving a collusion attack, so as to identify at least one of the hackers. For multimedia fingerprinting, robustness presents another need, namely the fingerprint's ability to remain traceable after intentional or unintentional modification of the fingerprinted media. A fingerprinting scheme for multimedia should also be robust to some amount of user-generated distortion. Examples of common mobile user-generated distortions include additive white Gaussian noise, linear filtering, JPEG compression, and geometric distortions among others.

In a fingerprinting scheme, each codeword from Δ is assigned to a different user. The goal of a malicious group of mobile users is to combine their codewords to produce a new codeword δ that cannot be traced back to the group. Formally, a collision attack made by a group of n users having their codewords in Δ can be defined by a function Col such that

$$\delta = Col(\Delta), A\lg(\delta) \notin P(\Delta) \setminus \emptyset.$$

In applications such as Video on Demand (VoD), it is impractical to send a unique fingerprinted video to each subscriber, because the bandwidth usage is excessive. The solution to this problem may be to send identical digital data to all subscribers, and then build a uniquely fingerprinted video at the user end. Such a scheme is referred to as fingerprinting in a broadcast channel environment. Let I be the multimedia object to be distributed to users (or buyers) and assume that it is a video sequence comprising of a set of s frames, say $I = \{I_1, I_2, ..., I_s\}$. Assume also that an encryption scheme E and a set of keys $K = \{K_1, K_2, ..., K_m\}$ is used to provide confidentiality of the multimedia during transmission. The object of E is to encrypt I to the set $\hat{I}$ so that any frame in $\hat{I}$ cannot be understood by eavesdroppers without the appropriate key. The encryption video $\hat{I} = E_K(I)$ is sent to m receivers who have legally purchased the multimedia object. Each key in the set K is assumed to belong to a specific receiver in the group. The decryption operator D maps $\hat{I}$ in a set of video sequences V_i, $i \leq m$ so that V_i is the fingerprinted version of video sequence I delivered to receiver i. This scheme is depicted in Figure 15.3.

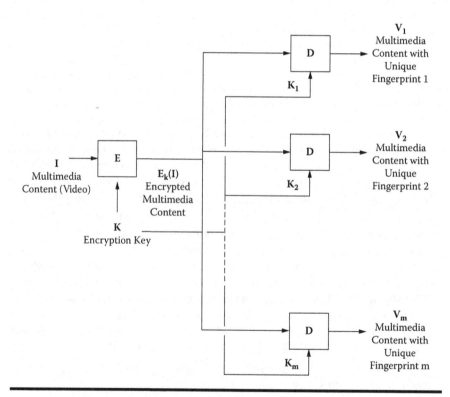

Figure 15.3 Fingerprinting in a broadcast channel.

To provide a robust fingerprinting scheme in a broadcast channel environment, the following requirements should be met:

- *Scrambled video signal*: The encrypted video sequence does not visually resemble the unencrypted video sequence I and should be made unintelligible.
- *Unique fingerprinted videos*: The decrypted versions (or fingerprinted videos) should be distinct and must use codewords that are collusion-resistant.
- *Watermarking scheme robustness*: The watermarking scheme should be robust to a set of feasible attacks.
- *Encryption security*: Without keys K_1, K_2, ..., K_m an eavesdropper cannot derive the fingerprinted version C in a reasonable period of time, given the encrypted video.

15.4 Major Watermarking Techniques

Many digital watermarking schemes have been proposed in the literature during the recent years for text, images, audio, and video streams. Some of these schemes are based on methods borrowed from radio communications. They are called spread spectrum watermarking techniques. Additive embedding of a pseudo noise watermark pattern and watermark recovery by correlation are among these techniques. The schemes can be distinguished in terms of the domain where the watermark is embedded, the real-time performance, and their resistance to particular types of attacks. In the sequel, we only discuss several techniques for the image and video watermarks.

15.4.1 Image Watermarks

Three classes of techniques have been identified. They are the spatial domain, the frequency domain, and MPEG coding structure based watermarks. Spatial domain watermark algorithms generally share the following characteristics: First, the watermark is applied in the pixel (or the coordinate) domain and the combination with the host signal is based on simple operations performed in the pixel domain. Second the watermark is derived from the message data via spectrum modulation. Finally, the watermark can be detected by correlating the expected pattern with the received signal. The pixel domain methods are simple to design and have low computational complexities. They have proven to be attractive for video watermarking applications where the real-time performance is important. On the other hand, the spatial domain methods present some limitations: First, the need for spatial synchronization leads to high vulnerability to certain attacks, called de-synchronization attacks (such attacks will be discussed in the following section). Second, the need to consider the temporal parameter leads to vulnerability to multiple frame collusion.

Several methods belong to the class of pixel domain methods. An example of such methods is given by a technique that can embed a watermark in an image by superimposing it over an area of the image and then add some fixed intensity value for the watermark to the varied pixel values of the image. The resulting watermark may be visible or invisible depending upon the value of the watermark intensity. Spatial watermarking can also be applied using color separation so that the watermark appears in only one of the color bands. This makes the watermark hard to detect under normal viewing. However, the watermark becomes visible when the colors are separated.

15.4.1.1 Least Significant Bit Modification (LSB)

The simplest method of watermark embedding is to embed the watermark into the least-significant bits of the cover object. Given the high channel capacity of using the entire cover for transmission in this method, a smaller object may be embedded multiple times. The LSB substitution, however, shows several drawbacks to this scheme, despite its simplicity. Although it may survive some transformations, any addition of noise or lossy compression is able to defeat the watermark. Another attack would be to simply set the LSB bits of each pixel to 1, fully defeating the watermark with negligible impact on the cover object. In addition, when the embedding algorithm is discovered, the embedded watermark could be easily modified by a third party placed on the transmission route.

An enhancement of the basic LSB substitution would be to use a pseudo-random number generator to determine the pixels to be used for embedding. The security of the watermark would be improved as the watermark could no longer be easily viewed by intermediate parties. The algorithm, however, would still be vulnerable to replacing the LSB's with a constant. Even in locations that were not used for watermarking bits, the impact of the substitution on the cover image would be negligible.

15.4.1.2 Correlation-Based Techniques

Another technique for watermark embedding is to exploit the correlation properties of additive pseudo-random noise patterns as applied to an image. A pseudo-random noise pattern $W(x,y)$ is added to the cover image $I(x,y)$ to obtain another image according to the following equation:

$$Iw(x, y) = I(x, y) + k \times W(x, y),$$

where k denotes a gain factor, and I_W the resulting watermarked image. Increasing k increases the robustness of the watermark, but may reduce the quality of the watermarked image. To retrieve the watermark, the same pseudo-random noise generation algorithm is seeded with the same key, and the correlation between the noise pattern and possibly watermarked image is computed. If the correlation exceeds a certain threshold, the watermark is detected, and a single bit is set. This method can easily be extended to a multiple-bit watermark by dividing the image into blocks and performing the above procedure independently on each block.

15.4.1.3 Frequency Domain Watermarks

Generally the discrete cosine transform (DCT), the fast Fourier transform (FFT), and the wavelet transform are used as the methods of data transformation. They allow a watermark to be embedded in a distributive way in the overall domain of an original data and be hard to be deleted, once embedded. The main strength offered by the frequency domain techniques is that they can take advantage of special properties of alternate domains to address the limitations of pixel-based methods or to support additional features. For instance, designing a watermarking scheme in the DCT domain leads to better implementation compatibility with popular video coding algorithms such as Moving Pictures Experts Group (MPEG)-2.

The DCT allows an image to be broken into different frequency bands, making it easier to embed watermarking information into the middle frequency bands of an image. The middle frequency bands are chosen such that they can avoid the most visual important parts of the image (low frequencies) without over-exposing themselves to removal through compression and noise attacks (high frequencies). One such technique utilizes the comparison of middle-band DCT coefficients to encode a single bit into a DCT block.

Several image watermarking techniques have been made available in the literature. While a large number of these techniques are classified as visible watermarking, a few of them are classified as invisible watermarking schemes. We discuss in the following three of these schemes:

- *Signature casting*: This technique is proposed signature casting on digital images. The signal embedding is done by addition to the luminance channel. The watermark consists of a binary pattern $S = \{s_{m,n}\}$ having a size equal to the original image I, where the number of 1's is equal to the number 0's. The image $I = \{I_{m,n}\}$, where $I_{m,n}$ is the luminance value at location (m,n), is divided into two sets Γ and Δ of equal size as follows:

$$\Gamma = \left\{ I_{m,n}: s_{m,n} = 1 \right\}, \ \Delta = \left\{ I_{m,n}: s_{m,n} = 0 \right\}.$$

The watermark is superimposed by changing the elements of the subset Γ by a positive integer factor k. Thus, Γ is mapped into a new set Γ' as follows:

$$\Gamma' = \left\{ I_{m,n} + k: s_{m,n} = 1 \right\},$$

and the watermarked image is then given by the union of Γ' and Δ. The watermark recovery is usually done by some sort of correlation method. To detect the presence of a watermark S in an image, the following hypothesis testing is applied. Let δ be the normalized difference between the mean of set Γ' and the mean of set Δ We have

$$\delta = \frac{\overline{\Delta} - \overline{r'}}{\sigma_\Gamma^2 - \sigma_\Delta^2},$$

where $\overline{X}$ and σ_x^2 denote the mean and variance of set X, respectively. The test is performed by comparing δ with a threshold to determine if there is a watermark.

■ *Cox et al watermarking*: This is an invisible robust watermarking technique. It inserts the watermark into the spectral components of the image using a technique analogous to spread spectrum communication. To do so, the watermark insertion integrates the following steps: First, the DCT of the entire original image is computed as one block. Second, the perceptually significant regions of the image are located. Third the watermark $S = s_1 \ldots s_n$ is computed. The values of s_i are assumed to follow a normal distribution with mean 0 and variance 1. Finally, the watermark is inserted in the DCT domain of the image by setting the frequency components v_i in the original image to $v_{i'}$ as follows:

$$v_i' = v_i(1 + kx_i),$$

where k is a scalar factor. The following expressions can also be used:

$$v_i' = v_i + kx_i \text{ and } v_i' = v_i e^{kx_i}.$$

The extraction of watermark consists of following steps: first, the DCT of the entire watermarked image is computed. Second, the entire original image is computed. Then the difference of the two is the watermark S^*. The extracted watermark S^* is compared with the original watermark S using appropriate similarity functions. It has been shown that this watermark is robust to common signal and geometric distortion such as compression, cropping, rotation, scaling, and translation.

■ *Kankanhalli method*: This is a visible watermarking technique. It typically divides the host image into different blocks and computes the discrete cosines transform (DCT) of each block. Then it classifies the blocks into six different classes in the increasing order of noise sensitivity, such as edge block, uniform with moderate intensity, uniform with high or low intensity, moderate busy,

busy, and very busy. Each block is then assigned two different values α and β. The host image blocks are then modified as follow:

$$\hat{X}_{i,j} = \alpha X_{ij} + \beta W_{ij},$$

where $\hat{X}_{i,j}$ is the DCT co-efficient of the watermarked image, X_{ij} is the corresponding DCT coefficient of the original image, and W_{ij} is the DCT coefficient of the watermark image.

15.4.2 Video Watermarking

Video streams consist of a series of consecutive and equally time-spaced images. Typically, the idea that image watermarking techniques are directly applicable to video sequences is obvious. This is partly true; however, there are also some important differences between images and video sequences that suggest specific approaches for video watermarking. One important difference is the available signal space. While the signal space (i.e., the number of pixels) is very limited for an image, the available signal space, for video, is excessively larger. In addition, video watermarking may require real-time constraints on the watermarking system. Consequently, it is less important, and for many applications even prohibitively complex, to use watermarking methods based on explicit models for image watermarking. The complexity of the computation of watermarking methods is more important for video applications than it is for image applications. Another concern is related to the security. In fact, the structure of video as a sequence of images gives rise to specific attacks, such as the frame averaging and frame dropping (Swanson, 1997b). A robust watermarking scheme should be able to resist to this type of attack by distributing watermark information over several consecutive frames, for example. However, it might be advantageous to retrieve the full watermark information from a short part of the sequence.

In the following, some watermarking methods exploiting uncompressed or compressed video properties are presented. In addition, some other methods that we describe are in fact image watermarking techniques applied to image sequences.

A first approach proposes to embed a spatial domain low-pass spread-spectrum watermark into 8x8 pixel blocks of video sequences (Darmstaedter, 1997). The blocks are first classified according to their activity. Blocks with low activity are not watermarked. A low-pass pseudorandom pattern is then added to each selected block. Typically, each block carries one bit watermark information; but, the bits are redundantly repeated over several blocks and several frames. Also, an error correcting code is applied. After embedding of the watermark, the video sequence is compressed using MPEG-2 compression. The extraction of the watermark is done in the spatial domain after decompression using a correlation function with thresholding. The embedding of one bit of watermark information into a total of 162,000 pixels allows to achieve an error-free watermark retrieval for compression at a video bit rate of 6 Mbit/s. It is worth noting that the approach has been verified, including real transmission over digital satellite links.

The second approach we discuss here applies watermarking of compressed video for fingerprinting applications. It employs a straightforward spread-spectrum approach and embeds an additive watermark into the video sequence. The watermark is generated using a pseudo-noise signal having the same dimensions as the video signal that is modulated with the information bits to be carried. Each information bit is redundantly embedded into many pixels. For each compressed video frame, the corresponding watermark signal frame is DCT-transformed on an 8 × 8 block-by-block basis, and the resulting DCT coefficients are added to the DCT coefficients of the video sequence as encoded in the video bit stream. This is done for I, P, and B frames. A rate control is realized by individually comparing the number of bits for each encoded watermarked DCT coefficient with respect to the corresponding encoded unwatermarked coefficient. Due to the inherent redundancy in the watermark, the watermark information can still be transported as long as a sufficient number of coefficients can be embedded. The scheme is compatible with all DCT-based hybrid compression schemes, such as MPEG-2, MPEG-4, and ITU-T H.263. The watermark is recovered from the decompressed video by correlation using the same pseudo noise sequence that was used during the generation of the embedded watermark.

A third approach proposes to embed a spread spectrum watermark into 3-D blocks of video by employing a 3-D DFT and adding to the transform coefficients (Deguillaume, 1999). The watermark is composed of the real watermark and an auxiliary pattern, called a template, which is easy to detect (especially under geometric attacks). The template can be used to undo geometric attacks to enable retrieval of the real watermark. The sequences that are handled consist of typically 16 or 32 frames. The template is embedded in a way that it is not affected by zoom and shift operations. Results are reported for an 88-bit watermark embedded into 3-D blocks of 352,288 pixels frames each (giving a watermark data rate of 1 bit per 36,864 pixels).

A fourth method applies an image watermarking method working on DCT blocks to video sequences. The watermarks are embedded into the luminance component of uncompressed video and retrieved after decompression. To improve the invisibility of the watermarks, blocks are selected for watermarking depending on the block activity. The method proposes to introduce additional temporal redundancy by embedding the watermark into several consecutive frames and averaging in the retrieval.

A fifth method develops a video watermarking method for video broadcast monitoring applications, called JAWS (just another watermarking system) method (Kalker, 1999). It performs the watermark embedding and detection in the spatial domain, meaning that embedding is performed before compression and detection is executed after decompression. The embedded watermark consists of watermark patterns of size 128 × 128 drawn from a white random process with Gaussian distribution that are repeated to fill the whole video frame. The activity measure is computed using a Laplacian high-pass filter. The same watermark is embedded into several consecutive video frames. To achieve watermark detection, a correlation detector is used after applying a spatial prefilter that reduces the cross talk between video signal and watermark. In order to embed arbitrary watermark information, the watermark signal is designed using several basic watermark patterns. The information is encoded in the choice of the basic patterns and their relative positions. The watermark can carry up to 50 bits/s; however, the watermark data rate is reduced to increase robustness for applications that require less watermark information per second. In addition, since the watermark detection must be achieved even in the presence of spatial shift operations, a search over all possible shifts is performed (typically in the FFT domain, to reduce the complexity of search and correlation).

Let us now notice that the above-mentioned watermarking methods present a few general observations. First, the proposed methods span a wide complexity range from very low complexity to considerable complexity including, for example, wavelet transforms and other models. In general, however, the more complex methods seem to embed the watermarks with higher robustness. Second, most methods operate on uncompressed video; only a few methods can embed watermarks directly into compressed video. For watermarking of compressed video watermarks can be embedded in the DCT coefficients, in the motion vectors, or inside information. Finally, the reported watermark data rates seem to range between a few hundred of bits per second and a few bits per second.

15.5 Attacks against Mobile Multimedia

A watermarked object is likely to be subjected to certain manipulations, some unintentional such as compression and transmission noise and some intentional such as cropping and filtering of images. We present in this section a list of common attacks that have been developed against watermarks. Most of these attacks have been discussed in the literature. The attacks can be classified into four categories (Hartung, 1999). Robustness of the watermarking techniques against these attacks is a major requirement. Several watermark attacks tend to show that the known watermarking techniques are vulnerable and are not robust.

15.5.1 Attack Classification

In several cases, the problems related to unauthorized embedding and detection are similar to those occurring in cryptography and can be solved in similar ways. One can notice for example that in symmetric cryptography, an encryption function, say $E_k(\cdot)$, that takes a known clear text m and provides a ciphertext $E_k(m)$, knowing the symmetric key *k*. Similarly, an embedding function $E_S(-)$, in watermarking, takes a multimedia object m and outputs a watermarked objects $E_S(m)$, knowing the watermark *S*. The addition of an encryption layer to a watermarking process ensures multimedia object security. In particular, someone who detects the watermark cannot decode it. In addition, it might prevent an attacker from detecting the presence of the watermark. However, in a wide range of watermarking systems, the encryption does not prevent the detection of the encrypted watermark. Therefore, in addition to attacks similar to those targeting cryptographic systems, specific attacks have been made available to target watermarks, even when they are encrypted.

For the sake of simplicity, we describe in this subsection attacks that do not damage the perceived quality of the host data. Six classes can be distinguished:

1. *Simple attacks*: These attacks are easy to design. They attempt to damage the embedded object by manipulating the whole watermarked data (i.e., the watermark and watermarked objects) and do not try to identify the watermark or extract it. Examples of simple attacks include linear and non-linear filtering operations, addition of noise, cropping, and quantization in the pixel domain.

2. *Unauthorized embedding attacks*: These attacks are closely similar of the attacks targeting sender authentication, which can be solved by the use of asymmetric cryptosystems.

3. *Detection disabling attacks*: These attacks attempt to break the correlation and to make the recovery of the watermark unachievable to a watermark detector. They typically use removal of pixels, insertion of pixels, pixel permutation, geometric transformation, geometric distortion shift in spatial or temporal direction, rotation, and cropping. Examples of such attacks include the synchronization attacks.

4. *Ambiguity attacks*: These attacks attempt to confuse by producing fake original data or fake watermarked data. Some attacks can delete the link to the author of a watermark by embedding additional watermarks such that the order of insertion of the watermarks (including the original one) becomes unclear. Examples of such attacks include confusion attacks, fake watermark attacks, and inversion attacks.

5. *Removal attacks*: These attacks attempt to analyze the watermarked data, estimate the watermark, separate the watermark and the host data, and then discard the watermark. Examples of removal attacks include the collision attacks, denoising attacks, and some non-linear filtering operations.

6. *Synchronization attacks*: These attacks target the synchronization method being used by the watermark detector. Two types of synchronization attacks can be distinguished: the template removal attacks and the auto correlation attack. The first class aims at removing peaks in the DFT domain and rotating the object. The second class applies a three-step process: apply an

autocorrelation function to the estimated watermark, detect significant peaks in the autocorrelation, and use this information to attack the watermark.

It is worthy noting the following facts related to the practice of watermarking and attacks:

- It has been shown that various watermarking methods are vulnerable to inversion attacks (Craver, 1998). Using such an attack, an attacker who receives the watermarked data can render the embedded watermark unclear.
- A multitude of attacks have been constructed to target digital versatile disk copy protection mechanisms. They usually include geometrical attacks using affine transformations, addition of noise, compression operations, collusion attacks, and attacks based on detector observations.
- Various spread spectrum watermarking techniques have been targeted by various attacks that analyze the vulnerability to watermark estimation via detector observations. The attacks proposed in Kalker (1998), for example, consider the schemes that provide a publically available black box watermark detector and propose methods that allow estimating the secret spread spectrum watermark by degrading gradually the watermarked data until the detector cannot detect the watermark in the degraded version.

15.5.2 Attacks Targeting Fingerprinting Schemes

Two classes of attacks targeting fingerprinting schemes can be distinguished, namely, the single-user attacks and the multimedia collusion attacks. In the first class of attacks, the attack involves one user. It is often found in the watermarking literature and is applied directly to the video sequence. The multimedia collusion attack is applied to a certain number of copies of a fingerprinted multimedia object.

15.5.2.1 Single-User Attacks

These attacks involve one copy of the multimedia and can be categorized into *unintentional attacks* and *intentional attacks* (Deguillaume, 2000). Unintentional attacks are attacks that occur due to bandwidth constraints, such as lossy copying and transcoding (i.e., compression, change in frame rate, format conversion, conversion in display format). Intentional attacks are user-generated attacks that aim to remove the watermark or fingerprint in the multimedia. Intentional attacks on video can be categorized into *single-frame attacks* and *statistical attacks*. Single-frame attacks can be categorized into *signal processing attacks* (i.e., band-pass filtering, adaptive Wiener denoising, etc.) and *desynchronizing attacks* (i.e., affine transformations, scaling, cropping, etc.). Statistical attacks for video are sometimes also called collusion. However, there is only one copy of the involved video and the term arises from the fact that consecutive frames in the video are used together to remove the watermark or fingerprint. A simple statistical attack on video is to average a small set of consecutive frames, so that this will remove the watermark. A more complex statistical attack is to first estimate the watermark in each individual frame and then average the estimated watermarks to obtain a final estimation, which is then subtracted from each frame.

15.5.2.2 Multimedia Collusion Attacks

Collusion attacks can be applied on multiple copies of multimedia. Figure 15.4 depicts a special classification of collusion attacks on fingerprinted multimedia. When fingerprints are embedded into multimedia, hackers can attempt to estimate the original non-fingerprinted multimedia. The collusion attack becomes an estimation problem and uses different techniques from the Estimation Theory. On the other hand, the hackers can attempt to scramble the fingerprint aiming at creating a non-compliant fingerprint.

For the sake of clarity, we consider that $\bar{I}_j^*(x,y)$ describes the jth frame, in the attacked sequence of frame I, at the (x,y)th pixel, using attack *, and that $\bar{I}_{j,i}(x,y)$ describes the jth frame at the (x,y)th pixel of the fingerprinted video of User i. The following equations define the simplest suboptimal estimation technique, which is simply to average the set of multimedia, and the min, max, minmax, median, modified negative, and randomized negative.

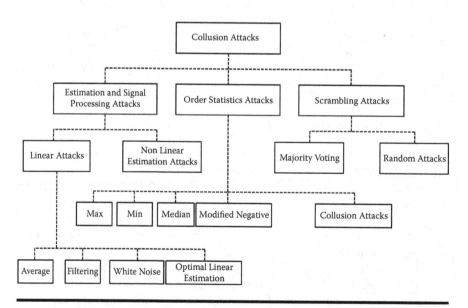

Figure 15.4 Types of collusion on fingerprinted multimedia.

$$\overline{I}_j^{Subopt}(x,y) = \frac{1}{m}\sum_{i=1}^{m}\overline{I}_{j,i}(x,y)$$

$$\overline{I}_j^f(x,y) = f\left(\overline{I}_{j,1}(x,y),\dots,\overline{I}_{j,m}(x,y)\right)$$

where f is equal to the operation min, max, minmax, median, modified negative, or randomized negative (Zhao, 2003). In particular for the random negative attack, we have

$$\overline{I}_j^{rand-neg}(x,y) = \overline{I}_j^{min}(x,y) \text{ with a given probability } \pi,$$

$$\overline{I}_j^{rand-neg}(x,y) = \overline{I}_j^{max}(x,y) \text{ with a probability } 1-\pi.$$

15.5.3 Attacks Targeting Watermarking Schemes

We present, in this section, some of the important attacks that have been developed against the common watermarking schemes.

15.5.3.1 Synchronization Attacks

Synchronization is the process of identifying the correspondence between the spatial and temporal coordinates of the watermarked signal and that of an embedded watermark. Preferably, the watermark detector receives a watermarked signal such that the coordinates of the embedded watermark have not been changed since the embedding process. If that is the case, one says that synchronization is true and the detector can proceed correctly. If the coordinates of the embedded watermark have been changed (changes occur, for example, when the watermarked signal is rescaled, rotated, and translated as shown on Figure 15.5), the detector must identify the coordinates of the watermark prior to detection. Synchronization is critical to achieve successful watermark detection. Many techniques have been developed to attack watermarked signals by simply desynchronizing the detector and not removing the watermark.

Synchronization is a problem that cannot be taken care of in video watermarking applications, even in the absence of a malicious attacker. In applications such as the secure digital television, the watermark detector may be expected to detect the watermark starting from any arbitrary temporal location within the video signal (as opposed to starting detection from the beginning). In other applications, such as the streaming video, the video signal arriving at the watermark detector may have been damaged or interrupted (intentionally or non-intentionally) to the extent that the detector loses synchronization and must resynchronize before watermark detection can resume. Establishing synchronization would involve an exhaustive

Watermarked   Attacked

Figure 15.5 An example of synchronization attack.

search over the space of all possible geometric and temporal transformations to find the watermark. This is not practical for video watermarking applications that require real-time watermark detection.

15.5.3.2 StirMark Attack

In its simplest version, StirMark introduces the same type of errors into an image as printing it on a high quality printer and then scanning it again with a high quality scanner. It applies a minor geometric distortion such as slightly stretching, shifting, and/or rotating by an unnoticeable random amount and then resampling it using bi-linear interpolation, for example. In addition, a transfer function that introduces a small and smoothly distributed error into all sample values is applied.

StirMark attack introduces a practically unnoticeable quality loss in the image if it is applied only once. However, after a few iterated applications, the image degradation becomes noticeable. With those simple geometrical distortions, one can confuse most marking systems available on the market. More distortions still unnoticeable can be applied to a picture.

One might try to increase the robustness of a watermarking system by trying to foresee the possible transforms used by the attackers; one might then use techniques such as embedding multiple versions of the mark under suitable inverse transforms. However, the general idea of the attacks is that, given a target marking scheme, one can develop a combination of distortions that will remove the watermark or at least make it unreadable, while leaving the perceptual value of the previously marked object undiminished.

15.5.3.3 Unauthorized Embedding Attacks

Such attacks attempt to forge a valid watermarked object for new host data and copy blocks of a valid object without understanding the content. A known example is the copy attack, which works as follows: Given a legitimately watermarked multimedia object O_1 and an unwatermarked object O_2, the method begins by applying a watermark removal attack to O_1, to obtain an approximation of the original mark by applying any method of estimation. The second step is to estimate the added watermark pattern by substracting the original from the watermarked object.

15.5.3.4 Scrambling Attacks

This is a system level attack in which the samples of a watermarked multimedia object are scrambled prior to presentation to a watermark detector and subsequently descrambled. The scrambling can be very simple (such as a permutation) or more complex (such a random scrambling of values in the object). To succeed, the scrambling should be invertible or near invertible. An example of scrambling attacks is given by the mosaic attack against images. Such an attack divides the attacked image into many small blocks. These small images are then displayed in a table so that they constitute an image "identical" to the attacked image. Scrambling attacks, however, require the receiver of the attacked object to obtain a descrambling scheme.

15.5.3.5 Unauthorized Removal

This is an operation performed by an attacker to remove the watermark from a watermarked multimedia object, no matter if the watermark is detected or not. Examples of unauthorized removal attacks include the filtering and noise removal attacks and the sensitivity analysis attack. A watermark with considerable energy in the high frequencies, for example, can be corrupted by the application of low-pass filter. Moreover, all watermarks that are noise-like added are susceptible to be removed by a noise removing technique. The sensitivity analysis attack is performed when the adversary is equipped with black-box detector. Three steps are required to launch this attack. The first step aims at finding a multimedia object O (deduced from the watermarked object) that lies close to the detection region and detecting the boundary of the detection region (this can be done by various methods including altering the watermarked object, and replacing samples of the object with the mean value of the watermarked object). The second step aims at approximating the direction of the normal to the surface of the detection region of O. The third step scales and substracts the normal form of the watermarked object.

15.6 Countermeasures against Watermarking Attacks

Spread spectrum watermarking systems can be defeated by a large set of the attacks mentioned in the previous section. To make these systems more resistant against attacks, the simple approach is to establish a set of countermeasures and rules for the design of watermarks. We discuss in the following some among the most important rules and countermeasures.

15.6.1 General Rules

Three major rules can be applied. First, the avoidance of cryptographic weaknesses and vulnerabilities should be observed. Obviously, watermark systems should be cryptographically secure. This implies that several features should be guaranteed, including the fact that all keys involved in the securing process (including those used in the embedding and retrieval) should be protected. In particular, if pseudo-random input generators are used, their structure and the seed they utilize should be impossible to determine. Second, the parameters of the watermarking system should be selected appropriately, since in that case filtering, compression, and similar operation do not represent any threat. In particular, the number of pixels that one bit of watermark information is distributed over should not be too small. Third, registration patterns that can be used to detect and reverse the geometrical transformation applied by the attack should be included as a safety measure that predicts detection-disabling attacks. However, since the registration patterns are easy to find, which could be exploited for attacks against the registration marks, the registration marks must provide sufficient security against removal and attacks.

15.6.2 Countermeasures against Collusion Attacks

Collusion attacks may be launched when a number of watermarked versions of the same multimedia object are distributed over a mobile network, for fingerprinting applications. It appears that, if spread spectrum watermarks are mean-free, they can be vulnerable to collusion attacks like averaging attacks. However, it has been shown that it is possible to construct collusion-secure watermark signals by applying a basic idea: to compose the watermarks out of static and dynamic components. The codes can be designed so that, for every possible combination of colluding parties, there are parts of the codes that do not average to zero. In fact, the static components do not vanish by averaging. Moreover, all colluding parties can be determined from the colluded (averaged) version. A limit, however, can be seen with the fact that the length of the proposed collusion-secure codes increases exponentially with the number of different distributed watermarks. To overcome this limit, the use of hierarchical codes is suitable.

15.6.3 Countermeasures against Ambiguity Attacks

Several solutions have been proposed against ambiguity attacks. There are two key principles in the design of non-invertible watermarks. A first solution is the use of signal-adaptive watermarks (Qiao, 1998) that depend on the host data in a one-way manner, by applying a hash function, for example. It has been shown that such watermarks are non-invertible. A second solution is the use of cryptographically secure timestamps provided by trusted third parties and encoded in the watermark. These time-stamps should typically be used for real-world applications, to avoid other drawbacks of watermarking and copyright protection systems.

15.6.4 Countermeasures against Embedding Attacks

An important preventative measure to avoid attacks is the selection of random spatial position of the embedded watermark bits. In fact, it has been shown that using regular arrangements, such as having k consecutive (or periodic positions) pixels to embed one watermark bit (as shown by examples (a) and (b) depicted in Figure 15.6), is not suitable. This would allow an attacker to target single bits of the watermark and modify the overall watermark, while leaving other portions of the data unmodified. For example, for fingerprinting applications, an attacker could move between various watermarked copies of the video sequence, and mix up the embedded watermark information. Consequently, it is required to distribute each bit of watermark information over pseudo-randomly selected pixels for a video sequence (as depicted by example (c) in Figure 15.6).

15.6.5 Use of Attack-Resilient Block-Based Watermark Decoder

Detection-disabling attacks, like StirMark, exploit the fact that the human visual system is not sensitive against shift and against small global modifications, provided that there are no severe and clear local modifications. As explained before, these attacks change the image or video globally, but local pixel neighborhoods are usually only shifted, rotated, or zoomed. This feature can be used to undo some

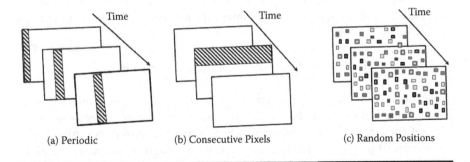

(a) Periodic   (b) Consecutive Pixels   (c) Random Positions

Figure 15.6 Examples of pixel positions used for watermark embedding.

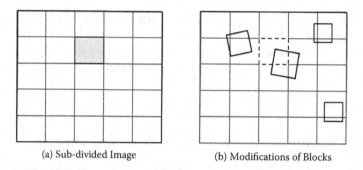

(a) Sub-divided Image   (b) Modifications of Blocks

Figure 15.7 Examples of modifications.

of the effects of the attack by applying a solution that extends the concept of the sliding correlator, as known in the radio spread spectrum. Such a method is used to re-synchronize 1D spread spectrum signals in the case of synchronization loss. This concept can be extended to images and video watermarking for counterattacking detection-disabling attacks. The idea of the solution is to re-synchronize small pixel clusters by testing different combinations of rotation, shift, and zoom, and possibly other distortions, then searching for the maximum correlation with respect to the original pseudo-noise signal used for watermark embedding.

For the sake of clarity, let us show how this solution works on the example of an image. The attacked image is divided into blocks of arbitrary size (such as 16x16). Then, all possible combinations of modifications (including shift, rotation, and zoom) are applied, for each block, as depicted by Figure 15.7. Then, the correlation between the modified block and the original pseudo noise signal is computed. The modification with the highest correlation for each block is supposed to be the one resulting from the attack. It is then used to compensate the effect of the attack on the block.

It can be noticed that, theoretically, the number of such modifications (combinations of shift, zoom, rotation, etc.) can grow very high with the number of bits in the block. In particular, for 256×256 image the number of modifications exceeds the following number:

$$512 \times 512 \text{ half-pel shifts} \times 360 \text{ rotations} \times 40 \text{ zooms}$$

In practice, however, the search space is often significantly smaller, since reasonable attacks cannot change the size and orientation of images too much. The default mode of the StirMark attack, for instance, introduces shifts by no more than 10 pixels, small rotations of less than 5 degrees, and magnifications by a factor less than 1:1. Thus, the search space needed to provide a countermeasure against the StirMark attack is notably smaller than the aforementioned number.

15.6.6 Preventing Unauthorized Detection

To prevent unauthorized detection, various countermeasures can be applied. A straightforward approach would be to use an encryption as follows: To embed the watermark the content provider performs the following steps:

1. Construct a description of the cover object based on information occurring in the object that would not change.
2. Obtain a digital signature of the message formed by the watermark and the description constructed by applying a hash function on it and encrypt it using a public key cryptosystem.
3. Embed the watermark message concatenated with signature in the cover object using an embedding algorithm that does not change the description computed in the first step.

To detect and check the validity of the watermark in the cover object, the receiver performs the following steps:

1. Detect the watermark and extract the watermark and signature.
2. Reconstruct the description of the cover object and compute the hash function of the watermark concatenated with the description.
3. Decode the signature and compare the result with the computed hash value. If the two values are identical, then the receiver of the watermarked object is guaranteed that the content provider is the one who has embedded the watermark.

15.7 Security of Mobile Multimedia Multicasting Schemes

In a near future, the video streaming is likely to become more widespread. Consequently, it is necessary to find a way of protecting digital video content during multicast and take care of the collusion problem. The traditional multicast security schemes encrypt the data using a group key and then multicast the encrypted multimedia object to the members of the group. When a mobile group member receives the encrypted data, he simply decrypts it using the group key. Therefore, the distribution of the group key in an efficient and secure manner is essential to the multimedia success.

15.7.1 Securing Multicasting Schemes

Most current schemes assume that the key distribution is performed during session setup or at least the key is not changed too frequently. These schemes present several limits in real time situations, where

- The multimedia data has strict time constraints and the group membership is changing to the mobility of users and the key must be delivered in a rapid way.
- The content sender and the mobile receiver are connected through an open network environment. Therefore, the multicast protocol does not have to rely on the security support of network component on the delivery paths.

Therefore, some rules need to be set up. Among these rules one can list the following:

- The key distribution scheme should cope with lossy and long delay paths of delivery. A rekey message may be lost, for example, for some of the group members.
- The copyright protection should be provided for multicast video sequences, so that the content provider (or owner) can be uniquely identified.
- The frequent changes of membership may generate a large overhead when a large number of rekey messages is needed for large groups.

In fact, the mobile customers should be allowed to join and leave a session at any time. They should be able to roam properly without affecting the quality of the received stream. In addition, when a customer leaves a group, he should not be able to decode the multicast stream, whereas when a new mobile customer joins a multicast session, he should be able to decode the multicast multimedia object. Moreover, if any entity on the network is serving as a group leader, it should be capable of blocking a customer from receiving a media stream.

Digital watermarking seems to be a good solution for the aforementioned needs (Lin, 2001). In several mobile multimedia applications, such as Video-on-Demand and Pay-Per-View, the digital watermarking has been used to enforce a copyright protection. In the following we discuss some schemes, based on watermarking, that provide copyright protection for mobile applications. A simple method for key distribution can be based on the following steps:

- A pairwise key k_i is created between the sender and the ith receiver during a receiver authentication procedure. The pairwise key can be symmetric or asymmetric. Then, the sender selects a data encryption key K, whenever it is needed, and encrypts the media object (or frame composing it) O with key K.
- The sender creates a packet with a special header to carry K and O the following structure:

$E_{k_1}(K)$	$E_{k_2}(K)$	$\cdots$	$E_{k_n}(K)$	$E_K(O)$

where n is the number of receivers and $E_{k_i}(K)$, $i \leq n$, stands for an encryption function using k_i. The following media objects are then encrypted using key K until there is a need to change K.

- When the mobile user m receives the encrypted stream, he looks first for the special header, if any, looks for the mth slot, decrypts the message $E_{k_m}(K)$, and uses K to decrypt the encrypted media document.

One can notice that the aforementioned method achieves the highest possible security level provided by the encryption utilized scheme. The security does not require any involvement of intermediate nodes. In addition, the method presents the following features:

- It can be applied to any multicast scheme;
- It achieves a minimal storage at each mobile device (since only k_m and K are stored, for the mth user); and
- It allows dynamic joining and leaving of mobile users.

However, the method assumes that all the keys should be received by the mobile users before the decryption process can be executed. In addition, the multicast channel is prone to packet loss and it is expensive to make multicast channel reliable in an open network (using TCP, for example). In a multicast group, some senders/ receivers may not get the key update message and keep encrypting/decrypting with the old key the multimedia data.

15.7.2 Authentication of Multimedia Content

A watermark can be used to support the verification and maintenance of the integrity of mobile multimedia. Mobile multimedia can easily be altered by intentionally malicious changes. Many non-watermarking techniques (such as the digital signature) have been provided to solve the problem of recognizing whether or not the transmitted multimedia object has been altered in a way or another. However, using watermarking to authenticate mobile multimedia content presents two advantages compared to these techniques. First, the watermark does not need to be stored in a separate data structure (such as a signature). Second, the watermark is subtle object; it undergoes the same processing as the object where it is embedded.

In the following we discuss a few issues related to the way watermarking can be used to check whether the transmitted multimedia object has been significantly altered and determine what parts of the object have been altered.

The application of cryptography-based authentication to multimedia has been well addressed. The authentication of an image can be based on the generation of a digital signature, which is able to identify the image, for example, and the transmission of that signature along with the image. If transmitted as a separate metadata, there may be risk that this metadata will be lost, particularly when the metadata experiences a number of format changes. Watermarking reduces this risk by embedding the signature within the image. This allows the format changes to occur without losing the authentication information. Consequently, representing the authentication as signature is as robust as the watermark would guarantee that the signature will be correctly extracted even if the watermarked object has been modified during transmission.

If the watermark is erasable, the following tasks can be performed to provide an interesting protocol: The provider of the media unit computes a signature that is able to authenticate the entire unit. On receiving the watermarked object, the recipient extracts the embedded signature, erases the watermark, and computes the one function on the resulting object; then he compares the hash value with the decrypted signature.

References

D. Boneh and J. Shaw, Collusion-Secure Fingerprinting for Digital Data, *IEEE Transactions on Information Theory*, Vol. 44, pp. 1897–1905, September 1998.

S. Craver, N. Memon, B. L. Yeo, and M. Yeung, Resolving Rightful Ownerships with Invisible Watermarking Techniques: Limitations, Attacks, and Implications. *IEEE Journal on Selected Areas in Communications*, 16(1998): 573–586.

V. Darmstaedter, J.-F. Delaigle, D. Nicholson, and B. Macq, A Block Based Watermarking Technique for MPEG-2 Signals: Optimization and Validation on Real Digital TV Distribution Links, in *Proc. European Conf. Multimedia Applications, Services, and Techniques—ECMAST '98*, Berlin, Germany, May 1998.

F. Deguillaume, G. Csurka, J. 'O Ruanaidh, and T. Pun, Robust 3D DFT Video Watermarking, in *IS&T/SPIE's 11th Annu. Symp., Electronic Imaging '99: Security and Watermarking of Multimedia Contents*, Vol. 3657, San Jose, CA, Jan. 1999.

F. Deguillaume, G. Csurka, and T. Pun, Countermeasures for Unintentional and Intentional Video Watermarking Attacks, in *Electronic Imaging 2000: Securityand Watermarking of Multimedia Content II*, 2000, Vol. 3971.

M. Gerla, B. K. F. Ng, M. Y. Sanadidi, M. Valla, and R. Wang, TCP Westwood with Adaptive Bandwidth Estimation to Improve Efficiency/Friendliness Tradeoffs, *Computer Communications*, 27(1): 41–58.

M. Hamdi and N. Boudriga, Four Dimensional Chaotic Ciphers for Secure Image Transmission, the *2008 IEEE International Conference on Multimedia & Expo (ICME 08)*, June 23–26, 2008, Hannover, Germany.

F. Hartung, J. K. Su, and B. Girod, Spread Spectrum Watermarking: Malicious Attacks and Counterattacks, in the *Proc. of SPIE, Security and Watermarking of Multimedia Contents*, San Jose, California, USA, Vol. 3657, pp. 147–158 SPIE.

A. Herrigel, J. Ó. Ruanaidh, H. Petersen, S. Pereira, and T. Pun, Secure Copyright Protection Techniques for Digital Images, *Information Hiding* 1998: 169–190.

T. Kalker, J. P. Linartz, and M. Van Dijk, Watermark Estimation Through Detector Observations, in *Proc. IEEE Benelux Signal Processing Symposium '98*, Leuven, Belgique, March 1998.

T. Kalker, G. Depovere, J. Haitsma, and M. Maes, A Video Watermarking System for Broadcast Monitoring, in *Proc. SPIE IS&T/SPIE's 11th Annu. Symp., Electronic Imaging '99: Security and Watermarking of Multimedia Contents*, Vol. 3657, Jan. 1999.

E. Lin, C. Podilchuk, T. Kalker, and E. Delp, Streaming Video and Rate Scalable Video: What Are the Challenges for Watermarking?, in *Security and Watermarking of Multimedia Contents III, Proceedings of SPIE*, Vol. 4314, pp. 116–127, 2001.

M. Mirhakkak, N. Schult, and D. Thomsom, Dynamic Bandwidth Management and Adaptive Applications for a Variable Bandwidth Wireless Environment, *IEEE JSAC*, 19(10)(2001): 1985–1997.

S. P. Mohanty, et al., A Dual Watermarking Technique for Images, *Proc. 7th ACM International Multimedia Conference, ACM-MM'99*, Part 2, pp. 49–51, Orlando, USA, Oct. 1999.

J. Padhye, V. Firoiu, D. Towsley, and J. Kurose, Modeling TCP Reno Performance: A Simple Model and Its Empirical Validation, *IEEE/ACM Trans. Networking*, 8(2)(2000): 133–145.

L. Qiao and K. Nahrstedt, Watermarking Schemes and Protocols for Protecting Rightful Ownership and Customer's Rights, *Journal of Visual Communication and Image Representation*, pp. 194–210, September 1998.

R. Rejaie, M. Handley, and D. Estrin, Rap: An End-to-End Rate-Based Congestion Control Mechanism for Real-Time Streams in the Internet, in *Proc. 18th Annual Joint Conference of the IEEE Computer and Communications Societies (INFOCOM '99)*, Vol. 3, pp. 1337–1345, New York, NY, USA, March 1999.

M. Swanson, B. Zhu, and A. Tewfik, Multiresolution Video Watermarking Using Perceptual Models and Scene Segmentation, in *Proc. IEEE Int. Conf. Image Processing 1997 (ICIP '97)*, Vol. 2, Santa Barbara, CA, Oct. 1997, pp. 558–561

M. D. Swanson, et al., Data Hiding for Video-in-Video, *Proc. IEEE International Conf. on Image Processing*, ICIP-97, Vol. 2, pp. 676–679.

F. Wu, G. Shen, K. Tan, F. Yang, and S. Li, Next Generation Mobile Multimedia Communications: Media Codec and Media Transport Perspectives, *China Communications*, October 2006, pp. 30–44.

T. Zhang and Y. Xu, Unequal Packet Loss Protection for Layered Video Transmission, *IEEE Trans. Broadcast.*, Vol. 45, No. 2, pp. 243–252, 1999.

H. Zhao, M. Wu, Z. J. Wang, and K. J. R. Liu, Nonlinear Collusion Attacks on Independent Fingerprints for Multimedia, in *IEEE International Conference on Acoustics, Speech, and Signal Processing*, 2003, pp. 613–616.

Index

605

Printed in the United States
by Baker & Taylor Publisher Services